TREATMENT OF EATING DISORDERS

TREATMENT OF EATING DISORDERS

Bridging the Research - Practice Gap

Edited by

MARGO MAINE

BETH HARTMAN MCGILLEY

DOUGLAS W. BUNNELL

AMSTERDAM • BOSTON • HEIDELBERG • LONDON
NEW YORK • OXFORD • PARIS • SAN DIEGO
SAN FRANCISCO • SINGAPORE • SYDNEY • TOKYO

Academic Press is an imprint of Elsevier

Academic Press is an imprint of Elsevier
32 Jamestown Road, London NW1 7BY, UK
30 Corporate Drive, Suite 400, Burlington, MA 01803, USA
525 B Street, Suite 1800, San Diego, CA 92101-4495, USA

First edition 2010

Copyright © 2010 Elsevier Inc. All rights reserved

No part of this publication may be reproduced, stored in a retrieval system or transmitted in any form or by any means electronic, mechanical, photocopying, recording or otherwise without the prior written permission of the publisher

Permissions may be sought directly from Elsevier's Science & Technology Rights Department in Oxford, UK: phone (+44) (0) 1865 843830; fax (+44) (0) 1865 853333; email: permissions@elsevier.com. Alternatively, visit the Science and Technology Books website at www.elsevierdirect.com/rights for further information

Notice
No responsibility is assumed by the publisher for any injury and/or damage to persons or property as a matter of products liability, negligence or otherwise, or from any use or operation of any methods, products, instructions or ideas contained in the material herein. Because of rapid advances in the medical sciences, in particular, independent verification of diagnoses and drug dosages should be made

British Library Cataloguing-in-Publication Data
A catalogue record for this book is available from the British Library

Library of Congress Cataloging-in-Publication Data
A catalog record for this book is available from the Library of Congress

ISBN: 978-0-12-810207-7

For information on all Academic Press publications
visit our website at www.elsevierdirect.com

Typeset by TNQ Books and Journals

Transferred to Digital Printing 2017

Contents

Biographies ix
Abbreviations xix
Introduction xxi

I
BRIDGING THE GAP: THE OVERVIEW

1. A Perfect Biopsychosocial Storm: Gender, Culture, and Eating Disorders 3
   MARGO MAINE AND DOUGLAS W. BUNNELL

2. What's Weight Got to Do with It? Weight Neutrality in the Health at Every Size Paradigm and Its Implications for Clinical Practice 17
   DEBORAH BURGARD

3. Neuroscience: Contributions to the Understanding and Treatment of Eating Disorders 37
   FRANCINE LAPIDES

4. Are Media an Important Medium for Clinicians? Mass Media, Eating Disorders, and the Bolder Model of Treatment, Prevention, and Advocacy 53
   MICHAEL P. LEVINE AND MARGO MAINE

II
BRIDGING THE GAP: DIAGNOSIS AND TREATMENT

5. The Assessment Process: Refining the Clinical Evaluation of Patients with Eating Disorders 71
   DREW A. ANDERSON, JASON M. LAVENDER, AND KYLE P. DE YOUNG

6. Medical Assessment of Eating Disorders 89
   EDWARD P. TYSON

7. Psychiatric Medication: Management, Myths, and Mistakes 111
   MARTHA M. PEASLEE LEVINE AND RICHARD L. LEVINE

8. Nutritional Impact on the Recovery Process 127
   JILLIAN K. CROLL

9. Science or Art? Integrating Symptom Management into Psychodynamic Treatment of Eating Disorders 143
   NANCY L. CLOAK AND PAULINE S. POWERS

10. New Pathways: Applying Acceptance and Commitment Therapy to the Treatment of Eating Disorders 163
    KATHY KATER

11. Outpatient Treatment of Anorexia Nervosa following Weight Restoration: Practical and Conceptual Issues   181
RICHARD A. GORDON

12. Recipe for Recovery: Necessary Ingredients for the Client's and Clinician's Success   197
BETH HARTMAN McGILLEY AND JACQUELINE K. SZABLEWSKI

III
BRIDGING THE GAP: SPECIAL POPULATIONS

13. Borderline Personality and Eating Disorders: A Chaotic Crossroads   217
RANDY A. SANSONE AND LORI A. SANSONE

14. Managing the Eating Disorder Patient with a Comorbid Substance Use Disorder   233
AMY BAKER DENNIS AND BETHANY L. HELFMAN

15. Comorbid Trauma and Eating Disorders: Treatment Considerations and Recommendations for a Vulnerable Population   251
DIANN M. ACKARD AND TIMOTHY D. BREWERTON

16. Healing Self-Inflicted Violence in Adolescents with Eating Disorders: A Unified Treatment Approach   269
KIMBERLY DENNIS AND JANCEY WICKSTROM

17. The Weight-Bearing Years: Eating Disorders and Body Image Despair in Adult Women   285
MARGO MAINE

18. Men with Eating Disorders: The Art and Science of Treatment Engagement   301
DOUGLAS W. BUNNELL

IV
BRIDGING THE GAP: FAMILY ISSUES

19. Mutuality and Motivation in the Treatment of Eating Disorders: Connecting with Patients and Families for Change   319
MARY TANTILLO AND JENNIFER SANFTNER

20. When Helping Hurts: The Role of the Family and Significant Others in the Treatment of Eating Disorders   335
JUDITH BRISMAN

21. The Most Painful Gaps: Family Perspectives on the Treatment of Eating Disorders   349
ROBBIE MUNN, DORIS AND TOM SMELTZER, AND KITTY WESTIN

V
BRIDGING THE GAP: MIND, BODY, AND SPIRIT

22. The Role of Spirituality in Eating Disorder Treatment and Recovery   367
MICHAEL E. BERRETT, RANDY K. HARDMAN, AND P. SCOTT RICHARDS

23. The Case for Integrating Mindfulness in the Treatment of Eating Disorders   387
KIMBERLI McCALLUM

24. The Use of Holistic Methods to Integrate the Shattered Self   405
ADRIENNE RESSLER, SUSAN KLEINMAN, AND ELISA MOTT

25. Incorporating Exercise into Eating Disorder Treatment and Recovery: Cultivating a Mindful Approach   425
RACHEL M. CALOGERO AND KELLY N. PEDROTTY-STUMP

26. Body Talk: The Use of Metaphor and Storytelling in Body Image Treatment   443
ANITA JOHNSTON

VI
BRIDGING THE GAP: FUTURE DIRECTIONS

27. The Research–Practice Gap: Challenges and Opportunities for the Eating Disorder Treatment Professional   459
JUDITH D. BANKER AND KELLY L. KLUMP

28. Call to Action   479

Index   491
Color Plates

Biographies

Senior Editor

Margo Maine, PhD, FAED, cofounder of the Maine & Weinstein Specialty Group, is a clinical psychologist who has specialized in eating disorders and related issues for 30 years. Author of: *Effective Clinical Practice in the Treatment of Eating Disorders: The Heart of the Matter,* co-edited with William Davis and Jane Shure (Routledge, 2009); *The Body Myth: Adult Women and the Pressure to Be Perfect* (with Joe Kelly, John Wiley, 2005); *Father Hunger: Fathers, Daughters and the Pursuit of Thinness* (Gurze, 2004); and *Body Wars: Making Peace With Women's Bodies* (Gurze, 2000), she is a senior editor of *Eating Disorders: The Journal of Treatment and Prevention* and vice president of the Eating Disorders Coalition for Research, Policy, and Action. A Founding Member and Fellow of the Academy for Eating Disorders and a member of the Founder's Council and past president of the National Eating Disorders Association, she is a member of the psychiatry departments at the Institute of Living/Hartford Hospital's Mental Health Network and at Connecticut Children's Medical Center, having previously directed their eating disorder programs. Dr Maine is the 2007 recipient of The Lori Irving Award for Excellence in Eating Disorders Awareness and Prevention, given by the National Eating Disorders Association. She lectures nationally and internationally on topics related to the treatment and prevention of eating disorders, female development, and women's health.

Editors

Douglas W. Bunnell, PhD, FAED, is a graduate of Yale University and received his doctoral degree from Northwestern University. He is a clinical psychologist and Vice President and Director of Outpatient Clinical Services for The Renfrew Center, overseeing the clinical programming and training for Renfrew's eight outpatient treatment centers. He is the editor of Renfrew's professional newsletter, *Perspectives,* and co-chairs their research committee. He serves on the editorial board of *Eating Disorders: The Journal of Treatment and Prevention.* A Fellow of the Academy for Eating Disorders, he is a former board president of the National Eating Disorders Association, a member of National Eating Disorder Association's Founders Council, and is the clinical advisor for the National Eating Disorder Association's Navigator program which trains parents and family members as resources for newly diagnosed patients and families. Dr. Bunnell also remains active in eating disorders advocacy and awareness. He has written and lectured, nationally and internationally, on eating disorders treatment, research, professional training, eating disorders in men, and the challenges of integrating science and practice. He is also a member of the Academy for Eating Disorders credentialing committee, working to develop practice standards for residential treatment of patients with eating disorders. In addition to his work with Renfrew, the Academy for Eating Disorders and National

Eating Disorder Association, Dr. Bunnell maintains a private practice in Wilton, Connecticut specializing in the treatment of eating disorders, chronic illness, and the psychological aspects of Lyme Disease.

Beth Hartman McGilley, PhD, FAED, Associate Professor, University of Kansas School of Medicine-Wichita, is a psychologist in private practice, specializing in the treatment of eating and related disorders, body image, athletes, trauma, and grief. A Fellow of the Academy for Eating Disorders, she has practiced for 25 years, writing, lecturing, supervising, directing an inpatient eating disorders program, and providing individual, family and group therapy. She has published in academic journals and the popular media, as well as having contributed chapters to several books. She is an editor for *Eating Disorders: The Journal of Treatment & Prevention*, and is working on her first book, a tribute to the patients she has served over the course of her career.

Dr. McGilley also specializes in applications of sports psychology and performance enhancement techniques with athletes at the high school, collegiate, and professional levels. She was the sports psychology consultant for the Wichita State University Women's Basketball team from 2005 to 2008, and serves as the co-chair of the Association for Applied Sports Psychology (AASP) Eating Disorders Special Interest Group.

Dr. McGilley co-founded and is the current President of the Healing Path Foundation, a non-profit foundation dedicated to the prevention and treatment of eating disorders in Kansas. She is a recent graduate of the Kansas Health Foundation Leadership Fellows Training program. Her hobbies include competitive cycling, hiking, and writing.

Contributors

Diann M. Ackard, PhD, LP, FAED, is passionate about helping us be the best that we can be. She is a licensed psychologist in private practice, and is an Adjunct Assistant Professor in the Division of Epidemiology and Community Health at the University of Minnesota, and a Research Scientist at Melrose Institute in St Louis Park, Minnesota. She sits on the Boards for the Academy for Eating Disorders and Break the Cycle, and co-founded the Trauma and Eating Disorders Special Interest Group of the Academy for Eating Disorders. She regularly publishes articles in peer-reviewed journals and frequently contributes at meetings and conferences.

Drew Anderson, PhD, is an Associate Professor in the Department of Psychology at the University at Albany, State University of New York. His research focuses on assessment and treatment of eating disorders, body image disturbance, and psychological and medical problems associated with obesity.

Amy Baker Dennis, PhD, FAED, is a clinical and research psychologist who has maintained a clinical practice over 36 years. She was the founding Board Secretary and served on the Board of the Academy for Eating Disorders (AED) for 11 years. She is also a founding member of the Eating Disorder Research Society (EDRS), founding Board President of the Eating Disorder Awareness and Prevention (EDAP) and a member of the Founders Council, and currently serves on the Board of the National Eating Disorder Association (NEDA). She has published and lectured extensively and received numerous awards for her contributions to the field, including the *Lifetime*

Achievement Award givn by NEDA. She is a certified cognitive therapist and has served on the faculties of University of South Florida, Department of Psychiatry and Behavioral Sciences, the Hamilton Holt graduate school at Rollins College in Orlando, Florida, and Wayne State University Department of Psychiatry in Detroit.

Judith Banker, MA, LLP, FAED, is the founder and executive director of the Center for Eating Disorders, a non-profit outpatient treatment center in Ann Arbor, Michigan. She is a Past President of the Academy for Eating Disorders and served as chair of the Academy for Eating Disorders Psychodynamic Psychotherapy Special Interest Group for 10 years. With over 35 years of clinical and training experience, Judith's teaching and writing focuses on the integrative clinical treatment of eating disorders and on research-practice integration in the eating disorders field.

Michael E. Berrett, PhD, received his PhD in Counseling Psychology in 1986 from Brigham Young University. He is CEO, Director, and Co-founder of Center For Change in Orem, Utah. Dr. Berrett has served as Chief of Psychology at Utah Valley Regional Medical Center and as Clinical Director of Aspen Achievement Academy. He has 25 years experience in the treatment of those struggling with eating disorders. He is co-author of the American Psychological Association book *Spiritual Approaches in the Treatment of Women With Eating Disorders* and multiple articles in professional journals.

Timothy D. Brewerton, MD, DFAPA, FAED, is Clinical Professor of Psychiatry and Behavioral Sciences at the Medical University of South Carolina in Charleston. He is triple board certified in general psychiatry, child/adolescent psychiatry and forensic psychiatry, Distinguished Fellow of the American Psychiatric Association and Founding Fellow of the Academy of Eating Disorders. Dr. Brewerton has published over 115 articles and book chapters, is editor of the book, *Clinical Handbook of Eating Disorders: An Integrated Approach*, and serves on the Editorial Boards of the *International Journal of Eating Disorders* and *Eating Disorders: The Journal of Treatment and Prevention.*

Judith Brisman, PhD, is Director and Co-Founder of the Eating Disorder Resource Center in New York City. She is co-author of *Surviving an Eating Disorder: Strategies for Family and Friends* (Collins Living, 2009, third edn), is an associate editor of *Contemporary Psychoanalysis* and is on the editorial board of the journal *Eating Disorders: The Journal of Treatment and Prevention*. Dr. Brisman is a supervisor of psychotherapy and a member of the teaching faculty of the William Alanson White Institute. She has published and lectured extensively regarding the interpersonal treatment of eating disorders and currently maintains a private practice in Manhattan, New York.

Deborah Burgard, PhD, specializes in the treatment of eating disorders and body image. She created www.BodyPositive.com and is one of the founding proponents of the Health at Every Size model. She co-wrote *Great Shape: The First Fitness Guide for Large Women,* and chapters in *Effective Clinical Practice in the Treatment of Eating Disorders: The Heart of the Matter, Feminist Perspectives on Eating Disorders*, and *The Fat Studies Reader*. Dr. Burgard is also a co-author of the Academy for Eating Disorder's "Guidelines for Childhood Obesity Programs" and co-leads the Sustainable Health Practices Registry, research on how people create ongoing practices that support their health.

Rachel Calogero, PhD, completed her M.A. at The College of William and Mary, and her doctoral and postdoctoral work in social psychology at the University of Kent in Canterbury, England. Currently, she is Assistant Professor of Psychology at Virginia Wesleyan College. Her primary interests cover a spectrum of socio-cultural factors that affect women's health and well-being, including the role of exercise in eating disorders treatment and recovery, the impact of sexual and self-objectification in girls' and women's daily lives, and the perpetuation of fat prejudice and stigmatization. She has published her research widely in peer-reviewed journals and book chapters, and is senior editor of the book, *Self-Objectification in Women: Causes, Consequences, and Counteractions* (APA, 2010). She presents her research frequently in Europe and North America, and offers workshops on mindful excercise in various clinical and community contexts.

Nancy Cloak, MD, attended medical school at the University of South Florida and did her psychiatric residency at the Menninger Clinic, where she was also a candidate in the Topeka Institute for Psychoanalysis. Following residency, she worked with eating disorder patients in a university health center, and then completed a fellowship in eating disorders at Sheppard-Pratt Hospital, after which she returned to Oregon to become the medical director of RainRock Treatment Center. Her professional interests include psychodynamic psychotherapy with eating disorder patients, the neurobiology of weight, appetite, and eating disorders, and medical complications of eating disorders.

Jillian Croll, PhD, MPH, RD, LD, is the Director of Communications, Outreach, and Research for the Emily Program. She is an Adjunct Assistant Professor in Department of Food Science and Nutrition at the University of Minnesota. She completed her MPH and PhD in Public Health Nutrition and Epidemiology at the University of Minnesota, and her MS in Nutritional Science at the University of Vermont. Her work in eating disorders includes program development, community education, teaching, research, clinical work, and advocacy.

Kimberly Dennis, MD, is the Medical Director at Timberline Knolls Residential Center for women with eating disorders and co-occurring disorders, and has a private practice with Working Sobriety Chicago. She specializes in group and individual treatment for patients with co-occurring eating and substance use disorders. She maintains a holistic perspective, and brings an awareness of the benefits of storytelling, creativity, and play in the recovery process. Dr. Dennis is a member of IAEDP, Academy for Eating Disorders, and ASAM. She is an editorial board member for *Eating Disorders: The Journal of Treatment & Prevention*.

Kyle P. De Young, MA, is currently an advanced graduate student in clinical psychology at the University at Albany, State University of New York. His research interests include the course and outcome of eating disorders, exercise, and assessment of eating and exercise-related constructs.

Richard A. Gordon, PhD, is Professor of Psychology at Bard College and a clinical psychologist in independent practice. He has treated patients with eating disorders for over 25 years. He is author of *Eating Disorders: Anatomy of a Social Epidemic*, Second Edition (Blackwell, 2000), and with Melanie Katzman and Mervat Nasser, *Eating Disorders and Cultures in Transition* (Brunner Routledge, 2001). He was made Honorary Fellow of the American Psychiatric Association for his contributions to the social understanding of eating disorders.

Randy K. Hardman, PhD, worked as a psychologist for 26 years. He was a co-founder and director of Center for Change, where he worked for 11 years. Dr. Hardman is currently working with students in the Counseling Center at Brigham Young University-Idaho in Rexburg, Idaho. He is a co-author of the book, *Spiritual Approaches in the Treatment of Women with Eating Disorders* (American Psychological Association, 2007). He has written and published articles on spirituality and other related eating disorder topics.

Bethany Helfman, PsyD, is a clinical psychologist who has practiced in the field for over 18 years. She is currently at Dennis & Moye & Associates in Bloomfield Hills, Michigan where she specializes in the treatment of adolescents, adults, and families affected by eating disorders and their comorbidities. She is a member of the Academy for Eating Disorders and the National Eating Disorder Association. Dr. Helfman supervises other professionals in the field, writes, lectures, and advocates for change related to the factors that make recovery from mental illness more difficult.

Anita Johnston, PhD, is Director of the Anorexia & Bulimia Center of Hawaii, which she co-founded in 1982, Clinical Director and Founder of Ai Pono Eating Disorders Programs in Honolulu, and Senior Advisor and Clinical Consultant for Focus Center for Eating Disorders in Tennessee. In 1986, she developed Hawaii's first in-patient eating disorders treatment program at Kahi Mohala Hospital. Dr. Johnston is the author of *Eating in the Light of the Moon: How Women Can Transform Their Relationships with Food Through Myth, Metaphor, and Storytelling* (Gurze, 2000) and an international speaker and workshop leader with a private practice in Kailua, Hawaii.

Kathy Kater, LICSW, is a St. Paul, Minnesota psychotherapist and an internationally known author, speaker, and consultant with over 30 years of experience specializing in the treatment and prevention of body image and eating-related disorders. Frustrated that progress in understanding these problems has not been matched by effective prevention, she authored *Healthy Body Image: Teaching Kids to Eat and Love Their Bodies Too,* a primary prevention curriculum for upper elementary school children, and *Real Kids Come in All Sizes; Ten Essential Lessons to Build Your Child's Body Esteem*, a companion guide for parents.

Susan Kleinman, MA, BC-DMT, NCC, is the dance/movement therapist for The Renfrew Center of Florida. She is a trustee of the Marian Chace Foundation, a past president of the American Dance Therapy Association, and a past Chair of The National Coalition for Creative Arts Therapies. She is a co-editor of The Renfrew Center Foundation's *Healing Through Relationship*, serves on the editorial board of the *Journal of Creativity in Mental Health*, and has published extensively on the use of dance/movement therapy in the treatment of eating disorders. She was the American Dance Therapy Association recipient of the 2009 Outstanding Achievement Award.

Kelly L. Klump, PhD, FAED, is an Associate Professor of Psychology at Michigan State University. Her research focuses on genetic and biological risk factors for eating disorders. Dr. Klump has published over 90 papers and has received a number of federal grants for her work. She has been honored with several awards including the David Shakow Award for Early Career Contributions to Clinical Psychology from the

American Psychological Association and New Investigator Awards from the World Congress on Psychiatric Genetics and the Eating Disorders Research Society. Dr. Klump is a Past President of the Academy for Eating Disorders.

Francine Lapides, MFT, writes and teaches from attachment and psychoneurobiological theories (including the arousal and regulation of affect) and their applications to relational and psychodynamic psychotherapy and adult romantic relationships. She supervised and taught psychotherapy through the 1970s and has been in private practice in Santa Cruz, California since 1980. She has trained with Daniel Siegel, is a member of Allan Schore's Berkeley study group, and has been strongly influenced by relational principles developed at The Stone Center at Wellesley College. She teaches workshops and conferences across the United States and provides an online seminar at www.PsyBC.com.

Jason M. Lavender, MA, is currently an advanced graduate student in clinical psychology at the University at Albany. His research interests include the functions of eating disorder behaviors, the course and outcome of eating disorders, and the assessment of body image and eating disorder behaviors.

Martha M. Peaslee Levine, MD, is Assistant Professor of Pediatrics, Psychiatry, and Humanities and the Director of the Partial Hospitalization and Intensive Outpatient Programs at Penn State Milton S. Hershey Medical Center.

Michael P. Levine, PhD, FAED, is Samuel B. Cummings Jr. Professor of Psychology at Kenyon College in Gambier, Ohio. His special interest is body image and eating problems and their links with preventive education, developmental psychology, and community psychology. His most recent book is Levine and Smolak's (2006) *The Prevention of Eating Problems and Eating Disorders: Theory, Research, and Practice* (Lawrence Erlbaum). Dr. Levine is a Fellow of the Academy for Eating Disorders. In June 2006 he received the Meehan-Hartley Award for Leadership in Public Awareness and Advocacy from the Academy for Eating Disorders.

Richard L. Levine, MD, is Professor of Pediatrics and Psychiatry and is the Chief of the Division of Adolescent Medicine and Eating Disorders at Penn State Milton S. Hershey Medical Center.

Kimberli McCallum, MD, CEDS, is a Fellow of the American Psychiatric Association and Associate Professor of Clinical Psychiatry at Washington University School of Medicine. She is a psychotherapist with a broad range of therapy skills, including dialectic behavior therapy, cognitive behavior therapy, family-based treatment, Family Systems Therapy, and psychoanalysis. She received her MD from Yale, general psychiatric training at UCLA, and child/adolescent training at Washington University. Dr. McCallum has co-founded several specialized eating disorders units, including inpatient, partial hospital, residential, and intensive outpatient programs. Her current programs include McCallum Place Treatment Center in St. Louis, MO, and Cedar Springs Treatment Center in Austin, TX.

Elisa Mott, MEd/EdS, a certified yoga teacher and graduate of University of Florida's Counselor Education program, also holds a Spirituality in Health Certificate. She was awarded an International Excellence in Counseling Research Grant from Chi Sigma Iota honor society for her study evaluating the use of yoga to improve wellness among females and presented this research at the 2010 ACA conference. She served as CSI's Wellness Committee chair and has presented on the use of yoga in the treatment of eating disorders at the International Association for Eating Disorder Professionals Conference and the University of Florida's Professional Development Day.

Robbie Munn, MA, MSW, is a clinical social worker who has spoken and written widely about the chaotic impact of eating disorders upon families and the challenges families face in obtaining appropriate treatment. Many women in her family have been affected by eating disorders, including her mother and daughter, nieces, and cousins. In 2000 she joined the Board of the National Eating Disorders Association (NEDA) as one of its first family members. In 2003 she helped to create and co-chair the first conference in the field to include families and individuals along with clinicians. This has become the esteemed annual conference hosted by NEDA.

Kelly N. Pedrotty-Stump, MS, is a high-school guidance counselor and an Exercise Consultant at the Renfrew Center. She co-developed the exercise program at Renfrew. Kelly is an experienced speaker on exercise and the treatment of eating disorders and has presented at national conferences including National Eating Disorder Association, Academy for Eating Disorders, and MEDA. She has taught workshops on various topics at West Chester University, Temple University and Philadelphia College of Osteopathic Medicine. She has published on the topic of exercise abuse and eating disorders. Kelly is also a certified yoga instructor.

Pauline Powers, MD, FAED, graduated from the University of Iowa College of Medicine and completed her residency at the University of California at Davis. She is Professor of Psychiatry and Behavioral Medicine in the Clinical and Translational Science Institute at the University of South Florida, Tampa, Florida. She was the Founding President of the Academy for Eating Disorders and was President of the National Eating Disorders Association 2005–2006. She has published three books on eating disorders and has reported research in several journals. She is currently Director of the University of South Florida Center for Eating and Weight Disorders and the Director of the USF Hope House for Eating Disorders.

Adrienne Ressler, MA, LMSW, CEDS, the National Training Director for The Renfrew Center Foundation, is the 2008–2010 president of the International Association for Eating Disorder Professionals board. She attended the University of Michigan and served as a faculty member in the School of Education. Her nationally renowned seminars reflect her background in gestalt, transactional analysis, psychodrama, bioenergetic analysis, and Alexander technique. She is published in the *International Journal of Fertility and Women's Medicine, Social Work Today* and authored the chapter *BodyMind Treatment* in *Effective Clinical Practice in the Treatment of Eating Disorders*. She is the featured body-image expert for documentaries on both cosmetic surgery and menopause.

P. Scott Richards, PhD, is a Professor of Counseling Psychology at Brigham Young University. He received his PhD in Counseling Psychology in 1988 from the University of Minnesota. He is the co-author of the book, *Spiritual Approaches in the Treatment of Woman with Eating Disorders* (American Psychological Association, 2007). He is also co-author of the book, *A Spiritual Strategy for Counseling and Psychotherapy*, which was published in 1997 and 2005 (2nd ed.) by the American Psychological Association. Dr. Richards has published on the topics of spirituality and eating disorders, religion and mental health, and spiritual issues in psychotherapy.

Jennifer Sanftner, PhD, is a Clinical Psychologist and tenured Associate Professor of Psychology at Slippery Rock University. She has been teaching in the areas of abnormal, clinical, health, and gender psychology, and directing the undergraduate practicum program at SRU for the last 8½ years. She has researched eating disorders for 19 years, resulting in publications in peer-reviewed journals and chapters. Her research focuses on the application of Relational Cultural Theory to understanding the etiology and maintenance of eating disorders. She is interested in using RCT to understand women's relationships with their bodies, with others, and with food, and to applying our understanding of RCT to treatment.

Lori A. Sansone, MD, is a civilian family medicine physician and the Medical Director for the Primary Care Clinic at Wright-Patterson Air Force Base in Dayton, Ohio. She has published over 100 refereed articles and 24 book chapters; co-authored the book, *Borderline Personality Disorder in the Medical Setting*; co-developed the *Self-Harm Inventory*; co-authors a professional column, *The Interface,* for the journal *Psychiatry,* and co-authors a local monthly newsletter, *Mental Health Issues in Primary Care.*

Randy A. Sansone, MD, is a professor at Wright State University School of Medicine in Dayton, Ohio, and Director of Psychiatry Education at Kettering Medical Center. He has published over 225 refereed articles and 33 book chapters; co-edited the books, *Self-Harm Behavior and Eating Disorders* and *Personality Disorders and Eating Disorders*; co-authored the book, *Borderline Personality Disorder in the Medical Setting*; co-developed the *Self-Harm Inventory*; and co-authors a professional column, *The Interface,* for the journal *Psychiatry.* Dr. Sansone is also the editor of the borderline personality module for the *Physician Information and Education Resource* and is on six journal editorial boards, including *Eating Disorders: The Journal of Treatment and Prevention.*

Doris and Tom Smeltzer, are career educators with master's degrees in education and counseling psychology, respectively. Tom is a college professor and Doris has taught throughout the K-12 spectrum. When their 19-year-old daughter Andrea died after 13 months of bulimic behaviors, Doris chose to leave her teaching position and has devoted her life to eating disorder prevention through Andrea's Voice Foundation, the non-profit organization she and Tom co-founded. Doris is the author of *Andrea's Voice: Silenced by Bulimia* and Gurze Books' "*Advice for Parents*" blog and is developing an educational curriculum for the ED field based on her Internet radio show.

Jacqueline Szablewski, MTS, MAC, LAC, is a psychotherapist and licensed addictions counselor who resides in Boulder, Colorado. Combining study in psychology, counseling, and world religions with a self-designed concentration in pastoral counseling, Jackie earned her Masters degree in Theological Studies from Harvard University. She has worked along the continuum of care in agency and hospital settings. Specializing in eating disorders, addiction recovery, and life transitions, particularly with individuals challenged by concomitant mood disorders, trauma, and grief issues,

Jackie has worked in the field for nearly two decades. She has maintained a private practice in Boulder, Colorado for the last 14 years.

Mary Tantillo, PhD, RN, CS, FAED, is the Director of the Western New York Comprehensive Care Center for Eating Disorders, an Associate Professor of Clinical Nursing at the University of Rochester School of Nursing, a Clinical Associate Professor in the Department of Psychiatry at the University of Rochester School of Medicine, and CEO/Clinical Director of a free-standing Eating Disorders Partial Hospitalization Program, The Healing Connection, LLC. She is a fellow of the Academy for Eating Disorders, as well as a previous board member, present chairperson for the Academy for Eating Disorders Credentialing Task Force, and co-chairperson for the Patient/Carer Task Force.

Edward P. Tyson, MD, has been treating eating disorders for more than 20 years and is board certified in both Family Medicine and Adolescent Medicine. After serving as Director of Adolescent Clinics for the Department of Pediatrics at Children's Hospital of Oklahoma, he opened a private practice in Austin, Texas specializing in eating disorders. Dr. Tyson is an active member and frequent presenter at the professional eating disorder organizations. He is an advocate for those with eating disorders and teaches residents and medical students, as well as undergraduate and graduate classes, at the University of Texas about eating disorders.

Kitty Westin is the founder and former President of the Anna Westin Foundation, which has now merged with the Emily Program Foundation. The Anna Westin Foundation was started by Anna's family after Anna died in 2000 as a direct result of anorexia. The Westins also started the first and only residential program to treat people with eating disorders in Minnesota. Kitty is also the past President of the Eating Disorders Coalition for Research, Policy & Action and she serves on the Academy for Eating Disorders Patient/Carer Task Force, and is the Co-chair of the Academy for Eating Disorders Advocacy/Communications Committee.

Jancey Wickstrom, AM, LCSW, is the Milieu Manager and DBT Specialist at Timberline Knolls Residential Center for women with eating disorders and co-occurring disorders. While a student at University of Chicago, she received training in DBT at the Emotion Management Program, and maintains a group and individual DBT practice there. Ms. Wickstrom firmly believes in the powerful effects of mindfulness meditation to help every person create a meaningful life.

Abbreviations

AA, Alcoholics Anonymous
ACC, anterior cingulate cortex
ACT, acceptance commitment therapies
ACTH, adrenocorticotropic hormone
ADHD, attention-deficit/hyperactivity disorder
AN, anorexia nervosa
ANBP, anorexia nervosa, binge purge subtype of anorexia nervosa
ANS, autonomic nervous system
BED, binge eating disorder
BMI, body mass index
BN, bulimia nervosa
BPD, borderline personality disorder
CAT, cognitive analytic psychotherapy
CBC, complete blood cells
CBT, cognitive behavior therapy
CPT, cognitive processing therapy
CRF, corticotrophin releasing factor
DBT, dialectical behavior therapy
DE, disordered eating
DEX, dysfunctional exercise
DEXA, dual energy X-ray absorptiometry
DMT, dance/movement therapy
DSM, Diagnostic and Statistical Manual
EBP, evidence-based practice
EBT, evidence-based treatment
ED, eating disorder
EDI, Eating Disorder Inventory
EDNOS, eating disorder not otherwise specified
EST, empirically supported treatments
FBT, family-based treatment
fMRI, functional magnetic resonance imaging
fNIRS functional near-infrared spectroscopy
FTT, failure to thrive
GABA, gamma-aminobutyric acid
GERD, gastroesophageal reflux disease
HPA, hypothalamic pituitary axis
IFT, internal family therapy
IPT, interpersonal psychotherapy
LH, left hemisphere
MAOI, monoamine oxidase inhibitors
MBCT, mindfulness-based cognitive therapy
MB-EAT, mindfulness-based eating disorder training
MBSR, mindfulness-based stress reduction
MET, motivational enhancement therapy
MI, motivational interviewing
MPC, medial prefrontal cortex
NA, narcotics anonymous
NES, night eating syndrome
OA, overeaters anonymous
OCD, obsessive-compulsive disorder
OFC, orbital frontal cortex
OTC, over the counter
PET, positron emission tomography
PFC, prefrontal cortex
PM, perceived mutuality
PPI, proton pump inhibitors
PTSD, post-traumatic stress disorder
RBC, red blood cells
R/M, relational/motivational approach
RCT, relational-cultural theory
RCTs, randomized controlled trials
RFS, refeeding syndrome
RH, right hemisphere
SD, standard deviation
SIV, self-inflicted violence
SMA, superior mesenteric artery

SOC, stage of change
SOCT, stages of change theory
SNRI, serotonin and norepinephrine reuptake inhibitors
SRED, sleep-related eating disorder
SSRI, selective serotonin reuptake inhibitor
SUD, subjective units of distress
TCA, tricyclic antidepressants
WBC, white blood cells

Introduction
Eating Disorders as Biopsychosocial Illnesses

> The point is that profound but contradictory ideas may exist side by side, if they are constructed from different materials and methods and have different purposes. Each tells us something important about where we stand in the universe and it is foolish to insist that they must despise each other. *Postman, 1995, p. 107*

The idea for this volume, *Treatment of Eating Disorders: Bridging the Research/Practice Gap*, emanated from our experiences as clinicians facing the challenge of helping patients and their loved ones back from the precipice of self-destruction brought on by eating disorders (EDs). While we are each very active in our professional development and ongoing education, every day we experience the impact of the significant gap between what the research in journals, books, and conference presentations provides and how our patients present clinically. Their needs rarely match the theories or studies intended to explain them.

For example, although Eating Disorders Not Otherwise Specified (EDNOS) is the most commonly diagnosed ED in clinical settings, ranging from 50 to 70% of all ED cases (Walsh & Sysko, 2009), research studies rarely include this diagnostic category. While more recent research is beginning to explore the complexities of EDNOS (Agras, Crow, Mitchell, Halmi & Bryson, 2010; Walsh, 2009; Wildes & Marcus, 2010), little is yet known about how this largest subgroup of ED patients progresses through the illness, responds to treatment, and fares in terms of outcome. Recent data seem to confirm what we have known clinically: many patients with EDNOS actually have poorer outcomes and higher mortality rates than patients with AN or BN (Crow, Peterson, Swanson, Raymond & Specker, 2009).

A multitude of other factors contribute to the research/practice gap. Despite the fact that many of our patients suffer from comorbid conditions, treatment research in our field tends to look at these problems more singularly (Haas & Clopton, 2003; Thompson-Brenner & Westen, 2005; Tobin, 2007). In clinical practice, it is often these comorbid factors, including depression, anxiety, and post-traumatic stress disorder, that dominate the process of therapeutic engagement. The process of engagement is known to be difficult in patients with ED, and adapting to the special demands of a patient's comorbidities makes each treatment relationship unique. This sense of uniqueness can create the perception that research does not easily, or often, apply to the individual patient with whom we sit. Furthermore, in clinical research trials, "relatively 'pure' groups of homogenous patients are selected for study, and are offered standardized treatment based on structured manuals. Everyone knows that therapy in the real world is far messier" (Herbert, Neeren & Lowe, 2007, p. 15). We designed this book with the clear intention of trying to bridge such gaps so that research can better inform clinical work, and clinical work can better inform the research agenda and process.

A historical view may help us to create the most-informed approaches to the field's current dilemmas. In her review of four decades of work, Hilda Bruch (1985), the pioneer to whom the field owes great respect and gratitude, identified the nature/nurture debate as a concerning gap. In her hopeful assessment, the two dimensions had finally found common ground. "Recent explorations of the neurochemical processes of the brain have revealed the close association of psychological experiences with alterations in brain metabolism, rendering the old dichotomy between physiological and psychological events untenable" (Bruch, 1985, pp. 8–9). The biopsychosocial model (Johnson & Connors, 1987; Lucas, 1981; Yager, 1982; Yager, Rudnick & Metzner, 1981) advanced this perspective and our understanding of ED, laying the groundwork for prolific empirical contributions in the subsequent decades. The field rigorously researched areas of pressing concern including, but not limited to: prevention; medical and psychiatric management; therapeutic tools and approaches; neuroscience and epigenetics; and the essential role of the family in the ED treatment and recovery process. In the clinical realm, innovative treatment approaches began to yield more positive outcomes.

The dialectic of the past decade, the science/practice gap, parallels, if not harks back to, that of Bruch's generation of ED specialists. Despite Bruch's prescient respect for the neuroscientific basis of psychological experience, integration of this work, and its implications for the therapeutic process, is relatively recent in the ED field. Although we cannot expect neuroscience to be the ultimate mediator for researchers and clinicians of discrepant viewpoints, it has undoubtedly provided a language and medium for professionals in both "camps" to appreciate the other's contributions to the understanding of the etiology and treatment of ED. Nearly 30 years have passed since Bruch's review, and the resurgence of interest in neuroscientific applications/understandings of ED, and in patients' subjective experiences, provides rich opportunities for collaboration between researchers and clinicians.

Today, we have the advantage of a knowledge base built on many more years of inquiry than Dr. Bruch and the other early writers had available to them. There are three scholarly journals dedicated solely to ED: *Eating Disorders: The Journal of Treatment and Prevention (EDJTP)*, the *International Journal of Eating Disorders (IJED)*, and the *European Eating Disorders Review*. Since the 1980s, approximately 1000 books have been published specifically regarding ED or closely related illnesses. *EDJTP* has published about 750 articles, and *IJED* has published approximately 1200 (L. Cohn, personal communication, January 28, 2010). Broadening the topic to body image, health psychology, obesity, or related areas, these numbers would vastly increase, but still do not reflect publications in a wide variety of basic science, psychiatric, medical, nutritional, and psychological journals. The point is that the ED field is relatively young and rapidly developing, with many talented clinicians and researchers whose contributions have the potential to bridge the current gaps, better serving the needs of our patients.

Helene Deutsch, the first psychoanalyst to specialize in the treatment of women, has been credited with saying, "after all, the ultimate goal of all research is not objectivity, but truth" (retrieved from: http://www.brainyquote.com/quotes/authors/h/helene_deutsch.html). *Treatment of Eating Disorders: Bridging the Research/Practice Gap* brings together the expertise of scientists and practitioners in an effort to further describe the truth about ED. Readers will find an unexpected irony: the effect of closing gaps also expands the realm of influence, information, and expertise across disciplines. Researchers will find accounts of the practiced experience and wisdom of clinicians who have been operating with skills and perspectives only partially informed by science. Likewise, clinicians will be exposed to scientific advances that have enriched our understanding of the biopsychosocial complexity of ED. Some of this research has substantiated the central role of the therapeutic relationship (American Psychiatric Association, 2006), and qualitative research is now giving the patient/subject an active voice and presence in the empirical process.

Readers will have access to chapters across a variety of topics where research and clinical work must come together to better shape the understanding, treatment, and outcome of ED. In light of the significant proportion of EDNOS cases, we encouraged our contributors to take a transdiagnostic approach (Fairburn & Cooper, 2007) when possible. We are also intrigued by the proposed alternative system for classification, Broad Categories for the Diagnosis of Eating Disorders (Walsh & Sysko, 2009). While the American Psychiatric Association refines its work on the DSM-V, many diagnostic issues are being considered, and it is premature to discuss the changes; however, we deeply appreciate the efforts of the ED work group.

The collaborative spirit of this book reflects our view that EDs are complex, multidetermined illnesses that must be understood and treated in the sociopolitical context. Effective treatment takes a *team* that includes the patient, the family, and a multidisciplinary group of clinicians working in concert. Successful recovery takes a *village*, interlocking communities of support (e.g. extended family, peers, team-mates, social networks, professional support) in which patients practice their recovery skills, and find vital sources of commonality, connection, optimism, and accountability. We hope that this book conveys respect for the daunting power of these illness processes, as well as the healing power of clinicians, researchers, patients, and families combining forces toward a common goal.

Readers will note recurring references to the importance of the clinical relationship, based on empathy, connection, compassion, respect, and affection, as well as the importance of using that relationship to best implement interventions that have demonstrated effectiveness (Zerbe, 2008). Furthermore, we hope a spirit of partnership emerges from this book—partnership between families and professionals, and between researchers and practitioners. Ideally, *Treatment of Eating Disorders: Bridging the Research/Practice Gap*, will help us to transcend the historical tensions and competitive relationships between researchers and practitioners in our field (Banker & Klump, 2007), and inspire us to proceed with collaborative efforts that appreciate and integrate the best from each domain's perspective. A paradigmatic shift of this magnitude, involving change in attitude and practice both within and between disciplines, will require more than an academic tome devoted to its necessity. As the final chapter of this book illustrates, we are *called to action* or we will remain a field destructively divided.

As editors, we also are aware of the limits of this volume. For example, the diversity, or the evolving face, of ED, is a critical issue beyond the scope of this book. Once the purview of young Caucasian women from higher socio-economic strata in the advanced technological nations, EDs are now global conditions occurring in over 40 countries, many of which are developing nations (Gordon, 2001). In their examination of how culture, ethnicity, difference, and EDs affect minority and non-western females, Nasser and Malson (2009) state:

> The spread of thinness as a master signifier of feminine beauty, promulgated by the mass media and the post-colonial operations of transnational capital, across all sections of western societies and across the world has been devastatingly effective in the 'globalisation' of 'eating disordered' subjectivities and practices...Thinness as a gendered body 'ideal' and a signifier of a multiplicity of positively construed 'attributes' can clearly no longer be considered exclusively western or white (p. 82).

Confirming this significant change in the face of ED, Grabe and Hyde (2006) conducted a meta-analysis of 98 studies, finding no significant differences in body dissatisfaction between Caucasian, Hispanic, and Asian women in the USA. Also, Bisaga et al. (2005) found similar rates of disordered eating (DE) across ethnicities in adolescent girls. Despite clinical impressions clearly confirmed by research, regarding the diverse presentation of ED, minority women experience worrisome barriers to their access to care, especially due to lack of recognition by providers (Cachelin & Striegel-Moore, 2006). Many of these same issues are factors in the underdiagnosis and treatment of men with ED. We must challenge these outdated stereotypes so all patients will be able to receive appropriate diagnosis and care.

Clearly, the field has much to learn about how EDs present across culture, country, ethnicity, and other divisions. We must begin to acknowledge that EDs no longer belong to a place, but instead inhabit many different and constantly evolving global social spheres. Nasser and Malson (2009) advise us to attend to both global and local factors in our attempts to understand ED. They explain that the "gendered aesthetics of thinness" are not always central to the DE or self starvation and that other "locally-specific discursive constructions of self-starvation may be more relevant" (p. 82).

The above findings remind us that our culture continues to drive vulnerable men and women into DE and ED. Although there seems to be a decreased appreciation for these sociocultural forces, enduring gender role stereotypes remain influential. Culture and diversity are enormously complex issues and, while we believe strongly in their importance in a discussion about ED, we could not do them full justice in this volume.

Despite this noted limitation, *Treatment of Eating Disorders: Bridging the Research/Practice Gap*, presents a range of topics critically illuminating the challenge of clinical work with ED patients. The informed clinician needs to be conversant with multiple literatures including research on the cultural, psychological, behavioral, medical, genetic, neurological, and spiritual dimensions of ED. If nothing else, this volume should put to rest the notion that there is any real dichotomy between the biology and the psychology of lived experience. We believe, also, that there is no validity to the dichotomy between clinical practice and research; it is, rather, the lack of resources, inadequate dialogue, disparate languages, and varied systems of inquiry that create this divisive impression (Banker & Klump, 2007). Clinicians collect data every day informing their sense of what does and does not help particular patients and families. Meanwhile, researchers are developing and refining methods of inquiry that allow for more relevant applications of evidence-based practices into naturalistic settings (Lowe, Bunnell, Neeren, Chernyak & Greberman, 2010). Historical differences between the two camps regarding what constitutes meaningful "evidence," or sources of information (e.g., clinical vs. empirical data) have impeded integrative, clinically driven investigations. Advances in qualitative and phenomenological research have begun to mediate this impasse and should be further incorporated into formal quantitative explorations (Jarman & Walsh, 1999; Kazdin, 2009). As Banker and Klump (2007) aptly state, it is time for a "researcher-clinician rapprochement" (p. 14).

Finally, the need to bridge the science/practice gap does not devalue either domain's distinct and relative merits, nor does it negate the necessity for interdisciplinary debate. In fact, as Nobel prize winner Ilya Prigogine has asserted, a certain degree of friction is vital for growth:

> It is precisely the quality of fragility, the capacity for being 'shaken up,' that is paradoxically the key to growth. Any structure—whether at the molecular, chemical, physical, social, or psychological level—that is insulated from disturbance is also protected from change (Levoy, 1997, p. 8).

Change, and exchange—in perspectives, attitudes, and practices—*is* the bridge this volume endeavors to create. It is no longer acceptable to rely on research that does not reflect clinical realities; thanks to the efforts of our authors and many other colleagues, we see promising signs that this gap is closing. Nor is it acceptable for therapists to base their treatment approaches solely on their own clinical intuition (Herbert et al., 2007). The research cited in this volume supporting innovative clinical work demonstrates the merits of Evidence Based Treatment (EBT) and the importance of incorporating EBT into treatment plans (Haas & Clopton 2003; Mussell, Crosby, Crow, Knopke & Peterson, 2000; Tobin, Banker, Weisberg & Bowers, 2007). Working from one theoretical perspective because that is how you were trained is no longer defensible. Clinicians need to be able to explain their rationale for their treatment approach and recommendations, and those explanations need to incorporate both science and clinical intuition. The following contributions seek to insure that researchers and clinicians are cross-trained in the best practices of ED treatment, building bridges that can withstand the inherent friction required for growth, and paving the way for future advances.

References

Agras, W. S., Crow, S., Mitchell, J., Halmi, K., & Bryson, S. (2010). A 4-year prospective study of eating disorder NOS compared with full eating disorder syndromes. *International Journal of Eating Disorders, 42*, 565–570.

American Psychiatric Association. (2006). Practice guidelines for the treatment of patients with eating disorders (3rd ed.). *American Journal of Psychiatry*, 1101–1185.

Banker, J., & Klump, K. (2007, Winter). Toward a common ground: Bridging the gap between research and practice in the field of eating disorders. *Perspectives*, 12–14.

Bisaga, K., Whitaker, A., Davies, M., Chuang, S., Feldman, J., & Walsh, B. T. (2005). Eating disorder and depressive symptoms in urban high school girls from different ethnic backgrounds. *Journal of Developmental and Behavioral Pediatrics, 26*, 257–266.

Bruch, H. (1985). Four decades of eating disorders. In D. Garner & P. Garfinkel (Eds.), *Handbook of psychotherapy for anorexia nervosa and bulimia*. New York, NY: Guilford Press.

Cachelin, F. M., & Striegel-Moore, R. H. (2006). Help seeking and barriers to treatment in a community sample of Mexican American and European American women with eating disorders. *International Journal of Eating Disorders, 39*, 154–161.

Crow, S. C., Peterson, C. B., Swanson, S. A., Raymond, N. C., Specker, S., Eckert, E. D., & Mitchell, J. E. (2009). Increased mortality in bulimia nervosa and other eating disorders. *American Journal of Psychiatry, 166*, 1342–1346.

Fairburn, C. G., & Cooper, Z. (2007). Thinking afresh about the classification of eating disorders. *International Journal of Eating Disorders, 40*, S107–S110.

Gordon, R. A. (2001). Eating disorders East and West: A culture-bound syndrome unbound. In M. Nasser, M. A. Katzman, & R. A. Gordon (Eds.), *Eating disorders and cultures in transition* (pp. 1–23). New York, NY: Taylor and Francis.

Grabe, S., & Hyde, J. S. (2006). Ethnicity and body dissatisfaction among women in the United States: A meta-analysis. *Psychological Bulletin, 132*, 622–640.

Haas, H., & Clopton, J. (2003). Comparing clinical and research treatments for eating disorders. *International Journal of Eating Disorders, 33*, 413–420.

Herbert, J. D., Neeren, A. M., & Lowe, M. R. (2007, Winter). Clinician intuition and scientific evidence: What is their role in treating eating disorders. *Perspectives, Winter,* 15–17.

Jarman, M., & Walsh, S. (1999). Evaluating recovery from anorexia nervosa and bulimia nervosa: Integrating lessons learned from research and clinical practice. *Clinical Psychology Review, 19*, 773–788.

Johnson, C., & Connors, M. E. (1987). *The etiology and treatment of bulimia nervosa: A biopsychosocial perspective.* New York, NY: Basic Books.

Kazdin, A. (2009). Bridging science and practice to improve patient care. *American Psychologist, 64*, 276–278.

Levoy, G. (1997). *Callings: Finding and following an authentic life.* New York, NY: Harmony Books.

Lowe, M. R., Bunnell, D. W., Neeren, A. M., Chernyak, Y., & Greberman, L. (2010). Evaluating the real-world effectiveness of cognitive-behavior therapy efficacy research on eating disorders: A case study from a community-based clinical setting. *International Journal of Eating Disorders.* (Advance online publication. doi: 10.1002/eat.20782).

Lucas, A. R. (1981). Toward the understanding of anorexia nervosa as a disease entity. *Mayo Clinic Proceedings, 56*(4), 254–264.

Mussell, M. P., Crosby, R. D., Crow, S. J., Knopke, A. J., Peterson, C. B., Wonderlich, S. A., & Mitchell, J. E. (2000). Utilization of empirically supported psychotherapy treatments for individuals with eating disorders: A survey of psychologists. *International Journal of Eating Disorders, 27*, 230–237.

Nasser, M., & Malson, H. (2009). Beyond western dis/orders: Thinness and self-starvation of other-ed women. In H. Malson & M. Burns (Eds.), *Critical feminist approaches to eating dis/orders* (pp. 74–86). London, UK: Routledge.

Postman, N. (1995). *The end of education.* New York, NY: Alfred Knopf.

Thompson-Brenner, H., & Westen, D. (2005). A naturalistic study of psychotherapy for bulimia nervosa: Comorbidity and therapeutic outcome: Part 1 & 2. *Journal of Nervous and Mental Diseases, 193*, 573–594.

Tobin, D. L. (2007, Winter). Research and practice in eating disorders: The clinician's dilemma. *Perspectives,* 8–10.

Tobin, D. T., Banker, J. D., Weisberg, L., & Bowers, W. (2007). I know what you did last summer (and it was not CBT): A factor analytic model of international psychotherapeutic practice in the eating disorders. *International Journal of Eating Disorders, 40*, 754–757.

Walsh, B. T. (2009). Eating disorders in DSM-V: Review of the existing literature (Part 1). *International Journal of Eating Disorders, 42*, 579–580.

Walsh, T., & Sysko, R. (2009). Broad categories for the diagnosis of eating disorders (BCD-ED): An alternative system for classification. *International Journal of Eating Disorders, 42*(8), 754–764.

Wildes, J. E., & Marcus, M. D. (2010). Diagnosis, assessment, and treatment planning for binge-eating disorder and eating disorder not otherwise specified. In C. Grilo & J. E. Mitchell (Eds.), *The treatment of eating disorders: A clinical handbook* (pp. 44–65). New York, NY: Guilford Press.

Yager, J. (1982). Family issues in the pathogenesis of anorexia nervosa. *Psychosomatic Medicine, 44*, 43–60.

Yager, J., Rudnick, F. D., & Metzner, R. J. (1981). Anorexia nervosa: A current perspective and some new directions. In E. Serafetinides (Ed.), *Psychiatric research in practice: Biobehavioral contributions* (pp. 131–150). New York, NY: Grune & Stratton.

Zerbe, K. (2008). *Integrated treatment of eating disorders: Beyond the body betrayed.* New York, NY: W. W. Norton & Company.

PART I

BRIDGING THE GAP: THE OVERVIEW

1 A Perfect Biopsychosocial Storm   3
2 What's Weight Got to Do with It?   17
3 Neuroscience   37
4 Are Media an Important Medium for Clinicians?   53

CHAPTER 1

A Perfect Biopsychosocial Storm
Gender, Culture, and Eating Disorders

Margo Maine and Douglas W. Bunnell

Although eating disorder(s) (ED) are multidetermined, biopsychosocial disorders, gender alone remains the single-best predictor of their risk (Striegel-Moore & Bulik, 2007). Most research asserts that anorexia nervosa (AN) and bulimia nervosa (BN) are 10 times more common in females than males, and binge-eating disorder (BED) is three times more common (Treasure, 2007). While some have argued that one in six cases occurs in males (Andersen, 2002), the gender disparity is still glaring. Furthermore, while ED is not the only gendered psychiatric condition, the degree of gender disparity is much greater than in most diagnoses (Levine & Smolak, 2006).

Now the third most common illness in adolescent females (Fisher et al., 1995), superseded only by diabetes and asthma, ED have become a major public health issue, affecting more and more women of all ages. Today they appear in every stratum of American culture and, with the impact of globalization, in more than 40 countries worldwide (Gordon, 2001). This exponential increase in a condition disproportionately affecting women must have its roots in the interplay of culture and gender, as a genetic mutation has not swept the globe. But media images of perfectly crafted female bodies and unprecedented role change have, in fact, swept the globe. The increased access to education and involvement in the workplace have transformed women's social roles dramatically, with rapid technological and market changes introducing a powerful global consumer culture and relentless expectations about appearance and beauty (Gordon, 2001). As the social changes accelerate, many women seek solace and mastery by controlling their bodies (Maine & Kelly, 2005).

Quite simply, gender creates risk. The World Health Organization's (WHO) evidence-based review of women's mental health (World Health Organization, 2000) concludes that gender is the strongest determinant of mental health, social position, and status, as well as the strongest determinant of exposure to events and conditions endangering mental health and stability. Furthermore, the WHO notes a positive relationship between the frequency and severity of social stressors and the frequency and severity of mental health problems in women. Despite the importance of gender disparities in mental health and risk for ED, the recent emphasis on biogenetic research risks minimizing the importance of the role of culture and gender in their etiology. As clinicians, we understand that the biopsychosocial whole is greater than the sum of its parts, despite the challenges this presents to the traditional research paradigms. This chapter explores the interplay of biopsychosocial factors contributing to the perfect storm of ED, especially examining culture and gender.

NATURE VERSUS NURTURE: A FALSE DICHOTOMY

Delineations between the biological, psychological, and social forces underlying ED are false distinctions, as nature and nurture always go hand in hand. Genes code RNA and DNA, the building blocks of cells, creating variations associated with risk. While they do not code behavior or disease, genes create vulnerabilities which will be tempered or intensified by other factors (Chavez & Insel, 2007), such as the family, early development, social experiences and expectations, physical conditions, and gender. Increasingly sophisticated research models investigate the complicated interactions in which environmental experience can alter gene expression (Hunter, 2005). Although they are not destiny, genes shape vulnerability and resilience, affecting how we perceive, organize, and respond to experiences, and contributing to the perfect storm of ED.

The rapid decline in the age at which girls enter puberty is an apt example of such a biopsychosocial storm. A century ago, the average age for menarche was 14.2 and now it is 12.3. In the 1970s, the average age of breast development was 11.5, but by 1997, it was less than 10 years old for Caucasian girls and 9 years old for African American girls, with a significant number developing even before age 8 (Steingraber, 2009). Girls who enter puberty earlier than peers have more self-esteem issues, anxiety, depression, adjustment reactions, eating disorders, and suicide attempts (Graber, Seeley, Brooks-Gunn & Lewinsohn, 2004). They are more likely to use drugs, alcohol, and tobacco, have earlier sexual experiences, be at increased risk of physical violence, and, due to prolonged estrogen exposure, have a higher incidence of breast cancer (Steingraber, 2009).

Early puberty may be best understood as an ecological disorder, an interaction of psychosocial, nutritional and environmental triggers, such as pollutants or chemical exposure; while family stress or trauma may also play a part. Aptly describing the false dichotomy between nature and nurture, Steingraber states: "The entire hormonal system has been subtly rewired by modern stimuli…female sexual maturation is not controlled by a ticking clock. It's more like a musical performance with girls' bodies as the keyboards and the environment as the pianist's hands" (2009, p. 52).

Sexual maturation brings increased attention to the body, sexuality, and the developmental pressures of adolescence, enhancing the impact of other ED risk-factors. Nature and nurture interact as girls' lives unfold.

GENDER: DIFFERENCE OR SIMILARITY?

Culturally constructed sexism has led to intense divisions between men and women, as expressed in common concepts such as "the war of the sexes," as if gender creates virtually different species with no hope of understanding each other. The media systematically promulgate gender differences, just as they have contributed to the objectification of women and sexism. Despite the popularity of books like *Men Are from Mars, Women Are from Venus* (Gray, 1995) and *You Just Don't Understand: Men and Women in Conversation* (Tannen, 2001), decades of psychological research suggest that men and women and boys and girls are much more alike than different (Hyde, 2005).

In their epic work, *The Psychology of Sex Differences*, Maccoby and Jacklin (1974) reviewed more than 2000 studies, dismissing many popular beliefs and identifying only four areas of difference: (i) verbal ability; (ii) visual-spatial ability; (iii) mathematical ability; and (iv) aggression. In 2005, Hyde's meta-analysis of the gender difference literature found that 78% of the differences are very small, actually close to zero, even in areas where gender differences have been consistently considered strong. The greatest gender difference is in motor performance, due to post-puberty differences in muscle mass and bone size. Measures of sexuality, especially the frequency of masturbation and attitudes toward "casual sex," also reveal significant gender differences, but virtually no difference in reported sexual satisfaction. The meta-analysis of aggression indicates a strong gender difference in physical parameters, but less so with verbal aggression. Despite the suggestion in the popular press and media that girls have a higher level of relational aggression, the evidence is mixed.

As gender differences fluctuate over the course of development, Hyde (2005) suggests that they are not as fixed as many believe. She also notes that the surrounding context, such as the written instructions, interactions between participant and experimenter, or expectations of gender differences, significantly affect results. The fact that both their strength and their direction depends on context challenges the notion of strong, stable gender differences.

NATURE, NURTURE, AND THE BRAIN

Research on the brain indicates important gender differences, despite the behavioral similarities noted above. In a thorough review of gender, Cahill (2006) noted significant gendered patterns in brain structure and neurochemistry associated with a wide range of emotional and cognitive functions including learning, emotional and social processing, memory storage, and decision-making. Male and female brains react differently to stress. Chronic stress is more damaging to the male brain, particularly to the hippocampal area thought to be central to memory and learning, while transitory interpersonal stressors result in a stronger adrenocortical response in women's brains (Stroud, 1999). At the neurochemical level, gender influences the ways in which our brains synthesize, metabolize, and respond to neurotransmitters such as serotonin, possibly helping to explain differential rates of mood disorders and substance addiction.

Brain differences have been disproportionately attributed to sex hormones, but research has now established that other distinctions exist. For example, the denser corpus callosum (the band of fibers bridging the brain's hemispheres) in the female brain allows greater connection between the two hemispheres, so women have less lateral specialization, whereas men have more of a division between the brain hemispheres. These neuroanatomical differences may explain women's superior language skills and men's superior visual-spatial skills. The neuroanatomy of the hypothalamus, instrumental in hormonal functions and reproduction, is also different, resulting in neurophysiological differences that in turn affect behavior.

The anterior cingulate gyrus, more active in women, is linked to nurturant social behaviors, while the amygdyla (more active in men) is linked to anger and rage. Although statistically significant, these differences are small (Solms & Turnbull, 2002). The environment and culture often intensify these differences with gender-laden messages, attitudes, and expectations, and thereby multiply their expression (Lee, 2007).

While the study of brain gender differences has been enhanced by technological developments, it has, perhaps, been retarded by the viewpoint that differences somehow imply deficiency. In a patriarchic culture, an androcentric bias may affect how scientific findings and models of psychopathology are interpreted.

GENDERING: A BIOPSYCHOSOCIAL PROCESS

Is it a boy or a girl? This is often the first question asked about the birth or pending birth of a baby. The answer shapes our reaction and expectations and impacts the child's life story and experience in countless ways. Simply put, the impact of gender occurs early and often.

According to social scientists, "gendering" is "the sum of all influences... that channel females and males into divergent life situations," which are then internalized into the self, leading to certain "sex-linked characteristics, cognitions, and interpersonal transactions" (Worell & Todd, 1996, p.135). By age 2, gender identity begins to emerge, with the child constructing a sense of self as either male or female (Worell & Todd, 1996).

Gender experiences interact with the gendered features of the brain to create a gendered self, and relationships with caregivers are the key arenas for these experiences to play out. Illustrating the intersections of culture, biology, and psychology, parental responses to an infant are driven by the parent's biology, by their cultural and psychological experiences, and by the gendered biology of the infant. According to Weinberg, Tronick, Cohn & Olson (1999), male and female infants display markedly different levels of emotional expressivity and arousal and evoke different parental reactions. Boys, who are less regulated, are actually more sociable than girls at this age. They seem to pull for more physical touch and perhaps greater relational involvement from their mothers due to the challenge of maintaining emotional regulation. Weinberg and her colleagues also found that boys and mothers stayed in attachment synchrony more than mother–daughter dyads but also took longer times to re-establish that synchrony after it had been disrupted. Girls, by comparison, require less soothing but also seem to present more subtle cues to their caregivers. Perhaps our belief that girls and women are more relational is rooted, at least in part, in the need for closer attention to these subtle expressions. Boys may be less relational because their emotions are so obvious.

In order to develop a sense of self, boys must psychologically separate from their primary attachment, usually their mother, and connect to the same sex figure, the father. Girls, on the other hand, must retain the connection to mother as the same sex identificatory figure but connect to the father in a new affective relationship. This developmental challenge places a premium on relational, as opposed to self-containing and separating, capabilities. Girls begin to explore who they are and who they want to be by comparing themselves with peers, parents, siblings, and the cultural images available to them (like characters in books or movies). Boys learn to harden themselves into self-sufficiency, fearful that dependence is shameful.

In optimal circumstances, these developmental challenges inter-twine with biological endowments and social values in ways that enhance and support healthy maturation. The environment can be more or less gendered, either reinforcing stereotypic behavior or allowing more room for difference or exploration. Individuals can also adjust their gender-typed behaviors in order to present in a certain way. For example, girls may act more typically feminine to get approval or attention pending the cues and demands they perceive. When psychopathology develops, it may reflect disruptions in this complicated process. When the psychopathology occurs at vastly different rates in men and women, the biopsychosocial construction of gender may be the source.

PSYCHOLOGICAL DEVELOPMENT IN A GENDERED ENVIRONMENT

Western culture is still androcentric, based on a patriarchy, as seen in our common language forms (think "chairman" of the board). Such gendered environments exert subtle, subliminal, but constant pressures on both sexes to act in certain ways. Gender stereotypes evolve based on a culture's belief systems regarding the attitudes, behaviors, and other characteristics that seem to differentiate the two sexes. This section focuses primarily on how gendering affects females, while Chapter 18 (Bunnell) examines the male experience and consequent risks for ED.

Frequent references to "the opposite sex" show our polarizing views of gender. Western culture usually emphasizes socio-emotional and body image (BI) issues when defining stereotypic femininity, and competence and autonomy when defining masculinity. These stereotypes prescribe certain behaviors: women are to take care of others and attend to their appearance, while men are to take risks, assume leadership, and focus on success and work. Such dichotomous views of gender give men public power and influence, while limiting them to women, with far-reaching consequences.

The impact of a gendered environment may intensify in the face of biopsychosocial developmental stressors. Puberty heralds both internal and external changes, clear markers of gender. For girls, it brings dramatic hormonal changes resulting in menstrual periods, breast development, and increased body fat. Between the ages of 10 and 14, in fact, the average girl gains 10 inches of height and between 40 and 50 pounds. Most double their weight by the time they finish puberty (Friedman, 1997). In addition to the physical events, puberty involves an increased attention to the demands and expectations of the dominant culture, as, emotionally ready or not, girls move from the safety of childhood into a universe increasingly driven by factors outside the family such as peers, school, and the media.

This heightened attunement to sociocultural demands or norms creates significant conflict for girls. Absorbing the external message that they need to control their weight and maintain an attractive, sculpted look, girls may feel unhappy about their body's natural changes. While body fat may be necessary to physical development, it contradicts the female ideals they have been taught, so it seems invalidating and frightening.

For some girls, this transition into puberty feels like the proverbial fall down a rabbit-hole, just like *Alice in Wonderland*, landing in a place where things may look the same but feel very, very different. One young woman in recovery from bulimia described that she went to bed at night after playing with dolls, then woke up with breasts, and everyone treated her differently.

With scant permission to explore complicated feelings about their bodies and maturation, and no rituals of celebration for this new life stage, many girls translate their distress and confusion into the language of fat (Friedman, 1997). Constant diet ads and messages about the dangers of obesity, and a weight-loss industry that now accounts for approximately $60 billion per year in the U.S. (Marketdata Enterprises, 2007), only reinforce the language of fat, making dieting normative. Again, the biopsychosocial whole is greater than the sum of the parts: the pressures to diet add to all the other developmental stressors of puberty, as "gendering" unfolds.

MEDIA IMAGES, GENDER, AND OBJECTIFICATION THEORY

Experimental and correlational studies, prospective research, and clinical accounts all validate that media images and influences are major factors in the etiology of ED (Levine & Smolak, 2006). American television and other popular media promulgate the strong message that women attract men through appearance. The media objectify women much more frequently than men and portray them as unrealistically thin (Engeln-Maddox, 2006). Meta-analytic reviews of research demonstrate that the contemporary cultural experiences of girls and women contribute to both BI dissatisfaction and to disordered eating (DE) (Murnen & Smolak, 2009). Exposure to mass media (Groesz, Levine & Murnen, 2002; Murnen, Levine, Smith & Groesz, 2007) increases the risk for ED, as do attempts to comply with traditional expectations regarding femininity (Murnen & Smolak, 1997). In essence, Western media systematically and relentlessly objectify and sexualize girls' and women's bodies, at great cost to their emotional and physical health (Maine, 2009).

Rapid social change, provocative media images, and pressures to attain the perfect body coalesce in the perfect storm of an ED. A dramatic example of the interplay of these factors occurred in Fiji after television was introduced. ED were basically non-existent there in 1995, but after less than three years of limited exposure to Western network television shows, they were rampant. Fijian girls had not spoken of diet or weight concerns previously but, by 1998, 11% used self-induced vomiting, 29% were at risk for ED, 69% had dieted to lose weight, and 74% felt "too fat." Watching popular female images on television seemed to have created a desire for the apparent life and presumed power of these stars and a commitment to change their bodies to get it. Once a culture where large female bodies were valued for their strength and contribution to the family and community life, and where food was celebrated and enjoyed with rich traditions and meanings, the Fijian experience demonstrates how rapidly cultural influences can overturn strong local cultural traditions and values, profoundly altering a woman's relationship to her body and to food (Becker, Burwell, Gilman, Herzog & Hamburg, 2002).

The media's objectification and sexualization of the female body has a lasting impact, persistently pressuring girls and women to assume an external view of themselves and of their value as people. In turn, they are less able to identify, express, process, or respect their emotions, thoughts, and instincts. Messages from the outside eclipse any inner life; in this way, social expectations regarding appearance, weight, and shape become pre-eminent, trumping women's appreciation for their natural bodies and paving the way for BI preoccupation and dissatisfaction, and eating pathology.

Contemporary culture's consistent sexualization of girls and women exerts a variety of individual harmful effects such as impaired cognitive functioning due to intrusive and negative thoughts, emotional distress, body dissatisfaction, negative self image, ED, and health problems (American Psychological Association [APA], 2007b). Objectification theory (Frederickson & Roberts, 1997) explains the lasting negative outcomes of such sexualization, especially the impact of media images and portrayals of women and the power of the sexual gaze. Girls come to see themselves as objects to be looked at and judged based on their appearance. Internalization of an external standard results in constant monitoring and self-scrutiny, so the individual has fewer resources for awareness of internal body states and experience. Thereby, the culture disrupts the connection to inner experience leading to a pervasive experience of disembodiment (Piran & Cormier, 2005), including denial of basic needs such as hunger and thirst.

GENDER DISTINCTIONS AND THE OUTCOME OF OBJECTIFICATION

Gender significantly affects the experience and outcome of objectification. According to Murnen, Smolak, Mills & Good (2003), children as young as 7 idealize objectified media images. Girls, however, are more likely to internalize and try to meet these idealized standards. Starting in childhood, girls are constantly exposed to criticisms and comments about their bodies, and not just in the media. Even in elementary school, boys feel free to disparage girls' bodies in their presence (Murnen & Smolak, 2000). With both anonymous and deeply personal messages conveying a constant stream of criticism, the average American girl struggles to simply feel safe in her own skin.

Studying children at age 11 and then at 13, Grabe, Hyde & Lindberg (2007) report that objectification appears to affect the emotional well-being of girls but not of boys. Adolescence and pre-adolescence are critical developmental periods for girls in their relationship to their bodies and their experience of objectification. Adolescent girls endure much more self-objectification, body shame, rumination, and depression than their male peers. In desperation, they internalize and try to meet the external standards, hoping to find safety and self-acceptance. Instead, they become deeply disconnected from their bodies, creating fertile ground for an ED.

Bandura's (1991) groundbreaking research demonstrates that when people fail to meet cultural ideals, intrusive thoughts can undermine cognitive functions. For women, such failure has a higher price, literally and figuratively. Cultural standards for appearance, weight, and beauty are much more rigid for women, so they experience more BI dissatisfaction than men (Ricciardelli & McCabe, 2004). In addition to their bodies being treated so differently, women also earn less, have less status and power, and are more likely to be victims of abuse and physical violence (Bordo, 1993). Intrusive and obsessive self-disparaging thoughts, absent from other areas in which they feel successful, can easily compromise their sense of self as well as their cognitive and emotional functioning.

According to Tiggemann and Slater (2001), the cumulative effect of self-objectification creates risk for three psychological issues: depression, sexual dysfunction, and ED. Self-objectification may become an internalized and long-lasting feature of the self, co-existing with many risk factors associated with ED, including diminished interoceptive awareness, poor self-esteem, body shame, and DE (Piran & Cormier, 2005). The internalization of the thin ideal, a predictable indirect result of self-objectification for girls in contemporary culture (Levine & Smolak, 2006), is one of the strongest predictors for BI distress and DE (Stice, 2002).

The effects of culture's objectification of girls and other prominent gender stereotyping not only harm girls; they foster a sexist or patriarchal culture that tolerates sexual harassment, violence, abuse, rape, and exploitation of girls and women (APA, 2007b). Sexism, however, is truly "a two-edged sword" (Sadker, Sadker & Zittleman, 2009, p. 208), victimizing boys and men as well. If girls must be objectified, boys must be objectifying, insensitive, and dismissive; if girls are to be harmed, then boys must harm through aggressive, insensitive, angry, and rejecting behaviors. Pushed to be independent, competitive, athletic, and disconnected from their own inner states much of the time, boys are also harmed by the culture's sexualization and objectification of women.

FEMINIST PERSPECTIVES: PROTECTION IN THE FACE OF OBJECTIFICATION?

Murnen and Smolak (2009) suggest that women who identify with feminism may be more resilient and less likely to succumb to concerns about BI or eating. Their hypothesis, that a feminist perspective may help girls resist objectification and other cultural forces that create risk for ED, has important implications for treatment and prevention. A feminist perspective may enable women to conceptualize objectification and cultural pressures related to weight, shape, and appearance as examples of the oppression of women, and may foster a critical perspective of media images for women, encouraging women to reject these messages or to limit exposure to them. When experiencing sexual harassment or any form of sexual discrimination, a feminist consciousness may help the individual to contextualize and externalize the experience rather than internalize or blame herself. This mindset also empowers a woman to define herself in areas of strength, rather than by focusing on imperfections, and to value internal attributes more than appearance. They also may be more capable of asserting themselves, expressing their beliefs, and avoiding the self-silencing that can lead to DE (Smolak & Munstertieger, 2002).

In focus groups, college-aged women who endorsed a feminist perspective were more likely to embrace body diversity and to be aware of the negative messages women receive about their bodies (Rubin, Nemeroff, & Russo, 2004). Feminism allowed them to resist the objectifying and sexualizing male gaze, to be more confident of their bodies, and to redefine beauty. Yet, despite such resilient attitudes, these women still felt pressured to comply with cultural values regarding women and weight. Other research using quantitative measures has been inconsistent. Tiggemann and Stevens (1999) reported a moderate positive association between feminist identity and BI, while Cash, Ancis & Strachan (1997) reported only a minimal association.

To shed light on the protective potential of feminism, Murnen and Smolak (2007) conducted a meta-analysis of 26 studies examining the link between feminism and BI, concluding that feminist identity and a lower drive for thinness go hand in hand. Adoption of a feminist perspective seemed to immunize women against the internalization of unhealthy attitudes about their bodies. Older feminists, lesbians, women's studies students, and activists derive the most benefit from their association with feminism, while younger women, or those less connected to feminism, struggle more with the pressures to be thin. A consolidated feminist identity, however, appeared to be a protective factor against dieting and ED symptoms, suggesting the value of a feminist-informed approach to treatment and prevention.

While the medical model focuses on the individual woman as sick or defective and attempts to subdue her disease, the feminist framework literally frames the woman's ED and behavior in the context of her cultural experience (Maine, 2009). Feminist-informed therapists recognize that the common issues prompting women to seek therapy are often related to powerlessness, experiences of trauma, low self-esteem, and the idealization of masculine qualities and the devaluation of feminine ones (Katzman, Nasser & Noordenbos, 2007). In other words, sexism and oppression bring women into treatment.

The feminist frame conceptualizes an ED as a solution to these problems, rather than being the problem itself. Moving far beyond symptom management, feminist-oriented treatment requires exploring the impact of gender-prescribed roles in contemporary culture and of the idealization of the masculine and devaluation of the feminine. Rather than pathologizing, a feminist approach is strength-based so women's needs for emotional connection and interdependence are seen as assets, not signs of weakness or dependency (Katzman et al., 2007; Sesan, 1994). These strengths are incorporated into treatment plans by encouraging interpersonal connections, utilizing group therapies, and including natural support systems such as families or friends. Furthermore, to counter the impact of objectification, involvement in collective social action may enhance recovery, restoring power and voice to the individual and a sense of connection to other like-minded people.

The feminist frame is also attuned to power, acknowledging the clinician's implicit power and the patient's likely chronic experiences of disempowerment (Sesan, 1994). Strategies to minimize the power differential include constant collaboration between the clinician and the patient and psycho-education, so the patient can make informed decisions about her treatment.

GENDER-INFORMED TREATMENT

Each patient brings her own rich constellation of biology, psychology, and culture into treatment, as does each clinician. Mature therapists learn to distinguish between acculturated reactions and their own idiosyncratic reactions by examining questions like: *"How am I as a therapist different with male and female patients?"* or, *"What biases color my reactions?"* and, *"What gendered expectations do patients have of me?"* Male therapists need to introduce these topics into their work with their female patients with ED. Female therapists must model how to work through their own gender biases and experiences and to be open to their patients' perceptions of body, maternality, sexuality, and experiences of power differentials (Piran, 2001).

Even in same sexed therapist–patient pairs, assumptions, values, and biases will shape the relationship. Psychotherapy research suggests that gender-stereotypes in diagnosis and treatment endure, although these may be more subtle than in the past. Women, for instance, are more likely to be diagnosed with Axis II disorders than men (APA, 2007b) and less likely to be diagnosed with PTSD. They also have higher rates of anxiety and depression and other internalizing disorders (Blatt, 2008). Clinicians may tend to attribute women's psychological disorders to endogenous or intrapsychic factors and minimize the centrality of external forces. On a more subjective level, many still label a female patient's emotional expressiveness as "needy" or "dependent."

These issues have particular power when treating women with ED. Culture and experience infuse the genetic vulnerabilities of these women with criticism and self-loathing, especially for the body. Consequently, feelings of fatness, fear of weight gain, the over-valuation of thinness, and the panic of fullness evoke a sense of being flawed or unacceptable. Experiences of helplessness, oppression, bias, violence, and trauma all affect self-concept (APA, 2007b). The internalization of the thin body ideal can be equally damaging, as seen in every treatment encounter with ED women.

The importance of gender is immediately apparent in the examination of the specific psychopathology of ED (Fairburn, 2008). Young girls are taught to use restraint, especially "dietary" restraint, and to monitor their value by body checking and constant self-scrutiny. While the over-evaluation of thinness, central in the cognitive model of eating pathology, appears to be relevant for both sexes, it is most consistent with the values of the female culture. Any effort to challenge this requires exploration of that patient's sense of her own gender. Other aspects of eating psychopathology (food avoidance, desire for emptiness, adherence to strict rules, fear of loss of control, avoidance of social eating, guilt and shame about eating) also reflect a gendered indoctrination regarding acceptable female behavior. Conforming to the thin body ideal provides a degree of stability and certainty and choosing to relinquish an ED comes at a high price including mourning, identity confusion, and ambiguity. As, historically, women have paid dearly for such non-conformity, time spent in psychotherapy on the exploration of the social context of gender and ED is time well spent.

Psychotherapy must help patients make sense of the biopsychosocial storm that led to their ED, through explicit discussion of what it means to be a woman with all its associated values, dreams, aspirations, hopes, fears, and anxieties. While directive, symptom-focused approaches have demonstrable power to help establish stability and normalization of eating, longer term recovery requires something more subjective. How can patients explore the significance of pleasure, desire, emotional intimacy, and safety if they only talk about eating, weight, and shape? Providing empathy and connection are rarely enough either. Effective treatment demands an integration that simultaneously sustains a collaborative alliance and an explicit push for change. The heart of the psychotherapeutic art conveys empathic acceptance while simultaneously pushing for things to be different. Although researchers continue to explore the power of the therapeutic alliance, we, to date, lack the quantitative sophistication to capture this core feature of therapeutic effectiveness (Maine, 2009).

Typically, clinical training and empirical treatment research have focused on the measurable behaviors and symptoms of DE. It is less clear how we can assess the role of relational connection or attachment, flexibility and adaptability in self evaluation, awareness of internal experiences, and issues such as trust, desire, intimacy, values, meaning, and spirituality. All these have a significant connection to gender and may have as much to do with full recovery as does the elimination of specific behaviors (see McGilley and Szablewski, Chapter 12).

ENHANCING GENDER COMPETENCE

As gender is a force in all relationships, including psychotherapy, and a core ingredient of the perfect storm of ED, we must appreciate, understand, and use it to enhance recovery.

Acknowledging its importance in psychological development and well-being, the American Psychological Association (APA, 2007a) has created guidelines for gender informed treatment, addressing eleven key features of gender "competence." Each of these dimensions has direct implications for both the specific aspects of eating psychopathology and more general aspects of psychological functioning such as self esteem, trust, safety, and intimacy. The APA stresses that clinicians must attend to these contextualizing factors and evaluate their conceptualizations, etiological models, and treatment approaches using these principles. This is especially critical when treating women with ED. Below we list the core competencies as proposed by APA and relate them to the treatment of ED.

- *Effects of socialization, stereotyping, and unique life events.* Case formulations need to include consideration of gender socialization and stereotyping. How does a woman's struggle with weight, shape, and eating reflect her sense of being female?
- *Effects of oppression and bias.* As noted earlier, we are at risk for minimizing the effects of oppression and disempowerment on the social, gender, and sexual identity development of our patients with ED. Therapists, too, must stay aware of how these forces impact our own sense of self and our therapeutic interactions
- *Bias and Discrimination.* These forces have demonstrable effects on physical and mental health. As we evaluate our patients, we need to explore the role of these experiences. Is the lack of research and treatment resources for ED an example of bias and discrimination?
- *Gender Sensitive and Affirming Procedures.* Therapists must incorporate explicit techniques that affirm and support girls and women. We should also be continually evaluating our theories and treatment models for negative gender bias
- *Impact on Therapist's Practice.* This guideline emphasizes the importance of therapist gender and self awareness. How do our own gender biases and experiences influence our work? Their influence is inevitable but we can monitor their effects and become more gender-sensitive
- *Evidence Based on Gender Sensitive Research.* The APA guidelines stress the importance of using gender sensitive research methods. Most of what we know about ED has, obviously, been based on research on women. But, are our methods free of bias? Perhaps the relative de-valuation of qualitative research reflects gender bias that undermines the field's knowledge and treatment approaches
- *Promote Initiative and Empowerment.* This is particularly relevant in that recovery from an ED requires a shift in self evaluation. Expanding the range of choices and opportunities for women enhances this shift away from a focus on weight, shape, and body as the foundations for self esteem
- *Assessment and Diagnosis.* Sociocultural values may skew assessment of things like emotionality, dependency, lability, and affective intensity. This may affect diagnosis of personality disorder in particular. As these features often accompany eating pathology, ED clinicians must be especially vigilant to their own biases, beliefs, and practices regarding assessment and diagnosis
- *Sociopolitical context.* What starts as a sociocultural attitude or trend can become an internalized constant. In addressing issues of body, eating, self-concept, femininity, or relationships, therapists need to keep their ears open for connections to their patients' external cultures. How have our patients taken in the messages from the culture and made

them their own? How can we best increase their awareness of the power of the sociocultural context?
- *Identifying Resources for Support.* In asking our patients to relinquish their ED, we need to help them incorporate other sources of support into their lives. This guideline pushes the gender sensitive clinician to identify these alternative sources in patient's relationships and community
- *Challenging Institutional Bias.* This final guideline encourages therapists to actively engage in challenging institutional bias and barriers. Women with ED are marginalized and diminished by the culture. Treatment is hard to access and difficult to afford. Research is underfunded. Even to this day, families, and mostly mothers, are still treated as the cause of ED (Maine, 2004). Therapists must look for opportunities to challenge these biases through advocacy, education, eating disorder prevention, or research. For some patients, the final phase of real recovery may require their own efforts to speak out and challenge the culture in some broader way.

CONCLUSION

Gender is a critical factor in ED, as being born female remains the greatest risk factor associated with these diagnoses. It may, however, be both a chicken and an egg, simultaneously creating critical biopsychosocial realities and reflecting them. As contemporary culture promotes an overevaluation of thinness, excessive self-criticism, and disembodiment in women, culture and gender create an essential context for women with ED; therapists must explicitly acknowledge and attend to that context.

To explore and acknowledge the power of external factors in the etiology and maintenance of eating pathology does not ignore the significance of biological and genetic vulnerabilities. We miss important aspects of our patients' lives and of their individual psychology if we minimize the power of the gendered context, including the effects of bias, marginalization, trauma, discrimination, and the roles and rights of women. Effective treatment requires a balance between individual adaptation and cultural gender awareness, and a deep appreciation for all the biopsychosocial elements of the perfect storm of eating disorders.

References

Andersen, A. E. (2002). Eating disorders in males. In C. G. Fairburn, & K. D. Brownell (Eds.), *Eating disorders and obesity* (pp. 188–192). New York, NY: Guilford Press.

American Psychological Association. (2007a). *Guidelines for psychological practice with girls and women.* Washington, DC: Author.

American Psychological Association. (2007b). *Report of the APA Task Force on the Sexualization of Girls.* Washington, DC: Author.

Bandura, A. (1991). Self-regulation through anticipatory and self-reactive mechanisms. In R. A. Dienstbier (Ed.), *Nebraska symposium on motivation: Vol. 38: Perspectives on motivation* (pp. 69–164). Lincoln, NE: University of Nebraska Press.

Becker, A. E., Burwell, R. A., Gilman, S. E., Herzog, D. H., & Hamburg, P. (2002). Eating behaviors and attitudes following prolonged exposure to television among ethnic Fijian adolescent girls. *British Journal of Psychiatry, 180,* 509–514.

Bordo, S. (1993). *Unbearable weight: Feminism, Western culture, and the body.* Berkeley, CA: University of California Press.

Blatt, S. (2008). Two primary configurations of psychopathology. In S. Blatt (Ed.), *Polarities of experience* (pp. 165–199). Washington, DC: American Psychological Association.

Cahill, L. (2006). Why sex matters for neuroscience. *Nature Reviews Neuroscience, 7,* 477–484.

Cash, T. F., Ancis, J. R., & Strachan, M. D. (1997). Gender attitudes, feminist identity, and body images among college women. *Sex Roles, 36,* 433–447.

Chavez, M., & Insel, T. R. (2007). Eating disorders: National Institute of Mental Health's perspective. *American Psychologist, 62*(3), 159–166.

Engeln-Maddox, R. (2006). Buying a beauty standard or dreaming of a new life? Expectations associated with media ideals. *Psychology of Women Quarterly, 30,* 258–266.

Fairburn, C. G. (2008). *Cognitive behavior therapy and eating disorders.* New York, NY: The Guilford Press.

Fisher, M., Golden, N. H., Katzman, D. K., Kreipe, R. E., Rees, J., Schebendach, J., ... Hobesman, H. M. (1995). Eating disorders in adolescents: A background paper. *Journal of Adolescent Health, 16,* 420–437.

Frederickson, B. L., & Roberts, T. A. (1997). Objectification theory: Toward understanding women's lived experiences and mental health risks. *Psychology of Women Quarterly, 21,* 173–206.

Friedman, S. S. (1997). *When girls feel fat.* Toronto, CA: Harper Collins.

Gordon, R. A. (2001). Eating disorders East and West: A culture-bound syndrome unbound. In M. Nasser, M. A. Katzman, & R. A. Gordon (Eds.), *Eating disorders and cultures in transition* (pp. 1–23). New York, NY: Taylor and Francis.

Grabe, S., Hyde, J. S., & Lindberg, S. M. (2007). Body objectification and depression in adolescents: The role of gender, shame, and rumination. *Psychology of Women Quarterly, 31,* 164–175.

Graber, J. A., Seeley, J. R., Brooks-Gunn, J., & Lewinsohn, P. M. (2004). Is pubertal timing associated with psychopathology in young adulthood? *Journal of the American Academy of Child and Adolescent Psychiatry, 43,* 718–726.

Gray, J. (1995). *Men are from Mars, women are from Venus.* New York, NY: Harper Collins.

Groesz, L. M., Levine, M. P., & Murnen, S. K. (2002). The effects of experimental presentation of thin media images on body satisfaction: A meta-analytic review. *International Journal of Eating Disorders, 31,* 1–16.

Hunter, D. J. (2005). Gene-environment interactions in human disease. *Nature Reviews Genetics, 6,* 287–298.

Hyde, J. S. (2005). The gender similarities hypothesis. *American Psychologist, 60*(6), 581–592.

Katzman, M. A., Nasser, M., & Noordenbos, G. (2007). Feminist therapies. In M. Nasser, K. Baistow, & J. Treasure (Eds.), *The female body in mind: The interface between the female body and mental health* (pp. 205–213). London: Routledge.

Lee, T. (2007). When the personal gets in the way of the interpersonal. In M. Nasser, K. Baistow, & J. Treasure (Eds.), *The female body in mind: The interface between the female body and mental health* (pp. 162–177). London: Routledge.

Levine, M. P., & Smolak, L. (2006). *The prevention of eating problems and eating disorders: Theory, research and practice.* Mahwah, NJ: Lawrence Erlbaum Associates.

Maccoby, E. E., & Jacklin, C. N. (1974). *The psychology of sex differences.* Stanford, CA: Stanford University Press.

Maine, M. (2004). *Father hunger: Fathers, daughters, and the pursuit of thinness.* Carlsbad, CA: Gurze.

Maine, M. (2009). Beyond the medical model: A feminist frame for eating disorders. In M. Maine, W. M. Davis, & J. Shure (Eds.), *Effective clinical practice in the treatment of eating disorders: The heart of the matter* (pp. 3–17). New York, NY: Routledge.

Maine, M., & Kelly, J. (2005). *The body myth: Adult women and the pressure to be perfect.* Hoboken, NJ: Wiley.

Marketdata Enterprises. (2007). *The U.S. weight loss & diet control market* (9th ed.). Market Research Study. Retrieved 5 June 2009 from. www.prwebdirect.com/releases/2007/4/preweb520127.php.

Murnen, S. K., & Smolak, L. (1997). Femininity, masculinity, and disordered eating: A meta-analytic review. *International Journal of Eating Disorders, 22,* 231–242.

Murnen, S. K., & Smolak, L. (2000). The experience of sexual harassment among grade-school students: Early socialization of female subordination? *Sex Roles, 43,* 1–17.

Murnen, S. K., Smolak, L., Mills, J. A., & Good, L. (2003). Thin, sexy women and strong, muscular men: Grade-school children's responses to objectified images of women and men. *Sex Roles, 43,* 1–17.

Murnen, S. K., Levine, M. P., Smith., J., & Groesz, L. (2007, August). *Do fashion magazines promote body dissatisfaction in girls and women? A meta-analytic review.* San Francisco, CA: Paper presented at the American Psychological Association.

Murnen, S. K., & Smolak, L. (2009). Are feminist women protected from body image problems? A meta-analytic review of the relevant research. *Sex Roles, 60,* 186–197.

Piran, N. (2001). A gendered perspective on eating disorders and disordered eating. In J. Worell (Ed.), *Encyclopedia of women and gender: Sex similarities and differences and the impact of society on gender* (pp. 369–378). San Diego, CA: Academic Press.

Piran, N., & Cormier, H. C. (2005). The social construction of women and disordered eating patterns. *Journal of Counseling Psychology, 52,* 549–558.

Ricciardelli, L. A., & McCabe, M. P. (2004). A biopsychosocial model of disordered eating and the pursuit of muscularity in adolescent boys. *Psychological Bulletin, 130,* 179–205.

Rubin, L. R., Nemeroff, C. J., & Russo, N. F. (2004). Exploring feminist women's body consciousness. *Psychology of Women Quarterly, 28,* 27–37.

Sadker, D., Sadker, M., & Zittleman, K. (2009). *Still failing at fairness.* New York, NY: Scribner.

Sesan, R. (1994). Feminist inpatient treatment for eating disorders: An oxymoron? In P. Fallon, M. A. Katzman, & S. C. Wooley (Eds.), *Feminist perspectives on eating disorders* (pp. 251–271). New York, NY: Guilford.

Smolak, L., & Munstertieger, B. F. (2002). The relationship of gender and voice to depression and eating disorders. *Psychology of Women Quarterly, 26,* 234–241.

Solms, M., & Turnbull, O. (2002). *The brain and the inner world: An introduction to the neuroscience of subjective experience.* London, England: Karnac.

Steingraber, S. (2009). Girls gone grown-up: Why are U.S. girls reaching puberty earlier and earlier? In S. Olfman (Ed.), *The sexualization of childhood* (pp. 51–62). Westport, CT: Praeger Publishing.

Stice, E. (2002). Risk and maintenance factors for eating pathology: A meta-analytic review. *Psychological Bulletin, 128,* 825–848.

Striegel-Moore, R. H., & Bulik, C. M. (2007). Risk factors for eating disorders. *American Psychologist, 62*(3), 181–198.

Stroud, L. R. (1999). Sex differences in adrenocortical responses to achievement and interpersonal stressors. *Dissertation Abstracts International, 60,* 1317B.

Tannen, D. (2001). *You just don't understand: Men and women in conversation.* New York, NY: Harper.

Tiggeman, M., & Stevens, C. (1999). Weight concerns across the lifespan: Relationship to self-esteem and feminist identity. *International Journal of Eating Disorders, 26,* 103–106.

Tiggemann, M., & Slater, A. (2001). A test of objectification theory in former dancers and non-dancers. *Psychology of Women Quarterly, 2,* 57–64.

Treasure, J. (2007). The trauma of self-starvation: Eating disorders and body image. In M. Nasser, K. Baistow, & J. Treasure (Eds.), *The female body in mind: The interface between the female body and mental health* (pp. 57–71). London, England: Routledge.

Weinberg, M., Tronick, E., Cohn, J., & Olson, K. (1999). Gender differences in emotional expressivity and self-regulation during early infancy. *Developmental Psychology, 35,* 175–188.

Worell, J., & Todd, J. (1996). Development of the gendered self. In L. Smolak, M. P. Levine, & R. Striegel-Moore (Eds.), *The developmental psychopathology of eating disorders: Implications for research, prevention, and treatment* (pp. 135–156). Mahwah, NJ: Lawrence Erlbaum Associates.

World Health Organization. (2000). *Women's mental health: An evidence based review.* Geneva, Switzerland: Mental Health Determinants and Populations, Department of Mental Health and Substance Dependence.

CHAPTER 2

What's Weight Got to Do with It?
Weight Neutrality in the Health at Every Size Paradigm and Its Implications for Clinical Practice

Deborah Burgard

My new client, who is fat, walks into the room, sits down across from me, and says, *"Well, as you can see, I'm a compulsive eater."*

I am still brought up short by this fairly common opening line. *"Ah, well, that is not something one can see with the naked eye,"* I say. *"Tell me about it."*

Our therapy is off and running, characterized from the start by the oppression my client has internalized about her body. She has felt the need to say first what she fears I am diagnosing simply by looking at her. Not knowing quite what to make of my comment, she goes on to tell the story of her problem as she understands it: that despite trying many diets, she has never been able to maintain any lost weight; that she despises herself for not having the control to keep restricting herself on a diet; that her body must be punishing her by making her fat; that she can't figure out why she can plan to have a "good" day with food and then find herself binging in front of the TV after her roommates have gone to bed; and that she is here because she hopes that understanding the reasons behind her eating habits will allow her to finally be thin and have the life she has dreamed of. She believes that there is something wrong with her self-control that therapy can fix, and her body and her life will be made over.

The time we spent initially speaking on the phone, when I told her about the Health at Every Size (HAES) model (Burgard, 2009), summarized in Box 2.1, and that I do not have a way to help her lose weight, has apparently not made a lasting impression. Indeed, we will experience this over and over during the course of our work together. The cultural beliefs about weight (including the meaning of a particular body size; the worth, health, and assumed behaviors of people who are thin versus fat; the power of thinness to confer blessings; the blame on fat for almost anything bad that happens) are almost seamless for most patients. People have their own internal versions of this story. Furthermore, they are surrounded by family, friends, news stories, medical personnel, and even workplace policies that reinforce the ideas that fat is bad, fat people make themselves fat by overeating, fat

> **BOX 2.1**
>
> ## WHAT IS HEALTH AT EVERY SIZE?
>
> - Accepting and respecting the diversity of body shapes and sizes
> - Recognizing that health and well-being are multi-dimensional and that they include physical, social, spiritual, occupational, emotional, and intellectual aspects
> - Promoting all aspects of health and well-being for people of all sizes
> - Promoting eating in a manner which balances individual nutritional needs, hunger, satiety, appetite, and pleasure
> - Promoting individually appropriate, enjoyable, life-enhancing physical activity, rather than exercise that is focused on a goal of weight loss.
>
> ASDAH, Health at Every Size (HAES) Principles, www.sizediversityandhealth.org

people will keel over and die, and right living will make everyone thin. Even though I am challenging those beliefs, and my patient is bemused, tolerant, or even agreeing with me while she is in my office, when she gets out the door she defaults back to the dominant cultural views.

Several months later, I discover my client's reaction to my initial comment, when she relates a recent conversation with her mother:

> *I told my mom, when we met that first time, you didn't assume that I overate. It was so weird, I went home wondering if you knew what you were doing. I mean, how could you not know that I must be overeating, after treating people with eating disorders? But I just felt kind of lighter—it was hard to explain. I guess it felt like maybe there was hope—like you saw me, not some Big Fat Couch Potato who eats all day.*

THE PANIC ABOUT FAT

The list of beliefs my patient voices is the same list most people, including health professionals, hold about fat and fat people. At various times, I have observed clinicians and researchers expressing the following:

- Body Mass Index (BMI) is a decent proxy for health, and existing categories of BMI accurately reflect risks for health problems and premature death
- Behavioral interventions (diet and exercise) will result in weight loss
- Health risks for reduced-fat people are the same as those who were never fat
- People can maintain weight loss if they try hard enough
- Fat people's bodies, or certainly their appetites, or surely their intake, must be pathological
- Fat people are walking time bombs for heart attacks and diabetes

- Fat people are weak and self-indulgent; or, in clinical language, they must be more impulsive or less able to inhibit themselves
- Therapy will allow people to achieve a "normal" weight by addressing their "issues"
- The most important intervention for a fat patient is to help her lose weight, so her life can be worth living.

When one begins from a neutral position on these statements, however, and asks for evidence, very little is available; in fact considerable evidence *contradicts* these assumptions. Many of the assumptions have not been adequately tested because people presume them to be true and not necessary to test in the first place. Others have been inadequately tested because of some common methodological problems: for example, attributing causality to a correlation; not understanding the magnitude of statistical relationships; or using clinical populations and generalizing to fat people in general. And most bewildering, many of the assumptions that have been proven wrong over and over never seem to die.

An adequate review of these issues is beyond the scope of this chapter, but Bacon (2008), Campos (2004), and Gaesser (2002) have provided comprehensive analyses of the relevant literature. Their work challenges us to face some important facts about "obesity" if we hope to provide ethical, effective, and safe health care:

- No known interventions lead to the long-term achievement of "normal" weight for the great majority of people classified as "obese," including weight loss surgery (Friedman, 2009; Mann et al., 2007)
- The people in the "overweight" and "mildly obese" BMI categories live the longest (Flegal, Graubard, Williamson & Gail, 2005; Orpana et al., 2009). For older adults (>69 years of age), higher BMI is a protective rather than a risk factor for mortality (Flegal et al., 2005; Tamakoshi et al., 2009)
- The diseases and conditions more highly correlated with higher weight occur across the weight spectrum and are known to vary with many confounding factors, such as: socio-economic status (SES) (Ernsberger, 2009); weight cycling (Kruger, Galuska, Serdula & Jones, 2004); stigma (Puhl & Heuer, 2009); disparities in access to medical care (Amy, Aalborg, Lyons & Keranen, 2006); and physical activity (Williamson et al., 1993)
- Most healthcare providers, including (especially!) clinicians who work with fat patients, are weight biased and feel frustrated by interactions with fat patients (Puhl & Heuer, 2009)
- Most fat people do not have eating disorders (ED) (Hudson, Hiripi, Pope & Kessler, 2007)
- Most fat people do not have psychiatric problems (Friedman & Brownell, 2002; Marcus & Wildes, 2009; Simon et al., 2006). A substantial proportion of fat people, especially women, do not have and never go on to develop diabetes, high blood pressure, high cholesterol, or heart disease (Sims, 2001; Wildman et al., 2008)
- When fat people make sustainable changes in their health practices, they reduce or eliminate the risk factors and health conditions associated with high BMI even when their weights do not change (Bacon, Van Loan, Stern, & Keim, 2005).

Weights have indeed risen in the past generation, by about 10–15 pounds on average (Flegal et al., 2005). The group of the very heaviest people (BMI>40; as an illustration, a 5'4" woman who weighs 234 pounds), although few in number, is now double the size it was in 1988 (5.9 vs. 2.9% of the adult population) according to the Centers for Disease Control

(2008). The percentiles established in the 1970s for children's weights, when the 95th percentile represented the heaviest 5%, have been continued in use even though now there is 14—19% (depending on the age group) in the "95th percentile" (Centers for Disease Control, 2010). Since 2004—2005, the weight gains seem to be leveling off (Centers for Disease Control, 2008) and we have no evidence for interventions that make lasting weight changes; still, proposals and policies for the "War on Obesity" continue to proliferate.

The conflating of "normal" BMI with health leads to statements like, "If all U.S. adults became nonsmokers of normal weight by 2020, we forecast that the life expectancy of an 18-year-old would increase by 3.76 life-years or 5.16 quality-adjusted years" (Stewart, Cutler & Rosen, 2009). But there is no evidence that: (1) people who have reduced their weight have the same health risks as those who were never heavier; (2) the higher health risks for higher-BMI people are caused by higher weight rather than the confounding variables; and (3) there is a way for fat people to become and remain "normal weight." It is similar to telling poor people that they really should consider making a US$1 million, because millionaires have better health profiles and live longer.

The neutral facts about weight changes in our population are not met with scientific curiosity. Instead, they are assumed to herald a growing (and impending) disaster for health, the economy (Finkelstein, Trogdon, Cohen & Dietz, 2009), even the global climate (Edwards & Roberts, 2009). Perhaps, like our ED patients, most people in our culture turn to trying to control weight when faced with dread and anxiety about forces that seem outside our control.

ABOUT THE DATA

One thread of my conversation with this patient concerns the nature of evidence. How do we know what (we think) we know about weight? And, what motivates us to maintain these beliefs?

Starting from scratch, we have just the fact of a person's weight, but not an assumption about it. Body size has acquired different meanings in different cultures, and at different times in history. At one point in time, fatness implied wealth; in our time, it implies being lower class (Ernsberger, 2009; Klein, 2001). These meanings are specific to time and place; they have no intrinsic relationship to fatness itself.

Examining my own beliefs and the evidence that shapes my thinking about weight, I have found several major sources. One is my personal experience growing up as a "90th percentile" child in height and weight. I know what it felt like to be weighed in front of the other kids, how I compared myself to the more delicately-built girls, and how it felt to have my parents fret about my weight. I also experienced the anguish of people in my family who had ED, with deeply held beliefs about weight. I was in the minority in rejecting the process of dieting at 19, though the other members of my family came to embrace a focus on health practices rather than weight in the decades that followed.

A second major source of evidence has been my clinical experience. The fact that I work with people with ED across the weight spectrum, as well as with people of all weights who are thriving and happy, has allowed me to test the hypothesis that fat is a proxy for ill health, depression, and overeating. I am privileged to see a greater range of health, moods, eating styles, and weights, than the vast majority of people do, and to get to know these people intimately. The truth is, people who binge come in all sizes. People who restrict come in all sizes. People who purge come in all sizes. Despite my decades of experience, I really cannot look at someone and know what they are going to tell me about what they are doing with food and exercise.

Moreover, as an ED specialist, I have been trained to challenge the beliefs of thin people who overvalue thinness and dread fatness. I have seen my patients' lives resume a normal developmental course when they can allow themselves to eat enough and let their bodies' genetic inheritance determine their "healthy" weight. I have seen that their weight-based beliefs were part of the problem, an uncontroversial and empirically supported aspect of ED treatment (Wilson & Fairburn, 1993).

Oddly enough, the very same beliefs about weight are somehow *not* seen as part of the problem if the patient is fat. In fact, some suggest that if fat people do *not* see themselves as sufficiently pathological, professionals should "educate" them to abhor being fat and to engage in the same weight loss attempts as the culturally dominant (white, professional) group (Bryner, 2009). Anyone who has worked with upper-middle class white women knows that this is not the group to use as role models for positive body image.

Finally, the third source of evidence is the vast number of studies from the ED, "obesity," and health fields that contribute to our understanding of weight dynamics, eating practices, and health. In my reading of this literature over 30 years, I have tried to capture and utilize the evidence that is solid, and maintain an agnostic stance toward the research that is too biased or too incomplete to provide answers.

THE VALUE OF A WEIGHT NEUTRAL STANCE: PART 1

Making the transition to a weight neutral world view is a daunting task. Such a perspective is a decidedly minority position, in opposition to broadly held cultural assumptions, and will most likely be under assault for years to come. Why even try to swim upstream against the cultural current?

The answer is that, over and over again, we have seen that members of oppressed groups make social progress and improve their mental and physical health by rejecting the negative and stigmatizing messages, paying attention to the truth of their own experience, comparing notes with other members of their group, and identifying the stereotypes, stigma, violence, and discrimination they face. Discrimination based on weight is as common as discrimination based on race and age, and more prevalent than that based on sexual identity, disabilities, and religious beliefs (Latner, O'Brien, Durso, Brinkman & MacDonald, 2008; Puhl, Andreyeva & Brownell, 2008). The correlation between higher weight and a lower SES seems to be causal in both directions (Ernsberger, 2009) in that poverty tends to restrict opportunities for safe physical activity and access to good nutrition, and discrimination against fat people in hiring and retention lowers their income. Public health interventions must address social justice issues like inequality, discrimination, and poverty, not just focus on individual choices. For people to do the hard work of caring for themselves, defending themselves, or loving and advocating for themselves, they have to believe they are worth caring for. Caring for our bodies is hard work, and some environments make it especially difficult. The first order of business is to understand that all bodies, of all sizes and all socio-economic classes, are precious and deserve care.

Historically we have witnessed oppressed groups move through a process where they transform shame into pride, and society's view changes from devaluing them to seeing the value of human diversity. Body size is a biological trait bound to vary widely in a population with a diverse ancestry. It is not due to any kind of pathology, but stems from having evolved from ancestors who survived by adapting to many different environments. Rather than understanding the global nature of our population, we still have a village-scale view of the issue of body size, where one group decides what is "normal" for "us," and the rest of the people ("they") are sick/wrong simply because they are fatter. But what if we humans did not have some people who are gifted at making fat from food? Are we so advanced that we can confidently predict that we will never need such a capacity for humans to survive in the future? Is such a capacity better labeled a pathology, to be dispensed with through "treatment," through surgery or medications or genetic manipulations, in order to satisfy the temporary esthetic ideals of the twenty-first century Western world? Would it not be better to preserve our capacities but continue to find ways of treating diseases?

We know that stigma itself causes ill health (Puhl & Latner, 2007), and that many of the heaviest people have weight cycled the most (Weiss, Galuska, Khan & Serdula, 2006). They are also poorer than thinner people (Ernsberger, 2009). But studies do not always control for SES, and none has controlled for weight cycling or stigma. Thus, we do not ever know when a health finding correlated to BMI is really caused by the numerous variables that co-vary with BMI. If, for example, weight cycling causes hypertension, and we conclude that because higher BMI is a risk factor for hypertension we should prescribe weight loss, and prescribing weight loss is, in essence, prescribing weight cycling, we are going to make people sicker and fatter with our "treatment."

In the realm of "obesity" treatment, the solution to being stigmatized is seen as weight loss: remove the person from the stigmatized group, and the assumption is that there is no longer a problem. But we do not understand the identification process with our bodies very well if we imagine that weight loss "removes" the "fat self" or the memories and expectations of being treated as fat. In fact, we should know from treating thin people with ED that a person may never have been fat at all to have a very strong identification with a "fat self" that is despised, feared, and rebelliously embraced by turns.

Cultural Meanings of Fat and Thin

Our culture associates fat with certain feeling states, causing us to act out complicated interactions with our bodies that really represent conversations and struggles we are having with disowned parts of ourselves.

Table 2.1 illustrates how our "fat" can denote literal evidence of vulnerable and shameful feeling states. Thin people are seen as free from these burdens, and more protected from being exposed, but the public relentlessly projects these stereotypes onto fat people. Whether they happen to feel depressed today or not, fat people do have a very real challenge, day in and day out, navigating these endless projections, assumptions, stereotypes, and objectifications based on body size. Because we all learn to associate certain emotional states with fat, fatness is "read" as tangible evidence of these vulnerable feelings. It is seen as shameful, and while such feelings are part of the human repertoire and thus experienced by all, it is the fat

TABLE 2.1 Associations to Thin and Fat

Thin	Fat
Successful	Loser
Confident	Insecure
In control	Out of control
Attractive	Ugly
Well-groomed	Blob
Graceful	Awkward
Conceited	Grateful for attention
Happy	Depressed, angry
Athletic	Couch potato
No eating problems	Overeater
Healthy	Impending illness, doom
Follows medical advice	Non-compliant
Good citizen	Scapegoat
Male	Female

people in our culture who bear the brunt of our projections. It is as if their innermost vulnerable feelings are being worn on the outside.

ED therapists know that thin people are every bit as likely to suffer from the internal experience of vulnerability; yet their social experience is different since they are more likely to live in fear, rather than in reality, that this vulnerability will be exposed. This fear leads to anxiety about eating in a way that someone else may view as out of control, being perceived as fat. Looking in the mirror, they recognize their flawed internal selves in the image before them.

And so, to be fat is to deal with people's projections that you are a loser, whether you feel like one or not; and, to be thin is to live in fear of exposure of your inner loser-ness. Clearly, changing one's body size provides no solution to the existential question of how to claim one's imperfection as a human being.

Part of my job is to see my fat patient without the assumptions. I need to learn from her who she is. I have seen that fat people have little in common with each other, with the exception that the majority of them have had these experiences of being stereotyped and projected upon. When meeting a new client, I try to understand how this unique person has met and managed this common experience.

THE VALUE OF A WEIGHT NEUTRAL STANCE: PART 2

Overhearing clinicians' stereotypical discussions about fat clients, I wonder why they so easily jettison the skills they use on the problems of people who are not fat. For the fat client, they seem to zero in on the "weight issue" and assume that the treatment plan must have weight loss as its central goal. Complaints that the clinician would interpret as depression or exposure to trauma are here deemed an expectable and predictable consequence of being fat, and the solution is weight loss. But what would we offer a thin person who is depressed or traumatized, who has been stereotyped, or who wants to lose weight as a fantasized solution to all problems? The treatment plan should not depend on the patient's body size.

But what about the fact that weight loss is not a *fantasized* solution to the problem of being fat? Isn't losing weight if you are fat a *real* solution? What does weight loss solve and what does it not solve?

Makeover Mind

Makeovers are today's version of fairy tales, with abundant before and after stories of weight loss, plastic surgery, lottery winners, and house remodels. In these narratives, all problems disappear with a makeover; indeed, they are almost always accompanied by the statement, *"I am a whole new person!"* The magical thinking of the makeover allows us to imagine that we are now going to live happily ever after. But in real life, there is an *after* before-and-after. After-the-"after" is when the reality sets in about what the makeover changes and what it does not change. When it comes to weight loss, the "after" is almost always the period of regaining weight, when people feel ashamed and like failures, avoiding people who have seen their weight loss. Weight rebound is the usual course of events, to some degree even with bariatric surgery (Sjostrom et al., 2007). This is true even among members of the Weight Control Registry, a group of several thousand people who are deemed to have "maintained weight loss" (McGuire, Wing, Klem, Lang & Hill, 1999). This fact does not seem to lessen the individual person's sense of personal inadequacy in not being able to maintain the weight loss. Regaining weight also reinforces the sense that, at one's core, the true self is the "loser" self, and everything associated with being fat (feeling vulnerable, inadequate, out of control, unable to make an impact, immobilized, and a binger) is reinforced rather than held compassionately as just one aspect of a person's personality.

One of the most important reasons for healthcare professionals to take a weight neutral stance is quite simply this: we do not have an intervention to permanently and safely change weight, at least not for the great majority of people (Mann et al., 2007). To prescribe weight loss is to in fact impose weight cycling on the patient, which can lead to poorer physical and emotional health itself (Blair, Shaten, Brownell, Collins & Lissner, 1993). The continued prescription of weight loss in the face of its continual failure is one of the most data-resistant features of healthcare.

But a second major reason for a weight neutral stance is the same reason we investigate the fantasized meanings of weight loss for our thin patients. Holding one's life hostage to the pursuit of weight loss is one way that many people get stuck developmentally. Disordered eating is almost always intimately bound to the goal of being a certain weight. Restricting-patients forced to eat more in treatment may start purging to keep from gaining weight, or their bodies may eventually compel them to binge in order to restore weight. Patients who are bulimic and trying not to purge in treatment may start restricting out of the terror of gaining weight. People who are binging without purging may restrict or excessively exercise in the name of dieting. Some may, with their healthcare provider's blessing, "try to get healthy," particularly if they are at a higher weight. Many people who are now fat have cycled their weights ever higher with each weight loss attempt, and the legacy of the restrictive phases (like that of all people who diet) is that they have lost touch with their body cues and eat in a highly psychologically reactive manner, where external cues like the presence of food or the opportunity to eat trigger eating whether one is hungry or full (Polivy, Herman & McFarlane, 1994).

If we had a culture where weight did not carry much meaning, what would provoke disordered eating? Perhaps we would still have the motive to soothe ourselves with food, but, if we were in touch with our bodies' needs, our physical experience of comfortable fullness would direct our eating, and the marketing and merchandising of food would have less influence. I attempt to make the therapeutic setting a weight neutral oasis, a subculture where weight has little inherent meaning, so that patients can sort out what truly troubles them and find effective solutions.

As people work through the loss of the fantasy of being thin, they examine the question: what does weight loss change and what does it not change? Theoretically, it should change the way someone is treated in a culture that discriminates on body size. But most people, when they regain weight, return to their former status and are faced with the original quandary of how to manage the social oppression. Even the minority of people who maintain weight loss, including those who have had bariatric surgery, usually remain in the "obese" BMI category (Bond, Phelan, Leahey, Hill & Wing, 2009) and, presumably, are still vulnerable to discrimination. We cannot improve health and prevent ED unless we make a culture that celebrates a diversity of body sizes.

Within the person's own internal world, the part of them that identifies as fat is still there, no matter what his or her body size, since the associated feeling states are not really about fat at all, but rather the human and existential feelings of vulnerability. Losing weight does not make you into a different person, one you can now like. It does not keep bad things from happening. It does not keep you from getting sick. It does not keep others from rejecting you or being mean.

Many people have described to me the ambivalent feelings they have when they are complimented on weight loss. *"I always think, what did you think of me before, if you are making this fuss now? And what will you think if I gain it back?"*

Our thin ED patients have taught us that you do not need to be fat to be haunted by the dread of being socially humiliated and rejected for being too fat. This quandary is never solved by being "thin enough." There is an entirely different solution—in which one's sense of self-trust and self-efficacy is sturdy enough to be an answer to the anxieties that we all have, no matter what our body size.

In the HAES approach, the therapist compassionately explores and challenges the person's fantasy of weight loss. For patients who are entering treatment with a weight loss goal, this sort of exploration usually requires a solid therapeutic alliance, since it will challenge the wish that is partly responsible for their seeking therapy in the first place. It also will challenge the patient who is aware that many other therapists and healthcare providers will promise weight loss, a perhaps more palatable option.

Perhaps it is a matter of self-selection and informed consent, but by the time someone is entering therapy with me, they usually have some sense of the disconnect between the fantasy of weight loss and their lived experience, and are in part seeking help in becoming more weight neutral themselves. The loss of the fantasy of being thin is a requirement for healing the adversarial relationship with one's body. It is filled with grief as well as unexpected gifts. Over time, progress is increasingly measured by movement towards a sense of being at peace within their own skin and in partnership with their imperfect but precious bodies. These goals gradually become more meaningful and more sustainable than weight loss.

When we are able to see what the patient associates with fatness and thinness, we begin to see the feelings states and parts of self that all need a home. The emerging view can often be a difficult experience. The shift from disowning to integrating these states gives a person some confidence that they can endure and care for themselves. When they experience themselves making a place for the disowned parts, they can believe that they can be loved fully by others as well. It is partly the experience of caring for these disowned parts of self that changes a person's actual experience and view of the universe in which they live—because most of our experiences are internal. When you take better care of yourself, there is a sense of "somebody here" to take care of you, demonstrated in the caring acts of feeding, reacting, advocating, etc. The whole universe seems filled with more parental love and respect.

THE VALUE OF A WEIGHT NEUTRAL STANCE: PART 3

Weight neutrality may actually be a necessary condition to enable people to develop sustainable health practices. We know that improved nutritional quality and more frequent physical activity lead to better health, whether weight is lost or not (Bacon et al., 2005; Sacks et al., 2001). But most people who start those practices with a goal of weight loss do not continue them. One hypothesis is that, because these changes are so difficult to make, the additional agenda of trying to do them in order to lose weight tends to be a trigger to *start* them but not to *sustain* them. If the purpose of these practices is weight loss alone, people give up their efforts to eat better and move more, as weight loss usually stalls or reverses at multiple points.

While this area deserves far more research, Bacon et al.'s study (2005), comparing a HAES model with a traditional weight loss intervention, is instructive. After 2 years, the HAES group was still practicing the health behaviors and had sustained the health benefits, even though their weight had not changed. Meanwhile, the weight loss group had lost, then regained weight, and, no longer practicing the healthier behaviors, had lost the health benefits.

We need to better understand the reasons for these motivational differences. People may need to find intrinsic motivators in order to maintain the practices; they may be demoralized by the added drama of the weight loss and gain; they may oscillate between identifications with their "good" (compliant) selves and their "bad" (rebellious) ones. It may take all of their available cognitive, logistical, and attentional resources to develop the capacity to come back to the health practices when inevitably forced out of their healthy routines by the vicissitudes of life. In any case, the more complicated the agenda becomes for a goal, the more that can go wrong. The goal to simply explore foods and physical activities that make you feel better, for the rest of your life, is as simple and direct as possible. We need a research agenda that includes the issues of motivation and sustainable behavioral goals to guide health care practice.

Weight Neutrality and Intuitive Eating

Many HAES clinicians use intuitive eating (Matz & Frankel, 2007; Tribole & Resch, 2003) as their primary skill-based eating intervention. As is clear to caregivers, infants are born with the ability to act upon hunger and satiety. The fact that many adults fail to use those cues does not mean they are not there. In fact, learning to diet can be seen as a tutorial in how to ignore physical hunger and satiety cues and instead eat according to external stimuli—exactly the opposite legacy we would wish, especially in today's environment that clamors with food marketing (Wansink, 2007).

The goal of intuitive eating is to return to a habit of using one's own body cues for decision-making about food, and to also increase the mindful awareness of the results of those decisions. Most people who consider themselves "compulsive eaters" believe they should be more abstinent and restrained because they have "over-indulged." In actuality, someone whose only pleasure or comfort comes from food is deprived of the full range of life's treats. In the context of HAES, we aim to *increase* the pleasure from food, to increase the number of sources of pleasure and satisfaction in the person's life, and to remove the moral associations of being "good" or "bad" from eating behavior. The rule-making of most diets is avoided, replaced with the simple goal of doing one's best to take care of oneself at the time. With absent deadlines, external "experts," and "allowed" or "forbidden" foods, the client begins the simple experiment of asking, "What am I hungry for?" and trying different ways of responding.

Some clinicians using intuitive eating do not particularly frame their approach as HAES, and promise weight loss as an outcome. They assume that, if someone has been eating more than their body cues demand, they are bound to lose weight. Little research supports this point, and, for all the reasons we have discussed above, it is critical to maintain an agnostic stance toward weight loss. I tell patients that I cannot predict what will happen as we treat their ED. Some people come to me at the bottom of the weight loss cycle (a "weight low-yo") and will gain weight when they practice intuitive eating, simply because their bodies are restoring their weight. Others lose weight if their bodies have been receiving food for all kinds of needs that were not just for food. Many simply stabilize in weight.

The important psychological stance is to put the practices of self-care first. Weight is the dependent variable. This proposal scares many people, who focus on weight as the "bottom line," ignoring the process. A central pathology in ED thinking is that what you weigh is more important than how you live your life. The agenda for therapy is to address dread, build skills, facilitate self-care, and engender hope and a sense of competence. These same tasks apply across the weight spectrum. Thus, as with any patient struggling with an ED, the focus is on getting the fuel one's body needs, holding an agnostic position about any weight outcome, and helping patients with their considerable anxiety about giving up the illusion of being able to both feed themselves well and choose their weight at the same time.

From Makeover Mind to Sustainable Eating

In keeping with the emphasis on "leaving behind Makeover Mind," my version of teaching intuitive eating skills emphasizes "sustainable eating." The question becomes: what works for you over time? If you have a regime that you can't keep doing through the ups and downs of your life, then it will lead to weight cycling.

One can begin to identify the important functions of food by building a matrix—for example, food as fuel (one axis) and a source of pleasure (the other axis), as seen in Figure 2.1. Using this matrix, therapist and patient can assess and explore the different experiences of food—the immediate taste experience as well as the functional benefits that food confers in the following hours or days.

Armed with a sense of their body's moderating influence, and its reliability compared to the swinging extremes of psychological and cognitive processes, much of the anxiety about the "permissiveness" of intuitive eating gradually resolves. Sometimes it is downright disappointing how little food one's body wants, and how dependent we are on our bodies to get hungry again in order to experience that degree of pleasure in eating.

In this stage of treatment, we begin to distinguish the great variety of sources for the desire to eat, and how many of them are not due to hunger. From the therapist's viewpoint, this information determines the next phase of the treatment plan. Some people learn the intuitive eating skills and take off with them, never to return to restricting or binging. Their difficulties seem to truly stem from losing track of those internal cues, most often from developing a dieting mindset. Other people may have all sorts of triggers for eating (or not eating), which take a long time to sort out and address. Among these are: regulating mood or emotions with food or hunger; physiological causes for binging, such as being below one's genetic weight range (Keys, 1953) even at a high BMI, or getting caught in vicious cycles of blood sugar dysregulation; more entrenched "dieting detox" reactions, encountering opportunities to eat formerly forbidden food, or reacting rebelliously to the perception that they should be

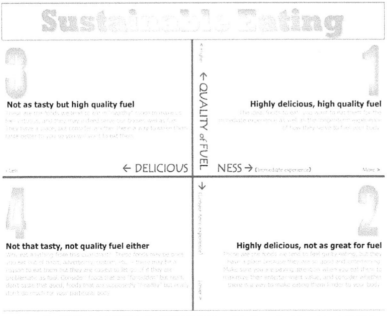

FIGURE 2.1 Sustainable Eating Chart.

dieting; and body image enactments, in which specific body sizes hold meaning for the person and any changes in weight (the "dependent variable") as a result of the process of intuitive eating can be problematic. Everyone also faces the daunting demands of breaking old habits, and needing to give energy and awareness to the tasks of developing new systems for self-care and solving everyday logistical problems.

We still have much to understand about the biological basis for ED. Clearly, our cultural associations with thinness and fatness often trigger dieting practices that can launch vicious cycles of distressing neurological, physiological, and emotional patterns. Treatment must help patients strip away the weight-related associations with foods and ways of eating so they may respond to their hungers in a straightforward, less driven, and more conscious manner.

Other Self-Care Behaviors

The HAES model does not focus solely on eating practices, examining other behaviors that can increase the experience of well-being, such as physical activity, obtaining restful sleep, and above all, strengthening social support.

The weight neutrality of the HAES approach is central to the process of integrating pleasurable physical activity into one's life. In both eating and exercise, basic human activities have been polarized into "good" (weight-reducing) and "bad" (weight-producing). Likewise, the foods eaten, or the exercise performed or avoided, trigger identification with the "good" and "bad" selves. As long as these selves are compartmentalized, patients may be caught up in rapid and disruptive shifts between overeating and compensation.

My clients usually have a mixed perception of exercise, depending on whether they remember it as recess or PE, recreation, or a weight control activity. Despite significant individual differences in how pleasurable movement is, most of my clients have at least some positive memories of playful activity. But for people who were not immediately gifted, or whose bodies were not seen as athletic, memories of movement may be quite painful. For anyone who has tried to exercise to lose weight, the association is usually more punitive than pleasurable. And so, as with eating, the legacy of pursuing weight loss is a barrier to the long-term integration of the very practices that will matter for well-being and health. *Great Shape* (Lyons & Burgard, 1988) examines the special issues that heavier people experience when exploring and integrating sustainable physical activity. For people of any size, the process of finding pleasurable movement can be the same as that of intuitive eating: becoming more aware of the hunger to move, and exploring the kinds of movement that satisfy and when. However, finding arenas for movement that do not further reinforce the association with "weight management" is a challenge in our culture; sadly, the weight loss agenda seems to dominate the experience in most locations.

Assessing the quality of sleep is a surprisingly overlooked aspect of healthcare, and sleep medicine is a relatively new field (Dement & Vaughan, 2000). Here, too, weight bias affects access to care, especially with diagnoses such as sleep apnea, which are present in people of all sizes but stereotypically associated with fatness. My thinner patients have had difficulty obtaining referrals to get a sleep study; and, oddly enough, after diagnosis of her sleep apnea, one patient with anorexia nervosa (AN) got a letter suggesting that she "consider weight loss." Both patients and doctors have presumed that sleep apnea resolves after bariatric surgery and do not even test for it at regular intervals, leaving many post-surgery patients with untreated apnea that has returned or never resolved (Hallowell et al., 2007; Lettieri, Elliasson & Greenburg, 2008).

Finally, the HAES model attends to social support as a key element of mental and physical well-being. Patients need strong and loving connections with others and freedom from stigma and violence. One way to bring this about is to connect patients to supportive communities—online, in their area, or in a support group. Because one source of distress is the weight bias in our culture, finding a new social niche, a healthier subculture, can nurture and strengthen weight diversity and minimize body disparagement.

THE VALUE OF A WEIGHT NEUTRAL STANCE: PART 4

The HAES model draws upon the general research on health and well-being to identify its important aspects: pleasurable physical activity, restful sleep, social support, safety, and freedom from stigma. Note that different mixtures of individual motivation and the broader environment are critical to each of these elements. The HAES model not only focuses on an individual's choices, but also on changing our cultural, social, and ecological environments to support well-being.

Because one of the central vehicles for the oppression of fat people is the contention that they cannot be healthy, HAES advocates argue that healthcare providers must recognize their role in designing and carrying out programs, albeit with the best of intentions, which are actually making people sicker (Isono, Watkins & Lian, 2009). The Academy for Eating Disorders initiated a statement in December 2009 along with all the other major eating disorders organizations, calling for a change in focus from weight-based programs to health, as stated by AED President Susan Paxton, PhD, FAED:

> There is concern that we have lost sight of the importance of avoiding harm in the process of addressing obesity. Further, we cannot ignore the opportunity to create a healthier environment, where people of all sizes are given the opportunity to lead healthy and productive lives, instead of singling out individual groups for reform based on weight alone. *(Academy for Eating Disorders, 2009)*

Another useful document addressing these issues is the Academy for Eating Disorder's *Guidelines for Childhood Obesity Prevention Programs* (Danielsdottir, Burgard & Oliver-Pyatt, 2009). The guidelines argue for health-based, rather than weight-based, interventions that focus on the health of all children and measure health outcomes rather than weight *per se*. They urge program developers to reduce stigma, teasing, and bullying, and to focus on reducing environmental barriers that prevent people from having access to nutritious foods and pleasurable and safe physical activity.

Individual healthcare practice, public health programs, workplace programs, and even diagnostic criteria, could all become more weight neutral. Current definitions of ED sometimes incorporate weight references or frank criteria that lead to harm. For example, a strict reading of the DSM-IV-R (American Psychiatric Assocation, 1994) dictates refraining from giving an AN diagnosis unless the patient is 15% below "minimally normal body weight." Despite some room for individual determination, most clinicians are forced to wait to give the diagnosis until the patient is well-entrenched in the disease. I have seen fat patients who really should be given an AN diagnosis, as they restrict, are intensely afraid of gaining weight, are rigidly focused on food and exercise, and may even have lost their periods. If we imagine the pathological process of AN as the problem, and the weight loss as the result, it makes little sense to have to wait until the disease is at a critical, life-threatening stage to name it a problem. If you are in a plane that loses an engine at 40,000 feet, you wouldn't want to have to wait to take action until you were at 10,000 feet.

These examples suggest that we need to address broader forces in order to improve public health. Many HAES practitioners are activists systemically challenging the policies and social norms that harm people across the weight spectrum. But while we are trying to change the culture we also have much to offer our individual clients.

STEREOTYPE MANAGEMENT SKILLS

For many clients, the acknowledgement that their health is linked to factors beyond their personal control is a new perspective. Many people have been focusing on specific food or exercise regimens to manage their less-conscious worries about mortality, and so many of them blame themselves for any physical or medical difficulty. Acknowledging that they are subject to stereotyping and stigma is oddly taboo. Perhaps it is part of the human condition to prefer to believe that aversive conditions are our own fault, rather than acknowledging that sometimes we cannot control broad, external forces that hurt us. I have found that giving patients tools to manage these broader forces generally feels empowering. In Box 2.2 is a summary of some of these tools.

INTERNALIZED OPPRESSION, EATING DISORDERS, AND HEALTH

Perhaps it is ironic that a "weight neutral" approach attends to the effects of being stereotyped according to body size. But research is beginning to show the importance of our beliefs about weight and self-worth and their impact on our health. Muennig, Jia, Lee & Lubetkin (2008) showed that the difference between actual and desired body weight was a stronger predictor of mental and physical health than was BMI. Puhl, Moss-Racusin & Schwartz (2007) noted that: "[Weight loss] participants who believed that weight-based stereotypes were true reported more frequent binge eating" and that such findings "challenge the notion that stigma may motivate obese individuals to engage in efforts to lose weight" (p. 1). In a well-designed prospective study of teens, Neumark-Sztainer et al. (2006) found that trying to lose weight resulted in both *higher* weight and greater likelihood of disordered eating. Mond et al. (2007) report that: "weight and shape concerns" were the mediating variable between obesity and psychosocial impairment and that "a greater focus on body acceptance in obesity treatment may be indicated" (p. 2769). In a more positive vein, Avalos & Tylka (2006) showed that "women who accept their bodies the way they are seem to be more likely to follow principles of healthy eating" (p. 486).

BOX 2.2
STEREOTYPE MANAGEMENT SKILLS

Know yourself:

- **Cultivate self knowledge** that challenges the stereotype
- **Invest in your own agenda**, as an antidote to being swayed to either act like or act unlike the stereotype
- **Foster your drive to be seen as your real self.** Remember, especially with negative projections: "if they knew me they would probably like me!".

Prepare yourself:

- **Expect stereotyping will happen**, as unpleasant as it is—budget it in
- **Know the specifics.** Understand the typical ways you are likely to be stereotyped, when, where, who. Understand the function of stereotyping in general
- **Blame the process, not your body.** Make sure you do not confuse the projection with your real self. Address internalized projections.

Impact the situation when possible:

- **Collect proven ways to "show up"** through the fog of the stereotype. What have others done? What has worked before? Think playfully
- **Participate in more personal settings** and situations where you can contribute, when you have a choice
- **Seek out or bring along allies** when you can.

Cultivate compassion:

- **Be kind to yourself** when you are hurt and when you are struggling. Stereotyping hurts no matter how strong your defenses are
- **Remember we all are prone to stereotyping.** Try to stay in touch with your belief in the humanity of the person stereotyping. It will make it more possible for them to snap out of it if you don't respond by stereotyping them back
- **Heal** with infusions of playfulness, righteous anger, passion, and humility.

Create new culture:

- **Bond** with other people who support your identity
- When you are facing stereotyping, think about telling your peeps about what is happening — **evoke them** and let them be present with you
- **Be an activist!** Do things that make you feel empowered. Take credit for artfully fighting the stereotype
- **Create new culture** with your words, pictures, and deeds.

Study history:

- **Research** resistance to oppression in all its forms
- **Feel connected** to all the humans who came before you and who created a new culture for you. Think about the people who will benefit in the future from your actions now.

AFTER

My fat client is still fat. She has spent time in therapy recognizing her own weight biases and their impact on her self-esteem and her motivation for taking care of the precious, flawed body she has. She has found other women who are trying to challenge the assumption that thinness is always better, and she has realized with a shock of recognition that she began life with the capacity to see beauty and worth in fat people. She is working on self-care, which is an ongoing project. She is surprised that she really doesn't like the taste of potato chips. She is surprised that she has both chocolate and fresh spinach in her house. She is walking her dog regularly and gearing up to try a belly dance class, maybe when she can find a buddy to go with her. She has told her physician that she knows he has to remind her about her BMI as a matter of procedure, but that, in case he is interested, it makes her feel worse and not really seen or appreciated. They are working on a better model for collaborating on her health.

She still has bad days. She still feels like a loser sometimes, but she can also hear the harshness in that appellation, and she is beginning to feel some compassion. She has met some friends who challenge her when she starts bashing her body—and they help her be curious about what is going on emotionally.

Every day she hears another news story about the "obesity epidemic" and forced weighing of children, or employees, or college seniors. Every day she sees another photo of some anonymous fat person whose body is deemed too shameful to show their face. But she is beginning to have a theory about all of this, fueled by insights from her own introspection. She is finally beginning to see the suffering among her thin friends, and is surprised not to envy them as much. She reports to me that one of them can't stand it that she has cookies in her house that she hasn't touched since the last time she visited and demands to know, "How are you doing this?" She is bemused.

As she takes her eggs out of the "Makeover" basket, she starts distributing them in the "go back to grad school," "go out on a few dates," "buy some really comfortable and really well-made clothes," "set a limit with my mother about diet talk," and "write a letter to the airlines" baskets. She is beginning to reliably experience a different way to take care of herself. She is mostly present now, in her body, and in her truth, and in her life. And so we carry on.

References

Academy for Eating Disorders (2009). Press Release, December 13. *Eating disorder organizations join forces to urge focus on health and lifestyle rather than weight.* (http://www.eatingdisordersmaine.com/id62.html); and message posted to aed-members@groups.aedweb.org

American Psychiatric Association (1994). *The diagnostic and statistical manual of mental disorders.* Washington, DC: Author.

Amy, N. K., Aalborg, A., Lyons, P., & Keranen, L. (2006). Barriers to routine gynecological cancer screening for White and African-American obese women. *International Journal of Obesity, 30,* 147–155.

Association for Size Diversity and Health (ASDAH), *HAES principles.* Retrieved 12/16/09 from www.sizediversityandhealth.org

Avalos, L. C., & Tylka, T. L. (2006). Exploring a model of intuitive eating with college women. *Journal of Counseling Psychology, 53*(4), 486–497.

Bacon, L., Van Loan, M., Stern, J. S., & Keim, N. (2005). Size acceptance and intuitive eating improve health for obese female chronic dieters. *Journal of the American Dietetic Association, 105,* 929–936.

Bacon, L. (2008). *Health at every size: The surprising truth about your weight.* Dallas, TX: Benbella.

Blair, S., Shaten, J., Brownell, K., Collins, G., & Lissner, L. (1993). Body weight change, all-cause mortality, and cause-specific mortality in the Multiple Risk Factor Intervention Trial. *Annals of Internal Medicine, 119*, 749–757.

Bond, D. S., Phelan, S., Leahey, T. M., Hill, J. O., & Wing, R. R. (2009). Weight-loss maintenance in successful weight losers: Surgical vs. non-surgical methods. *International Journal of Obesity, 33*(1), 173–180.

Bryner, J. (2009). *The obese don't always know it*. Retrieved 12/06/09 from http://www.livescience.com/health/091117-obesity-body-size-html

Burgard, D. (2009). What is health at every size? In E. Rothblum & S. Solovay (Eds.), *The fat studies reader* (pp. 41–53). New York, NY: NYU Press.

Campos, P. (2004). *The obesity myth*. New York, NY: Gotham Books.

Centers for Disease Control (2008). *Prevalence of overweight, obesity and extreme obesity among adults: United States, trends 1960–1962 through 2005–2006*. Retrieved 12/06/09 from http://www.cdc.gov/nchs/data/hestat/overweight/overweight_adult.htm

Centers for Disease Control (2010). *NHANES Surveys* (1976–1980 and 2003–2006). Retrieved from http://www.cdc.gov/obesity/childhood/prevalence.html

Danielsdottir, S., Burgard, D., & Oliver-Pyatt, W. (2009). *Guidelines for childhood obesity prevention programs*. Retrieved 12/13/09 from http://aedweb.org/media/Guidelines.cfm

Dement, W. C., & Vaughan, C. (2000). *The promise of Sleep*. New York, NY: Random House (Dell).

Edwards, P., & Roberts, I. (2009). Population adiposity and climate change. *International Journal of Epidemiology, 38*(4), 1137–1140.

Ernsberger, P. (2009). Does social class explain the connection between weight and health? In E. Rothblum & S. Solovay (Eds.), *The fat studies reader* (pp. 25–36). New York, NY: NYU Press.

Finkelstein, E. A., Trogdon, J. G., Cohen, J. W., & Dietz, W. (2009). Annual medical spending attributable to obesity: Payer- and service-specific estimates. *Health Affairs, 28*(5), 822–831.

Flegal, K. M., Graubard, B. I., Williamson, D. F., & Gail, M. H. (2005). Excess deaths associated with underweight, overweight, and obesity. *Journal of the American Medical Association, 293*(15), 1861–1867.

Friedman, J. M. (2009). Causes and control of excess body fat. *Nature, 459*(21), 340–342.

Friedman, M. A., & Brownell, K. D. (2002). Psychological consequences of obesity. In C. G. Fairburn & K. D. Brownell (Eds.), *Eating disorders and obesity: A comprehensive handbook* (2nd ed.). (pp. 393–398) New York, NY: Guilford Press.

Gaesser, G. (2002). *Big fat lies: The truth about your weight and your health*. Carlsbad, CA: Gurze.

Hallowell, P. T., Stellato, T. A., Schuster, M., Graf, K., Robinson, A., Crouse, C., & Jasper, J. J (2007). Potentially life-threatening sleep apnea is unrecognized without aggressive evaluation. *The American Journal of Surgery, 193*(3), 364–367.

Hudson, J., Hiripi, E., Pope, H., & Kessler, R. (2007). The prevalence and correlates of eating disorders in the national comorbidity survey replication. *Biological Psychiatry, 61*(3), 348–358.

Isono, M., Watkins, P. L., & Lian, L. E. (2009). Bon bon fatty girl. In E. Rothblum & S. Solovay (Eds.), *The fat studies reader* (pp. 127–138). New York, NY: NYU Press.

Keys, A. (1953). *The biology of human starvation*. Minneapolis, MN: University of Minnesota Press.

Klein, R. (2001). Fat beauty. In J. E. Braziel & K. LeBesco (Eds.), *Bodies out of bounds* (pp. 19–38). Berkeley, CA: UC Press.

Kruger, J., Galuska, D. A., Serdula, M. K., & Jones, D. A. (2004). Attempting to lose weight: Specific practices among U.S. adults. *American Journal of Preventive Medicine, 26*(5), 402–406.

Latner, J. D., O'Brien, K. S., Durso, L. E., Brinkman, L. A., & MacDonald, T. (2008). Weighing obesity stigma: The relative strength of different forms of bias. *International Journal of Obesity, 32*, 1145–1152.

Lettieri, C. J., Elliasson, A. H., & Greenburg, D. L. (2008). Persistence of obstructive sleep apnea after surgical weight loss. *Journal of Clinical Sleep Medicine, 4*(4), 333–338.

Lyons, P., & Burgard, D. (1988). *Great shape*. Palo Alto, CA: Bull.

Mann, T., Tomiyama, A. J., Wrestling, E., Lew, A., Samuels, B., & Chatman, J. (2007). Medicare's search for effective obesity treatments: Diets are not the answer. *American Psychologist, 62*(3), 220–233.

Marcus, M. D., & Wildes, J. E. (2009). Obesity: Is it a mental disorder? *International Journal of Eating Disorders, 42*, 739–753.

Matz, J., & Frankel, E. (2007). *Diet survivor's handbook*. Chicago, IL: Sourcebooks, Inc.

McGuire, M. T., Wing, R. R., Klem, M. L., Lang, W., & Hill, J. O. (1999). What predicts weight regain among a group of successful weight losers? *Journal of Consulting and Clinical Psychology, 67*, 177–185.

Mond, J. M., Rodgers, B., Hay, P. J., Darby, A., Owen, C., Baune, B. T., & Kennedy, L. (2007). Obesity and impairment in psychosocial functioning in women: The mediating role of eating disorder features. *Obesity, 15*(211), 2769–2779.

Muennig, P., Jia, H., Lee, R., & Lubetkin, E. (2008). I think therefore I am: Perceived ideal weight as a determinant of health. *American Journal of Public Health, 98*(3), 1–5.

Neumark-Sztainer, D., Wall, M., Guo, J., Story, M., Haines, J., & Eisenberg, M. (2006). Obesity, disordered eating, and eating disorders in a longitudinal study of adolescents: How do dieters fare 5 years later? *American Dietetic Association, 106*, 559–568.

Orpana, H. M., Berthelot, J., Kaplan, M., Feeny, D., McFarland, B., & Ross, N. (2009). BMI and mortality: Results from a national longitudinal study of Canadian adults. *Obesity* (18 June 2009, http://www.nature.com/oby/journal/v18/n1/abs/oby2009191a.html); doi:10.1038/oby.2009.191

Polivy, J., Herman, C. P., & McFarlane, T. (1994). Effects of anxiety on eating: Does palatability moderate distress-induced overeating in dieters? *Journal of Abnormal Psychology, 103*(3), 505–510.

Puhl, R. M., & Latner, J. (2007). Stigma, obesity, and the health of the nation's children. *Psychological Bulletin, 133*(4), 557–580.

Puhl, R. M., Moss-Racusin, C. A., & Schwartz, M. B. (2007). Internalization of weight bias: Implications for binge eating and emotional well-being. *Obesity (Silver Spring), 15*(1), 19–23.

Puhl, R. M., Andreyeva, T., & Brownell, K. D. (2008). Perceptions of weight discrimination: Prevalence and comparison to race and gender discrimination in America. *International Journal of Obesity* (http://www.nature.com/ijo/journal/v32/n6/abs/ijo200822a.html); doi:10.1038/ijo.2008.22

Puhl, R. M., & Heuer, C. A. (2009). The stigma of obesity: A review and update. *Obesity* (http://www.nature.com/oby/journal/v17/n5/full/oby2008636a.html); doi: 10.1038/oby.2008.636

Sacks, F. M., Svetkey, L. P., Vollmer, W. M., Appel, L. J., Bray, G. A., Harsha, D., … Lin, P. H. (2001). Effects on blood pressure of reduced dietary sodium and the Dietary Approaches to Stop Hypertension (DASH) diet. *New England Journal of Medicine, 344*, 3–10.

Simon, G., Von Korff, M., Saunders, K., Miglioretti, D., Crane, P., Van Belle, G., & Kessler, R. C. (2006). Association between obesity and psychiatric disorders in the U.S. adult population. *Archives of General Psychiatry, 63*, 824–830.

Sims, E. (2001). Are there persons who are obese, but metabolically healthy? *Metabolism, 50*(12), 1499–1504.

Sjostrom, L., Narbro, K., Sjostrom, C. D., Karason, K., Larsson, B., Wedel, H., … Carlsson, L. M. (2007). Effects of bariatric surgery on mortality in Swedish obese subjects. *New England Journal of Medicine, 357*, 741–752.

Stewart, S. T., Cutler, D. M., & Rosen, A. B. (2009). Forecasting the effects of obesity and smoking on U.S. life expectancy. *New England Journal of Medicine, 361*(23), 2252–2260.

Tamakoshi, A., Yatsuya, H., Yingsong, L., Tamakoshi, K., Kondo, T., Suzuki, S., … Kiruchi, S. (2009). BMI and all-cause mortality among Japanese older adults: Findings from the Japan Collaborative Cohort Study. *Obesity* (http://www.nature.com/oby/journal/v18/n2/full/oby2009190a.html); doi: 10.1038/oby.2009.190

Tribole, E., & Resch, E. (2003). *Intuitive eating*. New York, NY: St. Martin's Press.

Wansink, B. (2007). *Mindless eating*. New York, NY: Random House.

Weiss, E. C., Galuska, D. A., Khan, L. K., & Serdula, M. K. (2006). Weight-control practices among U.S. adults, 2001–2002. *American Journal of Preventive Medicine, 31*(1), 18–24.

Wildman, R. P., Muntner, P., Reynolds, K., McGinn, A. P., Rajpathak, S., Wylie-Rosett, J., & Sowers, M. R. (2008). The obese without cardiometabolic risk factor clustering and the normal weight with cardiometabolic risk factor clustering. *Archives of Internal Medicine, 168*(15), 1617–1624.

Williamson, D. F., Madans, J., Anda, R. F., Kleinman, J., Kahn, H., & Byers, T. (1993). Recreational physical activity and ten-year weight change in a U.S. national cohort. *International Journal of Obesity, 17*(5), 279–286.

Wilson, G. T., & Fairburn, C. G. (1993). Cognitive treatments for eating disorders. *Journal of Consulting and Clinical Psychology, 61*(2), 261–269.

CHAPTER

3

Neuroscience
Contributions to the Understanding and Treatment of Eating Disorders

Francine Lapides

With all the money, and all the glamour and all the fame and attention and success, it doesn't mean anything if you can't fit into your own clothes. It means the fat won. You didn't win. Here I am one of the most visible people in the world trying not to be visible, trying not to appear on the cover of my own magazine. I'm mad at myself. I'm embarrassed. I can't believe I'm still talking about weight **(Winfrey, 2008)**.

Oprah Winfrey's words illustrate how disordered relationships with food and weight can wreak devastating pain on even the strongest, most successful, and most visible. Most eating disorders (ED) have their onset during adolescence and arise from physical, psychological, cultural, and social factors that are dauntingly complex. The media and our thin-obsessed society certainly contribute. Also at play is the abundance of and ease of access to a staggering variety of foods, many of them "fast" and void of nutrition. But the origins of ED are much more intricate. They lie in the elaborate biology and neurobiology of our bodies and in the tender and earliest attachment relationships that shape the very development of our brains and nervous systems. Early histories of those with ED often include trauma, psychic injuries that occurred without secure containment or repair. Clear and impressive evidence supports a correlation between ED and these early attachment difficulties (Armstrong & Roth 1989; O'Kearney 1996).

While consensus is that the treatments for ED must be extensive, specialized, and multi-disciplinary, we currently know far too little about how to treat them effectively. This chapter translates findings from attachment research and from the field of neuroscience into a conceptual model of ED which emphasizes the unconscious, holistic, right hemisphere (RH), rather than the logical left hemisphere (LH) specifically targeted in approaches that utilize well-accepted techniques such as cognitive behavioral therapy (CBT). This neurodevelopmental, causal model of ED will draw from a multidisciplinary blending of developmental psychobiology and affective neuroscience to examine the role of the brain as the very *"substrate on which eating disorders are built"* (Treasure, Tchanturia & Ulrike, 2005, p. 194), and may well improve treatment outcomes as it generates innovative protocols.

UNCOVERING THE NEUROLOGICAL AND PSYCHOBIOLOGICAL CORRELATES OF EATING DISORDERS

The regulation of appetite and eating involves both peripheral (body–brain connections) and central (within the brain itself) pathways. Many neural center and biochemical irregularities are considered causal in the origins of ED. For example, altered brain metabolism in the locus coeruleus, a concentrated mass of norepinephrine-secreting neurons, may leave individuals vulnerable to surges of adrenaline and noradrenaline which in turn create a continual state of anxiety, arousal, and alarm (O'Donnell, Hegadoren & Coupland, 2004). Altered insula response to taste stimuli provides evidence that those with anorexia nervosa (AN) experience taste differently than a control group and that the insula and orbital frontal cortex (OFC) alter eating behavior by encoding changes in the value of food used as a reward (Wagner et al., 2008). Body image disturbances in those with AN may stem from underlying problems with right parietal lobe functioning (Grunwald, Etrrich, Assmann, Daehne & Krause, 2001; Ramachandran & Blakeslee, 1998). Reduced right dorsal ACC volume is observed in AN, and likely relates to deficits in perceptual organization and conceptual reasoning, (McCormick et al., 2008).[1]

Hypofunctioning dorsal striatum have been implicated in obesity (Stice, Spoor, Bohon & Small, 2008). In addition, the smaller size of the thalamus and thalamic perfusion changes in AN suggest that the thalamus plays an important role in mediating ED (Dusoir et al., 2005). Finally, OFC malfunctioning and the subsequent increases in impulsivity are a risk factor for ED (Wonderlich, Connolly & Stice, 2004), and changes in the reward circuitry (ventral tegmental area, nucleus accumbens and the prefrontal cortex) have been found in those diagnosed with ED (Bergen et al., 2005).

In spite of impressive growth in the research on ED, little relates to the most commonly diagnosed type, Eating Disorders Not Otherwise Specified (EDNOS), which accounts for the majority of ED patients seen in the community (National Alliance on Mental Illness, 1996). This review also demonstrates the lack of a unifying theory to make sense of the specific neural correlates at play in both the origin and treatment of ED.

A UNIFYING THEORY: EATING DISORDERS AND THE NEUROLOGICAL DEVELOPMENT OF AFFECT-REGULATION CAPACITIES

Affect-regulation refers to the way our nervous system manages the arousal in response to stress, or more simply, the way we manage energy. Brain researchers have examined why humans are so prone to dysregulated states, as in the chronic anxiety, anger, and depression

---

[1] **An aside on Language**: For the practicing clinician the foreign-sounding, multi-syllabic complex terminology of neuroscience and neuropsychology pose a barrier to understanding. These language barriers arise principally from the need for technical terms in order to precisely describe complex anatomical and functional neural systems. While the use of technical terms renders this field less accessible, and though it is beyond the scope of this chapter to define every term used, I have included them nevertheless, as they appear in the literature, along with occasional brief explanations, and a list of references for those wishing to gain a more fluid working vocabulary.

that often accompany and underlie unhealthy relationships to food (Fernandez-Aranda et al., 2007; Keel, Klump, Miller, McGue & Iacono, 2005). Through the lens of affect-regulation theory, ED can be thought of as strategies to manage excess and disruptive arousal.

Eating disorders are one of a subset of externalizing disorders, with a primary etiology being the failure of early attachment relationships to teach sufficient self-regulation and co-regulation capabilities. In the absence of a more direct or effective way to modulate arousal, those with ED rely on external regulators related to the seeking or avoidance of food. Some turn toward food as a source of comfort; others overeat in a futile attempt to "get enough" of what food can offer only fleetingly; while others turn just as obsessively away from food as the enemy. Thus, attempts to self-regulate externally can take the form of compulsive overeating, bulimic binging/purging to create a dissociative "numbing" state, anorexic avoidance of food, or a combination of all three.

Neural Pathways for Affect-Regulation

One of the basic psychobiological requirements of survival is the regulation of energy. Infants are born with a great deal of affect and little capacity to regulate their own rapid shifts between high and low arousal states. An infant's self-regulating capabilities are limited to sucking, gaze aversion, and dissociation (an unconscious, neurologically mediated way of shutting down when an experience is too powerful or otherwise unbearable). They learn everything else about effectively self-soothing or modulating affect from the adults who comfort and nurture them. In other words, attachment relationships are critical because they facilitate the development of the brain's self-regulatory mechanisms (Fonagy, Gergely, Jurist & Target, 2002). Thus, attachment and self-regulation go hand-in-hand.

Parents who are regulated themselves can be responsively attuned, moving toward the child to attend to them, or picking up the infant to stroke, rock, or walk the child, while cooing soothing sounds or singing lullabies. These are all right hemisphere (RH) nonverbal communications; a child has no left hemisphere (LH) verbal processing until the age of two. By introjecting these soothing moments, the infant learns to internally self-soothe. For insecurely attached children, whose bodies are left unregulated too often and for too long, affect-regulation is impaired. Subsequent dysregulation can take the form of chronic hyper- or hypoactivation and can leave the child predisposed to the eventual onset of psychiatric and psychosomatic disorders (Scaer, 2000; Schore, 1996, 1997). The child may grow to be a teenager who soothes himself or herself externally by using drugs, self-injuring, starving, gorging and then purging, or by eating when distressed.

Although our knowledge of functional neuroanatomy is far from complete, the connections between different areas of the cortex, and from the neocortex to lower brain structures, are well understood. Developmental studies elucidate how the brain structures involved in regulating energy are developed, and how they are impacted by our earliest relationships (Schore, 1994).

Neural Developmental Sequence

The capacity to regulate affect emerges in the very early months of life. An infant develops the ability to handle stressful changes in the external environment by experiencing the caretaker's regulation strategies and his or her own subsequent calming down.

The amygdala and insula, both deep in the temporal lobe, begin to develop during the last trimester of pregnancy. Working together, they register all sensory and propioceptive input and can trigger the body into high states of arousal by activating, through the hypothalamus, the pituitary and adrenal glands, and the autonomic nervous system (ANS), which signals danger or distress and can propel the body to fight, flight, or freeze.

These surges of energy begin to be internally modulated as the capacity for regulating affect is gradually transferred from the caretaker to the infant's brain through body to body, skin to skin contact, deep mutual gazing, facial expressivity, and the melodic prosody or "music of the voice" that we call "parentese." When adults consistently regulate the infant's shifting levels of arousal, they stimulate the infant's own regulatory centers to develop optimally. Neurological circuits of emotion regulation, the insula, then the anterior cingulate (AC), at 2 months, and the OFC at 12 months, come online and create cortical and subcortical pathways down into the amygdala. Activation of these pathways calms the child and inhibits the arousal and expression of emotional states (Schore 1994, 1997, 2001a). Extensive reciprocal connections exist between the amygdala and the prefrontal cortex (PFC) (Davidson, 1998), particularly the medial and orbital zones of the PFC, from which extend neurons of gamma-aminobutyric acid (GABA), a valium-like calming substance manufactured by the brain (Amaral, Price, Pitkanen & Carmichael, 1992). If early relationships are insecure, less than optimal connections are made within the OFC and AC and down into the insula and amygdala. The right brain is itself not as integrated; its regulating pathways may be too thin, its connections into the body not as good and greater amounts and higher frequencies of dysregulation will ensue (Schore, 1994, 1996, 2003a). This mis-wiring shows itself symptomatically as anxiety and affective disorders and in disordered relationships to food.

THE BODY AND A BIPHASIC SYSTEM OF AFFECT-REGULATION

A child must eventually learn to calm him- or herself when he or she is hyperaroused into excitement, anger, fear, or panic, and to bring themselves up from shame, despair, or boredom. Thus, a primary goal of these earliest attachment relationships is to instill in the infant a biphasic system of affect-regulation. To accomplish this, connections must form between the infant's own cortical regulating centers, and the rest of the child's body via the hypothalamic release of body-regulating hormones and through the autonomic nervous system (ANS), whose energizing sympathetic and calming parasympathetic branches move energy up from the body into the brain and back down again. When these neural connections have been wired well, the child can, by 24 months, begin to regulate their entire system when affect threatens to rise too high or fall too low for optimal functioning (Schore, 1994).

Chronic Hyperactivation via the Hypothalamic Pituitary Axis

The hypothalamic pituitary axis (HPA) is a major stress-regulating circuit from the brain into the body linking the hypothalamus, the pituitary, and the adrenal glands. During stress, this circuit triggers the release of corticotrophin releasing factor (CRF), the brain's major stress hormone that, in turn, causes the pituitary gland to release adrenocorticotropic hormone (ACTH) and, subsequently, the adrenal glands to release cortisol. Trauma seems to leave this brain system prone to oversecreting CRF which bathes the brain and body in excess levels of cortisol, resulting in chronic hyperarousal and explaining many of the symptoms of post traumatic stress disorder (PTSD). Once this system has been hyperactivated and hypersensitized, an individual's stress regulation capacities, especially in response to relational injuries, launch more quickly and involve deeper dysregulation (Schore, 2003a, 2001b). Individuals who are poorer at emotional regulation show higher evening levels of cortisol (Davidson, Fox & Kalin, 2007) and an exaggerated startle reflex.

Research (Treasure et al., 2005) utilizing the startle reflex as a measure of affective reactivity found that subjects with ED were not soothed by pleasant stimuli (such as pictures of baby animals) although those with bulimia nervosa (BN) were soothed by photographs depicting food. In contrast, when people with any type of ED viewed images of thin models, their defensive response increased. "Thus, salient emotion [sic] cues differentially produce either an exaggerated or attenuated automatic reaction, depending on the form of the eating disorder" (Treasure et al., 2005, p. 195).

Other studies have found hyperactive states of emotion, such as anxiety and anger, are particularly important in the triggering of binges and are more likely to precede binging than is depression (Agras & Telch, 1998; Arnow, Kenardy & Agras, 1992; Waller & Barter, 2005).

Chronic Hypoactivation as Dissociation

The brain is modular, comprised of well-defined systems, separate in structure and function, which must link up and work together in an integrated way to create our subjective experiences. The more traumatic an experience, the more energy is absorbed into the brain, resulting in a state of neural hyperactivation. The brain protects itself with a kind of circuit breaker system. When energy reaches a critical level, as in overwhelming trauma, neural connections are cut off between the prefrontal cortical and subcortical limbic structures, particularly in the right hemisphere (Schore, 2009), capping and trapping the intense energy beneath a deadened or numbed dissociative state (Porges, 1995).

Researchers have recreated this dissociative defense by using scanning technology and scripts of an individual's experiences, transcribed from their own verbal narratives. A 2007 study (Hopper, Frewen, van der Kolk & Lanius, 2007) observed this neural response to trauma in real time. Hearing their own trauma scripts read back to them, 70% of the subjects went into hyperarousal states, and 30% dissociated. In the latter group, instead of elevating, their heart rates decreased and the cortex became hypometabolic.

DISSOCIATION AND POST TRAUMATIC STRESS DISORDER IN THE DEVELOPMENT OF EATING DISORDERS

In neuroscience, ED is often viewed as a product of affect that is either under- or overregulated. Excessive bodily energy that cannot be regulated down using internal mechanisms can cause a person to reach out to food in order to feel calm. On the other hand, emotions that are overregulated, that are packed down or dissociated, leave the individual without connection to his or her own emotional and physiological states. Left in states of hypoarousal too often or for too long, individuals can lose the ability to sense their own corporeal selves. After years of this defensive overregulation, strong feelings in the body may be unable to come up the neural chain into consciousness and may instead be regulated by reaching for something external like food (Allan Schore, personal communication, September 27, 2008).

Some degree of dissociation underlies most forms of psychopathology, including those related to food. Eating disorders are characterized by both immediate state dissociation and by more chronic trait dissociation (De Berardis et al., 2009; Hallings-Pott, Waller, Watson & Scragg, 2005). For example, in people with AN, emotional blunting and apathy, a secondary effect of starvation, is more tolerable than arousal. "For such individuals, 'numbing' is preferable to the overwhelming experience of anger, jealousy, envy or regret" (Treasure et al., 2005, p. 197). There is a clearly and repeatedly established link between dissociation and bulimic behavior (Everill & Waller, 1995; Everill, Waller & Macdonald, 1995; Waller et al., 2003). Dissociation is both acute and chronic among bulimic women (Hallings-Pott et al., 2005), and binge eating and dissociation appear to serve the common function of blocking negative affect by distancing from it emotionally. Dissociative blunting also occurs during binges when sufferers eat without tasting and without the ability to stop. French researchers found alexithymic-like emotion-processing deficits (for example, the inability to identify and describe the subject's own feeling states) as well as mindsight deficits (impaired ability to understand other's emotional experience), in people with ED (Bydlowski et al., 2005). These test subjects also appeared to be easily overwhelmed by emotional circumstances, and appeared to use food to control intense affective experiences.

Dissociation is the neural mechanism of PTSD (van der Kolk & Fisler, 1995; van der Kolk, 2002). In a review of the ED and PTSD literature, Mantero and Crippa (2002) propose that ED may be a presenting complaint of an undiagnosed PTSD and suggest that clinicians who treat ED should be well-versed in the treatment of acute and chronic trauma. They point to the higher incidence of ED in the PTSD population, noting that ED and PTSD patients show similar trauma histories, which we know from the neuroscience literature are etched into the RH. Subjects with a history of PTSD-related dissociation demonstrate an increase in body dissatisfaction, when compared to the controls; thus, traumatized individuals may use ED as a coping strategy (Beato, Rodrıguez & Belmonte, 2003).

ATTACHMENT THEORY, AFFECT-REGULATION THEORY, AND EATING DISORDERS

When attachment theory, the collective body of research around this early development, is seen as the bonding process that not only creates a strong emotional link between parents and their infants that serves to protect the infant's life, but also the process that instills in the young the ability to self-regulate or modulate its own arousal, it becomes "affect-regulation theory" (Schore & Schore, 2008). As the name implies, Schore's affect-regulation model posits that a developmental deficit in the capacity to regulate powerful affect underlies the major forms of psychological pathology and clinical symptomatology. Thus, through the lens of affect-regulation theory, and the neuroscience that informs it, ED can be thought of as affect-regulation disorders unconsciously devised to externally manage excess arousal, stress, and anxiety, in order to compensate for inadequate wiring in the emotion-regulating circuits of the brain, mind, and body, and in the neural reward centers of the brain. In this way, food becomes an external regulator, a way to control intense affective experiences that otherwise threaten to overwhelm the individual.

Eating Disorders and Attachment

A review of the literature linking disturbances in attachment with ED (Ward, Ramsay & Treasure, 2000) found tentative but compelling evidence that children with dismissive parenting and subsequent avoidant attachment styles are more likely to develop AN, while those with preoccupied parenting, and the resulting anxious/resistant attachment styles, are more likely to develop BN.

A more recent study by Zhang, Li & Zhou (2008) examined brain responses to facial expression by adults with different attachment styles, and found attachment to be a powerful predictor of emotional processing. Anxious individuals were more, and avoidant individuals less, reactive in various areas of their brains to emotional stimuli. The human face is one of the most powerful emotional triggers. In the study, the anxiously attached subjects turned on more of their brain structures related to attentional processing than did secure subjects. Avoidantly attached subjects turned on fewer of those centers and with less metabolic activation, than did either their anxious or secure counterparts

When insecure attachment co-exists with trauma, especially in the first 18 months of life, neuroimaging shows impairment of the right hemisphere that predisposes children toward disorders of affect-regulation and thus may set the stage for the emergence of a disordered relationship to food (Schore, 2009).

Eating Disorders and Emotion Processing Deficits

Much of the research and clinical literature describes ED patients as easily overwhelmed by emotional circumstances. ED frequently coexist with anxiety disorders, and those who suffer from ED may either be less aware of emotions or have well-developed strategies to avoid experiencing their own affective states. Thus, people with ED often exhibit impaired ability to identify their own emotions (alexithymia), as well as deficits in judging others' emotional experience (mindsight) (Bydlowski et al., 2005; Speranza, Corcos, Guilbaud, Loas & Jeammet, 2005; Treasure et al., 2005; Troop, Schmidt & Treasure, 1995).

Davis and Marsh (1986) found that anorexic patients were less able to differentiate between hunger and satiety, and had difficulty differentiating their physical sensations from their emotions, which they also had difficulty describing. Their bulimic subjects often responded to stress with symptoms such as vomiting, even while they had difficulty correlating their ED behavior with any emotional stimulus.

Affect Disregulation and the Right Hemisphere

A dramatic contribution of neuroscience has been to demonstrate that a single human skull houses two virtually separate brains, distinct in both structure and functioning. There are major differences between our rational, logical, analytic, conscious left hemisphere (LH) and our largely unconscious, affective, nonlogical right hemisphere (RH). The latter is dominant for empathy, attachment, and affect-regulation, and stores an internal working model of the earliest attachment relationships and the strategies of affect-regulation that were learned in them (Allan Schore, personal communication, November 19, 2005). Mounting evidence (Lacey, 1986; Meyer, Waller & Waters, 1998; Schore, 2003a) supports an affect-regulation model of ED as a primarily RH phenomenon.

Although the etiology is complex (McManus & Waller, 1995), many, if not most, ED triggers originate in the unconscious subcortical RH. Experimental studies (Meyer & Waller, 1997, 1999; Waller & Mijatovich, 1998) have found that subliminal abandonment cues, presented too rapidly to be detected or reported consciously (Bornstein, 1990; Silverman, 1983), are processed unconsciously in the RH, and result in higher levels of food consumed. This effect was not observed when the presentation of the threat cue was longer, and perceived consciously. The subliminal perception of abandonment may lead to increased bulimic behavior, and binge eating may reduce negative affect through an "anesthetic" effect, a kind of spacing out, or escape behavior, that inhibits the learning of more mature coping (i.e., affect-regulation) strategies (Hallings-Pott et al., 2005). Inversely, the subliminal presentation of unification cues, such as "friendship," decreased the amount of food consumed (Waller & Barter, 2005).

THERAPEUTIC IMPLICATIONS

Such RH subliminal activation of the thoughts and affect underlying disordered eating implies the need for deeply relational and intersubjective therapeutic approaches, with an emphasis on RH resonance, rather than an exclusive reliance on LH cognitively constructed analysis and verbal interpretations (Waller & Barter, 2005). Thus, the goal of psychotherapy should be to enhance the ability for affect-regulation in our ED patients with approaches that activate the RH.

By the age of three, we largely rely on LH functions of thought, language, and the capacity to consciously analyze our experiences. But throughout life, cortical and deep subcortical RH information about what to expect, and "how to be" in relationships continues to exert powerful influence over both behavioral and emotional aspects of our lives, though it remains largely outside our conscious awareness, encoded in implicit (unconscious) procedural (body) memory (Schore, 2003b). In other words, we do not just think or reason our way through life; we also feel and sense our way, with our affective wisdom shaping the choices that we make (Demasio, 1999). Much of our behavior merely emerges from old patterns and attachment templates, rather than logical choice. In addition, traumatic experiences are stored largely in our RH and, if they remain unresolved, can emerge later in life as dysregulated energy and/or as compensatory patterns of behavioral acting-out that we label as psychopathology.

Effective treatment of ED requires the fostering of affect-modulating centers in the prefrontal cortices that developed relationally in infancy. This wiring of one brain through interaction with another brain extends well beyond childhood and is the psychoneurobiological basis for the healing potential of important relationships, including the deeply attuned dyad in psychotherapy (Safran & Muran, 2000). Evidence is accumulating that poorly wired circuits of interactive regulation, especially those between higher and lower centers on the right side of the brain, can literally be rewired in the therapist/patient dyad (Schwartz &

Begley, 2002; Linden, 2006). This can lead to an individual's increased ability to modulate their own emotional experiences.

The key is to connect with our patients' nonverbal, emotional RHs. For psychoanalytically oriented psychotherapists, this means more emphasis on "the primacy of affect" and less on cognitively constructed analysis and verbal interpretations (Allan Schore, personal communication, November 19, 2005). In general, psychotherapy must include more than verbal, investigative, and insight-based approaches to change; it must activate the deep subcortical recesses of the unconscious mind where our attachment templates have been laid down and trauma has been stored. The effective therapeutic dyad uses the same processes that occur between parents and infants in secure attachments (attunement, empathy, and resonance), to reactivate and rewire RH procedural templates from childhood. Through this exquisite attunement and timely coregulation within the therapeutic dyad, the ED individual's capacity for regulation, discernment, and impulse control improve, as higher centers become better at turning down lower centers of primitive arousal (Beer & Lombardo, 2007). To achieve this, psychotherapy must be an emotional, affective, somatic experience in the context of an intimate relationship and environment safe enough to enable experience of emotions and body sensations, even those that prove frightful or full of pain.

Left Hemisphere Cognitive, Conscious Strategies for Affect-Regulation

During its earliest, formative years, the infant has no verbal processing LH; this area of the brain does not begin to myelinate until the end of the second year of life. Once LH neural pathways are in place, given sufficient processing time, the slower LH can exert a regulating influence over the more emotional RH. We can use the verbal capacities of the LH to both dysregulate ourselves with anxious thoughts and imagined catastrophes, and to regulate ourselves and one another through calming soothing thoughts, and verbal reassurances. Other LH calming strategies include intentional distraction, deliberate suppression, reappraisal, or reframing.

Since Aaron Beck introduced cognitive psychotherapy in the 1970s (Beck, 1972; Beck, Rush, Shaw & Emery, 1979), clinicians have focused on the beliefs that individuals have about present or future events and the perceptions they have about their abilities to cope with these occurrences. LH-based cognitive therapy strategies have traditionally focused on evaluating the rationality of negative appraisals and beliefs, substituting more realistic, rational, or evidence-based viewpoints in their stead. In addition, the behavioral components of CBT for ED utilize patient education, self-monitoring, dietary records, shape and weight checking restraints, and verbal assessment questionnaires. Alone, or in combination with psychoactive medication, CBT has become a popular treatment model for ED (Fairburn, 2008; Waller et al., 2007).

Even psychoanalytic approaches have traditionally relied more on LH-based, cognitively-constructed analyses and verbal interpretations to encourage insight and the subsequent emergence of the unconscious realm into conscious thought. Neuroscience research, however, demonstrates that the neural mechanisms for affect-regulation exist primarily in the unconscious RH, and it is there that we must work to foster more permanent and automatic regulatory capabilities (Schore, 2003b). In short, cognitive strategies that encourage reappraisal of a stimuli triggering threat, and training awareness to accurately read "bottom-up," limbic system reactions of fear, anger, sadness, or shame, can improve the capacity to regulate, but a greater emphasis on RH modalities could well be the key to improving previously disappointing outcomes in ED.

RH modalities are already at play, at least peripherally, in LH-based therapeutic approaches like CBT. Ideally, all therapeutic contact, including CBT, occurs in a deeply affective, attuned, sensitive, and caring RH-to-RH bond between patient and therapist. Such synchronous, empathic, and transferential interactions create greater integration within the patient's own RH. Neural networks that had been dissociated connect again and eventually create stronger top-down, frontal lobe influences. As these high right neural centers send soothing extensions down into affective subcortical ones, affect-regulation and resiliency emerge.

Language of the Right Hemisphere

The conscious LH is not only structurally and functionally distinct from the mostly unconscious, affective RH, but it uses a different system of language as well. Information is communicated through words in the LH, while in the RH, affect is transmitted by the primary process communications of the face, gesture, posture, and prosody.

Beebe and Lachmann (1994) demonstrated that mothers engage in a synchronous dance of this primary-process attunement with their infants, yet are not aware, and cannot describe, what they are doing because it is happening too quickly for consciousness to track. Templates of secure attachment are built through repeated affect synchrony, as the child is met and amplified into positive peaks of joy and excitement and soothed or calmed out of states of fear or pain. In the absence of this attuned resonance, the child is left in extended states of disregulated affect and the attachment that evolves is insecure.

Psychotherapists who want to modify those early neural patterns must work simultaneously in both left and right hemisphere modalities. The RH synchrony of psychotherapy develops as we learn the unique rhythms of each patient and form a dyadic system with them. This requires listening carefully to track the patient's movements, to find their rhythm, to match their tempo and to reflect it back. We pace ourselves unconsciously to stay with them; we soften our voice, fall silent; join them in moments of humor or joy; fall into a hushed reverence. We speak with our posture, our eyes, our entire face. We are there to engage when they look up for us, wait patiently when they turn away, allow them time to recover and reengage, and sometimes reach out to them if they have moved too far away. In doing this, we form an empathic bond using our own bodies, our own somatic countertransference, and we match our own internal states with theirs so that they feel increasingly safe to drop defenses and open both to us and to their own deepening understanding of themselves. Affect gradually deepens as together we gain access to material stored deep in their emotional right brain. In this state of dyadic synchrony, interpretations or other interventions are experienced more deeply and their effects are more permanent (Allan Schore, personal communication, May 6, 2006).

Powerful empirical evidence suggests that the dyadic relationship contributes the majority of healing in psychotherapy, and it is now generally acknowledged that "…the therapeutic alliance is among the most robust predictors of treatment outcomes" (Karver, Handelsman,

Fields & Bickman, 2006, p. 35). When patients gradually learn to regulate, to stay more coherent, dissociating or fragmenting less during moments of heightened intensity, shifts occur in implicit relational patterns deep in the subcortical RH, which the LH may then later be able to translate into words. It is not, however, the words that make the changes. Rather, it is subcortical RH shifts that make the verbal awareness possible. Just moving emotional material into the left hemisphere does not suffice. If the affective material is regulated when it arrives in the LH, it becomes a coherent narrative. If it is still unregulated when it shifts left, it is disruptive and the individual may become confused, go blank or lose their train of thought (Allan Schore, personal communication, May 30, 2009).

The Body in Psychotherapy

The recognition that psychotherapy must be an affective experience in order to activate the subcortical limbic regions of the RH implicates the body as well in psychotherapy since the RH, far more than the left, is deeply connected into the body through the HPA axis and the ANS (Schore, 2003a; Allan Schore, personal communication, November 19, 2005). The neurological and trauma literature have thus expanded the definition of psychotherapy from a "mind-to-mind, thought-to-thought" connection to one that communicates "body-to-body" as well. Schore says repeatedly that trauma is about dissociation (Schore, 2009), neurologically disconnecting oneself as a means of self-defense from an experience that is too powerful or otherwise unbearable. Thus, trauma recovery is about coming out of dissociation and back into awareness of the external energy coming in and the energy arising up interoceptively from one's own corporeal self.

Healing from ED means improving the capacity to both individually and interactively regulate externally- and internally-arising energies, rather than reaching for external regulators like food, laxatives, or obsessive exercise. We differ in our capacities as therapists to notice our patients' subtle or micro facial expressions (faster than the 500 ms it takes to be consciously seen), changes in skin color, body posture, breathing patterns, or shifts in auditory prosody. The more accurately we track these and, simultaneously, our own somatic replies, the greater our capacity to affectively resonate and know when, and if, to move in and lend a regulating hand. Thus, a major outcome of good treatment in individuals with ED is a decrease in dissociation through an increased awareness of, tolerance for, and openness to, affect and sensation arising from the body in real time.

CONCLUDING THOUGHTS

The Therapist, More Than the Theory, Is the Agent of Change

Advances in neurological research have confirmed that a deep, safe, affective (and ideally affectionate) therapist/patient relationship is one of, if not *the*, central change element in psychotherapy; and, further, that the subcortical limbic system (that part of the brain that processes emotion, and is the center of the attachment relationship), is the key. Beneath the conscious models and techniques that distinguish the various schools of psychotherapy lies the therapist's ability to create the attachment intimacy of the patient/therapist bond, a relational process, a deep intersubjective being with that which is "beyond technique, frequently undervalued and mediated primarily by their and our more inexplicable and baffling RH" (Allan Schore, personal communication, September 27, 2008). The source of effective therapy is not so much what the therapist does for the patient, or says to the patient, as how to be with the patient, especially at moments of therapeutic rupture or when the patient's state of being is disintegrating in real time (Safran, 2003). When patients are in an overwhelmed state of terror, hopelessness, or despair, it is the therapist's job to be able to enter into, and hold that state with them. Although this can be enormously challenging, it is imperative, and so we do our best to stay present in our own bodies, to allow those feelings to be truly felt, and to stay connected in the relationship so the patient is not alone with them (Allan Schore, personal communication, February 11, 2006).

Significant gaps in our knowledge of the neurobiology underlying ED (as well as other mental disorders) persist, due to the difficulties of mapping the complex neural circuitry of higher brain functioning, and the complexity of sorting genetic and epigenetic developmental contributions (Hyman, 2007). Despite these limitations, neurobiological research will be an increasingly essential tool in helping us rethink our current approaches to the treatment of ED. In the meantime, it is safe to conclude that it is in relationships that we first emerge into our full neural complexity, and it is in relationships that we must heal.

References

Agras, W. S., & Telch, C. F. (1998). The effects of caloric deprivation and negative affect on binge eating in obese binge-eating disordered women. *Behavior Therapy, 29*, 491–503.

Amaral, D. G., Price, J. L., Pitkanen, A., & Carmichael, S. T. (1992). Anatomical organization of the primate amygdaloid complex. In J. P. Aggleton (Ed.), *The amygdala: Neurobiological aspects of emotion, memory and mental dysfunction* (pp. 1–66). New York, NY: Wiley-Liss.

Armstrong, J. G., & Roth, D. M. (1989). Attachment and separation difficulties in eating disorders: A preliminary investigation. *International Journal of Eating Disorders, 8*(2), 141–155.

Arnow, B., Kenardy, J., & Agras, W. S. (1992). Binge-eating among the obese: A descriptive study. *Journal of Behavioral Medicine, 15*, 155–170.

Beato, L., Rodrıguez, T., & Belmonte, C. A. (2003). Relationship of dissociative experiences to body shape concerns in eating disorders. *European eating disorders Review, 11*, 38–45.

Beck, A. T. (1972). *Depression: Causes and treatment*. Philadelphia, PA: University of Pennsylvania Press.

Beck, A. T., Rush, A. J., Shaw, B. F., & Emery, G. (1979). *Cognitive therapy of depression*. New York, NY: The Guilford Press.

Beebe, B., & Lachmann, F. (1994). Representation and internalization in infancy: Three principles of salience. *Psychoanalytic Psychology, 11*, 127–166.

Beer, J. S., & Lombardo, M. V. (2007). Insights into emotion regulation from neuropsychology. In J. J. Gross (Ed.), *Handbook of emotion regulation* (pp. 69–86). New York, NY: Guilford Press.

Bergen, A., Yeager, M., Welch, R., Haque, K., Ganjei, J. K., VandenBree, M. B., ... Kaye, W. H. (2005). Association of multiple DRD2 polymorphisms with anorexia nervosa. *Neuropsychopharmacology, 30*, 1703–1710.

Bornstein, R. F. (1990). Critical importance of stimulus unawareness for the production of subliminal psychodynamic activation effects: A meta-analytic review. *Journal of Clinical Psychology, 46*, 201–210.

Bydlowski, S., Corcos, M., Jeammet, P., Paterniti, S., Berthoz, S., Laurier, C., ... Consoli, S. M. (2005). Emotion-Processing deficits in eating disorders. *International Journal of Eating Disorders, 37*, 321–329.

Davidson, R. J. (1998). Affective style and affective disorders: Perspectives from affective neuroscience. *Cognition and Emotion, 12*(3), 307–330.

Davidson, R. J., Fox, A., & Kalin, N. H. (2007). Neural bases of emotion regulation in nonhuman primates and humans. In J. J. Gross (Ed.), *Handbook of emotion regulation* (pp. 47–68). New York, NY: Guilford Press.

Davis, M. S., & Marsh, L. (1986). Self-love, self-control and alexithymia: Narcissistic features of two bulimic adolescents. *American Journal of Psychotherapy, 15*, 224–232.

De Berardis, D., Serroni, N., Campanella, D., Carano, A., Gambi, F., Valchera, A.,... Ferro, F. M. (2009). Alexithymia and its relationships with dissociative experiences: Body dissatisfaction and eating disturbances in a non-clinical female sample. *Cognitive Therapy and Research, 333*(5), 471–479. Accessed via Superlink, published online: 6 May 2009.

Demasio, A. (1999). *The feeling of what happens: Body and emotion in the making of consciousness*. New York, NY: Harcourt Brace.

Dusoir, H., Owens, C., Forbes, R. B., Morrow, J., Flynn, P., & McCarron, M. (2005). Anorexia nervosa remission following left thalamic stroke. *Journal of Neurology Neurosurgery and Psychiatry, 76*, 144–145.

Everill, J., & Waller, G. (1995). Dissociation and bulimia: Research and theory. *European Eating Disorders Review, 3*, 129–147.

Everill, J., Waller, G., & Macdonald, W. (1995). Dissociation in bulimic and non-eating-disordered women. *International Journal of Eating Disorders, 17*, 127–134.

Fairburn, C. G. (2008). *Cognitive behavior therapy and eating disorders*. New York, NY: Guilford Press.

Fernandez-Aranda, F., Pinheiro, A. P., Tozzi, F., La Via, M., Thornton, L. M., Plotnicov, K. H., ... Bulik, C. M. (2007). Symptom profile and temporal relation of major depressive disorder in females with eating disorders. *Australian and New Zealand Journal of Psychiatry, 41*, 24–31.

Fonagy, P., Gergely, G., Jurist, E., & Target, M. (2002). *Affect regulation, mentalization, and the development of the self*. New York, NY: Other Press.

Grunwald, M., Etrrich, C., Assmann, B., Daehne, A., & Krause, W. (2001). Deficits in haptic perception and right parietal theta power changes in patients with anorexia nervosa before and after weight gain. *International Journal of Eating Disorders, 29*, 417–428.

Hallings-Pott, C., Waller, D. G., Watson, D., & Scragg, P. (2005). State dissociation in bulimic eating disorders: An experimental study. *International Journal of Eating Disorders, 38*, 37–41.

Hopper, J., Frewen, P., van der Kolk, B., & Lanius, R. (2007). Neural correlates of reexperiencing, avoidance, and dissociation in PTSD: Symptom dimensions and emotion dysregulation in responses to script-driven trauma imagery. *Journal of Traumatic Stress, 20*(5), 713–725.

Hyman, S. (2007). Can neuroscience be integrated into the DSM-V? *Nature Reviews/Neuroscience, 8*, 725–732.

Karver, M. C., Handelsman, J. B., Fields, S., & Bickman, L. (2006). Meta-analysis of therapeutic relationship variables in youth and family therapy. The evidence for different relationship variables in the child and adolescent treatment outcome literature. *Clinical Psychology Reviews, 26*, 50–65.

Keel, P. K., Klump, K. L., Miller, K. B., McGue, M., & Iacono, W. G. (2005). Shared transmission of eating disorders and anxiety disorders. *International Journal of Eating Disorders, 38*, 99–105.

Lacey, J. H. (1986). Pathogenesis. In L. J. Downey & J. C. Malkin (Eds.), *Current approaches: Bulimia nervosa* (pp. 17–26). Southampton, England: Duphar.

Linden, D. E. J. (2006). How psychotherapy changes the brain — the contribution of functional neuroimaging. *Molecular Psychiatry, 11*, 528–538.

Mantero, M., & Crippa, L. (2002). Eating disorders and chronic PTSD: Issues of psychopathology and comorbidity. *European Eating Disorders Review, 10*(1), 1–16.

McCormick, L. M., Keel, P. K., Brumm, M. C., Bowers, W., Swayze, V., Andersen, A., & Andreason, N. (2008). Implications of starvation-induced change in right dorsal anterior cingulate volume in anorexia nervosa. *International Journal of Eating Disorders, 41*, 602–610.

McManus, F., & Waller, G. (1995). A functional analysis of binge eating. *Clinical Psychology Review, 15*, 845–865.

Meyer, C., & Waller, G. (1997). *The impact of emotion upon eating behavior: The role of subliminal visual processing of threat cues*. United Kingdom: University of Southampton, Department of Psychology.

Meyer, C., Waller, G., & Waters, A. (1998). Emotional states and bulimic psychopathology. In H. Hoek, M. Katzman, & J. Treasure (Eds.), *The neurobiological basis of eating disorders* (pp. 271–289). Chichester, England: Wiley.

Meyer, C., & Waller, G. (1999). The impact of emotion upon eating behaviour: The role of subliminal visual processing of threat cues. *International Journal of eating disorders, 25*, 319–326.

National Alliance on Mental Illness. (1996). *Eating disorder not otherwise specified (EDNOS)*. Retrieved from http://www.nami.org/Content/ContentGroups/Helpline1/Eating_Disorder_Not_Otherwise_Specified_(EDNOS).htm

O'Donnell, T. O., Hegadoren, K. M., & Coupland, N. C. (2004). Noradrenergic mechanisms in the pathophysiology of post-traumatic stress disorder. *Neuropsychobiology, 50*(4), 273–283.

O'Kearney, R. (1996). Attachment disruption in anorexia nervosa and bulimia nervosa: A review of theory and empirical research. *International Journal of Eating Disorders, 20*(2), 115–127.

Porges, S. W. (1995). Orienting in a defensive world: Mammalian modifications of our evolutionary heritage. A polyvagal theory. *Psychophysiology, 32*, 301–318.

Ramachandran, V. S., & Blakeslee, S. (1998). *Phantoms in the brain*. New York, NY: Morrow.

Safran, J. D., & Muran, J. C. (2000). *Negotiating the therapeutic alliance: A relational treatment guide*. New York, NY: Guilford Press.

Safran, J. D. (2003). The relational turn, the therapeutic alliance and psychotherapy research. *Contemporary Psychoanalysis, 59*(3), 449–473.

Scaer, R. (2000). *The body bears the burden: Trauma, dissociation and disease*. Binghamton, NY: The Haworth Press, Inc.

Schore, A. N. (1994). *Affect regulation and the origin of the self: The neurobiology of emotional development*. New Jersey: Lawrence Erldbaum Associates.

Schore, A. N. (1996). The experience-dependent maturation of a regulatory system in the orbital prefrontal cortex and the origin of developmental psychopathology. *Development and Psychopathology, 8*, 59–87.

Schore, A. N. (1997). Early organization of the nonlinear right brain and development of a predisposition to psychiatric disorders. *Development and Psychopathology, 9*, 595–631.

Schore, A. N. (2001a). The effects of a secure attachment relationship on right brain development, affect regulation, and infant mental health. *Infant Mental Health Journal, 22*, 7–66.

Schore, A. N. (2001b). The effects of relational trauma on right brain development, affect regulation, and infant mental health. *Infant Mental Health Journal, 22*, 201–269.

Schore, A. N. (2003a). *Affect dysregulation and disorders of the self*. New York, NY: W.W. Norton & Company.

Schore, A. N. (2003b). *Affect dysregulation and the repair of the self*. New York, NY: W.W. Norton & Company.

Schore, A. N. (2009). Attachment trauma and the developing right brain: Origins of pathological dissociation. In P. Dell, & J. O'Neil (Eds.), *Dissociation and dissociative disorders* (pp. 107–141). New York, NY: Taylor & Francis Group.

Schore, J. R., & Schore, A. N. (2008). Modern attachment theory: The central role of affect regulation in development and treatment. *Clinical Social Work Journal, 36*, 9–20.

Schwartz, J. M., & Begley, S. (2002). *The mind and the brain: Neuroplasticity and the power of mental force*. New York, NY: Harper Collins.

Silverman, L. H. (1983). The subliminal psychodynamic activation method: Overview and comprehensive listing of studies. In J. Masling (Ed.), *Empirical studies of psychoanalytic theories, 1* (pp. 69–100). Hillside, NJ: Erlbaum.

Speranza, M., Corcos, M., Guilbaud, O., Loas, G., & Jeammet, P. (2005). Alexithymia, personality, and psychopathology. *American Journal of Psychiatry, 162*(5), 1029–1030.

Stice, E., Spoor, S., Bohon, C., & Small, D. M. (2008). Relation between obesity and blunted striatal response to food is moderated by TaqIA A1 allele. *Science, 322*(5900), 449–452.

Treasure, J., Tchanturia, K., & Ulrike, S. (2005). Developing a model of the treatment for eating disorder: Using neuroscience research to examine the how rather than the what of change. *Counselling and Psychotherapy Research, 5*(3), 191–202.

Troop, N. A., Schmidt, U. H., & Treasure, J. L. (1995). Feelings and fantasy in eating disorders: A factor analysis of the Toronto Alexithymia Scale. *International Journal of Eating Disorders, 18*(2), 151–157.

van der Kolk, B. A. (2002). Post-traumatic therapy in the age of neuroscience. *Psychoanalytic Dialogues, 12*(3), 381–392.

van der Kolk, B. A., & Fisler, R. (1995). Dissociation and the fragmentary nature of traumatic memories: Overview and exploratory study. *Journal of Traumatic Stress, 8*(4), 505–525.

Wagner, A., Aizenstein, H., Mazurkewicz, L., Fudge, J., Frank, G. K., Putnam, K., ... Kaye, W. H. (2008). Altered insula response to taste stimuli in individuals recovered from restricting-type anorexia nervosa. *Neuropsychopharmacology, 33*, 513–523.

Waller, G., & Mijatovich, S. (1998). Preconscious processing of threat cues: Impact upon eating upon women with unhealthy eating attitudes. *International Journal of Eating Disorders, 24*, 83–89.

Waller, G., Babbs, M., Wright, F., Potterton, C., Meyer, C., & Leung, N. (2003). Somatoform dissociation in eating-disordered patients. *Behaviour Research and Therapy, 41*(5), 619–627.

Waller, G., & Barter, G. (2005). The impact of subliminal abandonment and unification cues on eating behavior. *International Journal of Eating Disorders, 37*, 156–160.

Waller, G., Cordery, H., Corstorphine, E., Hinrichsen, H., Lawson, V., Mountford, V., & Russell, K. (2007). *Cognitive behavioral therapy for eating disorders: A comprehensive treatment guide.* Cambridge, England: Cambridge University Press.

Ward, A., Ramsay, R., & Treasure, J. (2000). Attachment research in eating disorders. *British Journal of Medical Psychology, 73*, 35–51.

Winfrey, Oprah (2008). *Oprah's weight loss confession. O, The Oprah Magazine.* Retrieved from http://www.oprah.com/slideshow/oprahshow/20081030_tows_bobgreene

Wonderlich, S. A., Connolly, K. M., & Stice, E. (2004). Impulsivity as a risk factor for eating disorder behavior: Assessment implications with adolescents. *International Journal of Eating Disorders, 36*(2), 172–182.

Zhang, X., Li, T., & Zhou, X. (2008). Brain responses to facial expression by adults with different attachment-orientations. *NeuroReport, 19*(4), 437–441.

CHAPTER 4

Are Media an Important Medium for Clinicians?
Mass Media, Eating Disorders, and the Bolder Model of Treatment, Prevention, and Advocacy

Michael P. Levine and Margo Maine

Advocacy and activism challenge us to take a moral and ethical stand with regard to the touchiest issues within our organizations, to publicly articulate our stand, and to risk the displeasure, if not the wrath, of those who hold power and authority. It may mean being unpopular, becoming a lightning rod for the anger and resistance of colleagues, and at times, it may mean being willing to put our jobs on the line in order to do the right thing **(Grieger & Ponterotto, 1998, p. 31, quoted in Vera & Speight, 2007, p. 377)**

The role of mass media in the development and maintenance of the spectrum of negative body image (BI) and disordered eating (DE) is a "touchy issue" in the field of eating disorders (ED) (Levine & Murnen, 2009; Smolak, Levine & Murnen, 2006). Many find the media's status as a cause of body dissatisfaction, a drive for thinness, ambivalence about food, and ED to be self-evident. Others, including an increasing number of parents and bio-psychiatric scientist-practitioners, see this contention as, *at best*, an irritating distraction from the need to acknowledge that ED are severe, self-sustaining psychiatric illnesses with a genetic, biochemical basis (Bulik, 2004; Klump, Bullik, Kaye, Treasure & Tyson, 2009). *At worst*, many research-oriented clinicians are concerned that a psychosocial or sociocultural perspective, in contrast to a biopsychiatric perspective, will add to the already troubling "blame-based stigma associated with anorexia nervosa" and other ED (Crisafulli, Von Holle & Bulik, 2008, p. 333).

This chapter is designed to educate and support clinicians who wish to integrate theory and research concerning mass media into the treatment and prevention of negative BI and DE, as well as into their own professional and personal development. After a brief review of the current status of research pertaining to mass media as a *causal risk factor* for clinically significant levels of negative BI and DE, we consider the implications of that work for treatment, prevention, and advocacy.[1]

As a researcher and a clinician collaborating on this chapter, we both find the scientific research supporting the role of the media in the risk factors associated with ED to be a remarkable validation of what some clinicians and many patients frequently discuss. When science and practice come together as these two areas do, it speaks to the tremendous work that both academicians and clinicians have contributed to this field.

SOCIOCULTURAL MODELS OF RISK

"Sociocultural" models of risk propose that macro-level cultural influences shape the content and the forms of those mass media, peer interaction, and parental variables that have considerable empirical support as risk factors in BI disturbances and eating problems (Becker & Fay, 2006; Smolak, 2009; Smolak et al., 2006). The challenge is to define the critical "content" and "messages," determine the extent of individual exposure and assimilation, and then establish how these contribute to the risk of ED (Anderson-Fye & Becker, 2004; Becker & Fay, 2006).

Substantial evidence suggests that the "Eating Disorders" recognized by the *DSM-IV-TR* (American Psychiatric Association, 2000) are composite expressions, to varying degrees, of dimensions such as negative emotionality, binge eating, and unhealthy forms of weight and shape management. In most cases, these features reflect a combination of negative BI and negative self-concept. For females, "body dissatisfaction" results from, and feeds, a schema easily found in the content of mass media. It is also comprised of four fundamental and culturally normative components: (i) a conviction that weight and shape are central determinants of one's identity; (ii) the idealization of slenderness and leanness; (iii) an irrational fear of fat; and (iv) the tyranny of "I must demonstrate my control over my body and its appetites, and I should be able to do this" (Gordon, 2000; Levine & Harrison, 2009; Levine & Smolak, 2006; Smolak & Levine, 1996).

Sociocultural models explain the multidimensional processes feeding the *development* of disorders. They do not deny that the disorders themselves typically reflect self-sustaining, chronic, extremely serious, and sometimes degenerative, and even life-threatening, processes that are rooted in a complex interplay between biopsychology, behavior, and interpersonal environments (Smolak et al., 2006). Nor do sociocultural models deny a role for genetics and/or neurobiology in either the individual risk for, or the progression of, ED. However, sociocultural models do argue that multiple pathways lead to similar outcomes. Moreover, in contrast to the mountain of evidence supporting the sociocultural impact, no evidence indicates that the majority of cases of DE are attributable to genetic vulnerability (Levine, 2009b; Smolak & Levine, 1994; Smolak et al., 2006). In fact, as behavior geneticists themselves have long argued, rare conditions can be explained by the multiplicative interaction of risk factors commonly distributed in a population (Levine & Smolak, 2009). Certain people will

---

[1] Although the relationships between mass media, negative body image, and unhealthy behaviors in males are receiving increasing attention (see, e.g., Pope, Phillips & Olivardia, 2000; Ricciardelli, McCabe, Williams & Thompson, 2007), this review concentrates on females.

have a genetic vulnerability placing them at high risk for an ED (Bulik, 2004), but the Rose Paradox demonstrates that the majority of cases of a disorder emerge from those at moderate to lower risk because they are simply more numerous than those at high risk (Austin, 2001). As stated in Chapter 1 (Maine and Bunnell), nature and nurture cannot be separated, with ED reflecting both individual and cultural risk factors.

Risk and Mass Media

A risk factor is a variable that is reliably and usefully associated with an increase over time in the probability of a subsequent outcome. Table 4.1 presents a summary of the argument for mass media as a causal risk factor by showing the necessary conditions (left-hand column), the status of the evidence, and representative citations for further reference. Readers who wish more detailed information are referred to literature reviews by Harrison and Hefner (2008), Levine and Harrison (2009), and Levine and Murnen (2009), and to a meta-analysis by Grabe, Ward & Hyde (2008).

Here are our seven basic conclusions from the voluminous and burgeoning research on mass media, BI, and DE. First, mass media are a major part of the lives of children, adolescents, and adults in many countries around the world (Anderson-Fye & Becker, 2004). Second, the content of mass media provides multiple, overlapping, and, all too often, unhealthy messages about gender, attractiveness, ideal body sizes and shapes, self-control, desire, food, and weight management. These messages indoctrinate developing girls and boys with insidious themes, such as: (i) women are "naturally" invested in their beauty assets; (ii) a slender, youthful attractive "image" is truly substantive, because it pleases males and demonstrates being "in control" to females; and (iii) the sources of ideals about attractiveness, style, and the best practices for becoming and staying beautiful are located outside the self (Levine & Harrison, 2009; Smolak & Levine, 1994, 1996; Smolak & Murnen, 2004, 2007). Thus, the slender beauty ideal, framed by other cultivated aspects of the feminine sex role, is not perceived as a mythic fantasy, but rather as ubiquitous, accepted by most people, and both normal and achievable (Thompson, Heinberg, Altabe & Tantleff-Dunn, 1999).

Third, cross-sectional surveys generally indicate modest but significant positive correlations between level of exposure to mass media and the important triad of body dissatisfaction, thin-ideal internationalization, and DE. However, we still have much to learn about the relationship between frequency and intensity of naturalistic media exposure and DE. Fourth, although there are surprisingly few longitudinal studies, and, therefore, much more research is needed, the available literature suggests that, for young people, the extent of media exposure predicts increases in negative BI and DE (Van den Berg, Neumark-Sztainer, Hannan & Haines, 2007), However, by early adolescence, the causal risk factor may not necessarily be media exposure, or even internalization of the slender beauty ideal, but rather the intensity and extent of "core beliefs and assumptions about the importance, meaning, and effect of appearance in an individual's life" (Tiggemann, 2006, p. 528).

Fifth, concentrated exposure to the media's slender ideal tends to produce an immediate and moderately large negative effect in how girls and women feel about their bodies and their selves. This appears to be true for content from magazines and television, as well as pro-anorexia web sites (Bardone-Cone & Cass, 2007) and video games featuring thin

TABLE 4.1 Status of Evidence for Mass Media as a Causal, Variable Risk Factor for Negative Body Image and Disordered Eating

Criterion	Status of Evidence	Further Reference
1. Content	Strong	
Slenderness is normative, attractive, attainable		Harrison & Hefner (2006); Thompson, van den Berg, Roehrig, Guarda & Heinberg (2004)
Fat is ugly and horrible: indicative of sloth, greed, extravagance, gluttony, and pride		Himes & Thompson (2007); Puhl & Latner (2007)
2. Exposure	Strong	Comstock & Scharrer (2007); Roberts, Foehr & Rideout (2005)
3. Positive correlation between SDE and level and/or type(s) of exposure:		
Cross-sectional studies		
Television	Moderate — Strong	Grabe et al. (2008); Murnen, Levine, Groesz & Smith (2007)
Magazines	Moderate — Strong	
Longitudinal studies		
Television	Moderate — Preliminary	Dohnt & Tiggemann (2006); Harrison & Hefner (2006); Tiggemann (2006)
Magazines	Weak — preliminary	
4. Controlled laboratory *experiments* assessing immediate effects of media content	Strong	Grabe et al. (2008); Groesz, Levine & Murnen (2002)
5. Subjective reports of desire to look like media "figures," perceived media pressures, and internalization of media ideals of beauty		
Cross-sectional studies	Strong	See reviews by Cafri, Yamamiya, Brannick & Thompson (2005); Levine & Harrison (2009); Levine & Murnen (2009); Murnen et al. (2007)
Longitudinal studies	Moderate — Preliminary	Stice (2002); see discussion by Levine & Murnen (2009)
Media literacy as prevention		
Controlled laboratory studies (analogs)	Strong	
Brief field studies	Moderate — Strong	See reviews by Levine (2009); Levine & Kelly (in press); Levine & Harrison (2009); Levine & Smolak (2006, Chapter 13)
Intensive field studies	Moderate — Preliminary	
Selective-targeted prevention	Moderate — Strong	

Note. In the Status of Evidence column, the designation of "preliminary" means that there are relatively few studies and/or the studies that have been done include variables that are, for various reasons, not direct reflections of the construct.

characters (Barlett & Harris, 2008). This "contrast effect" appears to be appreciably greater in females who are heavily invested in appearance concerns and/or who are already self-conscious and otherwise sensitive about their body shape (Groesz, Levine, & Murnen, 2002). A longitudinal study by Hargreaves and Tiggemann (2003) showed that adolescent girls whose BI was most negatively affected by experimental exposure to television commercials featuring the thin ideal had greater levels of body dissatisfaction and drive for thinness two years later, even when initial level of body dissatisfaction was controlled.

Sixth, if mass media are a causal risk factor in a psychological sense, then one should be able to determine their subjective influence in terms of ideals, motives, and experienced pressures. In general, studies of adolescent girls in various countries reveal that over two thirds feel that images in magazines do influence their conception of the "perfect body shape," while a significant minority report that such images motivate them to lose weight. Cross-sectional and longitudinal studies of adolescents and young women indicate that a clear desire to look like celebrities and models emerges as a strong predictor of weight concerns, dieting, binge-eating, and unhealthy weight management (see Levine & Harrison, 2009, and Levine & Murnen, 2009, for reviews).

Normative beliefs about the social *world*, as emphasized in mass media, become subjective beliefs and attitudes about the *self* and about what others believe and expect. Meta-analysis clearly shows that extent of self-reported media exposure correlates moderately with internalization of the slender beauty ideal (Grabe et al., 2008; Murnen, Levine, Groesz, & Smith, 2007). Cross-sectional research with middle and high school girls and undergraduate women (reviewed in Levine & Harrison, 2009, and Levine & Murnen, 2009) shows consistently that self-reported levels of perceived pressure from media and of internalization of the slender beauty ideal are highly correlated with each other. Individually, these two variables are each strong predictors of precursors of weight-and-body-dissatisfaction, unhealthy weight management, and DE. In support of these correlation-based findings, level of thin-ideal internalization has consistently been identified in experimental studies as a moderator of media effects (Levine & Murnen, 2009; Dittmar, Halliwell & Stirling, 2009).

Media Literacy

Thanks in large part to Stice (2002), the ED field now accepts that prevention studies are an *essential* component of "basic research" designed to clarify etiology. Thus, the seventh and final criterion: if media effects are a causal and variable risk factor, then reduction or elimination of negative media influences should reduce or prevent negative BI and other processes that, for some, eventually result in ED.

As is the case with most important phenomena in the behavioral sciences, "media literacy" is easy to talk about but difficult to define (Levine & Kelly, in press; Levine & Smolak, 2006). In general, "Media Literacy" refers to "the process of critically analyzing and learning to create one's own messages in print, audio, video, and multimedia" (Hobbs, 1998, p. 16). The first step in developing media literacy is establishing a collaboration between children or teens and their mentors that allows them to work together to understand and analyze the nature of mass media and one's relationships with them (Levine, 2009a; Levine & Harrison, 2004, 2009; Levine & Piran, 2004; Levine & Smolak, 2006).The next step involves active creative, well-planned efforts to translate the *knowledge* gained from a critical consciousness into *activism* toward offensive media and into *advocacy* of healthier messages. Learning how to *create* and *use* media, as well as how to "read" them, builds and reinforces advocacy skills involving research, problem-solving, assertion, and communication, as well as expanding the enjoyment and appreciation of media. These are the same "life skills" repeatedly found to play an important role in the long-term prevention of substance abuse (Botvin, 2000).

This approach to media literacy is part of a larger empowerment model of prevention developed by Piran (2001), Levine and Piran (2004), and Levine and Smolak (2006). It links media literacy directly to the "six Cs" of effective prevention work (Levine and Smolak, 2006): Connection, Consciousness-raising, Competence-building, the experience of Choice, participation in Change, and the Courage to be committed (or the Commitment to courage, defined as taking bold, existential action in the face of anxiety and doubt; May, 1958).[2] Body-relevant media literacy programs have the potential to improve culture by redefining the body as a site of public and effective action, not private self-consciousness, shame, and silence (Piran, 2001).

Systematic investigations of media literacy can be categorized into analog laboratory studies, brief interventions, and longer, more intensive programs (see Levine, 2009a; Levine & Kelly, in press; Levine & Smolak, 2006, 2009). Briefly, the findings from all three categories are promising but inconclusive. To date, there have been no direct, well-controlled, long-term studies of whether media literacy in particular can prevent development of negative BI and the spectrum of DE. We can with confidence state that: (i) brief training in media literacy as a critical social perspective can mitigate the immediate negative effects of exposure to the thin ideal; and (ii) more systematic, intensive interventions over days or weeks can significantly reduce one important risk factor: internalization of the slender ideal.

Are Mass Media a Causal Risk Factor?

It is nearly impossible to comprehend how appearance concerns, body dissatisfaction, and unhealthy eating and weight management could have become so prevalent, so influential, and, indeed, normative across age, socioeconomic status, cultures, and subcultures in the absence of *mass* communications. Yet, as typically happens in the development of an important research area, the empirical evidence is simultaneously supportive, ambiguous, occasionally negative, and, all too often, absent. Nevertheless, based on the weight of the empirical evidence (see Table 4.1), we conclude that the mass media, in and of themselves, and certainly in the social context of multiple, overlapping, synergistic messages and reinforcements (Gordon, 2000; Smolak & Levine, 1996; Thompson et al., 1999), are a *possible* causal risk factor. Yet, to honor the evidence-based "attitude" of the scientist—practitioner—advocate, we acknowledge that the following have not, to date, been demonstrated conclusively and persuasively: (i) direct engagement with mass media, or media effects mediated by parents and/or peers, *precedes* development of the more proximal risk factors such as negative BI; (ii) prevention programs can both increase media literacy *and* thereby reduce or eliminate negative media influences, and in turn reduce or delay development of proximal risk factors (e.g., internalization of the thin ideal, social comparison tendencies); and (iii) there is

---

[2]In the course of writing this chapter, it was discovered that the process of formulating the six Cs of prevention (see Levine & Smolak, 2006) occurred independently of the work of McWhirter and McWhirter (2007) in developing their five Cs of empowerment, even though the lists overlap in terms of collaboration, consciousness-raising, and competence.

a compelling conceptual framework for how media effects are shaped and moderated by characteristics such as age, developmental stage, race, ethnicity, gender, and macro-culture (Levine & Harrison, 2009; Levine & Smolak, 2006). To reiterate, "engagement with mass media is probably best considered a variable risk factor that might well be later shown to be a causal risk factor" (Levine & Murnen, 2009, p. 32).

That said, the field cannot afford to wait for full clarification of these issues (Levine & Smolak, 2008). The more proximal determinants of risk influenced by mass media, peers, families, and other sociocultural sources—including internalization of the slender or muscular beauty ideal, social comparison tendencies, sexual objectification of the female body, and implicit and explicit prejudice against fat people—are supported by a large confluence of research (Puhl & Latner, 2007; Smolak & Murnen, 2004; Thompson & Stice, 2001).

All the influential models of prevention agree that the "professional," the "political," and the "personal" are each significant aspects of the work (e.g., Levine, 1994; Levine & Smolak, 2006; Maine, 2000; Maine, Davis & Shure, 2009; Piran, 2001). This position is captured poignantly in the late Lori Irving's (1999) call for prevention specialists from all sectors of the academic, advocacy, and government worlds to transcend the venerable "scientist-practitioner" model (i.e., the "Boulder" model of clinical training) to embrace a broader and deeper model of professional work that embodies, figuratively and literally, a "Bolder" model of the *scientist-activist-and-practitioner of what she or he preaches*. The considerable research supporting the media as a risk factor for BI and ED reinforces the ability of clinicians to strongly attest to the role of cultural variables in the development of individual problems while providing specific findings that clinicians can use to augment their clinical insights regarding how patients can best mobilize their defenses against the toxic effects of our culture as they address their ED. Thus, clinicians owe it to their clients to develop and *embody* a critical social perspective (Maine, 2000; Piran, 2001; Steiner-Adair, 1994; Levine & Piran, 2004) and to embrace a Bolder Model of being in a culture that is so toxic for ED clients, their families, children at risk, and for the professionals who are devoted to them (Irving, 1999; Maine, 2000; Piran, 2001).

THE PERSONAL: EATING DISORDER PROFESSIONAL AS ROLE MODEL

The following is a partial list of what ED specialists can do to be "bolder" in taking their own stance against ED.

Be an Activist

Prevention and health promotion start with many "little things." Instead of receiving gifts for your birthday, ask that money be given to prevention organizations, or to local libraries for the purchase of relevant books and videos. Talk often about commercials, advertisements, billboards, jokes, etc., that objectify women or men, glorify slenderness, or treat fat people in disrespectful ways. Write to companies to protest such practices, or to praise positive images and statements in the media. Encourage service organizations or civic groups to raise money and to collaborate with local organizations to promote positive BI, healthy eating and activity levels, and youth leadership. The schools are an important focus of these efforts, as are Girl

Scouts, 4-H, equestrian clubs, theater programs (Haines, Neumark-Sztainer & Morris, 2008) and similar groups. Work with local newspapers, television, and radio to "cover" your innovative partnerships. And, each time you take such action, talk about your commitments and what they mean to you.

Promote "Media Literacy" and Other Forms of "Cultural Literacy"

Consciousness-raising is a key component of the Bolder Model. Renowned prevention specialist Niva Piran often talks about the importance of "cultural literacy" as she teaches how to analyze various cultural influences (e.g., weight-related teasing or jokes) in terms of what is present, what is absent, what the potential negative effects are, and who benefits as some suffer. Use your creative powers and those of others (Connection in the six Cs) to develop opportunities to discuss and critically evaluate society's obsession with images of what it considers beautiful, and how this propaganda expresses cultural pressures for girls and women to be thin, beautiful, and sexy. Consider how the increasing objectification of the male body as a symbol of "sleek, masculine, muscular power" may be contributing to men's ideas about women as the object of that power, to people's attitudes about "fat" and "thin" for both sexes, and thus to the emergence of BI disorders among males as well.

Enhancing "literacy" does not mean confronting and lecturing people on the evils of their preferred forms of mass media, nor does it mean substituting pseudo-intellectual cynicism for mindless consumption of media. An exception should be made for pornography. Mass media are a powerful tool in a democracy, but like all powerful tools they require safety glasses (E. Austin, personal communication, April, 1996). Sometimes cultural literacy means no more than reminding oneself and significant others of our values and the choices we make when we engage with (use) certain types of magazines, TV programs, and web sites. Other times it means cancelling magazine subscriptions, or writing an op-ed or letter to the editor in response to offensive images.

Take a Stand and a Stance Against Weight-ist Prejudice

Cultural literacy brings us back to the processes of Commitment and Change. We need to avoid making, and we need to confront, prejudicial comments and jokes about weight. Let family, friends, educators, and companies know that making fun of fat people is cruel and offensive, as is the implication that slender people are somehow "superior." Actively, assertively, and, if necessary, angrily convey your dislike for tasteless items, such as greeting cards, that proclaim things like "no fat chicks" or "nothing tastes as good as being thin."

Take a Stance Against "Object-ification"

Sexual "object-ification" is the set of processes whereby people are socialized to think of the female body in particular as an object intended primarily for the visual stimulation or other pleasures of males (Smolak & Murnen, 2004). Obvious sources of objectification include pornography, sexist "mainstream" movies, and teasing. Females *and* males committed to prevention of negative BI and DE should do whatever they can to criticize and reject such objectification, as well as other forms of sexual harassment and sexual violence (Levine, 1994, 2006).

THE PROFESSIONAL: BEING A BOLDER CLINICIAN

The Place of Media Literacy and Activism in Therapy

In the treatment process, most clinicians provide psycho-education about the multiple contributions to ED. Discussing the media and their impact on the individual, as well as on our collective attitudes toward women and their bodies, should be included in psycho-education because it contextualizes the individual problem. Certainly, therapeutic exploration of the context of the individual's symptoms helps the patient to feel less alone, less alienated, and less deviant. For young women, endorsement of a feminist perspective reduces the strength of the connection between awareness of the media's thin ideal and internalization of that ideal (Myers & Crowther, 2007). Perhaps, then, media literacy can actually promote recovery. Furthermore, knowing that their therapist has taken stands to challenge the objectification of women's bodies or the glorification of thinness, many patients feel empowered, validated, and hopeful.

Over the course of recovery, with an increasing understanding of the ongoing impact of media images on themselves and others, many patients decide to take a stand themselves. They may provide educational programs in their schools or communities, write letters to advertisers or to magazines, and voice their protest against media influences promulgating negative BI and DE. Patients usually find these experiences empowering. Therapists can support and validate these efforts, but must be sure not to use the patient's activism in any way to promote themselves or their practices, and must caution them to work on their own recovery before trying to change others. Occasionally, patients are not truly ready for the exposure and challenges involved with activism in such a deeply personal, but publicly misunderstood, area. Ongoing exploration of the impact of their activism is at such times an important part of therapy, to assure that patients are addressing their primary needs and not just taking care of others, a common dynamic in ED.

While therapeutic relationships are critical to recovery, patients also live in a culture that creates and sustains these disorders, so clarifying how they can manage these messages, translate them to a less toxic form, and stay centered on their own needs is a critical and ongoing discussion in treatment. Building resilience, resistance, and resources in order to navigate this culture—in the forms of family, peers, coaches, physicians, and so forth, as well as mass media—is a key aspect of treatment. Negative cultural values may present themselves even in the context of the therapy hour, and need to be acknowledged by exploring transference, shame, powerlessness, and other manifestations of our toxic culture's impact on women's selves-in-relationships (Maine et al., 2009).

Continuing Education

Clinicians working in this field are faced with many challenges in "keeping up" with the research literature and with advances in best practices (Maine et al., 2009). Knowledgeable therapists need to know about a dizzying array of topics, ranging from genetics and neurotransmitters, to categorical versus dimensional models of nosology, to sociocultural influences on nutrition and family meals. All clinicians working with ED must also learn, and somehow keep current with, the nature, content, and technologies of mass media (Roberts,

Foehr, & Rideout, 2005). A good, advanced general source on contemporary media effects can be found in Bryant and Oliver (2009).

The Office

Media literacy, and a critical social perspective in general, informs the structure of a clinician's office. The pictures on the walls, along with the magazines and books for clients and families, have tremendous meaning. We can reclaim the important role of a "model" by emphasizing images and actions of women of character who, regardless of appearance, did things such as working full time, raising a family, *and* caring for an ill parent, or who resisted negative cultural forces and went on to create laws, art, inventions, and other things of lasting value. The principal characters in these important narratives need not be famous people; they could just as easily be our ancestors and family members.

Assessment

Females with a negative BI or an ED are particularly vulnerable to media images of the slender ideal (Dittmar et al., 2009; Groesz et al., 2002). Clients may well actively seek out ideal-body media, or even pro-anorexia or pro-bulimia web sites, for the purposes of social comparison and motivation (so-called thinspiration) to "be good" and "stay in control." Often, the cost is, unfortunately, more negative feelings (Dittmar et al., 2009; Levine & Harrison, 2009; Trampe, Stapel & Siero, 2007). Clinicians need to determine if clients are engaged in a self-defeating cycle in which media use and "thinspirations" perpetuate their symptoms and psychopathology. Clinical assessment of media use, motivation, social comparison, and effects in relation to BI and DE is a developing area; we recommend assessments found in Thompson et al. (1999) and Thompson, Van den Berg, Roehrig, Guarda, and Heinberg (2004).[3]

Support of Parents and Loved Ones

ED treatment usually provides patients' loved ones with psycho-education about risk factors and the recovery process, including critical examination of media messages and the sociocultural role of women, as well as how they can help their loved one negotiate these messages. Often, men (fathers, brothers, spouses) are considering these issues for the first time; their sensitivity, and their actions (like a letter of protest to *Sports Illustrated* about the swimsuit issue) can be a powerful statement of support (Levine, 1994; Maine, 2004). Family members can begin to see that their role is one of many in a team effort to make health more attractive than illness, and that they are not to blame for the ED. Understanding and sharing the complex and powerful impact of the shared cultural environment may reduce their guilt, and strengthen their relationships.

[3] A search of titles and abstracts in EBSCO's on-line *Communication & Mass Media Complete* (CMMC) and in *PsycInfo*, using compounds such as *clinical + assessment + mass media* (or *media use*), yielded no published articles (CMMC and *PsychInfo*) and no papers presented at conferences (CMMC) on the topic of clinical assessment of media use for more effective understanding and treatment of eating disorders—or any other disorder.

THE POLITICAL: THERAPY, THERAPISTS, POWER, AND EMPOWERMENT

A sociocultural perspective, subsuming the effects of mass media, on the etiology of DE takes an ecological approach to prevention (Levine & McVey, in press; Levine & Piran, 2004; Levine & Smolak, 2006; Maine, 2000) and assumes that professionals committed to overcoming ED are necessarily involved in promoting social justice. Reflecting the substantial literature on the relationship between clinical practice and social justice (e.g., Aladarondo, 2007; Albee, 1983; Brown, 1997), Prilleltensky, Dokecki, Frieden & Wang (2007) argue for a direct relationship between: (i) enhancing wellness through a balanced satisfaction of personal, relational, and collective needs; and (ii) enhancement of justice in providing people with power, capacity, and opportunity. In their words, "personal, relational, and collective wellness can be neither studied nor pursued in isolation from each other" (Prilleltensky et al., 2007, p. 39; see also Piran, 2001).

This non-specific vulnerability-stressor model (Levine & Smolak, 2006) is consistent with the specific contentions of experts that the intersecting roles of gender inequality, sexual objectification, weight-based prejudice, race, and ethnicity make social justice a fundamental issue in "overcoming" ED (Levine, 1994; Levine & Smolak, 2006; Maine, 2000; Piran, 2001; Smolak & Murnen, 2004, 2007; Steiner-Adair, 1994). If, as the evidence demonstrates, various aspects of culture constitute risk for the development and maintenance of DE, then the "object" and "subject" of therapy is simultaneously the client with an ED and the culture (Maine, 2000; Steiner-Adair, 1994). "The challenge of therapy is to transform what appears as psychological resistance (a reluctance to know what one knows) to a political resistance (a refusal not to know what one knows)" (Steiner-Adair, 1994, p. 382). This transformation requires the aforementioned six C's, including courage and commitment modeled by a therapist who actively takes a stance against long-standing, powerful cultural beliefs and practices while reclaiming the power of what a "model" can and should be.

To be precise, then, *media* advocacy involves using the principles and forms of mass media in order to promote and market positive social changes (Wallack, Dorfman, Jernigan & Themba, 1993). We will get at the roots of the problem and sow the seeds of real, positive transformations only by becoming active advocates for *change* in the ways that mental health professionals think about mass media, for changes in the media themselves, and for *collaboration* with mass media to bring about other important changes in institutions, public policies, and cultural practices. Transforming mass media is daunting, but definitely possible, perhaps to an extent that even many professionals would find surprising (Kelly, 2002; Levine & Kelly, in press; Levine & Smolak, 2006; Maine, 2000; Maine & Kelly, 2005).

A FINAL REFLECTION AS AN EXAMPLE OF ADVOCACY

"Media" is one of those important concepts whose very ubiquity and familiarity constitute a formidable barrier to grasping its effects on people's lives and its implications for effective, humane clinical practice. The evidence in favor of the proposition that mass media are a causal risk factor for ED is solid but incomplete. Still, the evidence in favor of that proposition is much stronger than the current evidence for the theoretical and dogmatic contention that "eating disorders are genetic" (Smolak et al., 2006).

The mass media have had an impact on bodies, BI, gender identity, and eating for at least 125 years (Banner, 1983; Silverstein & Perlick, 1995), but the intersection(s) and transactions between individuals, family, and communities today are quite different than in the past. What constitutes "community" today is a culture increasingly saturated with instantaneous electronic connections and with extremely well-financed, extremely clever, profit-imperative marketing (Levine & Kelly, in press). Thus, it is extremely important, in personal, professional, and political terms, for clinicians to develop and keep current their media literacy. Clinicians need to marshal the six Cs of their own lives to understand how the immense, powerful, and continually evolving sociocultural "factor" called "mass media" works. And clinicians need to be a "Bolder Model" in challenging the medias' harmful effects by proactively expressing and asserting ourselves in, through, and with the media themselves. Media literacy is a promising area for prevention precisely because, whether target audiences are ED experts or 4th-grade boys and girls, it has the potential to foster and reinforce potent forms of consciousness-raising, connections, competence-building, choice, and cultural change (Levine, 2009a; Levine & Piran, 2004; Levine & Smolak, 2006; Piran, 2001).

References

Albee, G. W. (1983). Psychopathology, prevention, and the just society. *Journal of Primary Prevention, 4,* 5–40.
Aldarondo, E. (Ed.), (2007). *Advancing social justice through clinical practice.* Mahwah, NJ: Lawrence Erlbaum Associates.
American Psychiatric Association. (2000). *Diagnostic and statistical manual of mental disorders* (4th ed., Text Rev.; DSM-IV-TR). Washington, DC: Author.
Anderson-Fye, E. P., & Becker, A. E. (2004). Sociocultural aspects of eating disorders. In J. K. Thompson (Ed.), *Handbook of eating disorders and obesity* (pp. 565–589). Hoboken, NJ: John Wiley.
Austin, S. B. (2001). Population-based prevention of eating disorders: An application of the Rose prevention model. *Preventive Medicine, 32,* 268–283.
Banner, L. W. (1983). *American beauty.* New York, NY: Knopf.
Bardone-Cone, A. M., & Cass, K. M. (2007). What does viewing a pro-anorexia website do? Experimental examination of website exposure and moderating effects. *International Journal of Eating Disorders, 40,* 537–548.
Barlett, C. P., & Harris, R. J. (2008). The impact of body emphasizing video games on body image concerns in men and women. *Sex Roles, 59,* 586–601.
Becker, A. E., & Fay, K. (2006). Sociocultural issues and eating disorders. In S. Wonderlich, J. E. Mitchell, M. de Zwaan, & H. Steiger (Eds.), *Annual review of eating disorders: Part 2* (pp. 35–63). Oxon, UK: Radcliffe Publishing.
Botvin, G. (2000). Preventing drug abuse in schools: Social and competence enhancement approaches targeting individual-level etiologic factors. *Addictive Behaviors, 25,* 887–897.
Brown, L. S. (1997). The private practice of subversion: Psychology as Tikkun Olam. *American Psychologist, 52,* 449–462.
Bryant, J., & Oliver, M. B. (Eds.), *Media effects: Advances in theory and research* (3rd ed.). New York, NY: Routledge/Taylor & Francis.
Bulik, C. (2004). Genetic and biological risk factors. In J. K. Thompson (Ed.), *Handbook of eating disorders and obesity* (pp. 3–16). Hoboken, NJ: Wiley.
Cafri, G., Yamamiya, Y., Brannick, M., & Thompson, J. K. (2005). The influence of sociocultural factors on body image: A meta-analysis. *Clinical Psychology: Science and Practice, 12,* 421–433.
Comstock, G., & Scharrer, E. (2007). *Media and the American child.* Burlington, MA: Academic Press.
Crisafulli, M. A., Von Holle, A., & Bulik, C. M. (2008). Attitudes toward anorexia nervosa: The impact of framing on blame and stigma. *International Journal of Eating Disorders, 41,* 333–339.
Dittmar, H., Halliwell, E., & Stirling, E. (2009). Understanding the impact of thin media models on women's body-focused affect: The roles of thin-deal internalization and weight-related self-discrepancy activation in experimental exposure effects. *Journal of Social and Clinical Psychology, 28,* 43–72.

Dohnt, H., & Tiggemann, M. (2006). The contribution of peer and media influences to the development of body satisfaction and self-esteem in young girls: A prospective study. *Developmental Psychology, 42,* 929—936.

Gordon, R. A. (2000). *Eating disorders: Anatomy of a social epidemic* (2nd ed.). Malden, MA: Blackwell Publishers.

Grabe, S., Ward, L. M., & Hyde, J. S. (2008). The role of the media in body image concerns among women: A meta-analysis of experimental and correlational studies. *Psychological Bulletin, 134,* 460—476.

Groesz, L. M., Levine, M. P., & Murnen, S. K. (2002). The effect of experimental presentation of thin media images on body satisfaction: A meta-analytic review. *International Journal of Eating Disorders, 31,* 1—16.

Haines, J., Neumark-Sztainer, D., & Morris, B. (2008). Theater as a behavior change strategy: Qualitative findings from a school-based intervention. *Eating Disorders: The Journal of Treatment & Prevention, 16,* 241—254.

Hargreaves, D., & Tiggemann, M. (2003). Longer-term implications of responsiveness to thin-ideal television: Support for a cumulative hypothesis of body image disturbance? *European Eating Disorders Review, 11,* 465—477.

Harrison, K., & Hefner, V. (2006). Media exposure, current and future body ideals, and disordered eating among preadolescent girls: A longitudinal panel study. *Journal of Youth and Adolescence, 35,* 153—163.

Harrison, K., & Hefner, V. (2008). Media, body image, and eating disorders. In S. L. Calvert, & B. J. Wilson (Eds.), *The handbook of children, media, and development* (pp. 381—406). Malden, MA: Blackwell Publishing.

Himes, S. M., & Thompson, J. K. (2007). Fat stigmatization in television shows and movies: A content analysis. *Obesity, 15,* 712—718.

Hobbs, R. (1998, Winter). The seven great debates in the media literacy movement. *Journal of Communication,* 16—32.

Irving, L. (1999). A bolder model of prevention: Science, practice, and activism. In N. Piran, M. P. Levine, & C. Steiner-Adair (Eds.), *Preventing eating disorders: A handbook of interventions and special challenges* (pp. 63—83). Philadelphia, PA: Brunner/Mazel.

Kelly, J. (2002). *Dads and daughters: How to inspire, understand, and support your daughter.* New York, NY: Broadway Books.

Klump, K., Bullik, C. M., Kaye, W. H., Treasure, J., & Tyson, E. (2009). Academy for eating disorders position paper: Eating disorders are serious mental illnesses. *International Journal of Eating Disorders, 42,* 97—103.

Levine, M. P. (1994). "Beauty Myth" and the Beast: What men can do and be to help prevent eating disorders. *Eating Disorders: The Journal of Treatment and Prevention, 2,* 101—113.

Levine, M. P. (2006). Prevention is a necessity, not a luxury: Part 2: In search of a bolder model—What we can do and be in our everyday lives in order to prevent negative body image and disordered eating. June, *OUR Journey* [a publication of The Bronte Foundation, Melbourne, Australia], 3(1), 1, 4—12. Available by request from Levine@kenyon.edu

Levine, M. P. (2009a). Aportaciones desde el campo del la prevención: Implicaciones para la educación en comunición [Lessons from the field of prevention: Implications for Media Literacy Programs]. *Aula de Innovación Educativa* [Educational Innovations for the Classroom], *178,* 14—18.

Levine, M. P. (2009b). *Are media an important medium for clinicians? Mass media, eating disorders, and the Bolder Model of treatment, prevention, and advocacy.* April 18, Presentation at the conference "Eating Disorders: State of the Art Treatment Symposium," sponsored by The Center for Eating Disorders at Sheppard Pratt, Towson, MD. (Available by request, from Levine@kenyon.edu)

Levine, M. P., & Harrison, K. (2004). The role of mass media in the perpetuation and prevention of negative body image and disordered eating. In J. Kevin Thompson (Ed.), *Handbook of eating disorders & obesity* (pp. 695—717). New York, NY: John Wiley.

Levine, M. P., & Harrison, K. (2009). Effects of media on eating disorders and body image. In J. Bryant, & M. B. Oliver (Eds.), *Media effects: Advances in theory and research* (3rd ed.) (pp. 490—515). New York, NY: Routledge/Taylor & Francis.

Levine, M. P., & Kelly, J. (in press). A primer on media literacy's role in prevention of negative body image and disordered eating. In G. McVey et al. (Eds.), *Improving the prevention of eating-related disorders: Collaborative research, advocacy, and policy change.* Waterloo, ON, Canada: Wilfred Laurier University Press.

Levine, M.P., & McVey, G. (in press). The community-based prevention of negative body image and disordered eating: A framework for programs, research, and advocacy. In G. McVey et al. (Eds.), *Improving the prevention of eating-related disorders: collaborative research, advocacy, and policy change.* Waterloo, ON, Canada: Wilfred Laurier University Press.

Levine, M. P., & Murnen, S. K. (2009). Everybody knows that mass media are/are *not a cause* of eating disorders: A critical review of evidence for a causal link between media, negative body image, and disordered eating in females. *Journal of Social & Clinical Psychology, 29,* 9—42.

Levine, M. P., & Piran, N. (2004). The role of body image in the prevention of eating disorders. *Body Image, 1*, 57–70.

Levine, M. P., & Smolak, L. (2006). *The prevention of eating problems and eating disorders: Theory, research, and practice.* Mahwah, NJ: Lawrence Erlbaum Associates.

Levine, M. P., & Smolak, L. (2008). "What exactly are we waiting for?" The case for universal-selective eating disorders prevention programs. *International Journal of Child & Adolescent Health, 1*, 295–304.

Levine, M. P., & Smolak, L. (2009). Prevention of negative body image and disordered eating in children and adolescents: Recent developments and promising directions. In L. Smolak, & J. K. Thompson (Eds.), *Body image, eating disorders, and obesity in youth* (2nd ed.) (pp. 215–239). Washington, DC: American Psychological Association.

Maine, M. (2000). *Body wars: Making peace with women's bodies. An activist's guide.* Carlsbad, CA: Gürze Books.

Maine, M. (2004). *Father hunger: Fathers, daughters, and the pursuit of thinness* (2nd ed.). Carlsbad, CA: Gürze Books.

Maine, M., Davis, W. N., & Shure, J. (Eds.), (2009). *Effective clinical practice in the treatment of eating disorders: The heart of the matter.* New York, NY: Brunner-Routledge.

Maine, M., & Kelly, J. (2005). *The body myth: Adult women and the pressure to be perfect.* Hoboken, NJ: John Wiley & Sons.

May, R. (1958). To be and not to be: Contributions of existential psychotherapy. In R. May, E. Angel, & H. Ellenberger (Eds.), *Existence: A new dimension in psychiatry and psychology* (pp. 59–74). New York, NY: Basic Books.

McWhirter, E. H., & McWhirter, B. T. (2007). Grounding clinical training and supervision in an empowerment model. In E. Aldarondo (Ed.), *Advancing social justice through clinical practice* (pp. 417–442). Mahwah, NJ: Lawrence Erlbaum Associates.

Murnen, S. K., Levine, M. P., Groesz, L., & Smith, J. (2007, August). *Do fashion magazines promote body dissatisfaction in girls and women? A meta-analytic review.* Paper presented at the 115th meeting of the American Psychology Association, San Francisco, CA.

Myers, T. A., & Crowther, J. H. (2007). Sociocultural pressures, thin-ideal internalization, self-objectification, and body dissatisfaction. Could feminist beliefs be a moderating factor? *Body Image, 4*, 296–308.

Piran, N. (2001). Re-inhabiting the body from the inside out: Girls transform their school environment. In D. L. Tolman, & M. Brydon-Miller (Eds.), *From subjects to subjectivities: A handbook of interpretive and participatory methods* (pp. 218–238). New York, NY: NYU Press.

Pope, H. G., Jr., Phillips, K. A., & Olivardia, R. (2000). *The Adonis complex: The secret crisis of male body obsession.* New York, NY: The Free Press.

Prilleltensky, I., Dokecki, P., Frieden, G., & Wang, V. O. (2007). Counseling for wellness and justice: Foundations and ethical dilemmas. In E. Aldarondo (Ed.), *Advancing social justice through clinical practice* (pp. 19–42). Mahwah, NJ: Lawrence Erlbaum Associates.

Puhl, R. M., & Latner, J. D. (2007). Stigma, obesity, and the health of the nation's children. *Psychological Bulletin, 133*, 557–580.

Ricciardelli, L. A., McCabe, M. P., Williams, R. J., & Thompson, J. K. (2007). The role of ethnicity and culture in body image and disordered eating among males. *Clinical Psychology Review, 27*, 582–606.

Roberts, D. F., Foehr, U. G., & Rideout, V. (2005). *Generation M: Media in the lives of 8–18 year-olds.* A Henry J. Kaiser Family Foundation Study. Retrieved June 7, 2009, from: http://www.kff.org/entmedia/7251.cfm

Silverstein, B., & Perlick, D. (1995). *The cost of competence: Why inequality causes depression, eating disorders, and illness in women.* New York, NY: Oxford University Press.

Smolak, L. (2009). Risk factors in the development of body image, eating problems, and obesity. In L. Smolak, & J. K. Thompson (Eds.), *Body image, eating disorders, and obesity in youth: Assessment, prevention, and treatment* (2nd ed.) (pp. 135–155). Washington, DC: American Psychological Association.

Smolak, L., & Levine, M. P. (1994). Critical issues in the developmental psychopathology of eating disorders. In L. Alexander, & B. Lumsden (Eds.), *Understanding eating disorders* (pp. 37–60). Washington, DC: Taylor & Francis.

Smolak, L., & Levine, M. P. (1996). Adolescents' transitions and the development of eating problems. In L. Smolak, M. P. Levine, & R. Striegel-Moore (Eds.), *The psychopathology of eating disorders: Implications for research, prevention, and treatment* (pp. 207–234). Mahwah NJ: Lawrence Erlbaum Associates.

Smolak, L., Levine, M. P., & Murnen, S. K. (2006, June). *The scientific status of sociocultural models for eating disorders: A close look at controversy, theory and data.* Workshop presented at the International Conference on Eating Disorders of the Academy for Eating Disorders, Barcelona, Spain.

Smolak, L., & Murnen, S. K. (2004). A feminist approach to eating disorders. In J. K. Thompson (Ed.), *Handbook of eating disorders and obesity* (pp. 590–605). New York, NY: Wiley.

Smolak, L., & Murnen, S. K. (2007). Feminism and body image. In V. Swami, & A. Furnham (Eds.), *The body beautiful: Evolutionary and socio-cultural perspectives* (pp. 236–258). London: Palgrave Macmillan.

Steiner-Adair, C. (1994). The politics of prevention. In P. Fallon, M. A. Katzman, & S. C. Wooley (Eds.), *Feminist perspectives on eating disorders* (pp. 381–394). New York, NY: Guilford.

Stice, E. (2002). Risk and maintenance factors for eating pathology: A meta-analytic review. *Psychological Bulletin, 128*, 825–848.

Thompson, J. K., Heinberg, L. J., Altabe, M., & Tantleff-Dunn, S. (1999). *Exacting beauty: Theory, assessment, and treatment of body image disturbance*. Washington, DC: American Psychological Association.

Thompson, J. K., & Stice, E. (2001). Internalization of the thin-ideal: Mounting evidence for a new risk factor for body image disturbance and eating pathology. *Current Directions in Psychological Science, 10*, 181–183.

Thompson, J. K., van den Berg, P., Roehrig, M., Guarda, A. S., & Heinberg, L. J. (2004). The Sociocultural Attitudes Toward Appearance Scale-3 (SATAQ-3): Development and validation. *International Journal of Eating Disorders, 35*, 293–304.

Tiggemann, M. (2006). The role of media exposure in adolescent girls' body dissatisfaction and drive for thinness: Prospective results. *Journal of Social and Clinical Psychology, 25*, 523–541.

Trampe, D., Stapel, D. A., & Siero, F. W. (2007). On models and vases: Body dissatisfaction and proneness to social comparison effects. *Journal of Personality and Social Psychology, 92*, 106–118.

Van den Berg, P., Neumark-Sztainer, D., Hannan, P. J., & Haines, J. (2007). Is dieting advice from magazines helpful or harmful? Five-year associations with weight-control behaviors and psychological outcomes in adolescents. *Pediatrics, 119*, 30–37.

Vera, E. M., & Speight, S. L. (2007). Advocacy, outreach, and prevention: Integrating social action roles in professional training. In E. Aldarondo (Ed.), *Advancing social justice through clinical practice* (pp. 373–389). Mahwah, NJ: Lawrence Erlbaum Associates.

Wallack, L., Dorfman, L., Jernigan, D., & Themba, M. (1993). *Media advocacy and public health: Power for prevention*. Newbury Park, CA: Sage.

PART II

BRIDGING THE GAP: DIAGNOSIS AND TREATMENT

   5   The Assessment Process   71
   6   Medical Assessment of Eating Disorders   89
   7   Psychiatric Medication   111
   8   Nutritional Impact on the Recovery Process   127
   9   Science or Art?   143
  10   New Pathways   163
  11   Outpatient Treatment of Anorexia Nervosa following Weight Restoration   181
  12   Recipe for Recovery   197

CHAPTER 5

The Assessment Process
Refining the Clinical Evaluation of Patients with Eating Disorders

Drew A. Anderson, Jason M. Lavender, and Kyle P. De Young

Assessment is an often-overlooked but essential component of the treatment process. While assessments can be conducted for many reasons, this chapter will focus on assessments of clinical eating disorders (ED) conducted for the purpose of treatment planning and outcome.

To assist treatment planning, developing a symptom profile is often more important than obtaining a diagnosis *per se*. While treatments do exist for particular ED diagnoses (e.g., Fairburn, 1997; Fairburn, Marcus & Wilson, 1993; Lock, Le Grange, Agras & Dare, 2002), more recent research suggests that effective treatment can be conducted without a specific ED diagnosis, utilizing a transdiagnostic approach (Fairburn, 2008). Furthermore, the diagnostic criteria found in the current Diagnostic and Statistical Manual of Mental Disorders (DSM-IV-TR) (American Psychiatric Association, 2000) have been critiqued on a number of grounds (Fairburn & Cooper, 2007; Walsh, 2009a, b; Wonderlich, Crosby, Mitchell & Engel, 2007). Of particular relevance for clinical practice is the finding that the diagnosis of Eating Disorder Not Otherwise Specified (EDNOS) is at least as common as, if not more common than, anorexia nervosa (AN) and bulimia nervosa (BN) in clinical settings (Dalle Grave & Calugi, 2007; Machado, Machado, Goncalves & Hoek, 2007; Rockert, Kaplan & Olmsted, 2007; Turner & Bryant-Waugh, 2004). While the development of the latest edition of the Diagnostic and Statistical Manual of Mental Disorders (DSM-V) is underway, with an expected release in 2012, the resolution of these issues is unclear. Thus, this chapter will describe an assessment process that focuses on evaluating key symptoms rather than generating a specific diagnosis.

WEIGHT, MENSTRUATION, AND PHYSICAL CONDITIONS

Weight Status

Weight status is a key sign differentiating ED. In the current DSM-IV-TR (American Psychiatric Association, 2000) it represents the prime diagnostic difference between AN and BN. Weight status in ED can range from significantly underweight to obese. Weight itself, however, is not a behavior, but is the consequence of an interaction between behavior, the environment, and genetics.

A patient's weight also has important treatment implications. Due to serious medical consequences associated with being severely underweight (Keel et al., 2003; Mitchell, Pomeroy & Adson, 1997), weight regain is commonly the most pressing issue in treatment for underweight individuals (Fairburn, 2008). For non-overweight individuals, the goal is usually to stabilize weight and prevent any major weight fluctuations. For overweight and obese individuals, the role of weight change as a treatment goal is controversial. Some suggest that weight loss should not be encouraged because it promotes the very concerns that may drive individuals to develop ED (e.g., Bacon, 2008), while others suggest that weight loss may be appropriate in some cases (Wilfley, 2002).

Additionally, a comprehensive assessment of weight status provides a wealth of clinical data, offering an opportunity to learn how individuals view their weight and how they react to being weighed. For example, asking when, where, and how often they weigh themselves, how they would feel if they were asked to weigh themselves only once per week (as in the Eating Disorder Examination; Fairburn, Cooper & O'Connor, 2008), and whether they engage in any rituals associated with weighing reveals useful information about the relative importance of weight in their life.

Obtaining Current Weight

Obtaining an accurate weight is sometimes a sensitive topic for both patients and treatment providers. Directly weighing the client is the most accurate way to measure a patient's weight. An accurate weight is particularly important when a patient's weight is low and its restoration is a focus of treatment or when there are medical concerns related to low weight. A study by Kaplan et al. (2009) found that even a small weight loss following intensive treatment for AN was a powerful predictor of relapse, highlighting the need for frequent and accurate monitoring of weight in these cases.

Many patients experience substantial distress at the mention that they will be weighed, and the professional can take steps to alleviate this distress. For example, suggesting "blind" weight (i.e., with the patient's back to the scale or the numbers covered) alleviates discomfort for some, although it increases it for others. It may be clinically useful to have patients see the weight, however, even if it causes some discomfort. It is also an opportunity for clinicians to improve therapeutic relationships by demonstrating their sensitivity to how difficult this topic can be for individuals with ED. Unfortunately, it may become a point of disagreement between patient and clinician who do not see eye-to-eye regarding the benefits of obtaining detailed information pertaining to body weight.

Underweight ED patients may attempt to "add weight" by wearing multiple layers of clothes, placing heavy objects in pockets, etc. when being weighed. For that reason, many professionals prefer to weigh patients in minimal clothing (e.g., a hospital gown) to minimize this possibility. Excessive fluid-intake is also common and clinicians need to decide how to best approach these issues at assessment and throughout treatment, perhaps through monitoring before weighing or even testing of urine specific gravity.

Patients may also be asked their weight. The response often conveys much more than a number. Clients who respond without pause with an exact number may reveal the regularity of their weigh-ins and the importance they place on weight. Conversely, some may respond by claiming that they do not know their weight or by offering a rough estimate. These responses may indicate an avoidance of knowledge and stimuli pertaining to weight, perhaps because these stimuli are so distressing. Reluctance to report weight may also reflect resistance or ambivalence to the assessment process or even shame.

Although self-reported and measured weight are highly correlated, research indicates a tendency for individuals with a high weight to underestimate and for those with a low weight to overestimate, leading to systematic bias in self-reported weight at the extremes (Shapiro & Anderson, 2003; Villanueva, 2001).

Obtaining a Weight History

Obtaining a weight history from medical records can overcome the inherent limitations of memory, but records often have missing data. Some key pieces of information to be garnered from a weight history include the individual's highest weight, lowest weight, any significant weight gains or losses, and the ages during which these occurred. Developing a timeline, beginning in childhood and continuing to the present time, will help the clinician and patient to construct a model of the course of the ED over time. With younger patients, review of their pediatric growth charts can be especially helpful.

Research provides support for the prognostic value of obtaining an accurate weight history. In BN, weight suppression, or the extent to which current weight is below one's highest weight, has been found to predict subsequent weight gain as well as drop-out and poor treatment outcome (Butryn, Lowe, Safer & Agras, 2006; Lowe, Davis, Lucks, Annunziato & Butryn, 2006). Also, a history of AN in women with a diagnosis of BN has been shown to predict a longer course of illness and a higher probability of relapsing into AN compared to women without a history of AN (Eddy et al., 2007).

WEIGHT-RELATED PHYSIOLOGICAL MARKERS

Menstruation

While amenorrhea is currently one of the diagnostic criteria for AN in post-menarchial females (American Psychiatric Association, 2000), it does not appear to improve the specificity of the diagnosis (Mitchell, Cook-Myers & Wonderlich, 2005), and some have suggested that it be removed as a core diagnostic criterion (Attia & Roberto, 2009). Many women with very low body weight experience irregular menstruation (oligomenorrhea) rather than amenorrhea (Poyastro Pinheiro et al., 2007), while women who do not have very low weights may lose their menses due to strenuous physical activity (Glass et al., 1987). Also, the criterion is obviously irrelevant in certain subgroups (e.g., males and prepubescent and post-menopausal females).

Although the diagnostic utility of amenorrhea has been called into question, it remains an important indicator of the severity of low weight. Amenorrhea is associated with a number of other important physiological markers (Attia & Roberto, 2009). It has also been associated with self-induced vomiting in normal weight adolescent females (Austin et al., 2008), suggesting that it may be indicative of more than simply low body weight. Besides noting the presence or absence of menstrual regularity, additional questions can provide a clearer picture of hormonal status. For example, noting whether individuals have maintained a regular menstrual cycle since puberty may inform the clinical course of their ED. Importantly, the use of birth control pills and other hormonal supplements to stimulate or stabilize the menstrual cycle may mask underlying endocrine abnormalities. Finally, when menstrual irregularities have been present, assessing the individual's perception of their severity and implications can be useful.

Other Physical Conditions

Eating disorder patients often experience a host of related medical problems, some of them serious (American Psychiatric Association, 2000; Mitchell et al., 1997). Beyond the information obtained from medical records, a thorough evaluation by an experienced medical professional, throughout treatment, is strongly recommended (Carney & Andersen, 1996; Crow & Swigart, 2005; Work Group on Eating Disorders, 2006). Clinicians without medical training should be familiar with the basics of the medical issues commonly associated with ED.

Self report is also a source of data regarding the presence and severity of physical problems. For instance, individuals who abuse laxatives over extended periods of time may become dependent upon them to stimulate bowel movements (American Psychiatric Association, 2000). Thus, what may have begun as a method of purging can evolve into a method of self-medicating constipation. This and similar contextual information may not find its way into medical records but can be particularly informative during the treatment process.

DIETARY RESTRICTION AND DIETARY RESTRAINT

Dietary restriction refers to limiting the overall amount of food eaten, what foods are eaten, or both, while dietary restraint is the *intention* to restrict food intake, whether or not it is successful (Williamson et al., 2007). At initial assessment as well as throughout treatment, a thorough review of both actual and intentional daily eating patterns is necessary, including what meals, snacks, amounts, and types of food individuals typically eat and when. This information is extremely useful in treatment planning and progress-tracking, as dietary restriction may lead to weight loss but may also precede binge eating.

Assessing Quantity Of Meals And Snacks

The amount of food eaten (outside of objective binge episodes, when present) can vary substantially between and within different types of ED. Since many ED treatments (e.g., CBT, Fairburn, 2008; and Maudsley family therapy, Lock et al., 2002) involve establishing a regular meal pattern, an accurate initial assessment of current intake is essential. Asking how often patients eat breakfast, lunch, dinner, and snacks gives a good indication of overall patterns. They should also be asked whether they have any rules about the overall amount of food they eat, such as calorie limits, or rules about when they will eat, such as not eating after 8:00 p.m.

Arriving at an estimate of daily caloric intake involves a detailed assessment of consumption over a typical day, although many have difficulty reporting what constitutes a "typical" day. Most individuals make large errors in estimating food intake (Trabulsi & Schoeller, 2001), however, and individuals with higher levels of dietary restraint and eating-related pathology tend to be even less accurate (Bathalon et al., 2000; Hadigan, LaChaussee, Walsh & Kissileff, 1992). Thus, although many ED patients regularly track their caloric consumption closely, their estimates should not be assumed to be accurate. Even so, obtaining an approximate daily caloric intake is often helpful in estimating the trajectory of weight loss or gain and the degree of dietary restriction.

Assessing Diversity of Foods Eaten

In addition to attempts at limiting the overall quantity of food they eat, individuals suffering from ED often attempt to limit the types of food they eat, dividing food types into "good" and "bad" based upon their caloric value, fat or sugar (carbohydrate) content, or their propensity to trigger binges. Rules may include firm, inflexible statements, such as "no fried foods," in contrast to more flexible guidelines. Importantly, clinicians should pay attention to whether the rules are aimed at influencing weight or shape rather than being part of a well-balanced nutritional intake. Similarly, clinicians should exercise care when patients practice vegetarianism or veganism. As with fasting, people may choose these restricted dietary regimens for various reasons. If the motivation includes influencing body shape or weight, the dietary restriction is notable, but clinicians should avoid passing judgment about individuals' moral or health motives. Although some research has indicated that vegetarianism is associated with disordered eating (DE) (Robinson-O'Brien, Perry, Wall, Story & Neumark-Sztainer, 2009), this finding is not universal (Fisak, Peterson, Tantleff-Dunn & Molnar, 2006).

BINGE EATING

Although binge eating is a core feature of BN and binge eating disorder (BED), it is also common among individuals with AN, EDNOS, and other subclinical ED presentations (American Psychiatric Association, 2000). Binge eating is particularly difficult to assess, in part due to the complex and controversial nature of the definition of a binge. First, to be defined as a binge episode using the current DSM-IV-TR criteria (American Psychiatric Association, 2000), the individual must consume an unusually large amount of food in a discrete period of time and experience a sense of loss of control. However, many individuals who engage in binge eating do not describe it this way, which will be discussed later in this chapter. Also, occasional binge episodes are common in some cultures (Johnson, Schlundt, Barclay, Carr-Nangle & Engler, 1995), thus binge episodes alone do not necessarily reflect clinically significant eating problems. Because the definition of a binge episode remains problematic, a thorough assessment of the specific components and characteristics of a given eating episode is suggested.

Loss of Control

A subjective sense of having lost control over one's eating is perhaps the most salient defining feature of an episode of binge eating (Wolfe, Baker, Smith & Kelly-Weeder, 2009).

This loss of control may manifest in one or more ways. An individual may report feeling unable to stop eating once the binge commenced or unable to resist the urge to start binge eating. Also, some ED patients engage in planned binge eating episodes, which they may deny are "out of control." In such circumstances, the clinician can ask additional questions (e.g., could they have chosen not to engage in the episode?) in order to clarify the extent of loss of control.

Amount of Food

The size or amount of food consumed is a controversial component of the current definition of an eating binge. The DSM-IV-TR (American Psychiatric Association, 2000) does not specify an amount that must be consumed to qualify as a binge and some have argued that this requirement be modified or eliminated (Wolfe et al., 2009). Importantly, a substantial proportion of individuals with ED consider eating episodes to be binges even when the amount consumed is not objectively large (Beglin & Fairburn, 1992; Johnson, Boutelle, Torgrud, Davig & Turner, 2000; Telch, Pratt & Niego, 1998). Furthermore, individuals who experience loss of control during "binge" episodes that are not objectively large have similar levels of both ED psychopathology and general psychopathology (Keel, Mayer & Harnden-Fischer, 2001; Latner, Hildebrandt, Rosewall, Chisholm & Hayashi, 2007) as those individuals whose binge episodes are objectively large.

Still, obtaining information on the size of typical binge episodes is useful. For example, treatment strategies may differ substantially for an individual whose "binge episodes" consist of several thousand calories consumed over the course of two hours compared to an individual who consumes only a few hundred calories in a few minutes.

"Grazing" Behavior

Another eating pattern associated with ED is "grazing," a pattern characterized by frequently eating smaller amounts of food, with little time between episodes, over a longer period of time (e.g., throughout the course of a day). The individual may consume a large amount of food and experience a subjective sense of having lost control, making it difficult to distinguish these episodes from more traditional binge episodes. Because individuals may experience both traditional binge episodes and grazing behavior, assessing for both patterns is essential for designing an appropriate treatment plan.

INAPPROPRIATE COMPENSATORY BEHAVIORS

Individuals with ED can engage in both purging and non-purging methods of inappropriate compensatory behaviors (ICB). Common methods of purging are self-induced vomiting, laxative misuse, diuretic misuse, and enemas, while common non-purging methods include fasting, excessive exercise, and the misuse of insulin and thyroid medications (American Psychiatric Association, 2000). Assessment of ICB should include information about the variety of methods, their frequency, the situations in which they are employed, and their apparent functions. Although some ICB are rather straightforward (e.g., self-induced vomiting), considerable confusion exists regarding what other behaviors qualify as inappropriate, and many contextual variables must be taken into consideration. For instance, taking a leisurely walk after a large meal may be undertaken in part as a compensatory behavior but is unlikely to be considered inappropriate, as moderate levels of physical activity are health-promoting (Haskell et al., 2007). Exercising for several hours after an eating episode is much more likely to be viewed as inappropriate, even though exercise bouts lasting several hours are common in some sports. Although ICB may be present in AN, the diagnosis of BN and BED specifically rely upon the assessor's judgment regarding whether the ED is characterized by the *regular* use of ICB. This determination is important for making a differential diagnosis between BN and BED as well as for capturing the severity of these behaviors. Due to marked symptom frequency fluctuation that may occur in BN, considering the *average* frequency of ICB when making this diagnosis is extremely important; individuals can have "good" weeks during which they experience few or no episodes of ICB and "bad" weeks during which such behaviors are present nearly every day.

Self-Induced Vomiting

Self-induced vomiting is the most common form of purging in both outpatient ED (Favaro & Santonastaso, 1996) and community samples (Mond, Hay, Rodgers, Owen & Mitchell, 2006). Feelings of shame and embarrassment often accompany the disclosure of this behavior, so clinicians need to be sensitive when assessing a patient's experience with vomiting.

Assessing the frequency of self-induced vomiting is particularly tricky, depending upon the pattern of this behavior. For instance, some individuals induce vomiting once per day. Others induce vomiting after every binge eating episode they experience or even after regular meals or snacks. Thus, one should never assume that self-induced vomiting only follows episodes of binge eating.

Assessing the method by which individuals achieve vomiting is often useful. Some patients ingest syrup of ipecac or salt water to induce nausea and to facilitate vomiting. Since the use of syrup of ipecac has been linked to serious cardiac problems (American Psychiatric Association, 2000), it is important to note the use of this substance. Most often, however, individuals manually stimulate the gag reflex by inserting a foreign object (e.g., finger, eating utensil, or toothbrush) into their throats, increasing the risk of esophageal tears, accidental ingestion of the foreign body, and desensitization of the gag reflex (Mendell & Logemann, 2001). Over time, individuals may be able to induce vomiting with little or no manual stimulation, and may not report that they "induce" vomiting. Nevertheless, it should be considered an ICB.

Laxatives, Diuretics, and Enemas

The assessment of laxative, diuretic, and enema misuse should consider whether or not these behaviors are used to compensate for eating. Thus, if the only reason for laxative use is to relieve constipation and the recommended dosage instructions are followed, this should not be considered misuse. Noting the presence of tolerance to these substances aids treatment planning. Individuals who have not developed a tolerance may cease laxative use all at once, while those who have developed a tolerance may need to be weaned from their use.

Diuretic use presents additional complications for assessment including whether substances are used specifically for their diuretic effects and under what circumstances their use is appropriate. For instance, some "cleansing" routines and extreme diets include ingesting teas with diuretic properties. Also, some pharmacological agents, prescription and non-prescription, act as diuretics. One should exercise clinical judgment regarding whether the use of such substances should be considered ICB; the user's intent will be the prime determinant.

Fasting

Fasting as an ICB involves not eating food for extended periods of time due to body shape or weight concerns. Fasting may be a better predictor of future eating pathology than overall dietary restraint (Stice, Davis, Miller & Marti, 2008), underscoring the importance of assessing this behavior. Eight consecutive waking hours is probably the lower boundary for considering a period of time without eating as a fast (Fairburn, 2008). During this time, individuals may drink fluids without breaking the fast, but if they have eaten very small amounts of food, their behavior is better considered to be restriction.

The motivation behind fasting is important, but it is not uncommon to experience resistance when assessing this. Alternative explanations commonly include fasting for religious reasons, fasting for medical reasons, forgetting to eat, not having access to food, not having time to eat, and not feeling hungry. We will comment briefly on each.

Periodic fasting plays an important role in many religious and spiritual traditions, and it is important to avoid pathologizing religious fasting. If individuals indicate that *part* of the reason they fasted was to affect their weight or shape, however, the fast might be considered an ICB. Fasting for medical reasons has gained interest based on research which suggests that regular fasting can have health benefits even in the absence of weight loss, although this has not yet been firmly established in humans (Mattson & Wan, 2005; Varady & Hellerstein, 2007). As with religious fasting, if manipulating weight and shape underlies health-based fasting, it may be considered inappropriate. When individuals claim a lack of time to eat or access to food, the clinician should determine whether these were isolated, unpredictable instances. Predictable situations should not excuse fasting behavior. Finally, reports of frequently "not being hungry" or "forgetting to eat" should be viewed with some skepticism. Hunger generally returns within 3–4 waking hours after eating (e.g., Driver, 1988; Poortvliet et al., 2007), barring illness or other psychiatric disorders such as major depressive episode (American Psychiatric Association, 2000). A substantial proportion of ED patients do have co-occurring mood or anxiety disorders, which can make it difficult to determine the true reasons behind episodes of non-eating. Given the pathology of EDs, however, reports of frequent periods of fasting because of loss of hunger are suspect.

Exercise

The DSM describes exercise as excessive "when it significantly interferes with important activities, when it occurs at inappropriate times or in inappropriate settings, or when the individual continues to exercise despite injury or other medical complications" (American Psychiatric Association, 2000, pp. 590–591). Thus, to avoid pathologizing a behavior that is considered health-promoting for most individuals, the DSM suggests that exercise must be associated with impairment in functioning or increased risk of physical injury to be considered excessive. Because this criterion is vague, there is disagreement regarding how best to define and assess excessive exercise. Researchers have defined excessive exercise according to exercise type, number of days spent exercising, and/or the amount of time spent exercising per occasion (Davis, Fox, Brewer & Ratusny, 1995), the intensity of the exercise (Fairburn et al., 2008), whether the exercise is obligatory or compulsive in nature (Adkins & Keel, 2005; Pasman & Thompson, 1988), or the function of the exercise (De Young & Anderson, 2010).

It is important to determine whether individuals regularly exercise rather than engaging in other valued life activities. Frequently, however, patients report that exercise or sport competition is in itself a highly valued and important activity, complicating the clinical picture. Asking about any injuries that have been sustained while exercising and how the individual responded to these injuries can clarify this issue. If a medical professional was consulted as a result of an injury and the professional's recommendations were not followed, the exercise is more likely excessive. Additionally, asking whether friends or family have ever commented that they spend too much time exercising can be informative. Exercising in excess of one's sport-related practice time can also be an indicator of excessive exercise. Strong feelings of fear or anxiety related to missing practice or an exercise session can also suggest excessive exercise. Finally, asking directly whether the exercise is being used as a compensatory method or not can be useful.

Ultimately, clinical judgment should be used to determine whether exercise behavior is related to the ED. Clinicians must work to achieve a balance between discouraging ICB and encouraging health-promoting behaviors and adaptive coping strategies, while taking into consideration contra-indications for exercise such as emaciation and cardiac compromise. Excessive exercise following treatment has been shown to be a predictor of relapse in AN (Carter, Blackmore, Sutandar-Pinnock & Woodside, 2004), so achieving this balance is especially difficult.

Inappropriate Use of Medications

Occasionally, ED patients misuse medications (other than laxative, diuretic, and diet pill medications) for the purposes of influencing their shape or weight. These medications most commonly include insulin and thyroid medications. Thus, individuals who have insulin-dependent diabetes mellitus (i.e., type 1 diabetes) should be asked whether they follow their physician's recommendations regarding the administration of insulin, as this governs the metabolism of ingested sugars. Not doing so can be extremely dangerous (Rodin, 2002). Similarly, individuals with hyperthyroidism or hypothyroidism may either omit or overuse their medication, respectively. These behaviors are also extremely dangerous. Thus, a careful assessment of medical conditions and associated medication use is highly recommended.

BODY IMAGE AND SELF-EVALUATION

Body image (BI) is conceptualized as a multidimensional construct including cognitive, behavioral, and affective elements (Cash, 2004). BI disturbance is a core feature of ED, though its exact nature varies. BI assessments may address attitudinal (dissatisfaction with body weight or specific body areas) and/or perceptual (body size misperceptions) components (Cash & Deagle, 1997).

Attitudinal Body Image

ED sufferers usually have disturbances in attitudinal BI, including body dissatisfaction, fear of weight gain or becoming fat, and undue influence of BI on self-evaluation. Body dissatisfaction refers to the degree to which one dislikes one's body; assessments can address global elements (i.e., overall weight and shape satisfaction) and specific elements (i.e., satisfaction with particular body areas such as hips, abdomen, and arms). Determining the intensity of fear of gaining weight or becoming fat is also important, particularly if this fear drives extreme dietary restriction and inappropriate compensatory behaviors. Finally, for many with ED, BI plays a substantial role in overall self-evaluation. To assess this, clinicians may ask patients to rank order factors that contribute to their self-evaluation.

ED patients often exhibit preoccupations and behaviors associated with BI disturbance. They may frequently check themselves by measuring or scrutinizing a particular body part of concern. Other behaviors include camouflaging (e.g., concealing one's body) and avoidance of looking at one's body in the mirror or of situations in which one's body would be exposed (e.g., the beach). The frequency, severity, and associated psychosocial impairment of these behaviors vary widely, so the use of open-ended questions is especially important.

Perceptual Body Image

ED sufferers commonly experience distortions in body perception, frequently characterized by an overestimation of body shape and size. Assessing disturbance in perceptual BI is difficult due to the complex methodology and instrumentation required as well as the fact that it can fluctuate based on a number of factors (Garner, 2002; Gruber, Pope, Borowiecki & Cohane, 1999; Thompson & Gardner, 2002). Given these limitations, clinicians may opt to focus their assessment primarily on BI attitudes, particularly since attitudinal BI is more clinically relevant (Cash & Deagle, 1997).

COMORBID PSYCHOPATHOLOGY

Many ED patients experience one or more forms of comorbid psychopathology, including mood disorders, anxiety disorders, and substance use disorders (Hudson, Hiripi, Pope & Kessler, 2007). Given the high rates of comorbidity, assessing for the presence of other psychopathology is critical to treatment planning and case conceptualization.

Anxiety Disorders

A variety of anxiety disorders, including generalized anxiety disorder, social phobia, and obsessive-compulsive disorder (OCD), frequently co-occur with ED (Godart, Flament, Perdereau & Jeammet, 2002; Hudson et al., 2007). When assessing for the presence of an anxiety disorder in an ED sufferer, consider symptom overlap. For example, if obsessions and compulsions are entirely associated with eating and BI (e.g., obsessions about calories or body checking), these symptoms should not be interpreted as evidence of OCD. To clarify this issue, clinicians can ask whether the patient experiences anxious thoughts or behaviors unrelated to eating and/or BI.

Although classified as a somatoform disorder, body dysmorphic disorder (BDD) shares characteristics of an anxiety disorder and may occur among ED patients. To distinguish between ED and BDD, clinicians should assess the individual's primary areas of concern. In ED, the focus is on body weight and shape, while BDD concerns more often focus on specific body parts (Grant, Kim & Eckert, 2002; Phillips, Kim & Hudson, 1995).

Mood Disorders

Mood disorders are commonly associated with ED. Assessing for comorbid depression is essential, given the rate of suicidal ideation and related behaviors among ED patients (Fischer & Le Grange, 2007; Lacey, 1993; Wildman, Lilenfeld & Marcus, 2004). If the individual reports any thoughts of suicide or death, the clinician should ask about history of suicidal behaviors, current motivation and intent, and the presence of a plan (Sommers-Flanagan & Sommers-Flanagan, 1995). Patients may also report self-injurious behavior (e.g., cutting) without suicidal intent. If an individual reports such behaviors, assessing the frequency, severity, and potential lethality is critical. Assessments of comorbid depression should consider symptom overlap (e.g., weight change). If these symptoms seem primarily associated with the ED, they should not be considered to be symptoms of depression.

Substance Use Disorders

Many ED patients exhibit alcohol and other substance abuse (Root et al., 2010; von Ranson, Iacono & McGue, 2002). The clinician should assess the frequency and average amount of substance use, as well as associated problematic and high-risk behaviors, as these influence treatment planning. Alcohol use disorders also appear to be a strong predictor of mortality among persons with AN (Keel et al., 2003). See Baker Dennis and Helfman (Chapter 14) for more information about ED and substance abuse.

DENIAL AND MINIMIZATION

Given the social stigma associated with ED and the resulting embarrassment and shame, many sufferers deny or minimize the frequency or severity of their symptoms. Denial and minimization appear to be common in AN (Couturier & Lock, 2006; Vanderdeycken & Vanderlinden, 1983; Vitousek, Daly & Heiser, 1991), particularly denial of the seriousness of low body weight. Denial and minimization may also be manifested via under-reporting the frequency of certain ED behaviors. Vitousek and colleagues (1991) proposed three forms of denial: (i) deliberate distortion (i.e., intentionally underreporting or denying symptoms); (ii) inadvertent distortion (i.e., inaccurate reports resulting from a lack of awareness or insight); and (iii) overcompliance (i.e., the desire to please the clinician by providing inaccurate reports). Perceptions of anonymity and assessment conditions may also affect reporting (Keel, Crow, Davis & Mitchell, 2002; Lavender & Anderson, 2009).

Although a clinician may never be entirely certain that an individual is providing honest reports, developing strong rapport is important and may increase an individual's willingness to provide candid responses. A careful assessment can be part of the rapport-building process. Many patients have never spoken to anyone about the realities of their ED, so the initial assessment is often a critical moment for developing connection, trust, a language for treatment, and a rationale for explaining the illness and treatment.

Denial and minimization are often present when individuals feel ambivalent about, or do not see the need for, treatment. Instruments designed to assess readiness to change specifically in ED populations (Ackard, Croll, Richter, Adlis & Wonderlich, 2009; Geller, Cockell & Drab, 2001) can help the clinician explore this issue.

In summary, while there are no easy solutions to these issues, clinicians should remain aware of the possibility of denial, minimization, and ambivalence in their patients and address them as indicated.

ADDITIONAL AREAS TO ASSESS

Interpersonal and Family Functioning

Many individuals with ED experience difficulties in interpersonal functioning, and these may play an important role in the etiology and maintenance of their psychopathology (McIntosh, Bulik, McKenzie, Luty & Jordan, 2000). One useful assessment model is based on the four interpersonal problem areas identified in Interpersonal Psychotherapy (IPT) for ED: grief, role transitions, role disputes, and interpersonal deficits (Fairburn, 1997; Klerman & Weissman, 1993). It is also extremely important to assess family functioning and support (Work Group on Eating Disorders, 2006), particularly since some treatments include family involvement (Lock et al., 2002).

Trauma/Abuse History

A history of trauma, particularly childhood sexual abuse, may be associated with developing an ED and even impulsive behaviors, including substance abuse and self-injury (Corstorphine, Waller, Lawson & Ganis, 2007; de Groot & Rodin, 1999; Smyth, Heron, Wonderlich, Crosby & Thompson, 2008; Wonderlich et al., 2001). If the assessment reveals a trauma history, the clinician should assess for dissociation and other symptoms of post-traumatic stress disorder and formulate the treatment plan accordingly (see Ackard and Brewerton, Chapter 15; and Briere & Scott, 2007, for a discussion of trauma assessment in ED populations).

BRINGING IT ALL TOGETHER

The domains discussed in this chapter represent the key components of a comprehensive ED assessment. Based on these domains, a clinician can build a symptom profile that allows for a more fine-grained analysis of change than can be provided by simply achieving a diagnosis. Such a profile can be mapped onto diagnostic systems such as the DSM-IV-TR (American Psychiatric Association, 2000), but the reliance on diagnosis comes with a number of shortcomings. A symptom profile approach avoids these shortcomings and provides a high level of clinical utility.

All of these key domains can be assessed using an informal semi-structured or unstructured interview. While using only interviews may be common clinical practice (Anderson & Paulosky, 2004a), we recommend that clinicians conduct more formal assessments by supplementing clinical interviews with well-validated self-report inventories. While it is beyond the scope of this chapter to discuss specific instruments, several well-validated measures do exist, and comprehensive reviews of these measures are available (e.g., Allison, 2009; Anderson, Lundgren, Shapiro & Paulosky, 2004; Anderson & Paulosky, 2004b; Mitchell & Peterson, 2007). See also Chapter 15 (Ackard and Brewerton) and Chapter 27 (Banker and Klump). These measures have some advantages. First, they have well-established levels of reliability and validity. Second, most have norms, which allow the assessor to compare a given individual's scores to others. Third, as noted previously, individuals with ED might reply more honestly when not communicating face-to-face (Keel et al., 2002; Lavender & Anderson, 2009).

While most clinicians conduct a formal assessment at the beginning and ending of treatment (Anderson & Paulosky, 2004a), more frequent assessments provide valuable information about treatment progress and ultimate disposition. For example, early treatment response (after approximately 4 weeks of treatment) predicts longer-term treatment outcome (Doyle, Le Grange, Loeb, Doyle, & Crosby, in press; Grilo, Masheb & Wilson, 2006; Le Grange, Doyle, Crosby & Chen, 2008; Wilson, Fairburn, Agras, Walsh & Kraemer, 2002). Thus, an assessment at this point can assist clinicians in treatment planning. We suggest monthly formalized assessments, where feasible, to aid clinicians in tracking treatment. Weekly brief assessments using a brief paper and pencil instrument (e.g., the Eating Attitudes Test-26; Garner, Olmsted, Bohr & Garfinkel, 1982) can also be very helpful in tracking treatment progress.

Assessment is a critical component of effective treatment. A good assessment can help determine an initial level of care, decisions about changes in level of care (Work Group on Eating Disorders, 2006), and the ultimate treatment outcome. This chapter will hopefully serve as a guide for conducting more efficient and effective assessments in clinical practice.

References

Ackard, D. M., Croll, J. K., Richter, S., Adlis, S., & Wonderlich, A. (2009). A self-report instrument measuring readiness to change disordered eating behaviors: the Eating Disorders Stage of Change. *Eating and Weight Disorders, 14*, 66–76.

Adkins, E. C., & Keel, P. K. (2005). Does "excessive" or "compulsive" best describe exercise as a symptom of bulimia nervosa? *International Journal of Eating Disorders, 38*, 24–29.

Allison, D. B. (2009). *Handbook of assessment methods for eating behaviors and weight-related problems* (2nd ed.). Newbury Park, CA: Sage.

American Psychiatric Association. (2000). *Diagnostic and statistical manual of mental disorders* (4th ed., Text Revision). Washington, DC: Author.

Anderson, D. A., Lundgren, J. D., Shapiro, J. R., & Paulosky, C. A. (2004). Clinical assessment of eating disorders: Review and recommendations. *Behavior Modification, 28*, 763–782.

Anderson, D. A., & Paulosky, C. A. (2004a). A survey of the use of assessment instruments by eating disorder professionals in clinical practice. *Eating and Weight Disorders, 9*, 238–241.

Anderson, D. A., & Paulosky, C. A. (2004b). Psychological assessment of eating disorders and related features. In J. K. Thompson (Ed.), *Handbook of eating disorders and obesity* (pp. 112–129). New York, NY: Wiley.

Attia, E., & Roberto, C. A. (2009). Should amenorrhea be a diagnostic criterion for anorexia nervosa? *International Journal of Eating Disorders, 42*, 581–589.

Austin, S. B., Ziyadeh, N. J., Vohra, S., Forman, S., Gordon, C. M., & Prokop, L. A. (2008). Irregular menses linked to vomiting in a nonclinical sample: Findings from the National Eating Disorders Screening Program in high schools. *Journal of Adolescent Health, 42*, 450–457.

Bacon, L. (2008). *Health at every size: The surprising truth about your weight*. Dallas, TX: BenBella Books, Inc.

Bathalon, G. P., Tucker, K. L., Hays, N. P., Vinken, A. G., Greenberg, A. S., Mc Crory, M., & Roberts, S. B. (2000). Psychological measures of eating behavior and the accuracy of three common dietary assessment methods in healthy postmenopausal women. *American Journal of Clinical Nutrition, 71*, 739–745.

Beglin, S. J., & Fairburn, C. G. (1992). What is meant by the term "binge"? *American Journal of Psychiatry, 149*, 123–124.

Briere, J., & Scott, C. (2007). Assessment of trauma symptoms in eating-disordered populations. *Eating Disorders, 15*, 347–358.

Butryn, M. L., Lowe, M. R., Safer, D. L., & Agras, W. S. (2006). Weight suppression is a robust predictor of outcome in the cognitive-behavioral treatment of bulimia nervosa. *Journal of Abnormal Psychology, 115*, 62–67.

Carney, C. P., & Andersen, A. E. (1996). Eating disorders. Guide to medical evaluation and complications. *Psychiatric Clinics of North America, 19*, 657–679.

Carter, J. C., Blackmore, E., Sutandar-Pinnock, K., & Woodside, D. B. (2004). Relapse in anorexia nervosa: A survival analysis. *Psychological Medicine, 34*, 671–679.

Cash, T. F. (2004). Body image: Past, present, and future. *Body Image, 1*, 1–5.

Cash, T. F., & Deagle, E. A. (1997). The nature and extent of body-image disturbances in anorexia nervosa and bulimia nervosa: A meta-analysis. *International Journal of Eating Disorders, 22*, 107–125.

Corstorphine, E., Waller, G., Lawson, R., & Ganis, C. (2007). Trauma and multi-impulsivity in the eating disorders. *Eating Behaviors, 8*, 23–30.

Couturier, J. L., & Lock, J. (2006). Denial and minimization in adolescents with anorexia nervosa. *International Journal of Eating Disorders, 39*, 212–216.

Crow, S., & Swigart, S. (2005). Medical assessment. In J. E. Mitchell & C. B. Peterson (Eds.), *Assessment of eating disorders* (pp. 120–128). New York, NY: Guilford.

Dalle Grave, R., & Calugi, S. (2007). Eating disorder not otherwise specified in an inpatient unit: The impact of altering the DSM-IV criteria for anorexia and bulimia nervosa. *European Eating Disorders Review, 15*, 340–349.

Davis, C., Fox, J., Brewer, H., & Ratusny, D. (1995). Motivations to exercise as a function of personality characteristics, age, and gender. *Personality and Individual Differences, 19*, 165–174.

de Groot, J., & Rodin, G. (1999). The relationship between eating disorders and childhood trauma. *Psychiatric Annals, 29*, 225–229.

De Young, K. P., & Anderson, D. A. (2010). Prevalence and correlates of exercise motivated by negative affect. *International Journal of Eating Disorders, 43*, 50–58.

Doyle, P. M., Le Grange, D., Loeb, K., Doyle, A.C., & Crosby, R. D. (in press). Early response to family-based treatment for adolescent anorexia nervosa. *International Journal of Eating Disorders*. Published online 8 Oct 2008, doi: 10.1002/eat.20763

Driver, C. J. I. (1988). The effect of meal composition on the degree of satiation following a test meal and possible mechanisms involved. *British Journal of Nutrition, 60*, 441–449.

Eddy, K. T., Dorer, D. J., Franko, D. L., Tahilani, K., Thompson-Brenner, H., & Herzog, D. B. (2007). Should bulimia nervosa be subtyped by history of anorexia nervosa? A longitudinal validation. *International Journal of Eating Disorders, 40*, S67–S71.

Fairburn, C. G. (1997). Interpersonal psychotherapy for bulimia nervosa. In D. M. Garner & P. E. Garfinkel (Eds.), *Handbook of treatment for eating disorders* (pp. 278–294). New York, NY: Guilford Press.

Fairburn, C. G. (2008). *Cognitive behavior therapy and eating disorders*. New York, NY: Guilford Press.

Fairburn, C. G., & Cooper, Z. (2007). Thinking afresh about the classification of eating disorders. *International Journal of Eating Disorders, 40*, S107–S110.

Fairburn, C. G., Cooper, Z., & O'Connor, M. E. (2008). Eating Disorder Examination (Edition 16.0D). In C. G. Fairburn (Ed.), *Cognitive behavior therapy and eating disorders* (pp. 265–308). New York, NY: Guilford Press.

Fairburn, C. G., Marcus, M. D., & Wilson, G. T. (1993). Cognitive- behavioral therapy for binge eating and bulimia nervosa: A comprehensive treatment manual. In C. G. Fairburn & G. T. Wilson (Eds.), *Binge eating: Nature, assessment, and treatment* (pp. 361–404). New York, NY: Guilford.

Favaro, A., & Santonastaso, P. (1996). Purging behaviors, suicide attempts, and psychiatric symptoms in 398 eating disordered subjects. *International Journal of Eating Disorders, 20*, 99–103.

Fisak, B., Peterson, R. D., Tantleff-Dunn, S., & Molnar, J. M. (2006). Challenging previous conceptions of vegetarianism and eating disorders. *Eating and Weight Disorders, 11*, 195–200.

Fischer, S., & Le Grange, D. (2007). Comorbidity and high-risk behaviors in treatment-seeking adolescents with bulimia nervosa. *International Journal of Eating Disorders, 40*, 751–753.

Garner, D. M. (2002). Body image and anorexia nervosa. In T. F. Cash & T. Pruzinsky (Eds.), *Body image: A handbook of theory, research, and clinical practice* (pp. 295–303). New York, NY: Guilford.

Garner, D. M., Olmsted, M. P., Bohr, Y., & Garfinkel, P. E. (1982). The Eating Attitudes Test: Psychometric features and clinical correlates. *Psychological Medicine, 12*, 871–878.

Geller, J., Cockell, S. J., & Drab, D. L. (2001). Assessing readiness for change in the eating disorders: The psychometric properties of the Readiness and Motivation Interview. *Psychological Assessment, 13*, 189–198.

Glass, A. R., Deuster, P. A., Kyle, S. B., Yahiro, J. A., Vigersky, R. A., & Schoomaker, E. B. (1987). Amenorrhea in Olympic marathon runners. *Fertility & Sterility, 48*, 740–745.

Godart, N. T., Flament, M. F., Perdereau, F., & Jeammet, P. (2002). Comorbidity between eating disorders and anxiety disorders: A review. *International Journal of Eating Disorders, 32*, 253–270.

Grant, J. E., Kim, S. W., & Eckert, E. D. (2002). Body dysmorphic disorder in patients with anorexia nervosa: Prevalence, clinical features, and delusionality of body image. *International Journal of Eating Disorders, 32*, 291–300.

Grilo, C. M., Masheb, R. M., & Wilson, G. T. (2006). Rapid response to treatment for binge eating disorder. *Journal of Consulting and Clinical Psychology, 74*, 602–613.

Gruber, A. J., Pope, H. G., Borowiecki, J., & Cohane, G. (1999). The development of the somatomorphic matrix: A biaxial instrument for measuring body image in men and women. In T. S. Olds, J. Dollman & K. I. Norton (Eds.), *Kinanthropometry VI* (pp. 217–232). Sydney, Australia: International Society for the Advancement of Kinanthropometry.

Hadigan, C. M., LaChaussee, J. L., Walsh, B. T., & Kissileff, H. R. (1992). 24-hour dietary recall in patients with bulimia nervosa. *International Journal of Eating Disorders, 12*, 107–111.

Haskell, W. L., Lee, I. M., Pate, R. R., Powell, K. E., Blair, S. N., Franklin, B. A., ... Bauman, A. (2007). Physical activity and public health: Updated recommendation for adults from the American College of Sports Medicine and the American Heart Association. *Circulation, 116*, 1081–1093.

Hudson, J. I., Hiripi, E., Pope, H. G., & Kessler, R. C. (2007). The prevalence and correlates of eating disorders in the national comorbidity survey replication. *Biological Psychiatry, 61*, 348–358.

Johnson, W. G., Boutelle, K. N., Torgrud, L., Davig, J. P., & Turner, S. (2000). What is a binge? The influence of amount, duration, and loss of control criteria on judgments of binge eating. *International Journal of Eating Disorders, 27*, 471–479.

Johnson, W. G., Schlundt, D. G., Barclay, D. R., Carr-Nangle, R. E., & Engler, L. B. (1995). A naturalistic functional analysis of binge eating. *Behavior Therapy, 26*, 101–118.

Kaplan, A. S., Walsh, B. T., Olmsted, M., Attia, E., Carter, J. C., Devlin, M. J., ... Parides, M. (2009). The slippery slope: Prediction of successful weight maintenance in anorexia nervosa. *Psychological Medicine, 39*, 1037–1045.

Keel, P. K., Crow, S., Davis, T. L., & Mitchell, J. E. (2002). Assessment of eating disorders: Comparison of interview and questionnaire data from a long-term follow-up study of bulimia nervosa. *Journal of Psychosomatic Research, 53*, 1043–1047.

Keel, P. K., Dorer, D. J., Eddy, K. T., Franko, D., Charatan, D. L., & Herzog, D. B. (2003). Predictors of mortality in eating disorders. *Archives of General Psychiatry, 60*, 179–183.

Keel, P. K., Mayer, S. A., & Harnden-Fischer, J. H. (2001). Importance of size in defining binge eating episodes in bulimia nervosa. *International Journal of Eating Disorders, 29*, 294–301.

Klerman, G. L., & Weissman, M. M. (1993). Interpersonal psychotherapy for depression: Background and concepts. In G. L. Klerman & M. M. Weissman (Eds.), *New applications of interpersonal psychotherapy* (pp. 3–26). Washington DC: American Psychiatric Press.

Lacey, J. H. (1993). Self-damaging and addictive behaviour in bulimia nervosa. A catchment area study. *British Journal of Psychiatry, 163*, 190–194.

Latner, J. D., Hildebrandt, T., Rosewall, J. K., Chisholm, A. M., & Hayashi, K. (2007). Loss of control over eating reflects eating disturbances and general psychopathology. *Behaviour Research and Therapy, 45*, 2203–2211.

Lavender, J. M., & Anderson, D. A. (2009). Effect of perceived anonymity in assessments of eating disordered behaviors and attitudes. *International Journal of Eating Disorders, 42*, 546–551.

Le Grange, D., Doyle, P., Crosby, R. D., & Chen, E. (2008). Early response to treatment in adolescent bulimia nervosa. *International Journal of Eating Disorders, 41*, 755—757.

Lock, J., Le Grange, D., Agras, W. S., & Dare, C. (2002). *Treatment manual for anorexia nervosa: A family-based approach.* New York, NY: Guilford.

Lowe, M. R., Davis, W., Lucks, D., Annunziato, R., & Butryn, M. (2006). Weight suppression predicts weight gain during inpatient treatment of bulimia nervosa. *Physiology & Behavior, 87*, 487—492.

Machado, P. P., Machado, B. C., Goncalves, S., & Hoek, H. W. (2007). The prevalence of eating disorders not otherwise specified. *International Journal of Eating Disorders, 40*, 212—217.

Mattson, M. P., & Wan, R. (2005). Beneficial effects of intermittent fasting and caloric restriction on the cardiovascular and cerebrovascular systems. *Journal of Nutritional Biochemistry, 16*, 129—137.

McIntosh, V. V., Bulik, C. M., McKenzie, J. M., Luty, S. E., & Jordan, J. (2000). Interpersonal psychotherapy for anorexia nervosa. *International Journal of Eating Disorders, 27*, 125—139.

Mendell, D. A., & Logemann, J. A. (2001). Bulimia and swallowing: Cause for concern. *International Journal of Eating Disorders, 30*, 252—258.

Mitchell, J. E., Cook-Myers, T., & Wonderlich, S. A. (2005). Diagnostic criteria for anorexia nervosa: Looking ahead to DSM-V. *International Journal of Eating Disorders, 37*, S95—S97.

Mitchell, J. E., & Peterson, C. B. (2005). *Assessment of eating disorders.* New York, NY: Guilford.

Mitchell, J. E., Pomeroy, C., & Adson, D. E. (1997). Managing medical complications. In D. M. Garner & P. E. Garfinkel (Eds.), *Handbook of treatment for eating disorders* (2nd ed.). New York, NY: Guilford Press.

Mond, J. M., Hay, P. J., Rodgers, B., Owen, C., & Mitchell, J. E. (2006). Correlates of self-induced vomiting and laxative misuse in a community sample of women. *Journal of Nervous and Mental Disease, 194*, 40—46.

Pasman, L., & Thompson, J. K. (1988). Body image and eating disturbance in obligatory runners, obligatory weightlifters, and sedentary individuals. *International Journal of Eating Disorders, 7*, 759—769.

Phillips, K. A., Kim, J. M., & Hudson, J. I. (1995). Body image disturbance in body dysmorphic disorder and eating disorders obsession or delusions. *The Psychiatric Clinics of North America, 18*, 317—334.

Poortvliet, P. C., Berube-Parent, S., Drapeau, V., Lamarche, B., Blundell, J. E., & Tremblay, A. (2007). Effects of a healthy meal course on spontaneous energy intake, satiety and palatability. *British Journal of Nutrition, 97*, 584—590.

Poyastro Pinheiro, A., Thornton, L. M., Plotonicov, K. H., Tozzi, F., Klump, K. L., Berrettini, W. H., ... Brandt, H. (2007). Patterns of menstrual disturbance in eating disorders. *International Journal of Eating Disorders, 40*, 424—434.

Robinson-O'Brien, R., Perry, C. L., Wall, M. M., Story, M., & Neumark-Sztainer, D. (2009). Adolescent and young adult vegetarianism: Better dietary intake and weight outcomes but increased risk of disordered eating behaviors. *Journal of the American Dietetic Association, 109*, 648—655.

Rockert, W., Kaplan, A. S., & Olmsted, M. (2007). Eating disorder not otherwise specified: The view from a tertiary care treatment center. *International Journal of Eating Disorders, 40*, S99—S103.

Rodin, G. M. (2002). Eating disorders in diabetes mellitus. In C. G. Fairburn & K. D. Brownell (Eds.), *Eating disorders and obesity* (2nd ed.). (pp. 286—290) New York, NY: Guilford Press.

Root, T. L., Pinheiro, A. P., Thornton, L., Strober, M., Fernandez-Aranda, F., Brandt, H., ... Bulik, C. M. (2010). Substance use disorders in women with anorexia disorder. *International Journal of Eating Disorders, 43*, 14—21.

Shapiro, J. R., & Anderson, D. A. (2003). The effects of restraint, gender, and body mass index on the accuracy of self-reported weight. *International Journal of Eating Disorders, 34*, 177—180.

Smyth, J. M., Heron, K. E., Wonderlich, S. A., Crosby, R. D., & Thompson, K. M. (2008). The influence of reported trauma and adverse events on eating disturbance in young adults. *International Journal of Eating Disorders, 41*, 195—202.

Sommers-Flanagan, J., & Sommers-Flanagan, R. (1995). Intake interviewing with suicidal patients: A systematic approach. *Professional Psychology, Research and Practice, 26*, 41—47.

Stice, E., Davis, K., Miller, N. P., & Marti, C. N. (2008). Fasting increases risk for onset of binge eating and bulimic pathology: A 5-year prospective study. *Journal of Abnormal Psychology, 117*, 941—946.

Telch, C. F., Pratt, E. M., & Niego, S. H. (1998). Obese women with binge eating disorder define the term binge. *International Journal of Eating Disorders, 24*, 313—317.

Thompson, J. K., & Gardner, R. M. (2002). Measuring perceptual body image among adolescents and adults. In T. F. Cash & T. Pruzinsky (Eds.), *Body image: A handbook of theory, research, and clinical practice* (pp. 135—141). New York, NY: Guilford.

Trabulsi, J., & Schoeller, D. A. (2001). Evaluation of dietary assessment instruments against doubly labeled water, a biomarker of habitual energy intake. *American Journal of Physiology - Endocrinology and Metabolism, 281*, E891–E899.

Turner, H., & Bryant-Waugh, R. (2004). Eating disorder not otherwise specified (EDNOS): Profiles of clients presenting at a community eating disorder service. *European Eating Disorders Review, 12*, 18–26.

Vanderdeycken, W., & Vanderlinden, J. (1983). Denial of illness and the use of self-reporting measures in anorexia nervosa patients. *International Journal of Eating Disorders, 2*, 101–107.

Varady, K. A., & Hellerstein, M. K. (2007). Alternate-day fasting and chronic disease prevention: A review of human and animal trials. *American Journal of Clinical Nutrition, 86*, 7–13.

Villanueva, E. V. (2001). The validity of self-reported weight in US adults: a population based cross-sectional study. *BMC Public Health, 1*, 11.

Vitousek, K. B., Daly, J., & Heiser, C. (1991). Reconstructing the internal world of the eating-disordered individual: Overcoming denial and distortion in self-report. *International Journal of Eating Disorders, 10*, 647–666.

von Ranson, K. M., Iacono, W. G., & McGue, M. (2002). Disordered eating and substance use in an epidemiological sample: I. Associations within individuals. *International Journal of Eating Disorders, 31*, 389–403.

Walsh, B. T. (2009a). Special section/eating disorders in DSM-V: Review of existing literature (Part 1) [Special issue]. *International Journal of Eating Disorders, 42*(7), 579–580.

Walsh, B. T. (2009b). Special section/eating disorders in DSM-V: Review of existing literature (Part 2) [Special issue]. *International Journal of Eating Disorders, 42*(8), 673.

Wildman, P., Lilenfeld, L. R. R., & Marcus, M. D. (2004). Axis I comorbidity onset and parasuicide in women with eating disorders. *International Journal of Eating Disorders, 35*, 190–197.

Wilfley, D. E. (2002). Psychological treatment of binge eating disorder. In C. G. Fairburn & K. D. Brownell (Eds.), *Eating disorders and obesity: A comprehensive handbook* (2nd ed.). New York, NY: Guilford Press.

Williamson, D. A., Martin, C. K., York-Crowe, E., Anton, S. D., Redman, L. M., Han, H., & Ravussin, E. (2007). Measurement of dietary restraint: Validity tests of four questionnaires. *Appetite, 48*, 183–192.

Wilson, G. T., Fairburn, C. C., Agras, W. S., Walsh, B. T., & Kraemer, H. (2002). Cognitive-behavioral therapy for bulimia nervosa: Time course and mechanisms of change. *Journal of Consulting and Clinical Psychology, 70*, 267–274.

Wolfe, B. E., Baker, C. W., Smith, A. T., & Kelly-Weeder, S. (2009). Validity and utility of the current definition of binge eating. *International Journal of Eating Disorders, 42*, ,674–686.

Wonderlich, S. A., Crosby, R. D., Mitchell, J. E., & Engel, S. G. (2007). Testing the validity of eating disorder diagnoses. *International Journal of Eating Disorders, 40*, S40–S45.

Wonderlich, S. A., Crosby, R. D., Mitchell, J. E., Thompson, K. M., Redlin, J., Demuth, G., ... Haseltine, B. (2001). Eating disturbance and sexual trauma in childhood and adulthood. *International Journal of Eating Disorders, 30*, 401–412.

Work Group on Eating Disorders (2006). *Practice guidelines for the treatment of patients with eating disorders* (3rd ed.). Arlington, VA: American Psychiatric Press.

CHAPTER 6

Medical Assessment of Eating Disorders

Edward P. Tyson

This chapter intends to inform physicians and non-physician caregivers about the important medical complications of eating disorders (ED). Issues to be addressed include why and how they manifest, and how to quickly assess, understand, and treat them from all caregivers' perspectives. Family and other primary supporters (e.g., friends, teachers, coaches) have the most communication and contact with patients, and are typically their most trusted allies. However, they are sometimes left dangerously in the dark regarding their family member's critical medical care issues. Providing caregivers with a working understanding of these issues bridges this gap and promotes prompt, necessary, and collaborative intervention. Such knowledge may also circumvent or minimize medical complications, and/or allow for caregivers to be more appropriately supportive when adverse events occur.

A second goal of this chapter is to bridge a critical training gap among medical providers. Education on ED is variably included in health professionals' training programs. Eating disorders are not specifically listed as part of the Accreditation Council for Graduate Medical Education's (ACGME) required curriculum in a physician's training programs except in the adolescent medicine part of pediatrics, the combined medicine–pediatrics residency programs, or adolescent medicine fellowships. Neither general psychiatry, nor child and adolescent psychiatry requirements, specifically mandate training in ED. Therefore, medical practitioners who treat ED will usually have to seek out specialized education from the literature, from scientific meetings that address ED, and from colleagues in the field.

WOLVES IN SHEEP'S CLOTHING

Eating disorders are serious medical and psychiatric illnesses. With sometimes tragic outcomes, parents, coaches, and clinicians can be fooled into thinking that the bubbly and sweet teenager with anorexia nervosa (AN), who just won the state high school cross country meet, could not possibly be near death. Or, that the normal weight, straight-A physics graduate student who denies any complaints whatsoever may suffer a hypoglycemic seizure from her ED, crash her car, and kill herself and her passenger. Eating disorders are truly "wolves in sheep's clothing." Those with ED can look like they are doing "fine" (as patients often describe themselves even at their lowest points), even when all hell is breaking loose in their bodies. A teenager suffering from severe AN may be obviously ill-looking, but the looming dangers can still be missed or dismissed by those around them.

Eating disorder providers must always keep in mind that things can go terribly wrong and can be extremely complicated and expensive to treat. The worst outcome of an ED is death. It can come suddenly and without obvious warning. Eating disorders have the highest mortality of any psychiatric illnesses—higher than depression, schizophrenia, or bipolar disorder. Cardiac complications are the most common cause of death, with suicide being second in the younger population. Depending on which study one cites, lifetime mortality rates peak at 15–18%, approaching the death rates of certain serious cancers (Birmingham, Su, Hlynsky, Goldner & Gao, 2005; Neumärker, 1998; Sullivan, 1995). The mortality rate for AN is more than 12 times higher than the general population of 15- to 24-year-old females, and AN is the third most common chronic illness of adolescent females (Sullivan 1995; Lucas, Beard, O'Fallon & Kurland, 1991). The cost of medical care can be oppressive. The service and staffing capacities of college health services are being strained to accommodate increasing numbers of suffering students. According to the Agency for Healthcare Research and Quality, the problem is worsening for those in traditional and nontraditional groups (Zhao & Encinosa, 2009). From 1999 to 2006, ED admissions to nonpsychiatric community hospitals increased 18% overall, with related hospital costs increasing 61%. Rates of admissions for ED increased 119% for children under age 12, 48% for those aged 45 to 65, and 24% for those 64 and older.

While patients and their families typically are more distressed by purging behaviors than restriction, bulimia nervosa (BN) has an overall lower mortality rate than restricting AN. Although considered by some to be a non-specific term, Eating Disorder Not Otherwise Specified (EDNOS) is the most populated diagnostic group, and may carry a higher mortality and complication rate than either AN or BN (Crow et al., 2009). While overall death rates from ED approach 20%, those rates are averaged over all who have suffered and are not predictive of a particular patient. The more drastic, frequent, or chronic the behavior of the patient, the higher the risks, including the chance of death.

MENTAL ILLNESSES OR MEDICAL ILLNESSES?

Depending on one's perspective, an ED is either a medical illness with psychiatric features or a psychiatric illness with medical complications. In fact, biological and psychiatric issues impact each other. However, the tendency to utilize this deceptive dichotomy puts ED patients in a medical "no-man's land" that interferes with their getting appropriate assessment and care. Few medical hospitals or general psychiatric units are set up to manage the specific psychiatric needs and dietary management of patients with a serious ED. Therefore, specialized centers are often necessary for safe and adequate management.

The reciprocal relationship between a patient's medical and psychiatric status must be judiciously considered in the assessment and treatment of ED. Starvation, electrolyte abnormalities, hypoglycemia, caffeine toxicity, co-morbid medical conditions such as hypothyroidism, diabetes, autoimmune disorders, sleep deprivation, and other medical issues can all affect the onset, maintenance, and severity of the ED. Biological conditions, such as starvation, or exercise coupled with restriction, may be enough to trigger an ED or similar behaviors, even in otherwise seemingly low-risk individuals (Kalm & Semba, 2005). It becomes difficult to distinguish cause and effect.

The false dichotomy between medical and psychiatric aspects interferes with assessment and treatment in another important way. Health insurance has traditionally divided illnesses into either medical or psychiatric categories and "never the twain shall meet." Eating disorders are defined in the DSM-IV (American Psychiatric Association, 1994) and traditionally have been covered only by psychiatric benefits that usually have less reimbursement. They are classified as psychiatric illnesses in spite of the facts that: (i) medical complications are what usually instigate treatment and are a critical factor in determining proper levels of care; (ii) ED may be triggered by medical or biologic issues; and (iii) psychiatric residency programs are not required to specifically train about ED. With the significant exception of Adolescent Medicine units found in some academic medical centers, most ED treatment centers are considered psychiatric facilities. Medical units use traditional medical diagnoses (e.g., protein/calorie malnutrition, bradycardia, hypotension, hypoglycemia) as the primary codes rather than ED diagnoses, and are usually covered under medical benefits. In psychiatric centers, psychiatric codes are used as the primary diagnosis codes. Thus, even though the psychiatric and medical treatment centers are treating exactly the same illness, there is a dichotomy in reimbursement. In addition to the economic hardship imposed on families (see Munn, Smeltzer, Smeltzer and Westin, Chapter 21), limited reimbursement can translate into delays in assessment and intervention, interfere with ongoing medical treatment, and worsen outcomes. This issue is especially important in children and adolescents, who may ultimately suffer for many more years because of the earlier age of onset; the damage to their growing bodies may have geometric effects if not addressed early and aggressively.

WHAT PATIENTS DO IS MORE IMPORTANT THAN THEIR DIAGNOSTIC CATEGORY

Rather than discuss medical aspects and risks of each ED separately, it is more relevant to consider the patients' symptoms versus their diagnostic status. This is particularly true given that some ED patients move between diagnoses over the course of their illnesses, and that EDNOS accounts for the greatest ED population (Agras, Crow, Mitchell, Halmi & Bryson, 2010). Therefore, focus of medical concern should be on the degree of protein calorie malnutrition and dehydration rather than "AN" *per se*, or on the extent of vomiting or laxative abuse, rather than "BN." Clinicians using this approach will be more alert to consequences of patients' behaviors that may otherwise be missed if they do not recognize that patients with one form of an ED may develop symptoms usually found in another diagnostic category.

Knowing how their bodies have been injured may help patients turn around their behaviors and limit further complications (Zerbe, 2008). Focusing on medical conditions and consequences may feel more objective and less shameful to patients without using the more inflammatory diagnostic labels. Presenting and explaining the medical and physiological facts in a neutral, non-judgmental way enhances patients' acceptance that medical consequences are a major, but typically treatable, part of their ED. Providing this cautious reassurance of physical rehabilitation may instill hope and incentivize their early recovery efforts, whereas dealing with the enormity of the emotional and psychosocial aspects of their ED may, at least initially, seem beyond their grasp. Regardless, all involved should be mindful that medical stability is a necessary, but not sufficient, condition for recovery, and does not preclude the need for further or more intensive treatment.

GENERAL RECOMMENDATIONS ON EATING DISORDER ASSESSMENT

Box 6.1 provides a list of general recommendations to assist clinicians in circumventing the common pitfalls encountered in ED assessment and treatment. Physicians unfamiliar with ED are cautioned that these patients' presentations are often counterintuitive relative to those for other medical conditions (e.g., low resting heart rates in ED patients are signs of cardiac compromise versus athletic conditioning), and that their medical status is variable and apt to deteriorate rapidly if not addressed.

BOX 6.1

GENERAL PERSPECTIVES ON THE MEDICAL ASSESSMENT OF ED PATIENTS

1. Recheck any preconceived ideas or biases. Examples include:
   a. Presuming patients have to meet full DSM-IV criteria of AN or BN in order to be at serious risk. Patients with EDNOS have similar levels of psychiatric and medical comorbidity.
   b. Assuming that psychiatric complaints cannot be due to physical causes.
   c. Thinking that ED are not serious medical illnesses or that only very thin patients can be dangerously ill from an ED.
   d. Believing that ED do not occur in children, males, older persons, nonCaucasians or those from lower socioeconomic status.
   e. Suggesting to patients or their families that they will get over it on their own.
   f. Suggesting simplistic, misguided approaches, such as, "Just eat desserts" or "Eat snacks and come back in a month."

2. Treat all patients with respect. Most sufferers want to understand what is going on with their bodies. None asked for their eating disorder, neither do they know what to do about it.

3. Patients' general appearance/presentation coupled with their history and vital signs will give clinicians about 80% of the information related to the severity of their status and what is likely to be found on exam.

4. Eating disorders affect every organ system. Consider how each organ system will respond to the behaviors, and how those reactions should evidence themselves. This perspective should guide interpretation of the history, physical exam, laboratory results, ECG, and any radiographs done. See Birmingham and Beaumont (2004) for an exhaustive list of possible medical manifestations in ED patients.

(continued)

BOX 6.1—cont'd

5. Explain medical findings and their causes "as you go" during the exam. For example, acrocyanosis, lanugo, capillary refill delay, and low body temperature are all visual examples of how their body is trying to protect itself and conserve energy.
6. Make it real for them. Let the patient literally feel what is wrong. For example, let the patient feel the changes of pulse quality and rate when they go from lying to standing. Or, let them experience how hard it is to hear their BP when they stand up.
7. For those who claim they are "fine," challenge them to explain what they think causes the abnormal findings. Participation in treatment makes it harder for them to deny or doubt the obvious physical complications or medical recommendations.
8. Focus questions in the history towards medical issues specific to ED.
9. Look for the consequences that the patient's history would indicate. If they are not there or conflict with the history given, determine why. For example, if a patient indicates that she doesn't purge, yet has elevated salivary amylase and low potassium, calmly and directly explain how her medical findings "contradict her self report".
10. Denial is an integral part of ED; keep the patient's denials in perspective.
11. One cannot necessarily rely on the patient's or family's perceptions as to etiology or seriousness of the problems.
12. Coordinate treatment with other caregivers, including physicians, therapists, dietitians, coaches, and trainers.
13. Severely ill ED patients may be too cognitively impaired to provide informed consent or refuse treatment. Legal counsel may be necessary and treatment should be managed accordingly.

GATHERING THE HISTORY

No organ system is spared in malnutrition. Clinicians should therefore anticipate and evaluate for dysfunction of every organ system. Initial risk assessments should consider the following:

1. The types, timing, and pattern of ED behaviors (e.g., restriction, vomiting, laxatives, exercise, withholding insulin).
2. The duration of past and recent ED behavior (usually the longer the duration, the more damage is possible).
3. The severity of expression of the behavior (e.g., vomiting 10 times a day is more severe than twice a day; however, vomiting only once a week, but using ipecac syrup and a toothbrush to do so, may be more serious).

Evidence gathered by Crow et al. (2009) suggests that patients with subthreshold AN or BN (i.e., patients with EDNOS) may also have high rates of morbidity and mortality. In a study of 101 children diagnosed with ED between the ages of 5 to 13, 78% were hospitalized, 61% had potentially life-threatening complications, but only 37% of those met full diagnostic criteria, and only 51% met weight criteria (Madden, Morris, Zurynski, Kohn & Elliot, 2009). Other studies evaluating EDNOS in adolescents and adults have found similar trends (Dalle & Clugi, 2007; Eddy, Celio Doyle, Hoste, Herzog & Le Grange, 2008). Thus, a substantial number of ED patients will be dangerously overlooked if clinicians only consider at risk those who fulfill strict diagnostic criteria.

Taking the Medical History

The medical history must be detailed and specifically focused on ED in order for clinicians to identify the full range of existing problems. As denial and secrecy are prevalent and entrenched in ED, frequent surveying to assess for medical risk can prevent overlooking critical problems. This should be continued well into treatment, as medical problems can return with relapses or exacerbations of the ED. Using checklists and forms to be completed before the start of appointments is a way to do this efficiently. Box 6.2 provides an example of a patient checklist that enables both medical and nonmedical professionals to get a better sense of the patient's status and to see if more assessment or referral is needed. The emphasis is to be thorough and to check often.

In addition to assessing food and fluid intake, it is also important to estimate patients' use of caffeine. Caffeine intake can be alarmingly high and can cause cardiac problems, gastro-esophageal reflux, sleep dysfunction, anxiety, and dehydration. Paradoxically, when used "for energy," caffeine forces the body to burn calories that it may be trying to preserve, thus accelerating one's physical decline and fatigue.

Differential Diagnosis

It is important to distinguish ED from other illnesses. There are many medical and psychiatric illnesses that can present as, disguise, or be mistaken for ED. Incorrect diagnoses can have devastating consequences. Clinicians should screen for all possibilities early on and again if aspects of the ED are not responding as would be expected. Box 6.3 lists other illnesses that may mimic, obscure, or increase the risk of ED.

THE EXAMINATION: CORRELATE WITH THE HISTORY

Findings on the exam should be viewed in context with the history. Much can be learned from observing the person, gathering the history, and comparing expected findings to what is actually seen on examination (e.g., if a patient with low potassium indicates that she is not vomiting, the discrepancy between the history and findings needs to be resolved). Nonmedical professionals benefit from comparing the history to observations of the patient's interactions and appearance. For example, if the patient's hands or feet are blue, but she states that she is fine, it is reasonable to surmise that she is in a significant energy deficit, and probably has been for some time. If there is any concern, referral to an ED medical specialist is appropriate.

Vital Signs

It is impossible to overstate the importance of a patient's vital signs as indicators of overall health status. Vital signs consist of weight (and BMI), temperature, blood pressure (BP), and pulse. These are simple to assess even by non-medical professionals and can be very telling,

BOX 6.2

PATIENT CHECKLIST

Name: _____ Date: _____ Age: _____

The following is a list of symptoms or complaints you may or may not have. <u>Please read each one and check any and all items that apply,</u> even if they may not have changed since your last visit. **This is CONFIDENTIAL.** Please fold it over when done and hand it to the therapist or receptionist when you are ready to be seen.

_____ Feeling cold much of the time
_____ Fingers or toes turn blue at times*
_____ "Hot flashes" or sweating spells (at night or other times not related to exercise)*
_____ Dizziness or feeling like you're going to pass out at times*
_____ Mouth feels dry at times
_____ Chew gum frequently
_____ Heart beat going fast suddenly*
_____ Feeling your heart "skip beats" or like it "jumps" at times*
_____ Chest pain*
_____ Shortness of breath or trouble breathing recently*
_____ Difficulty thinking straight or remembering things as well recently
_____ Falling asleep or extreme fatigue during the daytime (e.g., at school or work)
_____ Trouble falling or staying asleep at night
_____ Loss or decrease of menstrual periods
_____ Hair falling out
_____ Constipation or difficulty having a bowel movement
_____ Swelling in your feet or hands*
_____ Stomach hurting*
_____ Blood when you have thrown up or gone to the bathroom*
_____ Noticing something that looks like coffee grounds when you have thrown up*
_____ Hurting yourself by cutting, scratching, burning, or other forms of self-inflicted injury*
_____ Pain in one or more of your bones (like your shin or feet) or joints*
_____ I have thrown up at least once recently*
_____ I have taken some laxatives recently*
_____ I have taken some diet pills recently*
_____ I have taken stuff to make me throw up*
_____ I have taken some water pills recently*
_____ I have been drinking alcohol enough to get drunk recently
_____ I have taken other stuff that I would rather not talk about
_____ I have had thoughts about hurting or killing myself recently*

About how many calories a day have you been eating for the last week (on average)?_____
About how much fluid have you consumed in the last day? _____ How much caffeine? ____
(Give the amount in ounces, cups, or # of water bottles)

*Indicates that the author suggests a physician be notified and involved.

> **BOX 6.3**
>
> ## DIFFERENTIAL DIAGNOSIS OF EATING DISORDERS
>
> **Medical illnesses that may mimic or present as an Eating Disorder**
>
> Any chronic disease or infection
> Gastrointestinal illnesses
>   Celiac Disease
>   Ulcerative colitis
>   Chronic parasitic, bacterial infections
>   Malabsorption
> Endocrine disorders
>   Diabetes mellitus
>   Addison's Disease (adrenal insufficiency)
>   Hyperthyroidism
>   Hypopituitarism
> Cancers
> Superior Mesenteric Artery Syndrome (can also be a consequence of an eating disorder)
>
> **Medical illnesses/conditions associated with an increased risk of an Eating Disorder**
>
> Diabetes mellitus
> Celiac Disease
> Gastric bypass
> Illnesses or conditions that require increased attention toward or regulation of food intake
> Attention Deficit Disorder

especially when put into perspective with the rest of the history. Because of their importance, the vital signs will be discussed in detail.

Weight and body mass index (BMI). Weight seems to be the primary focus of patients, their families, dietitians, and other professionals, but it's use is not specific, often not helpful, and can be misleading. There is debate about the clinical utility and impact of telling patients their weight. Given its lack of specificity and patients' ambivalence regarding their weight, consider redirecting the discussion to more meaningful determinants of their health, such as their body temperature, pulse, BP, hydration, and cardiac exam.

Clinicians should be cognizant that increases or decreases of weight do not give information as to what causes the weight change, except in the acute setting. Rapid and significant changes of weight are almost always from fluid shifts, and not from true body tissue increases or loss. Even as patients' weights change over time, it is hard to determine how much is due to muscle (including heart muscle), the other organs, or body fat. Since fat is primarily a form of stored energy and generally not stored by the body until there are excess calories in the diet, it will be one of the later causes of weight increase. However, patients requiring weight restoration usually believe that fat is the first source of weight gain, and fail to understand the vital role and functions of dietary fat and fat tissue in the body. Education and reassurance are a necessity for these patients during the refeeding process (see Croll, Chapter 8).

Without sufficient protein in their diet, even if patients take in enough calories, they will ultimately lose protein from organs (e.g., brain, kidney, liver), muscle (including heart muscle), tendons, ligaments, cartilage, neurotransmitters, hormones, and all other parts of the body that require protein. Patients who do not eat enough protein to compensate for losses due to routine activity and exercise tend to lose a disproportionate amount of muscle relative to body fat. When this occurs without enough substrate of protein to repair that breakdown, muscle will decrease both in mass and in its ability to function. Muscle weighs more than fat for the same volume; thus weight may go down faster as a result, but the percent body fat may actually increase. It is possible, then, that a patient who is severely malnourished and has very low body weight may have a relatively high percent body fat (the actual amount of fat though would still be low compared to peers). Clinicians should be careful in measuring body fat in ED as this fact may be misinterpreted and lead to misdirected recommendations from clinicians and calamitous reactionary behaviors by the sufferer.

Body mass index (BMI) may be a less triggering concept for patients than weight, although it is debatable whether it yields more clinically meaningful information (e.g., it does not account for natural age, ethnic, and/or body build sources of weight diversity). In fact, some ED specialists argue that over-emphasis on weight/BMI is "inappropriate, misleading and potentially harmful" (Lask & Frampton, 2009, p. 165). Measures of BMI attempt to put body size into perspective by factoring in weight as a function of height. Body mass index growth charts by sex and age can be downloaded from the Centers for Disease Control (see http://www.cdc.gov/growthcharts/clinical_charts.htm). Measurements from childhood can be extrapolated to reasonably predict where the young patient should be at a future date by following the curves. Rather than over-focus on the exact weight/BMI values, assessing their rate of change provides relevant clinical information. The lower the BMI, and the more rapid the rate of weight decline, the higher the likelihood of medical problems, especially cardiac issues. In pediatrics, the diagnosis of "Failure to Thrive (FTT)" can also be applied when a child's height for weight falls to less than 80% of predicted or when two major growth lines are crossed on a growth chart. Either should trigger aggressive evaluation. If FTT or its equivalent is diagnosed in adults, clinicians should consider an ED as a possible etiology.

Temperature. Patients' temperatures may reveal important clinical issues. As the nutritional state declines and the body is forced to use its reserves, the body will try to conserve itself in an effort to combat excessive losses. Since the body burns about 60% of its calories just to produce body heat, one of the most effective ways to conserve energy is to lower body temperature, sometimes to as low as 92–94°F. It may help to advise patients that their body is suffering just as much during temperature drops as with fever spikes of the same degree. Research by Nolan, Morley, Vanden Hoek & Hickey (2003) indicates that in cardiopulmonary resuscitation (CPR), lowering the body temperature improves morbidity and mortality. Thus, the hypothermia associated with declining body temperatures secondary to ED, may actually be cardioprotective, and a sign that the body is trying to protect the heart from serious consequence.

Hypothermia is accompanied by other findings. The body tries to prevent heat loss and tends to shunt blood more to the core and away from the periphery, such as fingers and toes, where the ability to lose heat quickly occurs. The fingers, hands, toes, and feet can be very cold to touch and be bluish in color (*acrocyanosis*). Often, delayed capillary refill occurs, where blood only slowly returns to the skin. One way to check this is to have both the patient and an observer squeeze their fists for 5 seconds and then simultaneously open them. Compare the speed at which the pink coloration and blood flow returns to their palms and fingers. Delays can be dramatic, revealing significant medical issues to the patient if explained sensitively.

Both acrocyanosis and delayed capillary refill may be signs of the heart having difficulty with pumping blood into the more distant body sites. As the hearts shrinks from malnutrition, its pumping capacity can decline. Therefore, physicians must consider whether these findings are from low body heat, poor cardiac capacity, or both.

Temperature, body heat, and calories in context. For ED patients who take in very few calories, even small levels of activity burn essential calories and contribute to weight decline. For example, a 20-year-old female who is 5'5" and weighs 115 lbs, burns about 1370 calories a day at rest with no added activity (i.e., her basal metabolic rate). To warm 3 liters of ice water (a reasonable daily intake of fluid), the body will need to burn about 150–180 calories. Standing burns about 30–50 more calories per hour depending on one's weight. Pacing or muscle co-contractions (or dynamic tension) burn even more calories depending on their duration. Even chewing gum can increase calories burned. These often overlooked behaviors are common in ED and insidiously contribute to weight decline. Understanding this will help patients and caregivers accept activity restrictions, even complete bedrest, when patients are acutely ill or severely underweight.

Pulse. Clinicians should note the heart rate (HR) in beats per minute and the quality of the pulse as strong, weak, or thready (like a thread sliding under the observer's finger). Check the pulse while the patient is laying down (supine) for a minute, preferably with the eyes closed, to mimic the resting pulse, when the pulse is at its slowest. Generally, a resting pulse of less than 50 is considered slow and termed *bradycardia*. A resting HR above 100 is termed *tachycardia*.

After the supine pulse is obtained, compare the speed and quality of the pulse upon standing. A significant increase in pulse upon standing is called *orthostasis* and is considered abnormal. Generally, an increase of 20 beats per minute for those with a normal baseline HR is considered significant, and is often, but not always, a sign of dehydration. In malnourished patients, the change may be due to the weakening capacity of the body's neurological and cardiac regulators needed for adaptation to changes in gravitational forces on the body. In patients with very low HR, using a percent of change of pulse rather than just a change of 20 beats may be more helpful, as a change of the same number of beats per minute in slower HRs is more significant (e.g., a resting HR of 40 that goes up to 60 with standing has increased by 50%). As malnutrition worsens, the body's ability to adapt even to positional changes can be impaired. If this minor challenge is taxing to the cardiovascular system, then more rigorous demands, such as exercise, may be dangerous. Orthostasis is a common problem in ED and can persist for weeks after nutritional rehabilitation has begun.

Understanding bradycardia in the context of ED is one area in which the gap in medical training on ED routinely manifests. It is critical to distinguish when bradycardia is a sign of fitness or a sign of dysfunction. The "athletic heart syndrome" is a term describing athletes' hearts that have adapted to aerobic exercise and developed the capacity to pump more blood with each beat, translating into a slower HR at rest. Clinicians should use caution in making this diagnosis, especially in athletes with signs of an ED. In a study by Bjørnstad, Storstein, Dyre Meen and Hals (1991) of 1299 athletics students in Norway, the average HR was 62, and for controls 68. In the complimentary study of 1450 athletes, the average HR was 61 for males and 64 for females compared to 68 for controls (Storstein, Bjørnstad, Hals & Dyre Meen, 1991). These studies, which are probably the largest ones looking at athlete HRs, indicate that resting HRs that are in the lower 50s or below are unlikely to be consistent with athletic training. In such cases, clinicians should be looking for other signs and causes of the body adapting to energy shortage.

Blood pressure (BP). Blood pressure is another fairly simple method that can give important signs of patients' cardiovascular status. The bottom limits of normal BP are usually considered to be 90/60 (readings for children can be somewhat different). As with pulse, low values or orthostatic BP changes are instructive when evaluating ED patients. An orthostatic drop of more than 20 points in the systolic (top number) and 10 of the diastolic (bottom number) is considered significant. As with HR, the lower the baseline pressure is, any change is considered more significant. Manual readings are recommended as the quality of the sound heard through the stethoscope can also be indicative of cardiovascular function.

The body can do a variety of things to increase or decrease the BP to adapt to circumstances. Adaptations that are weak, inappropriate, or do not occur as expected indicate at least some cardiovascular compromise. Usually in ED, especially if there is restriction, BP tends to decline and the heart sounds required to determine the BP decline can be harder to hear.

CARDIAC ASSESSMENT

Assessing the cardiovascular system is essential, as complications are common and are the primary cause of cause of death in ED. Box 6.4 lists possible cardiac findings in ED. While HR and BP readings are the most basic and necessary cardiac vital signs, feeling and listening to the heart by palpation and auscultation also provides essential data. Important information is gleaned from feeling the radial pulse in the wrist and the heart's pulse on the chest, called the point of maximal impulse (PMI). In normals, the PMI can usually be felt on the chest at about the 4th or 5th rib space (intercostal space), in line with the midpoint of the clavicle. In athletes, healthy adolescents and young adults, the PMI should be easily palpable because their heart beats strongly, especially if the chest wall is fairly thin. As the heart grows smaller from malnutrition, the PMI moves more to the midline and slightly upward. It also tends to get weaker and can even be absent to palpation.

The location and volume of the heart sounds can also suggest a smaller, less dynamic heart. As the heart muscle atrophies, the loudness of the heart sounds tends to decrease. The area on the chest where they can be heard best (the silhouette) also shrinks. Because of these architectural changes, valves may not fit quite as well, like a door in the winter, and slight murmurs may become more evident. They are usually not significant.

Physicians should also be alert to other cardiac complications, including pericardial effusions (fluid around the heart) and abnormal rhythms. There are differing reports on how common effusions occur. If in doubt, the physician should obtain an echocardiogram in addition to an electrocardiogram (ECG). Cardiac dysrhythmias are common. The most common is bradycardia; rates can be dramatically low, including into the teens. The Society for Adolescent Medicine and the American Academy of Pediatrics recommend hospitalization for those with HRs <50 beats per minute in the daytime or <45 at night (American Academy of Pediatrics, 2003; Golden et al., 2003). Despite these recommendations, it may be difficult to convince cardiologists, emergency medicine physicians, and hospitalists unfamiliar with ED to accept those criteria. Rarely, patients require pacemakers for extreme bradycardia. Generally, the HR will normalize with refeeding.

> **BOX 6.4**
>
> ## POSSIBLE CARDIOVASCULAR FINDINGS IN EATING DISORDERS
>
> ### ECG findings
>
> Bradycardia
> Low voltage
> Inverted T waves
> U waves
> Various degrees of heart block
> Prolonged QT interval
> Increased QT dispersion
> ST segment depression, elevation, and non-specific changes
> Ventricular premature complexes
> Ventricular tachycardia
> Torsades de pointes (a particularly ominous form of ventricular tachycardia)
> Ventricular fibrillation
> Asystole (cardiac arrest)
>
> ### Other cardiac findings/complications
>
> Decreased cardiac muscle mass
> Diminished cardiac output
> Weak, thready pulses
> Acrocyanosis
> Weak, medially displaced PMI
> Decreased heart sounds from decreased cardiac dynamics
> Increased heart sounds from decreased chest wall thickness
> Increased heart murmurs
> Friction rub (from pericardial effusion)
> Myofibrillar degeneration
> Decreased ventricular volume
> Decreased cardiac output
> Autonomic dysregulation
> Hypotension
> Increased peripheral resistance
> Mitral valve prolapse
> Pericardial effusions
> Heart failure
> Elevated cardiac enzymes (without coronary artery disease)

Two final caveats are important to remember. Cardiac findings in a resting patient may not be predictive of what will happen during exercise, or if electrolytes or other complications occur subsequent to the exam. Secondly, in those with ED, bad things tend to happen at the extremes of HR. The more significant the bradycardia, or as HR approaches maximum, the greater the risk of a life-threatening rhythm. Eating disordered patients with bradycardia may reach their HR maximum sooner and with less effort (Riggs, Harel, Biros & Ziegler, 2003). These issues are of special concern in athletes and heavy exercisers with ED, and may be used to differentiate a trained versus a wasting heart.

Electrocardiogram

An ECG is essential for a baseline ED cardiac assessment. Most ED patients, especially those with AN, will have some abnormal findings on an ECG. Common ECG findings include bradycardia and an electrical axis shifted to more vertical, indicative of a shrinking heart that is conserving energy. There can also be low voltage (amount of deflection of ECG tracing from baseline), T wave inversions, ST segment depression (usually from electrolyte imbalances), and abnormal rhythms including various degrees of heart block, tachycardia, and life threatening arrhythmias. Physicians should decide which of these abnormalities are significant and need further intervention. Low voltage in V6 may also indicate a lower left ventricular mass and is associated with a lower BMI. Another important measure is prolongation of the QT interval corrected for HR (QTc). Generally in ED, a QTc of >440 msec is considered prolonged. There is debate how well the QTc correlates to imminent risk, but physicians should also be aware that a normal QTc does not mean the patient is safe from a cardiac standpoint. Increased QT dispersion (the difference of QT length in different leads on the ECG) has been found in AN, and has also been linked to sudden death (Katzman, 2005).

Echocardiogram

An echocardiogram provides a more detailed picture of the anatomy and physiology of the heart. Although there is no clear guide as to when to perform an echo, specific indicators from the history or clinical exam warrant this consideration. For example, pericardial effusions and architectural problems (e.g., mitral valve prolapse) are best detected on an echocardiogram. Both of these are findings are usually minor, reversible, and not clinically significant.

Stress Tests

A stress test may be indicated for ED patients who are heavy exercisers and who have been determined *not* to be at risk from this procedure. Stress tests also provide data relevant for recovering patients' safe return to exercise. In these circumstances, the cardiologist must take the patient's age and compromised nutritional status into full account for an appropriate interpretation of the results. For example, the performance of a 19-year-old elite runner with AN should be compared with that of someone of that age, ability, and performance. Otherwise, the same runner's stress test compared with that of the cardiologist's usual patient (e.g., a 68-year-old overweight, diabetic with coronary artery disease), may be misinterpreted as normal, when it clearly does not reflect the anorexic's level of heart functioning prior to her malnutrition.

Reversibility

Fortunately, almost all cardiac complications reverse with normalization of nutrition. One exception is cardiomyopathy, which may develop from use of emetine (ipecac), a liquid that induces violent vomiting. Given its propensity for abuse by ED patients and its contraindication for poison management, its withdrawal from over the counter status has been strongly recommended (Silber, Maine & McGilley, 2008).

NEUROPSYCHIATRIC ASSESSMENT: IMPAIRED BRAIN MEANS IMPAIRED MIND

The effects of the psychiatric (mind function) and neurological (brain or biological function) aspects of ED are reciprocal (Siegel, 1999; see also Lapides, Chapter 3). The worse the

ED, the worse the malnutrition, and the more the brain is affected. The more the brain is affected, the more the mind is affected, and the worse the ED. A vicious cycle ensues, with each form of impairment aggravating the other. Impaired brain function can alter mood, affect, judgment, behavior, personality, cognition, response flexibility, autonomic nervous system control, and other mental and neurological aspects. Numerous brain studies of ED patients have demonstrated alterations in various areas that are probably intimately involved in triggering, aggravating, or prolonging an ED. Studies have detected alterations of neurotransmitter receptor activity and blood flow to different regions within the brain, and even dramatic loss of brain mass in severe AN, similar to that seen in Alzheimer's (Guido, Bailer, Henry, Wagner & Kaye, 2004). These changes may not be fully reversible.

As noted in Levine and Levine (Chapter 7) and Croll (Chapter 8) regarding the pharmacological and nutritional aspects of ED, medications for treatment of ED and comorbid illnesses may not be as effective until the brain is renourished. Psychotherapy can also be limited in effectiveness until all nutritional abnormalities are reversed. Fluctuating levels of glucose (the brain's primary food) may help explain some of the shifting personality, emotional, and other aspects in patient's behaviors. Because of these intertwining neurobiological, psychiatric, and nutritional aspects of ED, it is prudent to reverse malnutrition before formally diagnosing another psychiatric disorder in ED patients. Nutritional rehabilitation alone may clear the symptoms of other psychiatric disorders (see Box 6.3). Clinicians should also track how psychiatric features change with normalization of nutrition and adjust medications or interventions accordingly.

Food for Thought: Understanding the Neurological Impact of Hypoglycemia

The brain is the main user of glucose in the body. Neurons cannot store glucose, thus requiring a constant supply to the brain. When glucose levels are low, especially below 54 mg/dl (3 mmol/L), the brain starts to suffer. Typical hypoglycemic symptoms include shakiness, irritability, feeling hungry, weakness, increased pulse rate, and anxiety. With chronic malnutrition, it appears that the body's usual warning signs are blunted or absent. Instead, more drastic indicators of severely low glucose manifest, including hot flashes, sweating episodes (e.g. night sweats), seizures, and panic attacks. Patients with ED who experience panic episodes should be screened for hypoglycemia as an etiology, especially while an event is occurring.

During severe hypoglycemia the brain can use ketones (a breakdown product of fat metabolism) and lactic acid (a breakdown product of anaerobic metabolism) to keep going, misleading physicians that "all is well." There is some evidence that both may be responsible for blunting the hypoglycemic reactions normally seen (Amiel, Archibald, Chusney, Williams & Gale, 1991; Nagy, O'Connor, Robinson & Boyle, 1993). Ketones can be detected in the urine and indicate an ongoing starvation state. In patients with chronic or severe AN, there may be very little fat to use, even for the brain. These patients are particularly in trouble because the two preferred fuel sources are minimally available.

Even severely low glucose levels may not trigger obvious symptoms and may be overlooked by patients and clinicians. As with other medical problems in ED, appearing "normal" does not mean these patients' brains are not seriously suffering from low glucose. I have seen patients with glucose levels below 30 mg/dl (normal is about 70–100 mg/dl) who have been completely asymptomatic. Another patient had a random serum glucose that was zero, and the lab confirmed its findings on different analyzers and after recalibration. Glucose levels can drop to where seizures or coma develop, but that level is variable depending on individual circumstances. Hypoglycemia can persist up to a few weeks after initiation of refeeding. Therefore, physicians must maintain a high level of suspicion or consider rechecking glucose levels frequently in the initial treatment phase.

GASTROINTESTINAL (GI) ASSESSMENT

Gastrointestinal complications associated with ED are listed in Box 6.5. In general, these complications are largely dependent on the specific ED symptoms employed.

Restriction

Food restriction ultimately causes the GI tract to become malnourished. The stomach and intestines slow down (*gastroparesis*) and have a more difficult time digesting food. Gastroparesis can be severe. Even a small amount of food may take a very long time to move out of the stomach. Patients who already feel psychologically uncomfortable with food then have a physical complication which intensifies and prolongs their discomfort. An upper GI, or an upright abdominal X-ray showing an enlarged gastric bubble, can confirm the diagnosis. Superior mesenteric artery (SMA) syndrome can replicate postprandial discomfort. The SMA is an artery near the duodenum and is normally prevented from pushing on the duodenum by a fat pad. However, in chronic malnutrition, the fat pad may atrophy to the point where it cannot prevent the compression of the duodenum by the artery, resulting in abdominal pain. Clinicians must distinguish what complaints are from gastroparesis, SMA syndrome, or the ED itself. The treatment for all causes is still refeeding.

BOX 6.5

GASTROINTESTINAL COMPLICATIONS FROM EATING DISORDERS

- Dental cavities, soreness or loss of teeth
- Sialadenosis (enlarged salivary glands)
- Gastroesophageal Reflux Disease (GERD)
- Foreign bodies (from toothbrushes, etc. swallowed while inducing purging)
- GI bleeding (can be severe, including rupture of the esophagus and stomach)
- Precancer and cancer of the esophagus
- Gastroparesis (delayed emptying of stomach) and general slowed motility of intestine
- Elevated liver function tests
- Superior mesenteric artery (SMA) syndrome
- Chronic constipation
- Intestinal paralysis from stimulant laxative abuse
- Pancreatitis, usually from purging or high lipid levels from binging
- Sinusitis
- Asthma (from GERD)

Purging

For those who purge, GI complications include sore teeth, dental cavities, sialadenosis (enlarged salivary glands), sinusitis, asthma, GI bleeding, and Gastroesophageal Reflux Disease (GERD). Bleeding can come from instrumentation of the mouth and pharynx to stimulate a gag reflex (e.g., using a toothbrush) or tears of the stomach and esophagus. These are not uncommon and are usually mild, but they can be very serious and life-threatening. Chronic vomiting, or signs or symptoms of bleeding, require medical evaluation. Frequent vomiting can also lead to laxity of the gastroesophageal sphincter. This laxity means that stomach acid, which is very strong, can irritate acutely and chronically as it washes back up the esophagus. The common complication of this is GERD. Over time, this acid may cause precancerous and, eventually, cancerous changes to the esophagus. There are case reports of BN patients who developed cancer of the esophagus (Navab, Avunduk, Gang & Frankel, 1996; Shinohara, Swisher-McClure, Husson, Sun & Metz, 2007).

Treatment with antacids, H-2 blockers, and especially proton pump inhibitors (PPIs) can alleviate the symptoms and consequences of GERD. PPIs may also help prevent or reverse any bleeding or ulcers in the stomach as a result of vomiting. Because PPIs decrease the acidity dramatically, patients often have fewer stomach complaints and fewer complications from the acidity on other structures. Chronic use of PPIs may somewhat increase risk of osteoporosis, a condition not uncommon in ED, so prolonged use may require caution.

Pancreatitis, presenting as upper abdominal pain, can develop from purging. This is an uncommon but potentially severe and life-threatening complication. Serum amylase should be elevated, but that can also be elevated from salivary amylase in those who purge. Specifying salivary amylase to the lab will distinguish how much of the value is from salivary or pancreatic origin. An elevated serum lipase will usually pinpoint the pancreas as the cause of the abdominal pain.

Laxative Abuse

Those who abuse stimulant laxatives can develop chronic and severe constipation. Constipation can also occur as a result of the low fluid intake that often occurs in restrictors. Adequate hydration is a mainstay of treatment of constipation. Increasing fluid intake and adding a non-stimulant laxative, such as polyethylene glycol (Miralax™), may be helpful.

SKIN ASSESSMENT

In addition to acrocyanosis and capillary refill delay discussed above, other findings can be seen in the skin. Fine downy hair, called lanugo, can develop on the trunk, arms, and face in undernourished patients as a method to maintain body heat. It will resolve as nutrition improves. Scalp hair can fall out from malnutrition and as it starts to reverse (the new, healthy hair pushes out the old). Cuts, abrasions, and calluses from using the hands to stimulate vomiting occur on the dorsum of the hand, called "Russell's sign," in honor of the first author on BN. Skin can also have pigment changes from excess intake of certain foods. A high proportion of carrots and pumpkin can cause an orange tint, and squash can cause a yellow discoloration. Conjunctival hemorrhages and swelling of the face and eyes can occur in those who vomit forcefully. Fortunately, all of these skin changes will resolve in time.

ENDOCRINE ASSESSMENT

Although any aspect of the endocrine system can be affected by ED, the most common involve effects on the thyroid and the sex hormones, especially as they relate to menstrual periods and osteoporosis.

Thyroid Function

Palpation of the thyroid gland in the undernourished reveals it to be atrophic and thin. Laboratory testing usually shows a normal T_4 and TSH, indicating that the thyroid is still active enough in the low energy state. The T_4 and T_3 may be low and the TSH levels may be variable (euthyroid sick syndrome), but should not be treated with thyroid medication, as that will only cause the body to burn up more calories that it is already trying to preserve. After refeeding, thyroid values will return to normal.

Amenorrhea

Until recently, amenorrhea was generally considered to be a hallmark for AN, preceding actual weight loss in a significant percent (Golden & Shenker, 1993). However, some critically low weight patients meet all the criteria for AN, but never lose their periods. Pregnancy is possible for these patients if sexually active, thus birth control should be maintained. Although amenorrhea is most associated with restricting and over-exercising symptoms, menstrual dysfunction occurs in all ED subtypes (Pinheiro et al., 2007). The leutinizing hormone and follicle stimulating hormone usually decline in AN. Serum levels of estradiol (estrogen) may be low or undetectable even in patients having periods, or taking birth control pills containing estradiol. Low estradiol will cause breast atrophy and loss of menses (usually), but loss of bone density is a more serious long-term consequence (see below). Similarly, for males, low testosterone will result in bone loss, which may develop more quickly than in females (Mehler, Sabel, Watson & Andersen, 2008). Therefore, baseline estradiol and testosterone levels should be respectively assessed for female and male restricting patients.

BONE ASSESSMENT

Osteopenia and Osteoporosis

Loss of bone density is an ominous consequence of ED that may not fully recover after weight and nutritional restoration. It appears that bone loss increases as energy reserves decline along with weight, especially when coupled with inherited traits. Dual Energy X-ray Absorptiometry (DEXA) measures bone mineral density and should be a part of every patient's initial ED medical evaluation. A DEXA scan is one of the most accurate ways to diagnose osteopenia or osteoporosis. Osteopenia is bone loss that is between 1 and 2.5 standard deviations (SD) below the mean for age and sex. Osteoporosis is ≥ 2.5 SD below the mean. Fractures and stress fractures of the feet, lower legs, and spine are the primary concerns of loss of bone density in ED. Hip fractures are the most serious risk from osteoporosis and carry a death risk of ≥20% in 1 year, higher than for most cancers (Jensen & Tondevold, 1979; Moran, Wenn, Sikan & Taylor, 2005). Clinicians must educate sufferers about the risks of bone loss and be aggressive with assessment and management.

For children and adolescents, loss of bone density may carry geometrically increased risk for several reasons:

1. Bone density peaks in the late teens and early twenties. Patients with childhood onset ED may not get the benefit of that protective density, thus reaching osteoporosis more quickly.
2. Early loss of bone density can cause growth delays or failures to reach maximum predicted height, another one of the irreversible effects of ED.
3. Children have decades more exposure to the risk of consequences.
4. Osteopenia has been found within 3 months of loss of menses and is related to the total number of missed periods. Especially in early adolescence, the lack of regular menses in those with ED may be mistaken as normal.
5. Medications to treat adult osteoporosis, such as bisphosphonates and selective estrogen receptor modulators, may not be appropriate for those of child-bearing age because of concerns regarding pregnancy outcome, although they have been used for years in children with other orthopedic problems without associated complications (Golden et al., 2005). Estrogen, helpful in older, traditional populations with osteoporosis, does not appear to have a significant effect on osteoporosis in ED (Katzman, 2005; Mehler, 2003).

Treatment includes: weight restoration as soon as possible; calcium and vitamin D supplementation in adequate amounts; weight bearing exercise when appropriate; and certain medications to prevent or increase bone density when appropriate.

Finally, malnutrition also suppresses the function of the bone marrow, which produces white blood cells (WBC), red blood cells (RBC), and platelets. A complete blood count (CBC) will identify any such deficiencies. Generally, RBC levels decrease first, then WBC, and lastly, platelets. As WBC decrease, the ability to fight infection may be impaired, putting patients at risk of further medical compromise. If platelets get too low, bleeding can occur that is hard to stop. Anemia will also occur, but it is usually mild until malnutrition is severe. Vegetarians and vegans are at greater risk of anemia if they are not meeting their nutritional iron needs.

RENAL ASSESSMENT

The kidneys also are not spared. The more common issues involve a decreased ability to concentrate urine, kidney stones in those who are chronically dehydrated, and chronic renal insufficiency. These are apparently due to either low BP and volume depletion (dehydration) or chronic hypokalemia (low potassium), or both, all of which injure the kidney's ability to work appropriately. The paradox is that the patient may even seem to drink excessive volumes of fluids but measured urine output may still be greater than intake. Clinicians need to distinguish this characteristic in ED patients from those who have psychogenic polydipsia (drinking excessive amounts for psychological reasons).

Creatinine levels may be relatively elevated. Creatinine is a breakdown product of muscle and is usually maintained at a constant level by the kidneys. If the kidneys are injured, then they will not filter as well and creatinine levels will climb. However, those with low muscle mass should have a low creatinine. Therefore, physicians should interpret what the creatinine level should be in the individual patient, as levels in the "normal" range may not be normal for a patient with AN. If the injuries to the kidneys are not reversed, chronic renal failure and dialysis may follow. The key to prevention is to correct dehydration and avoid low potassium by stopping vomiting, laxative use, and diuretics.

MEDICAL AND LABORATORY TESTS

Laboratory studies and other tests are usually decided upon after the full history and physical are performed. However, certain ones are needed fairly consistently and others are ordered depending on specific findings. Box 6.6 shows lists of those tests that are commonly needed at initial assessment and others that are often necessary.

REFEEDING SYNDROME

Another dangerous consequence of ED is the Refeeding Syndrome (RFS). It occurs as patients are refed after a period of starvation, and is related to how fast nutrition is restored.

BOX 6.6

RECOMMENDED MEDICAL AND LABORATORY TESTS FOR INITIAL ED EVALUATION

Blood tests
- Chemistry panel, including electrolytes, glucose, calcium (consider an ionized calcium), liver functions, BUN, creatinine, albumin, total protein
- CBC with differential
- Thyroid function tests
- 25-OH vitamin D
- Magnesium and phosphorus, especially in low weight individuals
- Estradiol in low weight individuals or those who have lost their menstrual periods
- Testosterone in malnourished males

Urinalysis with specific gravity
Electrocardiogram
Bone density test (DEXA scan)

Other tests as indicated:
- Chest X-ray
- Urine pregnancy test
- Ionized calcium
- Amylase
- Lipase
- FSH, LH, prolactin
- MRI of the brain

It can develop even in those who are not severely underweight, but who have had acute malnutrition. As the body rebuilds from refeeding, it uses up stores faster than it can replace them. As that occurs, certain electrolytes (phosphorus, magnesium, potassium) can be depleted suddenly. Effects of RFS include edema, breakdown of skeletal muscle (rhabdomyolysis), WBC dysfunction (thus, increasing risk of infection), respiratory failure, heart failure, arrhythmias, delirium, seizures, coma, and sudden death. Cardiac complications can occur early, and delirium can occur even after the second week of refeeding. Edema and delirium may be the only obvious signs at first. Potentially fatal brain complications have been reported during refeeding including Wernicke's encephalopathy (from low thiamine) and central pontine myelinolysis (from low phosphorus or sodium) (Catini & Howells, 2007; Patel, Matthews & Bruce-Jones, 2008). Physicians must anticipate this complication and do repeat exams and lab for those at risk. While training in management of electrolytes is common in medical schools and training programs, management in ED patients can be especially complicated because of the many co-morbid medical conditions.

In those at risk, the primary rule to refeed is "start low and go slow." While no consensus on caloric refeeding guidelines have been yet developed, gains of 1–2 pounds of non-water weight per week are generally acceptable. An initial limit of between 1000–1200 calories per day of balanced nutrition, with twice weekly increases of 250–300 calories, will avoid most complications from refeeding. In those at greatest risk, prophylaxis with phosphorus, magnesium, potassium, and thiamine may be warranted (see also Croll, Chapter 8).

RISK OF LIFE-THREATENING INFECTION

In the well-nourished person, the body demonstrates infection by fever, elevated WBC, and an elevated sedimentation rate (a simple test; elevation indicates an infection or inflammation is likely). In the severely malnourished, the body temperature may be as low as 92°F, the marrow is severely suppressed and cannot put out many WBCs, and the sedimentation rate will be blunted. Thus, the usual indicators of infection (fever, elevated white count, and sedimentation rate) may all be "normal." If that is the case, then an overwhelming infection (sepsis) can develop with little warning and with devastating consequences. Clinicians should be aware that severely malnourished ED patients are at risk and even a two to three degree rise of their baseline temperature may be the only warning of impending life-threatening infection.

BRIDGING THE GAP FOR MEDICAL ASSESSMENT

As previously noted, significant gaps remain *within* medical training programs with regard to ED treatment. Considering the scope and severity of potential problems in these illnesses, advanced training is necessary for better assessment and intervention. With few exceptions, medical professionals and non-medical clinicians must seek training beyond their primary professional education to be suitably versed in anticipating and managing the possible complications. Eating disorder professional organizations, such as the Academy for Eating Disorders and the International Association of Eating Disorder Professionals, and other advocacy organizations, such as the National Eating Disorders Association, are trying to promote training in medical schools and residency programs about ED to reverse this inadequacy.

Bridging the gap *between* ED treatment providers, and between providers and caregivers, is also of paramount concern, as a multidisciplinary treatment approach is clearly the most effective. Providing clinicians and caregivers with the necessary knowledge, language, and tools for expert medical management of ED will not only facilitate early identification and assessment of medical complications; it will decrease the personal and financial cost, improve recovery and, ultimately, save lives.

References

Agras, W. S., Crow, S., Mitchell, J., Halmi, K., & Bryson, S. (2010). A 4-year prospective study of eating disorder NOS compared with full eating disorder syndromes. *International Journal of Eating Disorders, 42*, 565−570.

American Academy of Pediatrics. (2003). Committee on Adolescence. Identifying and treating eating disorders: A policy statement. *Pediatrics, 111*, 204−211.

American Psychiatric Association. (1994). *Diagnostic and statistical manual for mental disorders* (4th ed.). Washington, DC: Author.

Amiel, S. A., Archibald, H. R., Chusney, G., Williams, A., & Gale, E. (1991). Ketones lowers hormone responses to hypoglycaemia: Evidence for acute cerebral utilization of a non-glucose fuel. *Clinical Science, 81*, 189−194.

Birmingham, C. L., & Beumont, P. (2004). *Medical management of eating disorders: A practical handbook for health care professionals*. Cambridge, UK: Cambridge University Press.

Birmingham, C. L., Su, J., Hlynsky, J., Goldner, E., & Gao, M. (2005). The mortality rate from anorexia nervosa. *International Journal of Eating Disorders, 38*(2), 143−146.

Bjørnstad, H., Storstein, L., Dyre Meen, H., & Hals, O. (1991). Electrocardiographic findings in athletic students and sedentary controls. *Cardiology, 79*, 290−305.

Catini, M., & Howells, R. (2007). Risks and pitfalls for the management of refeeding syndrome in psychiatric patients. *The Psychiatrist, 31*, 1209−1211.

Crow, S., Peterson, C., Swanson, S., Raymond, N., Specker, S., Eckert, E. D., & Mitchell, J. E. (2009). Increased mortality in bulimia nervosa and other eating disorders. *American Journal of Psychiatry, 166*, 1342−1346.

Dalle, G. R., & Clugi, S. (2007). Eating disorder not otherwise specified in an inpatient unit: The impact of altering the DSM-IV criteria for anorexia and bulimia nervosa. *European Eating Disorders Review, 15*(5), 340−349.

Eddy, K. T., Celio Doyle, A., Hoste, R. R., Herzog, D. B., & Le Grange, D. (2008). Eating disorder not otherwise specified in adolescents. *Journal of the American Academy of Child and Adolescent Psychiatry, 47*(2), 156−164.

Golden, N. H., Iglesias, E. A., Jacobson, M. S., Carey, D., Meyer, W., Schebendach, J., … Shenker , I. R. (2005). Alendronate for the treatment of osteopenia in anorexia nervosa: A randomized, double-blind, placebo-controlled trial. *Journal of Clinical Endocrinology & Metabolism, 90*(6), 3179−3185.

Golden, N., Katzman, D., Kriepe, R., Stevens, S., Sawyer, S., Rees, J., … Rome, E. S. (2003). Eating disorders in adolescents: Position paper of the Society for Adolescent Medicine. *Journal of Adolescent Health, 33*, 496−503.

Golden, N., & Shenker, R. (1993). Amenorrhea in anorexia nervosa. Neuroendocrine control of hypothalamic dysfunction. *International Journal of Eating Disorders, 16*(1), 53−60.

Guido, F., Bailer, U., Henry, S., Wagner, A., & Kaye, W. (2004). Neuroimaging studies in eating disorders. *CNS Spectrums, 9*(7), 539−548.

Jensen, J. S., & Tondevold, E. (1979). Mortality after hip fractures. *Acta Orthopaedica Scandinavica, 50*(2), 161−167.

Kalm, L. M., & Semba, R. D. (2005). They starved so that others be better fed: Remembering Ancel Keys and the Minnesota experiment. *Journal of Nutrition, 135*, 1347−1352.

Katzman, D. (2005). Medical complications in adolescents with anorexia nervosa: A review of the literature. *International Journal of Eating Disorders, 37*, S52−S59.

Lucas, A. R., Beard, C. M., O'Fallon, W. M., & Kurland, L. T. (1991). 50-year trends in the incidence of anorexia nervosa in Rochester, Minn: A population-based study. *American Journal of Psychiatry, 148*, 917−922.

Lask, B., & Frampton, I. (2009). Anorexia nervosa: Irony, misnomer and paradox. *European Eating Disorders Review, 17*(3), 165–168.

Madden, S., Morris, A., Zurynski, Y., Kohn, M., & Elliot, E. (2009). Burden of eating disorders in 5–13-year-old children in Australia. *Medical Journal of Australia, 190*(8), 410–414.

Mehler, P. (2003). Osteoporosis in anorexia nervosa: Prevention and treatment. *International Journal of Eating Disorders, 33,* 113–126.

Mehler, P., Sabel, A., Watson, T., & Andersen, A. (2008). High risk of osteoporosis in male patients with eating disorders. *International Journal of Eating Disorders, 41*(7), 666–672.

Moran, C. G., Wenn, R. T., Sikan, M., & Taylor, A. M. (2005). Early mortality after hip fracture: Is delay before surgery important? *The Journal of Bone and Joint Surgery (American), 87,* 483–489.

Nagy, R., O'Connor, A., Robinson, B., & Boyle., P. J. (1993). Hypoglycemia unawareness results from adaptation of brain glucose metabolism. *Diabetes, 42*(Suppl. 1), 133A.

Navab, F., Avunduk, C., Gang, D., & Frankel, K. (1996). Bulimia nervosa complicated by Barrett's esophagus and esophageal cancer. *Gastrointestinal Endoscopy, 44*(4), 492–494.

Neumärker, K.-J. (1998). Mortality and sudden death in anorexia nervosa. *International Journal of Eating Disorders, 21*(3), 205–212.

Nolan, J., Morley, P., Vanden Hoek, T., & Hickey, R. (2003). Therapeutic hypothermia after cardiac arrest. An advisory statement by the advanced life support task force of the International Liaison Committee on Resuscitation. *Circulation, 108,* 118–121.

Patel, A., Matthews, L., & Bruce-Jones, W. (2008). Central pontine myelinolysis as a complication of refeeding syndrome in a patient with anorexia nervosa. *Journal of Neuropsychiatry and Clinical Neuroscience, 20,* 371–373.

Pinheiro, A. P., Thornton., L. M., Plotonicov, K. H., Tozzi, F., Klump, K., Berrettini, W. H., … Bulik, C. M. (2007). Patterns of menstrual disturbance in eating disorders. *International Journal of Eating Disorders, 40,* 424–434.

Riggs, S., Harel, D., Biros, P., & Ziegler, J. (2003). Cardiac impairment in adolescent girls with anorexia nervosa: What exercise stress testing reveals. *Journal of Adolescent Health, 32*(2), 126.

Shinohara, E., Swisher-McClure, S., Husson, M., Sun, W., & Metz, M. (2007). Esophageal cancer in a young woman with bulimia nervosa: a case report. *Journal of Medical Case Reports, 1*(160), 1–4.

Siegel, D. (1999). *The developing mind: Toward a neurobiology of interpersonal experience.* New York, NY: Guilford Press.

Silber, T., Maine, M., & McGilley, B. (2008). Academy for Eating Disorders position statement: Over-the-Counter status of ipecac should be withdrawn. Retrieved from. http://aedweb.org/policy/ipecac.cfm

Storstein, L., Bjørnstad, H., Hals, O., & Dyre Meen, H. (1991). Electrocardiographic findings according to sex in athletes and controls. *Cardiology, 79,* 227–236.

Sullivan, P. F. (1995). Mortality in anorexia nervosa. *American Journal of Psychiatry, 152,* 1073–1074.

Zerbe, K. (2008). *Integrated treatment of eating disorders: Beyond the body betrayed.* New York, NY: W.W. Norton & Company.

Zhao, Y., & Encinosa, W. (2009, April). *Hospitalizations for eating disorders from 1999 to 2006. (Statistical Brief #70.) Healthcare Cost and Utilization Project (HCUP).* Rockville, MD: Agency for Healthcare Research and Quality. http://www.hcup-us.ahrq.gov/reports/statbriefs/sb70.pdf

CHAPTER 7

Psychiatric Medication
Management, Myths, and Mistakes

Martha M. Peaslee Levine and Richard L. Levine

Eating disorders (ED) can be considered "jigsaw puzzles" with multiple contributing factors and ensuing health issues. Treatment can be its own puzzle especially with the diverse range of psychiatric medications typically used to address these conditions. While medications are only part of the necessary multidisciplinary treatment, they are an important component, with the potential to minimize symptoms, improve recovery, and limit relapse.

Facets of a patient's ED dictate which classes of medications should be used; comorbidities and family history also influence choices. Decisions can change if patients migrate from one diagnostic category to another. While research and clinical experience provide insights into treatment decisions, there is often a gap between these fields. This chapter works to bridge that gap. We will examine each diagnosis and its psychiatric comorbidities, and offer a guide so practitioners can use all available and applicable medications to craft the picture of recovery.

ANOREXIA NERVOSA: OVERVIEW OF PSYCHIATRIC MEDICATION MANAGEMENT

Diagnostic Characteristics

Anorexia nervosa (AN) is defined by severely low weight, intense fear of gaining weight, disturbed perception of body shape and weight, and amenorrhea (American Psychiatric Association, 2000). AN is potentially life-threatening (Attia & Schroeder, 2005) with increased death rates, as compared to the general population, ranging from a six-fold (Papadopoulos, Ekbom, Brandt & Ekselius, 2009) to a ten-fold increase (Button, Chadalavada & Palmer, 2010). Unfortunately, no medication has been proven to have a definitive impact on weight gain or the psychological features of AN (Attia & Schroeder, 2005; Bulik, Berkman, Brownley, Sedway & Lohr, 2007; Guarda, 2008).

Psychiatric Comorbidities

Anorexia nervosa has many associated psychiatric comorbidities; treatment of these conditions can facilitate recovery. In one study, 97% of female inpatients with ED exhibited one or more Axis I comorbid diagnoses, with the most common associations being mood (94%), anxiety (56%), and substance use (22%) (Blinder, Cumella & Sanathara, 2006). Papadopoulos et al. (2009) reported 20 to 30% of deaths from AN resulted from suicide. Alcohol misuse has also been identified as a potentially fatal issue in patients with AN (Papadopoulos et al., 2009; see also Baker Dennis and Helfman, Chapter 14).

Depression, with decreased appetite and weight loss, may become so entwined with AN that it is difficult to distinguish which behavior initiated the downward spiral and how both perpetuate it. Depression adds to the risk for osteoporosis, a concern in AN (Konstantynowicz et al., 2005).

Medication Guidelines: Antidepressants

For the sake of readers unfamiliar with the generic names of psychiatric medications, Table 7.1 provides a list of the medications included in this chapter with their associated brand name.

Selective serotonin reuptake inhibitors (SSRIs). SSRIs offer a reasonable therapeutic profile for depression and anxiety without significant side effects. Choosing an SSRIs can be based on many factors—past response, family history, or insurance coverage. Cost is important; we know patients who have been discharged from inpatient stays with such large co-pays, could not afford the medication. Of the SSRIs, fluoxetine can be more activating and needs to be used cautiously in anxious patients (Stahl, 2005). Other medications in this class include sertraline, citalopram, and escitalopram. The US Food and Drug Administration (http://www.fda.gov/ForConsumers/ConsumerUpdates/ucm048950.htm) warns that antidepressants can trigger suicidal thinking and behavior in young adults, the principal population of AN, during the initial two months of treatment.

Serotonin and norepinephrine reuptake inhibitors (SNRIs). Venlafaxine and duloxetine can be as effective as SSRIs (Ricca et al., 1999).

Tetracyclic antidepressants. Mirtazapine has been used in treating adolescent AN with the thought that the anxiolytic or antidepressant effect improves patient compliance and facilitates psychotherapy (Hrdlicka, Beranova, Zamecnikova & Urbanek, 2008).

Tricyclic antidepressants (TCAs) and monoamine oxidase inhibitors (MAOIs). TCAs can place patients at risk for medical complications. They lower blood pressure, which can affect orthostatic hypotension (Fluoxetine Bulimia Nervosa Collaborative [FBNC] Study Group, 1992) and can prolong the QT interval in a population already experiencing conduction abnormalities (Tousoulis et al., 2009; Lock & Fitzpatrick, 2009). MAOIs are often more effective in atypical or treatment-resistant depression (Fiedorowicz & Swartz, 2004). However, the food restrictions associated with a low tyramine diet may fuel ED behaviors in AN. Adverse reactions to TCAs and MAOIs are more pronounced in malnourished patients (American Psychiatric Association, 2006).

Bupropion. The FDA has warned against using bupropion in any formulation in patients with a current or previous diagnosis of AN or BN because of a higher incidence of seizures

TABLE 7.1 Generic and Trade Names of Medications

Generic	Trade name
Fluoxetine	Prozac
Sertraline	Zoloft
Citalopram	Celexa
Escitalopram	Lexapro
Venlafaxine	Effexor
Duloxetine	Cymbalta
Mirtazapine	Remeron
Bupropion	Wellbutrin
Fluvoxamine	Luvox
Paroxetine	Paxil
Buspirone	Buspar
Sildenafil	Viagra
Quetiapine	Seroquel
Olanzapine	Zyprexa
Risperidone	Risperdal
Aripiprazole	Abilify
Topiramate	Topamax
Lamotrigine	Lamictal
Ondansetron	Zofran
Sibutramine	Meridia

noted in patients with BN treated with this medication (PDRHealth, 2009). Adhering to this warning is important, as evidence shows that the majority of women with AN experience diagnostic crossover to other diagnostic categories, including BN (Eddy et al., 2008).

Atypical Antipsychotics

While the term "antipsychotic" can frighten some patients and families, the reasons for this treatment become clearer if the distorting nature of ED thoughts is explained. Studies demonstrate the benefit of atypical antipsychotics in AN. Improvement includes decreased delusional thoughts related to food and weight, and decreased eating-related anxiety (Bosanac et al., 2007; Couturier & Lock, 2007; Powers, Bannon, Eubanks & McCormick, 2007). Choosing an atypical antipsychotic can be based on side effects. If patients have difficulty sleeping, a more sedating medication, such as quetiapine can be helpful (Stahl, 2005). Low dose olanzapine (La Via, Gray & Kaye, 2000) and risperidone (McElroy et al., 2006) have been effective in patients with AN, especially with treatment-refractory illness.

Aripiprazole is less sedating and has a lower weight gain side effect profile (Stahl, 2005). It can be useful in patients who would benefit from an atypical antipsychotic but refuse to take medications that might trigger weight gain. It is often given in the morning because it can be more activating (i.e., energizing) than other atypical antipsychotics (Stahl, 2005). Patients should be monitored for any increase in jitteriness. Excess exercise is a hallmark of AN, with patients pacing the halls, taking the stairs, and jiggling during group sessions, all in an attempt to burn off calories. If this behavior increases, akathesia (an uncomfortable restless feeling) should be assessed.

Atypical antipsychotics are associated with an increased risk of type II diabetes. This can be related to increased food intake, particularly carbohydrate craving, and excessive weight gain (Bailey, 2003). Hyperlipidemia and hyperglycemia, unrelated to weight gain, may occur (Gardner, Baldessarini & Waraich, 2005); blood sugar should be monitored. One patient's hyperglycemia resolved (despite staying on the medication) after weight restoration, suggesting a role of malnutrition (Yasuhara, Nakahara, Harada & Inui, 2007).

Alternative Pharmacologic Agents

Anxiolytics. Short-acting benzodiazepines before meals have been tried. The literature supports our clinical experience regarding the risks of this practice because of the long-term nature of the illness and the comorbidity of substance abuse in this population (Krüger & Kennedy, 2000). Buspirone, while not useful for acute anxiety, can augment antidepressant and antianxiety effects for patients on SSRIs (Trivedi et al., 2006).

Dronabinol. Rarely clinicians put patients on the oral formula of Δ^9-THC, dronabinol, to stimulate appetite. One study (Gross et al., 1983) found this did not enhance weight gain and led to severe depressive reactions with paranoid ideation.

Micronutrient Supplementation

The majority of patients, even with significant nutritional and hormonal abnormalities, do not demonstrate severe vitamin deficiencies (Castro, Deulofeu, Gila, Puig & Toro, 2004). However, one of our patients with substantial weight loss developed symptoms of Wernicke's encephalopathy—confusion, ataxia, and nystagmus—related to thiamine deficiency (Peters, Parvin, Petersen, Faircloth & Levine, 2007). Pellagra, with skin changes related to sunlight, has been reported (Prousky, 2003). Although these conditions are rare, they are potentially fatal. In our clinic, we recommend a multivitamin/multi-mineral preparation, thiamine 100 mg daily (for severe malnutrition), and calcium with vitamin D.

Combining Treatments

Antidepressants and atypical antipsychotics often need to be combined. The SSRIs treat depression, anxiety, and both obsessive-compulsive and bulimic symptoms. Atypical antipsychotics maximize treatment by affecting resistance to weight gain, severe obsessional thinking, and denial that becomes delusional (American Psychiatric Association, 2006).

Recovery from AN is best viewed as a two-stage process of weight restoration followed by relapse prevention (Guarda, 2008).

Treatment Considerations

High dropout rates from treatment are often seen with AN (Berkman et al., 2006). This may be related to the entrenched mindset that becomes the hallmark of the ED. Clinical practice suggests that patients who have some motivation to relinquish the ED respond better to medications. They have, on some level, begun to acknowledge the extent of their illness. If, however, their anorexic mindset and behaviors are ego-syntonic, and their illness becomes integrated into their identity, recovery is a significant challenge (Crow, Mitchell, Roerig & Steffen, 2009a). Patients' people-pleasing tendencies may make it nearly impossible for them to admit that they are not taking their medication. While patients may not admit noncompliance to the prescriber, they often will tell another team member whom they perceive as having less of a stake in the issue.

In our experience, SSRIs are useful to treat associated depressive, anxiety, or obsessive-compulsive disorder (OCD) symptoms. Atypical antipsychotics appear to help with ED thoughts. We find it most effective to start at low doses, and gradually increase the dose depending on clinical assessment, side effects, continued symptoms, and treatment response. Unlike patients with schizophrenia who require higher doses of the antipsychotics, low doses are often all that are required to affect the ED beliefs. We usually start by halving the lowest pill in order to minimize side effects. Clinically, AN patients appear sensitive to side effects because of their intense focus on their body.

Some studies suggest that fluoxetine can help prevent relapse (Kaye, Weltzin, Hsu & Bulik, 1991; Kaye, Nagata, Weltzin, Hsu, Sokol, McConaha, 2001), while other studies contradict these results (Walsh et al., 2006). The antidepressant effect is limited if a patient is very malnourished (Ferguson, La Via, Crossan & Kaye, 1999). The SSRIs achieve their benefit by preventing reuptake of serotonin. Without enough raw materials in the body to form the neurotransmitters (e.g., tryptophan, the precursor for serotonin), the medications cannot work effectively.

Length of treatment. Patients with a positive response to antidepressant treatment should remain on it for a minimum of 6 to 12 months, and longer if relapse has occurred in the past (Heiden, De Zwaan, Frey, Presslich & Kasper, 1998). This prevention goal should be discussed with patients. Frequently, when mood and anxiety symptoms remit, and patients are feeling better, they may minimize symptoms and prematurely stop the medication. It is important to inquire about medication compliance when patients undergo precipitous declines in functioning. Lastly, AN patients often have difficulties with change. It is not advisable to discontinue their medication during times of significant transition (e.g., going away to school).

BULIMIA NERVOSA: OVERVIEW OF PSYCHIATRIC MEDICATION MANAGEMENT

Diagnostic Characteristics

Bulimia nervosa (BN) is defined by recurrent episodes of binge eating occurring on a regular basis with compensatory behavior to prevent weight gain (American Psychiatric

Association, 2000). Elevated mortality related to suicide and other medical causes is seen in this group (Crow et al., 2009b).

Psychiatric Comorbidities

Lifetime histories of major depression (68%), OCD (25%), generalized anxiety disorder (11%), alcohol dependence (32%), and other substance dependence (21%) were found in a group recovered from BN (Kaye et al., 1998). Clinically, we have seen many patients who suffer from the "triple Bs" of BN, bipolar disorder and borderline personality disorder.

Medication Guidelines: Antidepressants

Selective serotonin reuptake inhibitors. The low side effect profile makes the SSRIs the most commonly prescribed class of antidepressants. Effective treatment of bulimic symptoms requires higher doses of SSRIs than typically used for depression (Krüger & Kennedy, 2000). Fluoxetine is the most widely studied SSRIs, and is the only one which has received FDA approval for the treatment of BN. Even in the absence of depressive symptoms, fluoxetine is beneficial (American Psychiatric Association, 2006). Fluoxetine 60 mg per day decreases core bulimic symptoms (i.e., binging and purging) and associated psychological symptoms (Shapiro et al., 2007). Depression, carbohydrate craving, and pathological eating attitudes and behaviors improve significantly with fluoxetine (FBNC Study Group, 1992). This higher dose also contributes to reduced relapse at the 1 year mark (Shapiro et al., 2007).

Clinically, we start with fluoxetine at 20 mg and adjust to 60 mg depending on response. If patients or their families have history of bipolar illness, a slow titration may be advisable due to the risk of potentially triggered hypomanic symptoms. Patients with BN and mild depression may experience symptom improvement (i.e., reduced vomiting) taking only 20 mg of fluoxetine (Goldstein, Wilson, Ascroft & Al-Banna, 1999).

Sexual dysfunction associated with SSRIs usually occurs early in treatment, and rarely remits spontaneously (Nurnberg et al., 2008). Patients requiring higher doses of SSRIs are more apt to experience sexual side effects. Sildenafil (50–100 mg), taken 1 to 2 hours before anticipated sexual activity, demonstrated a reduction in adverse sexual effects in women taking SSRIs (Nurnberg et al., 2008).

Serotonin and norepinephrine reuptake inhibitors. Venlafaxine has been helpful with depressive symptoms and binging and purging. Clinically, one patient on venlafaxine for depression was switched to fluoxetine when her bulimic symptoms were identified. She developed suicidal ideation on the fluoxetine and had to be hospitalized. When she was placed back on her venlafaxine, her suicidal ideation resolved. Her venlafaxine was increased to a higher therapeutic dose and her bulimic symptoms improved.

Tricyclic antidepressants and monoamine oxidase inhibitors. Tricyclic antidepressants have been shown to work in the treatment of BN, but their use is not recommended because of the side-effect profile (Krüger & Kennedy, 2000). Anticholinergic, sedative, and postural hypotensive effects are commonly described in studies with TCAs (FBNC Study Group, 1992). MAOIs are not recommended as first-line treatments (Krüger & Kennedy, 2000). Depending on the binge food, if high tyramine foods are eaten, MAOIs could place patients at risk for a hypertensive crisis (Fairburn, 2008).

Bupropion. As previously noted, bupropion is contraindicated because of associated seizures in patients with BN (American Psychiatric Association, 2006; Horne et al., 1988; PDRHealth, 2009). Some clinicians interpret this guideline as not recommending bupropion for BN unless it is combined with anticonvulsant medication (Krüger & Kennedy, 2000). The challenge comes when BN patients come into treatment and have had resolution of their depression only with bupropion, and experience side effects with the other classes of antidepressants. Clearly, the clinician needs to weigh all the evidence, fully inform the patient of the heightened risks and implications of experiencing a generalized seizure, consider safer alternatives, and make a collaborative decision in the overall best interests of the patient.

Anticonvulsants

Topiramate. This anticonvulsant is effective in the treatment of binge eating disorder (BED) and BN (Arnone, 2005). Weight loss is a potential drawback; topiramate is not indicated in patients who are at a low weight, already struggle with restricting behaviors, or have a clear pattern of restricting with significant weight loss in their history (Nickel et al., 2005). For example, a patient was struggling with binging and purging even on fluoxetine 60 mg. When topiramate was added, her symptoms shifted to significant restricting, and she dropped a concerning amount of weight. Topiramate is best used with patients suffering from BN who are average to slightly above average in the weight range.

Topiramate, by report and through observation, has tolerable side effects, especially with slow titration. Clinically, we have had patients complain of "feeling spacey" if the medication is increased too quickly. Another patient demonstrated word-finding difficulties. These side effects resolved when the dose was decreased and titration slowed. Other observed adverse events are paresthesias and headache (Arnone, 2005).

We typically start topiramate 25 mg per day or, at the most, 25 mg twice a day and increase by 25 mg per day every 5 to 7 days until symptoms resolve. Improvement has been demonstrated at doses between 50 and 100 mg twice a day. Nickel et al. (2005) used up to 250 mg per day. As with all medications, it is important to make sure patients are not purging soon after taking the medication. The therapeutic effect of topiramate may not occur until weeks into therapy.

Concerns have been raised about topiramate possibly compromising the efficacy of oral contraceptive agents (Rosenfeld, Doose, Walker & Nayak, 1997). It appears to only have an effect on oral contraceptive agents at doses greater than 200 mg/day (Crawford, 2009). Control of bulimic symptoms typically occurs at or below 200 mg/day.

Mood Stabilizers

Mood stabilizers are thought to be useful in BN because of the impulsive nature of the illness (McElroy, Kotwal & Keck, 2006). Mood stabilizers can be especially helpful when bipolar disorder is associated with BN. Lithium and valproic acid tend to induce weight gain in patients, which may not be acceptable. Lithium is particularly not recommended for patients with BN because of potential toxicity if patients continue with purging behavior (American Psychiatric Association, 2006). We treated one patient who had AN with purging features and became toxic on a relatively low dose of lithium. If lamotrigine is used, slow titration is important to reduce the risk of life-threatening rash (Stahl, 2005).

Atypical Antipsychotics

These medications can be used to treat mood swings and impulsivity related to bipolar disorder and so may have an occasional role in this population. As noted in the section about AN, aripiprazole has a lower weight gain profile (Stahl, 2005), and may be the best choice. Clinically, we have seen individuals with bulimia who have experienced increased binging related to atypical antipsychotic use.

Alternative Pharmacologic Agents

Naltrexone. This medication has been studied in AN (bulimic subtype) and BN because of its effects on opiate blockade. Marrazzi, Bacon, Kinzie & Luby (1995) view ED in an auto-addiction model; naltrexone allows patients to control their addictive eating behaviors with minimal side effects. Although the APA practice guidelines (American Psychiatric Association, 2006) question the efficacy of naltrexone for bulimic symptoms, they support its efficacy in treating substance abuse (opiate and alcohol), which co-occurs in one-third of the ED population. Our clinical experience is consistent with research demonstrating that naltrexone is also effective for reducing self-injurious behavior associated with ED (Griengl, Sendera & Dantendorfer, 2001; Sonne, Rubey, Brady, Malcolm & Morris, 1996; Svirko & Hawton, 2007; see also Dennis and Wickstrom, Chapter 16). The typical naltrexone dose is 50 mg/day, although a higher dose (up to 200 mg/day) may be needed to treat self-harm (Casner, Weinheimer & Gualtier, 1996). Dose-related hepatotoxicity has been shown, and thus liver function tests should be monitored (Williams, 2005).

Ondansetron. Fung and Ferrill (2001) studied ondansetron, a selective serotonin receptor antagonist used to prevent nausea and vomiting associated with chemotherapy, radiotherapy, and postoperative care. The proposed mechanism of action is correcting abnormal vagal activity which influences meal termination and satiety (Faris et al., 2000). One challenge is the dosing schedule because of the medication's short half-life (Faris et al., 2000). This, and the expense of the medication, may be reasons why it is not typically used clinically with the ED population.

Stimulants. Bulimia nervosa patients with concurrent attention-deficit/hyperactivity disorder (ADHD) respond well to stimulants, such as methylphenidate (American Psychiatric Association, 2006). One must be clear about the diagnosis and distinguish it from substance abuse and bipolar disorder. One patient who presented with BN often binged as a way to focus her attention at work. With a trial of lisdexamfetamine, her concentration and bulimic symptoms improved.

Micronutrient Supplements

Little is known about micronutrient supplements in the treatment of BN. Inositol, unofficially referred to as "vitamin B_8" has been studied in ED, panic disorder, and depression with improvement in symptoms and limited side effects (Gelber, Levine & Belmaker, 2001).

Combining Treatments

Combining medications may be necessary to help treat not only the ED symptoms, but also associated mood and attention symptoms. If binging and purging symptoms do not fully resolve on higher doses of antidepressants, topiramate may need to be added. Naltrexone can be useful in patients with combined bulimic symptoms and substance abuse or self-injury. Mood swings or irritability may suggest the addition of a mood stabilizer.

Treatment Considerations

While studies demonstrate the benefits of medication, the best results are with medication and psychotherapy, specifically cognitive behavior therapy (CBT) (Fairburn, 2008; Goldbloom, 1997). In direct comparisons of their relative therapeutic merits, patients receiving CBT showed more improvement in their ED symptoms than those receiving only medication (American Psychiatric Association, 2006; Berkman et al., 2006; McGilley & Pryor, 1998). Conversely, medication has been effective in patients who have not responded to psychotherapy. Walsh et al. (2000) assigned patients who had failed either CBT or interpersonal psychotherapy (IPT) to a trial of either placebo or fluoxetine with statistically significant results favoring fluoxetine.

Sometimes patients do not remain on medication as they start to recover. When their level of suffering decreases, their commitment to pharmacologic treatment also decreases. However, if BN symptoms persist or resume, the physiologic effects of disordered eating appear to maintain the core features of the disorder, resulting in a self-perpetuating cycle (McGilley & Pryor, 1998). Discussing the potential for non-compliance throughout treatment can be helpful.

Length of Treatment

With response to medication, antidepressant therapy should continue for a minimum of 9 months and probably a year for most patients with BN (American Psychiatric Association, 2006). This time period would start after resolution of ED symptoms; patients need to understand the importance of staying on their medications even if they are no longer symptomatic. When treating with antidepressants, it is important to remember that most antidepressants take weeks to demonstrate complete effectiveness. Side effects often occur early and then decrease, but the therapeutic effect requires more time (Stahl, 2005). If patients are tolerating the medication, a trial should not be discontinued prematurely or before maximum dosage has been reached. However, clinically we have had success in changing antidepressants if, even after achieving the full therapeutic dose, a patient has not demonstrated a decrease of binging and purging after 3 to 6 months.

EATING DISORDERS NOT OTHERWISE SPECIFIED: OVERVIEW OF PSYCHIATRIC MEDICATION MANAGEMENT

Diagnostic Criteria

Eating Disorders Not Otherwise Specified (EDNOS) is the category for disorders that do not meet the criteria for any other specific ED (American Psychiatric Association, 2000). Some patients may have symptoms similar to those of AN or BN, but which do not meet all the criteria. Other specific symptoms fall within EDNOS. One is BED, characterized by binges without the compensatory behaviors characteristic of BN (American Psychiatric Association, 2000). Sleep-related eating disorder (SRED) involves recurrent episodes of eating after an arousal from sleep and can lead to not only bingeing, but ingesting bizarre substances (Howell, Schenck & Crow, 2009). Another variant of BED is night eating syndrome (NES), characterized by patients overeating in the evening and subsequent morning anorexia. In NES, food intake is lower in the first half of the day and greater in the evening and nighttime (O'Reardon et al., 2006). In EDNOS, mortality rates are elevated from suicide and other causes (Crow et al., 2009b).

Psychiatric Comorbidities

A sizeable number of overweight teens who struggle with binge eating have higher negative mood scores (Glasofer et al., 2007). Patients with BED demonstrate elevated risk for depression, generalized anxiety disorder, panic attacks, past suicide attempts, and probable alcohol use disorder (Wonderlich, Gordon, Mitchell, Crosby & Engel, 2009). Obesity without BED does not have significant associations with these disorders (Grucza, Przybeck & Cloninger, 2007). Yet studies note that major depression and mood disorders are common in patients with obesity, occurring in 20% to 60% of women 40 years or older with a BMI >30 kg/m^2 (Lau et al., 2007). The presence of a mood disorder may negatively affect adherence to weight management interventions (Lau et al., 2007). Allison, Grilo, Masheb & Stunkard (2007) found a higher proportion of both BED and NES groups reported neglect and emotional abuse compared with overweight/obese controls. Comorbidity rates in BED are similar to those of other ED, and are not explained by obesity alone (Wonderlich et al., 2009).

Medication Guidelines: Antidepressants

Selective serotonin reuptake inhibitors. With the high comorbidity of mood disturbance and depression in BED, antidepressant treatment should be considered. As described in the section on BN, fluoxetine is useful in reducing binges and can be helpful in BED. Fluvoxamine has been found to reduce the frequency of binges in BED and to decrease body mass index regardless of the presence of depression (Hudson et al., 1998). The SSRIs led to greater reductions in binge eating and psychiatric symptoms in BED patients when compared to placebo (Brownley, Berkman, Sedway, Lohr & Bulik, 2007). Sertraline has been found to be effective in treating NES (O'Reardon et al., 2006). Some patients demonstrated a rapid response (within 2 weeks) and others a more gradual response (between 4 to 8 weeks). Sertraline is also effective for post-traumatic stress disorder (PTSD) symptoms (Davidson, Rothbaum, Van der Kolk, Sikes & Farfel, 2001). Many ED patients have experienced trauma in their lives and often have intrusive thoughts, flashbacks, and nightmares. Finally, of the SSRIs, paroxetine is associated with the highest weight gain, and thus it may not be the ideal front line treatment choice for BED (Stahl, 2005).

Tricyclic antidepressants (TCAs) and monoamine oxidase inhibitors (MAOIs). The TCAs may help in the short-term reduction of binge eating episodes (Krüger & Kennedy, 2000), but the effect may not endure. Indeed, some studies show weight increases and increased preference for sweets in long-term treatment with TCAs (Berken, Weinstein & Stern, 1984). Antidepressant action did not appear essential for the benefit of TCAs on reducing binge eating (McCann & Agras, 1990). The therapeutic action may be in reducing appetite rather than elevating mood. The MAOIs are similarly contraindicated with BED due to issues of dietary restrictions and risks for a hypertensive crisis if the diet is not followed.

Anticonvulsants

Topiramate. Topiramate decreases binge eating behavior, obesity, and other associated features in BED (McElroy et al., 2007a). We found it effective in reducing bingeing in a morbidly obese woman. As previously discussed in BN, slow titration should be followed to limit side effects. Topiramate may be effective in the treatment of SRED (Howell et al., 2009). Sleep-related EDs are more common than typically realized, especially in patients who suffer from daytime ED (Winkelman, Herzog & Fava, 1999).

Zonisamide. Zonisamide was found to decrease binge eating episodes in BED patients, but was not well-tolerated (McElroy et al., 2007b).

Alternative Pharmacologic Agents

Sibutramine. Sibutramine is a schedule IV controlled serotonin and norepinephrine reuptake inhibitor and has demonstrated effectiveness in decreasing binge eating, eating pathology (disinhibition and hunger), and depressive symptoms (Appolinario et al., 2003; Wilfley et al., 2008). The medication was well tolerated in studies, but some patients experienced small but significant increases in pulse and blood pressure. For this reason, the authors recommended trying the lower dose of 10 mg as opposed to the studied dose of 15 mg (Wilfley et al., 2008). Yager (2008) advocates using the studied dose of 15 mg/day, but carefully monitoring blood pressure and modifying the treatment as needed.

Other agents. Baclofen, a GABA-B receptor agonist, has been shown in some small samples to reduce the binge eating by at least 50%, to reduce food cravings and to be well-tolerated (Broft et al., 2007). Memantine, an NMDA receptor antagonist, demonstrated reduced binge behavior, improved global functioning, and was well-tolerated (Brennan et al., 2008).

Combining Treatments

Topiramate might be considered a first-line therapy for BED patients with obesity, along with CBT, IPT, SSRIs, and possibly sibutramine and orlistat (McElroy et al., 2007a). Evaluation of BED treatment studies suggest that there is a greater reduction in binge eating symptoms when receiving targeted ED psychotherapies such as CBT or IPT (Wonderlich et al., 2009). Clinically, we start with SSRIs combined with psychotherapy as our initial treatment. We are using topiramate more frequently as adjunctive treatment if bingeing has not resolved with the therapeutic dose of SSRIs. Frequently, we see patients whose bingeing did not improve after treatment with a number of SSRIs, in which case we often go directly to topiramate.

Treatment Considerations

Leombruni, Gastaldi, Lavagnino & Fassino (2008) questioned whether BED made patients less tolerant of side effects. This raises an interesting question as to whether patients who work to numb emotional discomfort with food may find elements of treatment intolerable. It is helpful to discuss common side effects with patients, advising them that they often diminish over time and that their clinician will adjust medications if the side effects remain intolerable.

For SRED, patients' current medications should be reviewed to see if they are affecting sleep or increasing appetite. Zolpidem (as well as other sedative hypnotics) has been associated with this condition; patients are either amnesic or partially aware of the eating when reflecting on a disrupted sleep (Najjar, 2007). In such cases, the offending medication should be stopped.

MEDICATION MANAGEMENT: BRIDGING THE PUZZLING GAPS

This chapter has "pieced together" the salient psychopharmacological literature on ED. To summarize, the most important puzzle pieces to remember include:

- In AN, antidepressants, particularly SSRIs, can be helpful in treating the psychiatric comorbidities and may help prevent relapse
- Atypical antipsychotics can help patients to manage ED thoughts
- In BN, higher doses of SSRIs are needed to treat bulimic symptoms. Unresolved bulimic behaviors can be treated with topiramate. Mood stabilizers may be necessary because of comorbid bipolar disorder
- Naltrexone is helpful for both ED behavior and self-injury, which is often seen in this impulsive group
- For EDNOS, guidelines should be followed for AN or BN depending on the predominant symptoms
- For BED, SSRIs or anticonvulsants may be beneficial
- Bupropion should not be used with any of these conditions because of the risk of seizures.

Eating disorder treatment is a challenging psychiatric puzzle, requiring management of both eating symptoms and other psychiatric comorbidities. Luckily, some medications used to treat specific ED also target mood disorders, anxiety, PTSD, and other related conditions. Within this work, it is vital to develop good rapport with patients, take an in-depth history, and be open to adjusting the pieces of the puzzle as new issues come to light. These are complicated conditions, but piecing together these many disparate elements into a picture of recovery is well-worth the effort.

References

Allison, K. C., Grilo, C. M., Masheb, R. M., & Stunkard, A. J. (2007). High self-reported rates of neglect and emotional abuse, by persons with binge eating disorder and night eating syndrome. *Behavior Research and Therapy*, 45(12), 2874–2883.

American Psychiatric Association (2000). *Diagnostic and statistical manual of mental disorders. Washington DC* (4th ed., text rev). Author.

American Psychiatric Association (2006). *Practice guideline for the treatment of patients with eating disorders.* Available online at: http://www.psychiatryonline.com/pracGuide/pracGuideTopic_12.aspx doi:10.1176/appi.books.9780890423363.138660

Appolinario, J. C., Bacaltchuk, J., Sichieri, R., Claudino, A. M., Godoy-Matos, A., Morgan, C., ... Coutinho, W. (2003). A randomized, double-blind, placebo-controlled study of sibutramine in the treatment of binge-eating disorder. *Archives of General Psychiatry, 60,* 1109–1116.

Arnone, D. (2005). Review of the use of topiramate for treatment of psychiatric disorders. *Annals of General Psychiatry, 4*(5), 1–14. doi:10.1186/1744-859X-4-5

Attia, E., & Schroeder, L. (2005). Pharmacologic treatment of anorexia nervosa: Where do we go from here? *International Journal of Eating Disorders, 37,* S60–S63.

Bailey, R. K. (2003). Atypical psychotropic medications and their adverse effects: A review for the African-American primary care physician. *Journal of the National Medical Association, 95*(2), 137–144.

Berken, G. H., Weinstein, D. O., & Stern, W. C. (1984). Weight gain. A side-effect of tricyclic antidepressants. *Journal of Affective Disorders, 7*(2), 133–138.

Berkman, N. D., Bulik, C. M., Brownley, K. A., Lohr, K. N., Sedway, J. A., Rooks, A., & Gartlerhner, G. (2006). *Management of eating disorders (RTI-UNC Evidence report/Technology assessment No. 135).* Retrieved from Agency for Healthcare Research and Quality website: http://www.ahrq.gov/downloads/pub/evidence/pdf/eatingdisorders/eatdis.pdf

Blinder, B. J., Cumella, E. J., & Sanathara, V. A. (2006). Psychiatric comorbidities of female inpatients with eating disorders. *Psychosomatic Medicine, 68,* 454–462.

Bosanac, P., Kurlender, S., Norman, T., Hallam, K., Wesnes, K., Manktelow, T., & Burrows, G. (2007). An open-label study of quetiapine in anorexia nervosa. *Human Psychopharmacology: Clinical and Experimental, 22,* 223–230.

Brennan, B. P., Roberts, J. L., Fogarty, K. V., Reynolds, K. A., Jonas, J. M., & Hudson, J. I. (2008). Memantine in the treatment of binge eating disorder: An open-label, prospective trial. *International Journal of Eating Disorders, 41*(6), 520–526.

Broft, A. I., Spanos, A., Corwin, R. L., Mayer, L., Steinglass, J., Devlin, M. J., Attia, E., & Walsh, B. T. (2007). Baclofen for binge eating: An open-label trial. *International Journal of Eating Disorders, 40*(8), 687–691.

Brownley, K. A., Berkman, N. D., Sedway, J. A., Lohr, K. N., & Bulik, C. M. (2007). Binge eating disorder treatment: A systematic review of randomized controlled trials. *International Journal of Eating Disorders, 40*(4), 337–348.

Bulik, C. M., Berkman, N. D., Brownley, K. A., Sedway, J. A., & Lohr, K. N. (2007). Anorexia nervosa treatment: A systemic review of randomized controlled trials. *International Journal of Eating Disorders, 40,* 310–320.

Button, E. J., Chadalavada, B., & Palmer, R. L. (2010). Mortality and predictors of death in a cohort of patients presenting to an eating disorders service. *International Journal of Eating Disorders.* E-pub ahead of print, http://www.ncbi.nlm.nih.gov/pubmed/19544558; 10.1002/eat.20715

Casner, J. A., Weinheimer, B., & Gualtier, T. C. (1996). Naltrexone and self-injurious behavior: A retrospective population study. *Journal of Clinical Psychopharmacology, 16*(5), 389–394.

Castro, J., Deulofeu, R., Gila, A., Puig, J., & Toro, J. (2004). Persistence of nutritional deficiencies after short-term weight recovery in adolescents with anorexia nervosa. *International Journal of Eating Disorders, 35,* 169–178.

Couturier, J., & Lock, J. (2007). A review of medication use for children and adolescents with eating disorders. *The Journal of the Canadian Academy of Child and Adolescent Psychiatry, 16*(4), 173–176.

Crawford, P. M. (2009). Managing epilepsy in women of childbearing age. *Drug Safety, 32*(4), 293–307.

Crow, S. J., Mitchell, J. E., Roerig, J. D., & Steffen, K. (2009a). What potential role is there for medication treatment in anorexia nervosa? *International Journal of Eating Disorders, 42*(1), 18.

Crow, S. J., Petersen, C. B., Swanson, S. A., Raymond, N. C., Specker, S., Eckert, E. D., & Mitchell, J. E. (2009b). Increased mortality in bulimia nervosa and other eating disorders. *American Journal of Psychiatry, 166*(12), 1342–1346.

Davidson, J. R., Rothbaum, B. O., Van der Kolk, B. A., Sikes, C. R., & Farfel, G. M. (2001). Multicenter, double-blind comparison of sertraline and placebo in the treatment of posttraumatic stress disorder. *Archives of General Psychiatry, 58,* 485–492.

Eddy, K. T., Dorer, D. J., Franko, D. L., Tahilani, K., Thompson-Brenner, H., & Herzog, D. B. (2008). Diagnostic crossover in anorexia nervosa and bulimia nervosa: Implications for DSM-V. *American Journal of Psychiatry, 165* (2), 245−250.

Fairburn, C. G. (2008). *Cognitive behavior therapy and eating disorders*. New York, NY: The Guilford Press.

Faris, P. L., Kim, S. W., Meller, W. H., Goodale, R. L., Oakman, S. A., Hofbaver, R. D., ... Hartman, B. K. (2000). Effect of decreasing afferent vagal activity with ondansetron on symptoms of bulimia nervosa: A randomized, double-blind trial. *The Lancet, 355*, 792−797.

Ferguson, C. P., La Via, M. C., Crossan, P. J., & Kaye, W. H. (1999). Are serotonin selective reuptake inhibitors effective in underweight anorexia nervosa? *International Journal of Eating Disorders, 25*, 11−17.

Fiedorowicz, J. G., & Swartz, K. L. (2004). The role of monoamine oxidase inhibitors in current psychiatric practice. *Journal of Psychiatric Practice, 10*(4), 239−248, July.

Fluoxetine Bulimia Nervosa Collaborative Study Group (1992). Fluoxetine in the treatment of bulimia nervosa. *Archives of General Psychiatry, 49*, 139−147.

Fung, S. M., & Ferrill, M. J. (2001). Treatment of bulimia nervosa with ondansetron. *The Annals of Pharmacotherapy, 35*, 1270−1273.

Gardner, D. M., Baldessarini, R. J., & Waraich, P. (2005). Modern antipsychotic drugs: A critical overview. *Canadian Medical Association Journal, 172*(13), 1703−1711.

Gelber, D., Levine, J., & Belmaker, R. H. (2001). Effect of inositol on bulimia nervosa and binge eating. *International Journal of Eating Disorders, 29*, 345−348.

Glasofer, D. R., Tanofsky-Kraff, M., Eddy, K. T., Yanovski, S. Z., Theim, K. R., Mirch, M. C., ... Yanovski, J. A. (2007). Binge eating in overweight treatment-seeking adolescents. *Journal of Pediatric Psychology, 32*(1), 95−105.

Goldbloom, D. S. (1997). Evaluating the efficacy of pharmacologic agents in the treatment of eating disorders. *Essential Psychopharmacology, 2*(1), 71−88.

Goldstein, D. J., Wilson, M. G., Ascroft, R. C., & Al-Banna, M. (1999). Effectiveness of fluoxetine therapy in bulimia nervosa regardless of comorbid depression. *International Journal of Eating Disorders, 25*, 19−27.

Griengl, H., Sendera, A., & Dantendorfer, K. (2001). Naltrexone as a treatment of self-injurious behavior—A case report. *Acta Psychiatrica Scandinavica, 103*, 234−236.

Gross, H., Ebert, M. H., Faden, V. B., Goldberg, S. C., Kaye, W. H., Caine, E. D., ... Zinberg, N. (1983). A double-blind trial of delta-9-tetrahydrocannabinol in primary anorexia nervosa. *Journal of Clinical Psychopharmacology, 3*(3), 165−171.

Grucza, R. A., Przybeck, T. R., & Cloninger, C. (2007). Prevalence and correlates of binge eating disorder in a community sample. *Comprehensive Psychiatry, 48*(2), 124−131.

Guarda, A. S. (2008). Treatment of anorexia nervosa: Insights and obstacles. *Physiology & Behavior, 94*, 113−120.

Heiden, A., De Zwaan, M., Frey, R., Presslich, O., & Kasper, S. (1998). Paroxetine in a patient with obsessive-compulsive disorder, anorexia nervosa and schizotypal personality disorder. *Journal of Psychiatry & Neuroscience, 23*(3), 179−180.

Horne, R. L., Ferguson, J. M., Pope, H. G., Jr., Hudson, J. I., Lineberry, C. G., Ascher, J., & Cato, A. (1988). Treatment of bulimia with bupropion: A multicenter controlled trial [Abstract]. *Journal of Clinical Psychiatry, 49*(7), 262−266.

Howell, M. J., Schenck, C. H., & Crow, S. J. (2009). A review of nighttime eating disorders. *Sleep Medicine Reviews, 13*, 23−34.

Hrdlicka, M., Beranova, I., Zamecnikova, R., & Urbanek, T. (2008). Mirtazapine in the treatment of adolescent anorexia nervosa. *European Child and Adolescent Psychiatry, 17*, 187−189.

Hudson, J. I., McElroy, S. L., Raymond, N. C., Crow, S., Keck, P. E., Jr., Carter, W. P., ... Jonas, J. M. (1998). Fluvoxamine in the treatment of binge-eating disorder: A multicenter placebo-controlled, double-blind trial. *American Journal of Psychiatry, 155*, 1756−1762.

Kaye, W. H., Greeno, C. G., Moss, H., Fernstrom, J., Fernstrom, M., Lilenfeld, L. R., ... Mann, J. J. (1998). Alterations in serotonin activity and psychiatric symptoms after recovery from bulimia nervosa. *Archives of General Psychiatry, 55*, 927−935.

Kaye, W. H., Nagata, T., Weltzin, T. E., Hsu, L. K., Sokol, M. S., McConaha, C., ... Deep, D. (2001). Double-blind placebo-controlled administration of fluoxetine in restricting- and restricting-purging-type anorexia nervosa. *Biological Psychiatry, 49*(7), 644−652.

Kaye, W. H., Weltzin, T. E., Hsu, L. K., & Bulik, C. M. (1991). An open trial of fluoxetine in patients with anorexia nervosa. *Journal of Clinical Psychiatry, 52*(11), 464−471.

Konstantynowicz, J., Kadziela-Olech, H., Kaczmarski, M., Zebaze, R. M., Iuliano-Burns, S., Piotrowsfka-Jastrzebska, J., & Seeman, E. (2005). Depression in anorexia nervosa: A risk factor for osteoporosis. *The Journal of Clinical Endocrinology & Metabolism, 90*, 5382–5385.

Krüger, S., & Kennedy, S. H. (2000). Psychopharmacotherapy of anorexia nervosa, bulimia nervosa and binge-eating disorder. *Journal of Psychiatry & Neuroscience, 25*(5), 497–508.

La Via, M. C., Gray, N., & Kaye, W. H. (2000). Case reports of olanzapine treatment of anorexia nervosa. *International Journal of Eating Disorders, 27*, 363–366.

Lau, D. C., Douketis, J. D., Morrison, K. M., Hramiak, I. M., Sharma, A. M., & Ur, E. (2007). For members of the Obesity Canada Clinical Practice Guidelines Expert Panel. 2006 Canadian clinical practice guidelines on the management and prevention of obesity in adults and children (summary). *Canadian Medical Association Journal, 176*(8), S1–S13.

Leombruni, P., Gastaldi, F., Lavagnino, L., & Fassino, S. (2008). Oxcarbazepine for the treatment of binge eating disorder: A case series. *Advances in Therapy, 25*(7), 718–724.

Lock, J. D., & Fitzpatrick, K. K. (2009, March). Anorexia nervosa. *Clinical Evidence, 3*(1011), 1–19.

Marrazzi, M. A., Bacon, J. P., Kinzie, J., & Luby, E. D. (1995). Naltrexone use in the treatment of anorexia nervosa and bulimia nervosa. *International Clinical Psychopharmacology, 10*, 163–172.

McCann, U. D., & Agras, W. (1990). Successful treatment of nonpurging bulimia nervosa with desipramine: A double-blind, placebo-controlled study. *American Journal of Psychiatry, 147*, 1509–1513.

McElroy, S. L., Kotwal, R., & Keck, P. E., Jr. (2006). Comorbidity of eating disorders with bipolar disorder and treatment implications. *Bipolar Disorders, 8*, 686–695.

McElroy, S. L., Hudson, J. I., Capece, J. A., Beyers, K., Fisher, A. C., Rosenthal, N. R., & Topiramate Binge Eating Disorder Research Group. (2007a). Topiramate for the treatment of binge eating disorder associated with obesity: A placebo-controlled study. *Biological Psychiatry, 61*, 1039–1048.

McElroy, S., Kotwal, R., Guerdjikova, A., Weige, J., Nelson, E., Lake, K. A., ... Hudson, J. (2007b). Zonisamide in the treatment of binge eating disorder with obesity: A randomized controlled trial. *Journal of Clinical Psychiatry, 68*(1), 1897–1906.

McGilley, B. M., & Pryor, T. L. (1998). Assessment and treatment of bulimia nervosa. *American Family Physician, 57*(11), 2743–2753.

Najjar, M. (2007). Zolpidem and amnestic sleep related eating disorder. *Journal of Clinical Sleep Medicine, 3*(6), 637–638.

Nickel, C., Tritt, K., Muehlbacher, M., Pedrosa Gil, F., Mitterlehner, F. O., Kaplan, P., Lahmann, C., ... Nickel, M. K. (2005). Topiramate treatment in bulimia nervosa patients: A randomized, double-blind, placebo-controlled trial. *International Journal of Eating Disorders, 38*, 295–300.

Nurnberg, H., Hensley, P. L., Heiman, J. R., Croft, H. A., Debattista, C., & Paine, S. (2008). Sildenafil treatment of women with antidepressant-associated sexual dysfunction: A randomized controlled trial. *The Journal of the American Medical Association, 300*(4), 395–404.

O'Reardon, J. P., Allison, K. C., Martino, N. S., Lundgren, J. D., Heo, M., & Stunkard, A. J. (2006). A randomized, placebo-controlled trial of sertraline in the treatment of night eating syndrome. *American Journal of Psychiatry, 163*, 893–898.

Papadopoulos, F. C., Ekbom, A., Brandt, L., & Ekselius, L. (2009). Excess mortality, causes of death and prognostic factors in anorexia nervosa. *The British Journal of Psychiatry, 194*, 10–17.

Physicians' Desktop Reference Health (2009). *Prescription Drug Information and Health Information. Wellbutrin / Prescription Drug Information, Side Effects/PDRHealth.* Available at: http://www.pdrhealth.com/health/clinical-trials-index.aspx

Peters, T. E., Parvin, M., Petersen, C., Faircloth, V. C., & Levine, R. L. (2007). A case report of Wernicke's encephalopathy in a pediatric patient with anorexia nervosa-restricting type. *Journal of Adolescent Health, 40*, 376–383.

Powers, P. S., Bannon, Y., Eubanks, R., & McCormick, T. (2007). Quetiapine in anorexia nervosa patients: An open label outpatient pilot study. *International Journal of Eating Disorders, 40*, 21–26.

Prousky, J. E. (2003). Pellagra may be a rare secondary complication of anorexia nervosa: A systematic review of the literature. *Alternative Medicine Review, 82*, 180–185.

Ricca, V., Mannucci, E., Paionni, A., Di Bernardo, M., Cellini, M., Cabras, P. L., & Rotella, C. M. (1999). Venlafaxine versus fluoxetine in the treatment of atypical anorectic outpatients: A preliminary study. *Eating and Weight Disorders, 4*(1), 10–14.

Rosenfeld, W. E., Doose, D. R., Walker, S. A., & Nayak, R. K. (1997). Effect of topiramate on the pharmacokinetics of an oral contraceptive containing norethindrone and ethinyl estradiol in patients with epilepsy. *Epilepsia, 38*(3), 317–323.

Shapiro, J. R., Berkman, N. D., Brownley, K. A., Sedway, J. A., Lohr, K. N., & Bulik, C. M. (2007). Bulimia nervosa treatment: A systematic review of randomized controlled trials. *International Journal of Eating Disorders, 40*, 321–336.

Sonne, S., Rubey, R., Brady, K., Malcolm, R., & Morris, T. (1996). Naltrexone treatment of self-injurious thoughts and behaviors. *Journal of Nervous and Mental Disease, 184*(3), 192–195.

Stahl, S. M. (2005). *Essential psychopharmacology: the prescriber's guide.* Cambridge, UK: Cambridge University Press.

Svirko, E., & Hawton, K. (2007). Self-injurious behavior and eating disorders: The extent and nature of the association. *Suicide & Life-Threatening Behavior, 37*(4), 409–421.

Tousoulis, D., Antonopoulos, A. S., Antoniades, C., Saldari, C., Stefanadi, E., Siasos, G., ... Stefanadis, C. (2009). Role of depression in heart failure—Choosing the right antidepressive treatment. *International Journal of Cardiology.* Advance online publication, http://www.ncbi.nlm.nih.gov/pubmed/19501922?itool=EntrezSystem2.PEntrez

Trivedi, M. H., Fava, M., Wisniewski, S. R., Thase, M. W., Quitkin, F., Werden, D., ... Rush, A. J. (2006). Medication augmentation after the failure of SSRIs for depression. *New England Journal of Medicine, 354*(12), 1243–1252.

Walsh, B., Agras, W., Devlin, M. J., Fairburn, C. G., Wilson, G., Kahn, C., & Chally, K. (2000). Fluoxetine for bulimia nervosa following poor response to psychotherapy. *American Journal of Psychiatry, 157*(8), 1332–1334.

Walsh, B., Kaplan, A. S., Attia, E., Olmsted, M., Parides, M., Carter, J. C., ... Rockert, W. (2006). Fluoxetine after weight restoration in anorexia nervosa. *The Journal of the American Medical Association, 295*(22), 2605–2612.

Wilfley, D. E., Crow, S. J., Hudson, J. I., Mitchell, J. E., Berkowitz, R. I., Blakesley, V., & Walsh, B. T. (2008). The Sibutramine Binge Eating Disorder Research Group. Efficacy of sibutramine for the treatment of binge eating disorder: A randomized multicenter placebo-controlled double-blind study. *American Journal of Psychiatry, 165,* 51–58.

Williams, S. H. (2005). Medications for treating alcohol dependence. *American Family Physician, 72*(9), 1775–1780.

Winkelman, J. W., Herzog, D. B., & Fava, M. (1999). The prevalence of sleep-related eating disorder in psychiatric and non-psychiatric populations [Abstract]. *Psychological Medicine, 29*(6), 1461–1466.

Wonderlich, S. A., Gordon, K. H., Mitchell, J. E., Crosby, R. D., & Engel, S. G. (2009). The validity and clinical utility of binge eating disorder. *International Journal of Eating Disorders, 42*(8), 687–705.

Yager, J. (2008). Binge eating disorder: The search for better treatments [Editorial]. *American Journal of Psychiatry, 165* (1), 4–6.

Yasuhara, D., Nakahara, T., Harada, T., & Inui, A. (2007). Olanzapine-induced hyperglycemia in anorexia nervosa. *American Journal of Psychiatry, 164*(3), 528–529.

CHAPTER 8

Nutritional Impact on the Recovery Process

Jillian K. Croll

Eating disorders (ED) are not just about food and eating. A complex constellation of factors is implicated in their development and the recovery process. Nonetheless, the establishment or restoration of a positive, peaceful relationship with food and eating is essential for ED recovery. Currently, clinical experience, more than empirical research, informs the nutrition restoration process (Albers, 2003; Koenig, 2005; Satter, 2007; Tribole & Resch, 2003). Hence, the gap in the nutritional treatment of ED is not one of translating empirical findings into clinical practice, it is the paucity of research in this area. There is, however, a significant body of literature detailing the effects of nutritional intake that informs ED treatment in conjunction with practitioners' clinical experience.

PHYSICAL AND PSYCHOLOGICAL IMPACTS OF NUTRITIONAL INTAKE

Inadequate, imbalanced nutrition in ED can impact the entire physiological and psychological condition of the body. Food restriction contributes to potentially devastating physical consequences including amenorrhea, compromised bone health, nutrient deficiencies, and significant alteration of mood (Mitchell & Crow, 2006; Williams, Goodie & Motsinger, 2008; see also Tyson, Chapter 6). Inadequate intake may impair gut motility and inhibit appetite (Nakahara et al., 2008). Even in the face of adequate or excessive energy intake, unbalanced nutrient intake can have a significant impact. For example, patients with bulimia nervosa (BN) or binge eating disorder (BED) may take in sufficient calories, but, because of restrictive dietary rules, may consume few foods high in calcium, placing themselves at risk for bone related complications. The nutritional chaos and imbalance that comes with restricting, binging, purging, overeating, and/or weight cycling, can result in adverse physiological states, regardless of diagnosis (Mitchell & Crow, 2006; Williams et al., 2008).

The landmark Minnesota Semi-Starvation Study extensively informed our knowledge of the physiological and psychological effects of undernutrition (Keys, Brozek, Henschel,

Mickelsen & Taylor, 1950). Maintaining the body's metabolic engine, and the structure and integrity of tissues, requires proper intake of essential nutrients and adequate energy (Keys et al., 1950). Additionally, nutrition and eating behaviors impact how the body can handle physical and emotional stressors, operate known cognitive pathways, generate new cognitive processes, and manufacture and replenish biochemical and neurochemical modulators (Keys et al., 1950; Spinella & Lyke, 2004). Lastly, ED patients' misuse of food and disordered eating interferes with the body's intricate, natural hunger and satiety systems, in addition to compromising their abilities to emotionally and physiologically connect with these systems (Dalle Grave, Di Pauli, Sartirana, Calugi & Shafran, 2007; Geliebter, Yahav, Gluck & Hashim, 2004; Latner, Rosewall & Chisholm, 2008; Waters, Hill & Waller, 2001). These factors have distinct implications for the ED recovery process.

The goals of nutritional restoration across the ED spectrum focus on achieving balanced, adequate, and fulfilling eating, while enabling patients to recognize and honor their internal hunger and fullness signals, and enhance their emotional well-being. This process requires great patience, creativity, and skill on the part of the nutrition practitioner, and hard work, risk-taking, and practice on the part of the recovering patient. It is best done in concert with a multidisciplinary team of ED specialists and the patient's support system.

Effects of Unbalanced Nutrition on Cognitive Processes

The impact of nutritional status on cognitive and emotional processes demonstrated by Keys et al. (1950) has crucially informed the field of ED. Basic knowledge of this study is critical knowledge for all ED clinicians. Briefly, 36 male subjects were selected from over 400 volunteers assessed to have the strongest mental and physical health (Keys et al., 1950). The year-long study involved an initial 3-month control period during which the men were fed adequate diets, a 6-month semi-starvation period during which each man was fed in a manner that induced a weight loss of approximately 25% of his body weight, a 3-month period of restricted rehabilitative feeding that restored weight and nutritional status at varying rates and with various nutrients, and a final 2-month period of unregulated nutritional restoration and rehabilitation during which participants could eat whatever and as much as they wished.

In addition to the significant physical effects experienced by the participants during the semi-starvation phase, including decreased metabolic rate, decreased heart rate, slowed respirations, dizziness, hair loss, and lowered body temperature, participants reported dramatic changes in mood, cognition, comprehension, and judgment during both the semi-starvation phases and the rehabilitative phases. Some of the participants withdrew from academic courses because they felt unable to participate as the semi-starvation period progressed. Interestingly, standardized cognitive tests administered as part of data collection during this phase did not show diminished cognitive abilities. Participants reported that they experienced continued difficulties in cognition and judgment in the rehabilitative phases as well. One participant reported that the rehabilitative phase was even more difficult cognitively. Expecting to feel relief with refeeding, he was dismayed by continued obsessive thoughts of food even when the strict eating rules of the semi-starvation phase were lifted (Kalm & Semba, 2005).

These 60-year-old accounts of the rapid and ill effects of semi-starvation on otherwise healthy research subjects sound tragically similar to those heard by contemporary clinicians treating unwitting victims of ED. It is not uncommon in our ED practices to work with students needing to drop a course or a whole semester, or a sales executive suffering a drop in productivity, due to their inability to concentrate or complete the required work.

Other data support these early findings of disturbances in brain functioning and dysregulated eating behaviors in ED. Complex brain imaging studies have described changes in ED patients in areas involved in the complex appetite, food regulatory, and impulse regulation systems (Frank, Bailer, Wagner, Henry & Kaye, 2004). Dietary restraint, disinhibited eating or overeating have also been associated with a decrease in decision-making and other cognitive capabilities (Spinella & Lyke, 2004). Stealing food and diet aids occurs at a higher incidence among ED patients than in the general population, and may be partially related to disturbances in cognitive functioning resulting from imbalanced nutritional status (Goldner, Geller, Birmingham & Remick, 2000).

In addition to compromised cognitive abilities, Bailer and Kaye's (2003) empirical review of neuropeptides involved in food regulatory systems, mood, and behavior modulation described alterations in neuropeptides across the spectrum of ED behaviors. This burgeoning body of research may provide some insight into ED behaviors. Since neuropeptides play an important role in appetite (the desire to eat) and hunger (the biological need to eat), alterations due to bingeing, purging, and/or restricting may contribute to the cyclical nature of ED behaviors. For example, in a patient with anorexia nervosa (AN) with low total calorie and protein intake, restrictive eating may lead to decreased neuropeptide production, compounds that are created from the amino acid building blocks of dietary proteins. This in turn could influence mood and appetite regulation. If this perturbation is experienced as decreased mood and dulled appetite signals, it may reinforce continued restriction. In a patient with BN, bingeing and purging may also lead to a disruption in neuropeptide levels, which could contribute to mood dysregulation and intensified hunger, setting someone up for the next binge/purge cycle. While some of these differences persist after recovery, many resolve with adequate, balanced nutrition (Bailer & Kaye, 2003).

Significant interest is being directed to compulsive eating behaviors and the interaction between food intake, and cognitive features such as reward sensitivity, dopamine pathways, impulsivity, and decision-making (Davis & Carter, 2009; Munsch, Michael, Biedert, Meyer & Margraf, 2008). In their review of the concept of compulsive overeating as an addiction disorder, Davis and Carter (2009) suggested that compulsive overeating may be conceptualized as a unique interplay of a substance dependence (i.e., nutrients in the food) and a behavioral addiction (i.e., the activity of eating) that is both promoted and maintained by high reward sensitivity in the dopamine system. Their conceptualization lends itself to establishing treatment goals, including learning more helpful strategies around food consumption and emotion regulation. For example, patients experiencing anxiety, intense food cravings, and strong rewards from eating foods high in sugars and fats, may benefit from education about the brain's dopamine reward system contributing to their experiences. This information can help them decide how and when to consume these foods.

While the relationship between brain functioning and behavior is extraordinarily complex, it is clear that alterations in nutrition and eating patterns significantly impact concentration and judgment. Eating disorder behaviors can result in dramatic alterations in eating patterns and subsequently, nutritional intake. Normalizing eating patterns is a strong step towards resolving some of the cognitive effects of ED.

Effects of Unbalanced Nutrition on Emotional and Behavioral Processes

Unbalanced nutritional intake can also severely impact psychological functioning. Similarly to ED patients, nearly all of the men in the Minnesota Starvation Study reported severe emotional distress and depression, and a dramatic obsession with all things food-related (Keys et al., 1950). Food preoccupations, including reading and collecting cookbooks and recipes, and eating rituals became commonplace, just as they frequently do among ED patients. Maladaptive behaviors around food arise when food intake is altered. A brain restricted from adequate nutrition, whether imposed by diet, famine, or ED, reacts by focusing thoughts and energy on food in hopes of motivating its procurement.

Levels of emotional distress regarding these behaviors vary widely among patients. Distressed by the mere contemplation of grocery shopping, many report spending hours agonizing over which peanut butter, fruit snack, or bread to purchase that fits all their "ED requirements." Some report feeling utterly detached from the experience of shopping, yet find themselves purchasing a range of items either to binge on or that fit within restrictive ED rules. Nutritional therapy interventions address planning and accomplishing less overwhelming shopping trips.

Those who restrict food may chew multiple packs of gum each day, drink enormous amounts of water and diet soda, chew ice, or eat high volumes of very low energy density, high-fiber foods in an attempt to fill up and quell hunger pangs. Dietary restriction and the associated feelings of control may be experienced as rewarding by the patient (Dalle Grave et al., 2007; Shafran, Fairburn, Nelson & Robinson, 2003). Interventions must help patients learn more adaptive ways of experiencing control and empowerment.

Patients struggling with bingeing describe trying to fill an endless void with food. During recovery, they discover that the void is actually related to emotional, not physical, hunger. Nutritional intervention for bingeing establishes regular eating patterns so that emotional hunger can be addressed in therapy. In the context of BN, this interrupts restricting and prolonged delays between meals, while in BED, this contributes to regulation of mood states.

In sum, the impact of food intake on cognition, physiology, and emotional states is wide ranging and powerful, presenting clear opportunities for interventions that can yield marked changes. In order for ED to remit, eating must become an activity that is free of dread, anxiety, and determined alteration, and distinguished by self-care, flexibility, and positive coping.

FINDING NUTRITIONAL BALANCE: NUTRITIONAL REHABILITATION

Nutritional rehabilitation is a dynamic process that proceeds in stages. Typically, this entails progressing from symptom interruption with a highly structured eating plan, to teaching a more intuitive, mindful way of eating, and ultimately, to the patient mastering self-regulated eating connected to the body's cues of hunger and satiety. Instruction and direction through these stages is done in conjunction with teaching alternative coping skills.

Recovered patients describe the changes in their thoughts about food as moving from irrational, dominant, fearful, compulsive thoughts, to thoughts of flexibility, openness, acceptance, exploration, and freedom (Björk & Ahlström, 2008). The actions and behaviors that go along with these cognitive shifts include: eating a wide range of foods and eating regularly; eating without fear or limitations based on the feared effects of the food; eating comfortably with others; and eating in response to the body's natural signals. Thus, this new "healthy" relationship to food involves the body and mind operating in concert, rather than in internal conflict.

Duration of Nutritional Rehabilitation

Recovery takes copious amounts of time, practice, and patience on the part of patients and their supporters. These dramatic changes in thoughts and orientation to food rarely occur in a linear fashion. Even in the case of the imposed undernutrition of healthy men, as demonstrated in the Minnesota Semi-Starvation study (Keys et al., 1950), rehabilitation is protracted and challenging. Many of the men reported that the rehabilitation period was the most difficult part of the study, as the intense effects of food restriction lingered well after the reintroduction of adequate nutrition (Kalm & Semba, 2005; Keys et al., 1950). Subjects reported that it took from 2 to 24 months after the 5 month nutritional rehabilitation to feel fully recovered from the experience (Kalm & Semba, 2005). Similarly, research examining time to ED recovery report durations of 3 to 5 years or longer (Grilo et al., 2007; van Son, van Hoeken, van Furth, Donker & Hoek, 2010).

The length, structure, and path of the nutritional rehabilitation process depends a great deal on the specific struggles of the patient, the ED behaviors in question, the available supports (e.g., access to specialized treatment; financial resources; involvement of family, friends, co-workers), and the internal motivation of the patient.

Interrupting Symptom Use

For patients with low body weights, initial nutritional rehabilitation is more prescriptive and directed in an effort to restore well-being and weight, so that they are more physically equipped to make decisions, develop new coping skills, and handle the psychological work involved in the process of treatment and recovery (American Dietetic Association, 2006; Cockfield & Philpot, 2009). The early phase of refeeding focuses on the intake of nutrients and energy in a regulated and adequate manner.

Prevention of possible medical complications of increased food intake is critical (Tresley & Sheehan, 2008; see also Tyson, Chapter 6). Refeeding syndrome can occur when food intake, particularly carbohydrate, is increased too dramatically, leading to a sudden rise in insulin production and movement of phosphate, magnesium, and potassium into cells. Low blood concentrations of these nutrients involved in muscle contraction, together with fluid overload from decreased water and sodium excretion in the face of the energy influx, can lead to cardiac complications and death (Tresley & Sheehan, 2008). Because of these risks, caloric increases are done gradually and systematically with close medical monitoring of labs and vital signs.

Early fullness is a common complaint among patients with AN, likely due to a combination of gastric slowing, appetite hormone dysregulation, and anxiety with eating (Cockfield & Philpot, 2009; Nakahara et al., 2008). This uncomfortable experience can be managed proactively with education and suggestions of creative solutions for lower volume, energy dense foods, and liquids (Cockfield & Philpot, 2009). For example, one alternative is a meal of 3 ounces of chicken, 1 cup of brown rice with butter, 1 cup of salad with dressing, 1 cup of berries, 1 cup of skim milk, and a slice of chocolate cake. However, a meal of chicken and vegetable stew, a biscuit with butter, and a milkshake with 1 banana, $^3/_4$ cup of whole milk, and 1 cup of premium ice cream accomplishes delivery of similar calories and nutrients in a lower volume of food.

For patients struggling with chaotic, compulsive, and/or out of control eating behaviors, with or without purging, bringing structure to eating is critical to interrupting symptoms. Establishment of regular meals and snacks focused on adequate, balanced energy and nutrient intake brings a sense of predictability and regularity to eating that can help prevent or mitigate dramatic swings in hunger and satiety, improve cognitive processes by decreasing obsession with food, and minimize symptom use (American Dietetic Association, 2006).

Patients' participation in food choices varies depending on the treatment setting. Clinically, some patients are not initially well-equipped to make choices most supportive of their health, and will need more guidance. Others are able and eager to be more involved in food selection, even when quite compromised. Clinical skill, empathy, and flexibility are key components of navigating this early stage of rehabilitation in a way that allows the patient to feel heard, supported, respected, and involved.

Eating plans. As noted, the early stages of nutritional rehabilitation involve providing guidance on establishing regular patterns of eating. These recommendations range from having three meals and two to three snacks per day, going no more than four hours between eating episodes, and no skipping meals (Fairburn, 2008), to a detailed meal plan that offers guidance on the types and amounts of food to be eaten at each meal and snack. The latter approach is common in many practice settings. The overall emphasis is on eating meals and snacks regularly throughout the day, in adequate amounts, comprised of foods that will provide nourishment appropriate to the undernourished or nutritionally unbalanced state of the patient. Simply put, but not easily done, the job of the patient is to try to eat according to the established guidelines.

The structure of a meal plan contributes a sense of order, accountability, acceptability, and nutritional adequacy to eating. Often broadly based on food groups (protein, calcium-rich foods, grains, vegetables, fruits, fats, desserts, and other), the meal plan will illustrate how much of each food group is recommended and how these amounts could be distributed across three meals and multiple snacks. The "other" category may include supplementary foods, such as drinks, bars, and powders that are condensed energy sources for those patients needing to consume a larger amount of energy for weight restoration. Table 8.1 illustrates an example of a weight restoration meal plan with menu ideas incorporating supplementary foods and ways to reduce volume when possible.

All foods can be accommodated, although certain ones considered too initially overwhelming can be reintroduced more slowly over time. For example, patients struggling with restricting and strong food fears may have to start with low amounts of fats and work to add additional servings over time. Patients struggling with bingeing and purging may need to pay particular attention to eating meals and snacks that include a mixture of carbohydrate, fats, and proteins, rather than primarily carbohydrate, to help the meal be more physiologically satisfying and to prevent bingeing resulting from inadequate intake. For patients trying to decrease binge eating, common binge foods, such as desserts high in carbohydrate and fat, may need to be briefly avoided to allow for normalized eating before being reintegrated later in the recovery process. Patients may experience the meal plan as a relieving

TABLE 8.1 Sample Weight Restoration Meal

Serving Size	Food Group	Breakfast	Snack	Lunch	Snack	Dinner	Snack	Total number of exchanges
1 oz meat, fish, poultry; 1 egg, 1 oz cheese; 1TBSP peanut butter; 3 oz tofu; 1 oz nuts	Protein	2 scrambled eggs		2 slices cheese/meat pizza (3 oz protein in 2 slices)		3 oz hamburger with 1 oz cheese		9
1 cup milk, soy milk, or yogurt; 1 oz cheese	Dairy	1 oz cheese		8 oz milk	1 cup yogurt	8 oz milk	8 oz milk with instant breakfast powder	6
1 slice bread, 1 cup flake cereal, ½ cup dense cold cereal; ½ cup hot cereal; ½ cup rice, pasta, potato; small tortilla; 4–6 crackers; ½ bagel, bun	Grain	2 small tortillas	1 bagel	Crust of pizza	½ cup granola cereal	1 cup oven roasted potato cubes; 1 bun		11
½ cup cooked; 1 cup raw; 4 fl oz veg juice	Veg			2 cups mixed greens		1 cup steamed broccoli		4
1 medium piece; ½ cup canned; 1 cup fresh; 4 oz juice; ¼ cup dried fruit	Fruit	1 cup strawberries	4 fl oz orange juice		½ cup canned fruit		1 large banana	5
1 tsp butter, margarine, cream cheese; 1 tbsp regular mayo, salad dressing, whipped butter or margarine; 2 tbsp light dressing or mayo; ½ oz nuts	Fat	2 tbsp sour cream on breakfast burritos	2 tbsp cream cheese on bagel	1 tbsp dressing on salad; ½ oz nuts on salad		1 tsp olive oil on potatoes	1 tbsp peanut butter	7
1 medium slice cake, pie, brownie; regular size candy bar; 2–3 homemade cookies; 3–4 sandwich cookies; 1 cup ice cream, pudding	Dessert			Medium brownie			1 cup premium ice cream	1–2 per day
Powder addition to milk; sports bar, drink, or shake mix; liquid supplement	Other (energy dense options for extra kcal)						Make shake with peanut butter, ice cream, banana, milk, and breakfast powder	

sort of "permission to eat normally." The meal plan and its structure may also function as a way to move ahead from a less successful day. A wise patient once said that her meal plan helped her to remind her: "No matter how bad yesterday was, **always** eat breakfast."

The meal plan is not intended to be a maximum amount of food that can be eaten in a day, nor should it be rigid or inflexible. It should be designed with the patient's ED symptoms, height, weight, gender, activity, and energy needs in mind and planned in a manner that increases the likelihood of the patient's success. For example, some patients find it easier and prefer to consume more energy earlier in the day. Others find it significantly more challenging to start the day with a large meal, and will find it more manageable to break it into multiple smaller meals across the morning. Other factors to consider may include time, finances, cooking ability, and living situation. Thus, this process is ideally personalized and the plan is revised over time to meet the patient's changing needs.

Self-monitoring. Frequently, self-monitoring of food consumption and symptom use is recommended in conjunction with the meal plan. Self-monitoring, a cornerstone of cognitive behavioral treatment (Fairburn, 2008), can be a partner to the meal plan, helping patients compare how their eating for the day matches the intended structure and content. As shown in Figure 8.1, the food log details food eaten, amount, time, place, thoughts, emotional state, and symptom use. Reviewing a self-monitoring log can help clarify when and which components of the meal plan are missing or in excess, how symptoms interplay with food intake, and what emotional states are associated with presence or absence of symptom use.

Willingness and interest regarding completing self-monitoring logs varies across patients and is likely dependent on motivation for recovery and on symptom patterns. Self-monitoring might be more acceptable to patients in active stages of change (Nichols & Gusella, 2003). These patients typically find self-monitoring to be a helpful, insightful means to behavior change. Food logs may show how irregular eating with long periods between meals lead to episodes of bingeing and purging. When experiences regarding emotions and binge eating can be clearly linked, emotional triggers, and strategies for self-soothing and increasing emotional tolerance can be addressed.

Among patients less motivated for change, self-monitoring may feel like more of a chore than a tool. They may feel so emotionally detached that they have difficulty even naming feelings that arise in relation to food and symptom use. In these situations, simply recording food intake and symptom use may be all they are able to do. Discussion of the logs in session may help them become more aware of connections between emotions and eating behaviors, or long gaps between meals, and ravenous hunger later or even the next day.

Shame may also interfere with a patient's ability or willingness to complete self-monitoring logs, given that it can be a core struggle that affects eating behaviors, thoughts, and attitudes (Goss & Allan, 2009). For those patients who experience significant shame in relation to eating in general or to symptom use, self-monitoring can be a powerfully negative experience that keeps them stuck, rather than a tool to enhance self-awareness of behavior. In these situations, it may be prudent to avoid food logs, at least initially, and work closely with the therapist to help the patient identify alternatives to symptom use.

In addition to these emotional obstacles to self-monitoring, other barriers must be explored as well. When asked to self-monitor, patients will often say: "I don't want people to see me doing that," or "I don't want to carry something like that with me; I might lose it." Patients should be encouraged to complete self-monitoring records as thoroughly, beneficially, and realistically as possible. Logs may be handwritten or completed electronically, in a journal format—any manner that works best for the patient. One family devised a system of colored chips in which each color represented a serving of a particular food group. Once the

Food Record and Hunger Ratings
DATE: _____

Time	Item Eaten	Amount	PR	GR	VG	FR	FT	ML	DS	Other	Hunger/ Fullness (see scale below)	Symptoms (binge, purge, diet pill, laxative, restrict, other)	Activity (type, length, intensity)	Thoughts/ Feelings
		Intake												
		Goal												

PR: Protein; GR: Grain; VG: Vegetable; FR: Fruit; FT: Fat; ML: Dairy; DS: Dessert.
Hunger/Fullness Scale: 0=extremely hungry, 1=very hungry, 2=quite hungry, 3=fairly hungry, 4=getting hungry, 5=neither full nor hungry, 6=getting full, 7=fairly full, 8=quite full, 9=very full, 10=extremely full.

FIGURE 8.1 Sample Food Record with Hunger and Satiety Ratings

adolescent consumed a meal or snack, a chip for each serving was moved from the *"to have"* to the *"had"* jar, with the goal being an empty *"to have"* jar at the end of the day. A quick glance was all that was needed for the parents to assess how the day was going.

Meal plan duration. The meal plan should be conceptualized from the beginning as a temporary tool used to navigate the process to mindful, nurturing, self-regulated eating, much as a map is a temporary tool used to navigate trips to new destinations. Properly and attentively used, both the map and the meal plan eventually become unnecessary. The timing of meals and snacks, and the amounts of food that are outlined in the meal plan, are intended to become internalized and matched with patient's internal signals of hunger and satiety. Ideally, their acquired knowledge and competence with food, used in conjunction with this inner compass, becomes their guide to eating.

On-going Nutritional Rehabilitation

The middle phase of nutritional restoration is marked by practice, experimentation, and repeated attempts to reincorporate previously feared foods, and previously avoided food-related situations. In this phase, the goal is change rather than compliance. This approach is not linear; but a process of back pedaling or side stepping intermit with forward motion.

With progression, physical health complications of unbalanced nutrition resolve, cognition and focus improve, and obsessions and compulsions with food begin to shift. The main focus of the nutrition-related work in this phase is reincorporation of regular, normal eating with the goal of developing a mindful or intuitive nutritional relationship with one's body. Once structured eating is mastered, awareness of hunger and satiety cues, flexibility, and balance become the key factors in nutritional rehabilitation.

Hunger and satiety. Research in hunger, satiety, appetite, and food intake regulation continues to inform the process of improving dysfunctional eating patterns (Hetherington & Rolls, 2001). An explanation of the concepts of hunger, satiety, and appetite can aid patients in attending to internal signals. Hunger, the biological need for food, may be expressed by a growling stomach, a headache, fatigue, or distractedness. If ignored, it will likely disappear, only to reappear later in a different form such as stronger fatigue, emotional disconnectedness, cognitive impairment, or thirst. Satiety may be a quiet fullness, a feeling of calm, a slight decrease in the palatability of the food being eaten, a sensation in the belly. Appetite, or desire for food with or without hunger, may be experienced as an insatiable, frightening state, because emotional appetites can be confused with physical, food-related appetites. Perception of hunger, satiety, and appetite cues are typically so significantly impacted by ED behaviors that a substantial portion of nutritional rehabilitation work is done around understanding, recognizing, and reconnecting with the intricate internal signals related to food intake.

Monitoring hunger and satiety. When monitoring logs include ratings of hunger and satiety, patients can learn to discern and understand how symptoms interact with these cues. As in the example shown in Table 8.1, hunger and satiety rating scales often range from 1 to 10, with 1=extremely hungry and 10=extremely full. In order to avoid getting too hungry and being at risk for overeating, it is helpful to encourage eating prior to reaching a 3 or 4 on the hunger scale. Similarly, in order to prevent or recognize overeating, patients are encouraged to stop eating prior to reaching 8 or 9. Eating disorder symptoms can interfere with attempts to respond to hunger and fullness in the middle area of the range, and draw patients to the extremes of the range of hunger and fullness.

Flexibility. A patient who has mastered eating regular meals and snacks from an expanded realm of food choices, and has begun understanding hunger/satiety cues, has significantly progressed in nutritional recovery. The next step is developing flexibility (Satter, 2007). Once a difficult eating related task is mastered, expanding the boundaries of comfort and competence can be a challenging process.

For example, a patient's meal plan is structured such that breakfast involves the following servings: two grains, a fruit, a fat, and a calcium food. The patient has gradually become comfortable with this breakfast, typically having a bowl of cereal or oatmeal with milk and fruit and a slice of toast with peanut butter. While these foods well represent what the meal plan outlines, it is rarely varied. However straightforward it seems to have something else for breakfast, it may have taken the patient months to comfortably incorporate an adequate fat serving or second grain serving. Most likely, great progress was made to consume this entire breakfast, building up from one small apple, or in lieu of an hour-long binge and purge session.

Alterations in the patient's schedule due to work, traveling, social events, etc., are inevitable. Formulating plans in advance allows the patient to be more prepared and flexible when situations demand it. A number of typical meal scenarios can be planned in session and practiced in familiar surroundings, so that once in the actual situation, the patient feels more confident and calm. This type of practice expands patient's eating competence, developing their flexibility with an array of food possibilities and circumstances (Satter, 2007).

Off-limits foods and challenging situations. Often, across all types of EDs, certain foods are abandoned due to fear, misinformed nutritional beliefs, or their potential for symptom triggering. Patients are best assisted by discussing their concerns, thoughts, and/or beliefs about these foods, and distinguishing between nutritional myths and truths. It is also important to understand the situations in which these foods were typically eaten, how that became problematic, and how the patient would like to consume them in recovery. Reincorporation of these foods utilizing an internally motivated regulation of food intake, as opposed to cognitively or externally regulated eating, promotes health, well-being, and balance (Kratina, 2003).

For example, a patient decides that she would like to reincorporate pizza, a previously favorite food typically eaten out with friends. Formulating this plan in stages is most helpful. The first stage might involve having pizza in a nonpublic private setting that is familiar, safe, and supportive. This helps to minimize her anxiety, and to experience how pizza can fit into normal, healthy eating. Once this has been successfully accomplished, the next step may be going out for pizza with safe, supportive others. The patient is encouraged to choose the kind of pizza truly desired. Being in public adds the element of the outside world to the experience, which, depending on the person, may increase anxiety or may be calming by adding a sense of normalcy. The last step would be to have pizza out, with friends, as she did prior to her ED. Having two or more previous experiences focused on reintegrating pizza into her diet, the combination of the previously feared food and the social situation is less overwhelming than it would have been otherwise. As illustrated, with practice, support, and successive approximation to the ultimate goal, an array of foods can be reclaimed and reintegrated into the patient's routine intake.

Life events, such as wedding, graduations, traveling, holidays, new jobs, relationships, schedules, or other significant occasions offer real-life opportunities to practice new ways of interacting with food. Acclimating to these situations may require trial and error until the patient feels confident with multiple food related activities. Common occurrences also become manageable. For example, with planning and practice, the previous ordeal of grocery shopping involving copious label reading and few food purchases may shift to a shorter, less label-bound, more tolerable event involving moderate amounts of food purchasing. Teaching cooking skills may also be necessary. This can be done in a group, ideally in an actual kitchen with hands-on food preparation, or individually in the office if no facilities are accessible. Cookbooks of quick, easy meals can be helpful teaching tools. Eventually, these can become normal life responsibilities that stir up minimal emotional turmoil, and yield a navigable kitchen stocked with adequate, nourishing foods.

Shift away from structure and monitoring. Inevitably, at some point in the treatment process, patients will ask: "How long will I be on this meal plan?" "How long do I have to do these food records?" "Do 'normal' people follow meal plans?" or "How do 'normal' people feel when they are full?" These are typically signs that they may be ready to make the transition away from structure. Conversely, other patients may cling to it, stating: "As long as I follow my meal plan or write down my food, I do okay," or "How will I know how to eat if I don't have my meal plan?" When and how quickly to shift away from the meal plan and self-monitoring process is dependent on numerous factors, including treatment setting, motivation, and health status.

At times, patients may need gentle, compassionate prodding to loosen their grip on the meal plan or food record tools. These can become too comfortable in a way that precludes patients from expanding their food repertoire or keeps them too tightly tied to a structure that continues to limit normal flexibility with eating. It may be necessary to provide safe challenges to help them expand their latitude with eating. One challenge is to have patients not look at the meal plan or complete food records for a few days, and then describe in a session how this impacted what they ate and how they felt. Letting go of the meal plan may have been anxiety producing or freeing. They may notice that not recording intake changes thoughts about the structure of eating, e.g., "If I don't have to write things down, then I'm not going to eat it!" or "I liked not having to write things down; I felt more normal." These observations, and possible changes in behavior, are gauges of a patient's readiness to transition to less structure.

Another goal for those slower to relinquish reliance on external tools is to guide eating with internal hunger and fullness cues without relying on the meal plan for a day or more. This may or may not still include written self-monitoring. If hunger and fullness cues are still difficult for the patient to discern, this type of experiment can be helpful in steadily strengthening those skills. Generally, these prompts encourage the patient to push beyond a safe set of foods or settings to try more challenging situations or foods, observe how it went, and report back with successes and areas for continued practice.

The internal compass. Novel or re-engaged competence with life skills involving food and eating is critical to replacing ED symptoms with positive, self-care behaviors. A new type of map is in order: the internal compass. Internally directed and regulated eating is described as mindful eating (Albers, 2003; Smith, Shelley, Leahigh & Vanleit, 2006), intuitive eating (Tribole & Resch, 2003), normal eating (Koenig, 2005), and competent eating (Satter, 2007). External factors may still impact the internal compass, such as food availability, resources, time, facilities, presence or absence of others, and situation. While these external factors may influence what and how much food is available, they do not have to unduly influence which and how much food is eaten, when eating begins and ceases, or the state of mind experienced by the patient at the time of eating.

In practice, patients often question this part of recovery, doubting that they can eat in this internally regulated manner. Reassurance that it is possible helps them stay the course. A common theme among recovered patients is that they now possess the ability to eat whatever is wanted, without avoiding foods based on calorie or fat content, and without focusing on externally regulated portions or cues (Björk & Ahlström, 2008). Developing skills around emotional recognition and regulation and stress management is essential, as they allow the patient to forego ED behaviors to deal with life events, thoughts, and feelings. It is important to remember that eating behaviors are not only connected to negative emotions, but positive emotions as well (Dingemans, Martijn, van Furth & Jansen, 2009). For example, birthday parties or other celebrations can result in binge eating even though the patient is happy and having fun, if they fail to follow internal hunger and satiety signals.

Turning up the volume on internal regulation. Strengthening internal regulation of eating begins by listening to and observing thoughts, feelings, and physical sensations in relationship to food, eating, and the body. Clinically, ED patients often express an inability to relax and may be resistant when asked to develop mindfulness skills that request observation of the body's feelings (Douglass, 2009). Despite feeling like they "can't do it," research shows that mindfulness can be learned, developed, and cultivated as a skill and tool even among more highly anxious patients (Shapiro, Oman, Thoresen, Plante & Flinders, 2008; see also McCallum, Chapter 23). Extending these skills to eating practices is encouraged. Indeed, programs designed for the general public to develop or enhance mindfulness typically include a mindful eating component (Kabat-Zinn, 1990; Shapiro et al., 2008). By exploring and observing (i.e., being mindful of) and coming to understand their own sensations of hunger, satiety, and desires for particular foods or aversions to others, patients can start to hear more of their body's signals regarding food intake. Learning the language of internal signals makes reading the messages possible.

The End of the Journey

The final stage of nutritional rehabilitation is focused on maintaining internally regulated, mindful eating skills with attention to any specific nutritional needs. For example, a patient recovering from AN may have high calcium and vitamin D needs to address compromised bone health. A patient with BED can practice mindful eating and also be attentive to cholesterol and saturated fat intake to address heart disease risk. Research demonstrates that internally regulated eating leads to health improvements, in lieu of restrictive dieting (Bacon, Stern, Van Loan & Keim, 2005; Hawley et al., 2008; Smith et al., 2006). Body acceptance and self-care in conjunction with internally regulated eating become the focus of life in recovery mode (Björk & Ahlström, 2008).

Letting go. The process of normalizing food intake and ED recovery requires the patient to let go of perfectionistic ED rules and structure in favor of bodily connection. In practice, patients working on detangling from their ED sometimes struggle with having to "do recovery perfectly" or in some "right" way. This particularly includes the manner in which they eat, the foods eaten, and the acceptability of and satisfaction with these foods. In their exacting ED world, foods need to be "just right, just what they wanted, cooked just right, presented just right, and enjoyed in just the right manner."

Patients must recognize and move through this struggle with "food perfectionism." Eating and food experiences are not perfect; in fact, they are highly variable. They can be wonderful, boring, messy, predictable, surprising, satisfying, disappointing, tasty, bland, exciting, or unmemorable. As patients allow themselves to experience this full range of possibilities, and become more mindful of their responses to situations, they are better able to let go of dissatisfying eating. If a meal or snack proves disappointing, their perfectionistic food orientation bears reminding them that there will be opportunities for satisfaction the next time.

Mindful eating in a less than mindful world. Flexibility with food and eating, accepting and cooperating with the body, and allowing for a more peaceful relationship with weight, are characteristics of recovered patients (Björk & Ahlström, 2008). In many ways, however, the sociocultural environment does not support mindful, joyful, internally regulated eating. Patients developing competence with internal regulation skills may feel counter-cultural, as if they have skills not possessed by most people. This feeling may be quite accurate, as internally guided eating is sadly lacking today, but can be learned (Satter, 2007). Despite lack of support for internally regulated eating in the culture, patients must continue to develop skills in this area to flourish in recovery.

The tenants of recovery eating. This approach to eating, regardless of context, is summarized by the following principles:

1. Begin eating in sync with hunger, satiety, and appetite cues rather than in response to emotional state, external motivators, or fear.
2. Choose foods based on internal cues, desires about what is really wanted, and with consideration of availability.
3. Experience eating as it is happening.
4. Stop eating in accordance with hunger, satiety, and appetite cues.
5. Reflect on the degree of satisfaction with the experience to inform future eating choices and opportunities.
6. Honor your right to listen to your body and allow your internal compass and wisdom to guide you.

CLOSING THE RESEARCH/PRACTICE GAP IN THE NUTRITIONAL TREATMENT OF EATING DISORDERS

While a gap remains in ED research examining specific nutritional interventions, clinical experience across disciplines highlights the importance of including a nutrition professional on the treatment team. The dietitian plays an integral role in ED recovery by aiding patients in establishing positive, internally guided approaches to eating while working with other team members on the physical, psychological, and spiritual components of their recovery. Ongoing communication and consultation between providers regarding patients' successes and challenges is critical for their progress and the team's effectiveness.

Empirical research is needed to both support and innovate specific nutritional interventions in ED treatment. The dietitian is often at odds with patients at the beginning of treatment, facing the force of their resistance to weight gain or symptom interruption. Over time, this typically dissipates and develops into a strong, trusting relationship that helps patients expand their eating repertoire. Given the dearth of ED nutrition research, qualitative research focused on the patient's and dietitian's experiences throughout the course of treatment would add critical information regarding the process and outcomes of nutritional treatment. Additionally, increased ED nutrition intervention research can serve to support improved insurance coverage of nutrition services; a challenging financial issue faced by many patients.

Clinicians new to the ED field are advised to thoroughly immerse themselves in ED nutrition therapy resources, as well as the related psychological treatment literature. Typical academic nutrition training will equip new dietitians with robust clinical nutrition skills, but not necessarily the therapeutic and counseling skills required to effectively navigate the psychological nuances of ED treatment. Mentorship with an experienced ED dietitian or psychotherapist can be invaluable in developing competency in the field. The effort is worth it; helping patients to achieve freedom from the tyranny of food and experience joy in life is extremely fulfilling.

References

American Dietetic Association. (2006). Position of the American Dietetic Association: Nutrition intervention in the treatment of anorexia nervosa, bulimia nervosa, and other eating disorders. *Journal of the American Dietetic Association, 106,* 2073−2082.

Albers, S. (2003). *Eating mindfully. How to end mindless eating and enjoy a balanced relationship with food.* Oakland, CA: New Harbinger Publications.

Bacon, L., Stern, J., Van Loan, M., & Keim, N. (2005). Size acceptance and intuitive eating improve health for obese, female chronic dieters. *Journal of the American Dietetic Association, 105,* 929−936.

Bailer, U. F., & Kaye, W. H. (2003). A review of neuropeptide and neuroendocrine dysregulation in anorexia and bulimia nervosa. *Current Drug Targets: CNS and Neurological Disorders, 2,* 53−60.

Björk, T., & Ahlström, G. (2008). The patient's perception of having recovered from an eating disorder. *Health Care for Women International, 29,* 926−944.

Cockfield, A., & Philpot, U. (2009). Feeding size O: The challenges of anorexia nervosa. Managing anorexia from a dietitian's perspective. *Proceedings of the Nutrition Society, 68,* 281−288.

Dalle Grave, R., Di Pauli, D., Sartirana, M., Calugi, S., & Shafran, R. (2007). The interpretation of symptoms of starvation/severe dietary restraint in eating disorder patients. *Eating and Weight Disorders, 12,* 108−113.

Davis, C., & Carter, J. (2009). Compulsive overeating as an addiction disorder. A review of theory and evidence. *Appetite, 53,* 1−8.

Dingemans, E., Martijn, C., van Furth, E., & Jansen, A. (2009). Expectations, mood, and eating behavior in binge eating disorder. Beware of the bright side. *Appetite, 53,* 166−173.

Douglass, L. (2009). Yoga as an intervention in the treatment of eating disorders: Does it work? *Eating Disorders: The Journal of Treatment and Prevention, 17,* 126−139.

Fairburn, C. (2008). *Cognitive behavior therapy and eating disorders.* New York, NY: Guilford Press.

Frank, G. K., Bailer, U. F., Wagner, A., Henry, S., & Kaye, W. H. (2004). Neuroimaging studies in eating disorders. *CNS Spectrums, 9,* 539−548.

Geliebter, A., Yahav, E. K., Gluck, M. E., & Hashim, S. A. (2004). Gastric capacity, test meal intake, and appetitive hormones in binge eating disorder. *Physiology & Behavior, 81,* 735−740.

Goldner, E. M., Geller, J., Birmingham, L., & Remick, R. (2000). Comparison of shoplifting behaviors in patients with eating disorders, psychiatric control subjects, and undergraduate control subjects. *Canadian Journal of Psychiatry, 45,* 71−475.

Goss, K., & Allan, S. (2009). Shame, pride and eating disorders. *Clinical Psychology and Psychotherapy, 16,* 303−316.

Grilo, C., Pagano, M., Skodo, A., Sanislow, C., McGlashan, T., Gunderson, J., & Stout, R. (2007). Natural course of bulimia nervosa and of eating disorder not otherwise specified: 5-Year prospective study of remissions, relapses, and the effects of personality disorder psychopathology. *Journal of Clinical Psychiatry, 68,* 738−746.

Hawley, G., Horwath, C., Gray, A., Bradshaw, A., Katzer, L., Joyce, J., & O'Brien, S. (2008). Sustainability of health and lifestyle improvements following a non-dieting randomized trial in overweight women. *Preventive Medicine, 47,* 593−599.

Hetherington, M., & Rolls, B. (2001). Dysfunctional eating in the eating disorders. *Psychiatric Clinics of North America, 24,* 235−248.

Kabat-Zinn, J. (1990). *Full catastrophe living: Using the wisdom of your body and mind to face stress, pain, and illness.* New York, NY: Dell Publishing.

Kalm, L. M., & Semba, R. D. (2005). They starved so that others be better fed: Remembering Ancel Keys and the Minnesota experiment. *Journal of Nutrition, 135,* 1347−1352.

Keys, A., Brozek, J., Henschel, A., Mickelsen, O., & Taylor, H. L. (1950). *The biology of human starvation* (2 volumes). Minneapolis, MN: University of Minnesota Press.

Koenig, K. (2005). *The rules of "normal" eating: A commonsense approach for dieters, overeaters, undereaters, emotional eaters, and everyone in between.* Carslbad, CA: Gurze Books.

Kratina, K. (2003). Health at every size: Clinical applications. *Healthy Weight Journal, 17,* 19−23.

Latner, J. D., Rosewall, J. K., & Chisholm, A. M. (2008). Energy density effects on food intake, appetite ratings, and loss of control in women with binge eating disorder and weight-matched controls. *Eating Behaviors, 9*, 257–266.

Mitchell, J., & Crow, S. (2006). Medical complications of anorexia nervosa and bulimia nervosa. *Current Opinions in Psychiatry, 19*, 438–443.

Munsch, S., Michael, T., Biedert, E., Meyer, A. H., & Margraf, J. (2008). Negative mood induction and unbalanced nutrition style as possible triggers of binges in binge eating disorder. *Eating and Weight Disorders, 13*, 22–29.

Nakahara, T., Harada, T., Yasuhara, D., Shimada, N., Amitani, H., Sakoguchi, T., … Invi, A. (2008). Plasma obestatin concentrations are negatively correlated with body mass index, insulin resistance index, and plasma leptin concentrations in obesity and anorexia nervosa. *Biological Psychiatry, 64*, 252–255.

Nichols, S., & Gusella, J. (2003). Food for thought: Will adolescent girls with eating disorders self-monitor in a CBT group? *Child and Adolescent Psychiatry Review, 12*, 37–39.

Satter, E. (2007). Eating competence: Definition and evidence for the Satter Eating Competence Model. *Journal of Nutrition Education, 39*, S142–S153.

Shafran, R., Fairburn, C. G., Nelson, L., & Robinson, P. H. (2003). The interpretation of symptoms of severe dietary restraint. *Behaviour Research and Therapy, 41*, 887–894.

Shapiro, S., Oman, D., Thoresen, C., Plante, T., & Flinders, T. (2008). Cultivating mindfulness: Effects on well-being. *Clinical Psychology, 64*, 840–862.

Smith, B., Shelley, B., Leahigh, L., & Vanleit, B. (2006). A preliminary study of the effects of a modified mindfulness intervention on binge eating. *Complementary Health Practice Review, 11*, 133–143.

Spinella, M., & Lyke, J. (2004). Executive personality traits and eating behavior. *International Journal of Neuroscience, 114*, 83–93.

Tresley, J., & Sheehan, P. M. (2008). Refeeding syndrome: Recognition is the key to prevention and management. *Journal of the American Dietetic Association, 108*, 2105–2108.

Tribole, E., & Resch, E. (2003). *Intuitive eating: A revolutionary program that works* (2nd ed.). New York, NY: St. Martin's Griffin.

van Son, G. E., van Hoeken, D., van Furth, E. F., Donker, G. A., & Hoek, H. W. (2010). Course and outcome of eating disorders in a primary care-based cohort. *International Journal of Eating Disorders, 43*(2), 130–138.

Waters, A., Hill, A., & Waller, G. (2001). Bulimics' responses to food cravings: Is binge-eating a product of emotional hunger or emotional state? *Behavior Research and Therapy, 39*, 877–886.

Williams, P., Goodie, J., & Motsinger, C. (2008). Treating eating disorders in primary care. *American Family Physician, 77*, 187–195.

CHAPTER 9

Science or Art?
Integrating Symptom Management into Psychodynamic Treatment of Eating Disorders

Nancy L. Cloak and Pauline S. Powers

...what we call their symptoms they call their salvation (**Boris, 1984**)

INTRODUCTION

As one veteran clinician in the field succinctly observed, "Treating eating disorders is not for the faint of heart" (Zerbe, personal communication, April, 2009). Not only are the eating disorder symptoms themselves frequently intransigent and ego-syntonic, but most patients—90% in one naturalistic study (Thompson-Brenner & Westen, 2005a)—suffer from co-morbid psychiatric disorders that complicate management and impede recovery (Hudson, Hiripi, Pope & Kessler, 2007). Although many researchers advocate the use of manual-based treatments that have demonstrated efficacy in randomized controlled trials, both clinical experience and an increasing number of outcome studies indicate that while these treatments often produce symptomatic improvement, they rarely lead to full recovery. For example, while Cognitive-Behavioral Therapy (CBT) is considered the evidence-based treatment of choice for bulimia nervosa, only seven of 43 outcome measures in 15 trials of CBT showed that patients' post-treatment binge-purge frequency or Eating Disorder Examination (EDE) scores were equivalent to those of a normal sample (Lundgren, Danoff-Burg & Anderson, 2004). In another analysis of 19 studies, purge abstinence rate after CBT ranged from only 35 to 55% (Richards et al., 2000). For anorexia nervosa (AN), evidence-based treatment is essentially non-existent at this time. Other than family therapy for younger adolescents, there is little evidence to support any specific intervention for anorexia (Bulik, Berkman & Brownley, 2007); time to recovery is protracted (Strober, Freeman, & Morell, 1997) and less than half of patients are described as recovered at long-term follow-up (Steinhausen, 2002). Additional research on interventions for eating disorders is urgently needed but, in the interim, clinicians must find ways to treat their patients effectively.

A common approach is to use a mixture of interventions, and this is supported by treatment guidelines, especially for AN (Work Group on Eating Disorders, 2006). Given the tenacity of the symptoms, substantial co-morbidity, and limitations of evidence-based treatments, it is not surprising that most therapists identify something other than CBT as their primary approach to treatment (Simmons, Milnes & Anderson, 2008). A substantial minority of these clinicians describe themselves as primarily psychodynamic in orientation (Thompson-Brenner & Westen, 2005b) and there is a large body of literature about psychodynamic approaches to treating eating disorders. While psychodynamic psychotherapy is sometimes dismissed as not being evidence-based, a growing body of sophisticated research supports its effectiveness in a broad range of psychiatric conditions (Knekt et al., 2008; Leichsenring, Rabung & Leibing, 2004; Leichsenring & Leibing, 2007). In populations with chronicity and complex co-morbidity, long-term psychodynamic psychotherapy is significantly superior to shorter-term methods of psychotherapy with regard to overall outcome, target problems, and personality functioning (Leichsenring & Rabung, 2008). Furthermore, psychodynamic therapy has been effective in a number of controlled studies of eating disorder patients, with short-term outcomes equivalent to CBT in studies where the comparison was made (Bachar, Latzer, Kreitler & Berry, 1999; Dare, Eisler, Russell, Treasure & Dodge, 2001; Garner et al., 1993; Treasure et al., 1995). Thus, a long clinical tradition combines with more recent outcomes literature to provide a solid evidence base for using a psychodynamic approach to the treatment of eating disorders.

However, there are some difficulties inherent in using an approach that focuses primarily on relationship, exploration, and interpretation when treating patients with eating disorders. Though it is true that short-term, behaviorally-oriented treatments often result in limited improvement and symptom-substitution, it is equally true that long-term psychodynamic therapies can flounder due to intractable symptoms. Symptoms can become health- or even life-threatening, creating a sense of urgency (and sometimes, actual emergency) rarely experienced with other disorders. Frightened family members and managed care pressures add to this recipe for therapist angst, making it extremely difficult for clinicians to retain a reflective stance and maintain a treatment frame that facilitates psychodynamic work. Therapists are often torn between the need for interventions for symptoms and a desire to understand, or at least not re-enact, the patient's problematic family relationships. As pressures build and objectivity diminishes, therapists become vulnerable to unproductive enactments unless they have a framework for introducing symptom-directed interventions. On the other hand, implementing symptom-directed interventions without understanding relevant psychodynamic factors can be problematic, as Case 1 illustrates.

CASE 1

Alan was a very disturbed teenager with severe binging-purging type AN who was quite expressive in his art work during art therapy, though he also had very low self-esteem and did not recognize that his work was expressive, symbolic, and very well executed. The treatment team suggested that he have an art show of his work in the hopes that it would improve his self-esteem. He reluctantly agreed. When the art work was displayed, it was quite revealing and indicated a marked sexual preoccupation and sexual conflict that hadn't been previously apparent. After the art show, he became less cooperative and ran away from the hospital. Alan's core conflicts had not been fully elucidated and understood in his psychodynamic psychotherapy prior to introducing art therapy, and the art show was a premature confrontation that resulted in his leaving treatment.

The example in Case 1 illustrates the importance of psychodynamic understanding and the hazards of premature integration of active interventions. Alan's lack of enthusiasm for the art show should have been a clue to his team that he was not ready for this intervention. A psychodynamic framework helps the therapist determine the best nature and timing of an active intervention, and to understand and make corrections in the process when symptom-directed interventions are resisted. When active interventions are integrated within a psychodynamic psychotherapy process, patients and therapists can benefit from the greater efficacy of longer-term psychodynamic treatment in addressing chronic and complex mental disorders, and premature terminations can often be prevented.

INTEGRATING SYMPTOM-FOCUSED INTERVENTIONS WITHIN PSYCHODYNAMIC THERAPY: THEORY

Traditionally, classical psychoanalysts have avoided incorporating behavioral interventions into psychodynamic therapy processes as much as possible, believing that activity by the therapist could forestall the development and elaboration of transference and foster acting on the counter-transference. However, even the founder of psychoanalysis acknowledged that free association and interpretation sometimes had to be supplemented by other interventions when he advocated a form of exposure therapy for agoraphobia: "One can hardly master a phobia if one waits till the patient lets the analysis influence him to give it up...One succeeds only when one can induce the patients by the influence of the analysis...to go into the street and to struggle with their anxiety while they make the attempt" (Freud, 1919, p. 166). Most of Freud's early followers were less flexible, perhaps because they were uncertain about how to use behavioral interventions while remaining faithful to psychoanalytic tenets.

More recently, psychoanalysts with relational and social constructivist perspectives have since argued that it is preferable to understand the idiosyncratic meanings of the therapist's actions to the patient rather than to try to avoid activity altogether (Mitchell & Black, 1995). Indeed, some have emphasized the potential for classical, "abstinent" analytic behavior to recapitulate the patient's painful childhood relationship with a schizoid, depressed, or neglectful parent, with the result that the patient either flees treatment or acts out in order to draw the therapist into greater involvement (Hoffman, 1994; Searles, 1999). Relational theory allows considerable freedom for active interventions in psychodynamic therapy by defining transference as the patient's idiosyncratic perspective on the real activity of the therapist. Thus, rather than obscuring the transference, introduction of behavioral techniques may highlight previously unrecognized aspects of the transference (Tobin & Johnson, 1991). For example, patients with narcissistic traits may avoid CBT homework because they feel it is somehow "beneath" them, while others may experience it as punitive or controlling. Still others may welcome it as an opportunity to be effective and independent of the therapist. If explored sensitively, the patient's reactions to the behavioral intervention often bring up new historical material that can be used to understand and modify problematic object relations.

INTEGRATING SYMPTOM-FOCUSED INTERVENTIONS WITHIN PSYCHODYNAMIC THERAPY: PRACTICE

General Considerations

Patient selection, timing, and the therapist's expertise with various interventions are important factors that affect decisions about including active interventions within psychodynamic psychotherapy. Although probably both treatments (psychodynamic psychotherapy and symptom-focused interventions) include aspects of each other (whether acknowledged or not), the presentation of the patient often determines which methods will be primary. At one extreme, for example, is the patient who comes to treatment indicating that he or she needs help, that their eating behavior is out of control, and that they are willing to do whatever it takes to recover. If such a patient has little evidence of character disorder, a supportive healthy family, and a brief history of bulimia nervosa, straightforward educational or cognitive behavioral approaches may suffice. One might expect that such a patient could be given a self-help manual for cognitive behavioral therapy and recover without additional treatment. However, even in this situation, meeting periodically with a therapist who provides guidance and support while the patient utilizes the CBT manual results in a much greater likelihood of response and recovery (Perkins, Murphy, Schmidt & Williams, 2006). At the other extreme, patients with a classic neurotic issue (for example, an unresolved oedipal conflict) that is entwined with an eating disorder might be considered ideal candidates for longer-term psychodynamic psychotherapy. However, even these patients can benefit greatly from incorporation of appropriately timed active interventions. Finally, some patients with multiple, serious, chronic illnesses (e.g., a patient with alcohol dependence, AN, and narcissistic personality disorder) might seem untreatable with any method. These are patients who may initially require draconian methods, including involuntary hospitalization and court ordered medication. Even so, an empathic doctor—patient relationship, in which underlying psychodynamic conflicts are considered during the course of a set of very active interventions, is likely to lead to a more positive outcome. In time, the patient may recover sufficiently to be able to participate in meaningful insight-oriented psychotherapy with strategically integrated active interventions.

The hallmarks of successful integration are flexibility and creativity. We liken the treatment to putting together a puzzle rather than following a road map in a manual. Therefore, instead of drawing a map, we will outline a few principles that may serve as pieces for the puzzle that each therapist—patient dyad must put together to promote healing.

Respect the Symptoms

The first puzzle piece represents an attitude rather than an activity. Faced with frightening medical complications, distressed family members, and impatient insurance reviewers, even psychodynamically oriented clinicians can quickly begin to view symptoms only within the frame of "things to be gotten rid of." This is especially true when behavioral interventions are implemented. We must bring the wise words of Harold Boris (1984) into our work by recognizing that the illness, though destructive, represents a psychological achievement that has saved the patient from something worse. As another well known clinician (Craig Johnson, personal communication) remarked, "There are worse things than eating disorders."

There are several reasons why respecting symptoms forms an important foundation for subsequent interventions directed at ameliorating them. First, symptoms themselves represent an important part of the patient's experience—respecting them conveys respect for the patient. Second, exploring symptoms can open up a wealth of associations and historical information that may not have come to light otherwise. A third reason is that often it is only in a respectful, safe atmosphere that the patient will reveal the details that are necessary to plan good behavioral interventions. Finally, as seen in the following case example, one common function of eating disorder symptoms is to preserve autonomy. Respecting symptoms can facilitate a subtle shift in the therapeutic relationship that paradoxically makes it easier for the patient to give them up.

CASE 2

Anita, a 22-year-old college student with a history of anorexia in adolescence, entered psychodynamic therapy for anxiety and depression related to her upcoming graduation. However, she appeared to be losing weight and was not discussing this in therapy. One day when Anita referred to her "anorexic episode" while discussing recent events in her family, the therapist remarked, "*It was kind of ingenious for a 14-year-old to come up with something* [the anorexia] *that would show your folks they couldn't control you, while at the same time keeping them close and involved.*" Anita was momentarily quiet and thoughtful while this sank in, then said, "*But you know, that didn't really work—it got me locked up. And I'm afraid I'm going back to some of the same habits more recently.*"

Discuss the Symptoms in Detail

While exploring the meaning and functions of symptoms is a basic ingredient of all psychodynamic psychotherapy, details about the eating disordered patient's symptomatic thoughts and behaviors often get overlooked, for several reasons. First, the patient is ashamed of the symptoms, or, on the other hand, is protective of them and equates revealing them to giving them up. Second, therapists may assume that they know what a patient means when in fact they don't. What constitutes a binge? What does it mean when the patient says he or she is restricting? This information might be obtained on intake, but not routinely re-explored. Third, therapists may have their own countertransference discomfort with details about the symptoms, especially binging, vomiting, and laxative abuse. Finally, other topics, such as relationships or childhood events, may seem more interesting or important. Sometimes this is true, but it can also be an expression of resistance or denial on the part of the patient or therapist (Gutwill, 1994).

We believe that talking with patients about symptomatic behavior in detail is important, not only because it conveys respect for the symptom (and the patient), but also because it elicits two kinds of important information. First, we learn things that are important for cognitive-behavioral interventions, such as antecedents, consequences, and maladaptive cognitions that can sustain the behaviors. Secondly, we often hear important themes about pieces of history, and family interaction patterns that might not have been revealed otherwise. Detailed discussions of symptoms deepen the alliance and facilitate the development of a "shared language" between patient and therapist. With practice, clinicians can develop skills for "listening with both ears" when the patient describes her symptoms—that is, attending both to the behavioral factors and to the psychodynamic meanings.

To explore symptoms fully, we must be willing to ask very specific questions. Talking about the behaviors in detail ameliorates the patient's shame and expectation that the therapist will view symptoms as *"just stupid, something I need to get over."* This expectation may serve as a defense against knowing what the eating and body image disturbances reveal about the core of the patient's most profound and intimate struggles (Bloom, Kogel & Zaphiropoulos, 1994). For example, if the patient says that she had a hard day at work, binged and purged, and ended up hating herself, ask questions like, "What was hard about work?" "When did you start thinking about binging?" "What food did you think about binging on?" "Did you try to stop it? How?" "What did you do when you left work…and then…and then…etc?" "What did you do after binging?" "What did you find so hateful about yourself?" "Did you have other feelings?"

Deconstruct the Meanings of Food and Weight

In addition to talking about symptomatic behaviors, therapists should be prepared to discuss specifics about food, weight, and size. Again, this is not just important for planning cognitive-behavioral interventions—it also attends to a significant and emotionally charged aspect of the patient's experience (Bloom & Kogel, 1994a). One clinician likens discussions about food to explorations of dreams. In the appropriate context, she will ask patients to describe what comes to mind about a particular food item, what childhood experiences the patient might have had with different foods, or what she imagines the therapist does or does not eat (Zerbe, 2008). This can become part of a dialectic in which educational and behavioral interventions are woven together with psychodynamic explorations, as the following vignette (case 3) illustrates.

CASE 2, CONTINUED

Several months later, Anita was telling her therapist about challenging herself to go out to breakfast with some fellow students. Afterwards she had intense urges to purge and exercise, but she was at a loss to explain this resurgence of symptoms. In the ensuing discussion, the therapist asked Anita to tell her about what she ate. Anita replied, *"Orange juice, eggs, and 2 pieces of toast—not a big deal for me anymore."* When the therapist asked her to associate to these particular food items, however, Anita said, *"Now I remember. I couldn't put any jelly on my toast, even though everyone else was. I remember that was how my anorexia started: I wouldn't put anything on my bread. My dad used to make fun of me for it—he didn't understand."* The therapist encouraged Anita to work on adding appropriate condiments to her meals, while facilitating a fruitful discussion about how her father's responses continued to trigger her symptoms.

Be Aware of Common Functions of Symptoms.

Understanding the meanings and functions of symptoms is a major goal for psychodynamic psychotherapy in itself. It also is an essential foundation for integrating non-dynamic interventions within the therapy, because knowing the functions served by a particular patient's symptoms allows us to better plan the type, timing, and potential reactions to any kind of non-dynamic intervention In addition to suggesting clarifications or interventions that we might otherwise overlook; another important role of theoretical literature is to help contain our anxieties and foster reflective thinking, much as a good supervisor would do (Almond, 2003).

A thorough review of the meanings and functions of eating disorder symptoms could be the subject of another book. We have chosen to list a number of them in Box 9.1. We will explore, in more detail, one of the most common and vexing uses of symptoms: to test the therapist within the transference.

A common example of a test of control and structure (Stern, 1986) is a patient with AN who continues to lose weight, yet does not acknowledge his or her condition or refuses to enter residential treatment. In childhood, many such patients were effectively coerced into complying with overly strict rules that met their caregivers' narcissistic needs, yet lacked more benevolent forms of discipline such as having limits set in their best interests. These patients present therapists with a dilemma: should the therapist aggressively pursue the higher level of care, and risk being perceived as an autocratic, narcissistic caregiver, or should they allow the patient to remain as he or she is, and thus express the uninvolved, neglectful characteristics of the caregiver? In general, a therapist passes a test when he or she is able to attend to both sides of the conflict. Often the best approach is to turn the dilemma back to the patient (Hoffman, 1992) by saying something like, *"I'm extremely concerned about your weight loss. Part of me wants to try to make you go to residential treatment, but I'm afraid you'd feel controlled and pushed. Yet I'm afraid you'd experience me as not really understanding your illness or caring about your well-being if we do nothing. How can we deal with this?"* Additional guidelines for responding to tests involving control and structure are listed in Box 9.2.

While transference tests are often painful for therapists, we must realize that the patient is from the outset presenting us with two of his or her most fundamental dilemmas: how to become one's true self, and how to get one's dependency needs met without being controlled. One positive aspect of these quandaries is that they highlight another important piece of the puzzle: the importance of understanding transference implications of symptoms as a foundation for designing non-psychodynamic interventions.

Relate the Symptom to the Transference

Another common function of symptoms is as a communication to the therapist, often about transference issues (Hamburg, 1989). Symptoms can communicate feelings or ideas that the patient cannot yet put into words, and thus worsening symptoms may represent "speaking louder so that the therapist can hear" (Brisman, 1996).

BOX 9.1

SOME COMMON PSYCHODYNAMIC FUNCTIONS OF EATING DISORDER SYMPTOMS

- Rebel against a strict caregiver/superego and express autonomy (Zerbe, 1998)
- Test the therapist to see if s/he will respond in the same way as early caregivers (Weiss & Sampson, 1986)
- Displace anxiety or shame onto fears of weight gain or hatred for the body (Mintz, 1985b)
- Avoid recovery because it is perceived as resulting in overwhelming demands for performance (Mintz, 1992)
- Avoid recovery because it is equated with becoming a narcissistic extension of the therapist (Reich & Cierpka, 1998)
- Substitute self-destructiveness related to the eating disorder for suicide (Mintz, 1985a)
- Rid the self of an intrusive and/or abusive internalized caregiver (Zerbe, 1993)
- Preserve a relationship with an internalized caregiver (Zerbe, 1996)
- Elicit caregiving that was absent in earlier life, including limit-setting (Zerbe, 2001)
- Maintain a pathological identity (Dennis & Sansone, 1990)
- Avoid recovery because it is equated with losing the therapist (Garner, Garfinkel & Bemis, 1982)
- Distract both patient and therapist from painful topics (Thompson & Sherman, 1989)
- Avoid mourning (particularly of lost youth in older patients) (Zerbe, 2008)
- Precipitate rejection by the therapist, because it is less painful to be rejected for unacceptable behavior than for intrinsic unacceptability (Garner, Garfinkel, & Bemis, 1982)
- Regulate feelings and protect the self from fragmentation or emptiness (Krueger, 1997)
- Express emotional pain for which the patient cannot yet find words (Bloom & Kogel, 1994b)
- Express anger or exact revenge (Stern, 1986)
- Communicate feelings to the therapist via projective identification (Mintz, 1992)
- Illustrate problematic object relations via metaphor (e.g., the relationship with food parallels relationships with important others) (Rozen, 1993)
- Re-enact dissociated traumatic memories (Gutwill & Gitter, 1994)
- Protect the self against re-traumatization by becoming unattractive (Gutwill & Gitter, 1994)
- Undo or compensate for positive steps in other areas of life out of unconscious guilt (Reich & Cierpka, 1998)
- Obtain nurturance and connection while avoiding the risks of intimacy (Rozen, 1993)
- Communicate transference feelings that are too frightening to express in words (Hamburg, 1989)
- Defend against feelings of powerlessness and ineffectiveness (Bruch, 1982)

BOX 9.2

RESPONDING TO PATIENTS' TESTS OF CONTROL AND STRUCTURE (STERN, 1986)

- Keep the patient's needs and interests primary at all times
- Provide only as much structure as the patient needs, and only in the areas he or she needs it
- Take a clear, firm therapeutic stance, emphasizing the concern for the patient's well-being
- State your rationale, but don't insist that the patient agree with you
- Set clear behavioral limits and apply consequences consistently
- Reassess the needs for structure frequently as treatment progresses

---

CASE 3

Barbara, a 35-year-old woman with EDNOS (Eating Disorder Not Otherwise Specified), cancelled several sessions after her therapist gave her a self-monitoring assignment, then arrived in great distress, reporting worsening symptoms. Her therapist then inquired, *"When we talk, you say you appreciate me and like my suggestions for getting better control of the binging and purging, but your cancelling sessions says something else. Are there perhaps other parts of our relationship that you find more difficult to express in words?"* Barbara replied that she was ashamed that she could not control her symptoms and felt put down by the therapist's suggestions of *"oversimplified behavioral stuff."* Yet she did not want to risk losing the therapist by being open about her feelings, stating *"You just did not do that in my family."* Her therapist realized that he would have to consider Barbara's tendency to be narcissistically injured by interventions whenever he made concrete suggestions to Barbara about managing her symptoms.

Even when explicit transference interpretations are not made, clinicians must be aware of transference issues when designing and introducing non-psychodynamic interventions. Before intervening, the therapist should ask him or herself, "What role is the patient placing me in? Given this patient's relationship history, how are they likely to experience my recommendation for this particular intervention at this particular time?" Sometimes an intervention must be implemented before the answers to these questions are clear, but they can be answered retrospectively in collaboration with the patient, a process that is often very beneficial see Chapter 19).

Monitor the Countertransference

Therapists should also ask themselves a third, critical question: "What part, if any, of the motivation for my intervention is based on my countertransference rather than the patient's needs at the moment?"

CASE 3, CONTINUED

Barbara continued to struggle with binging despite some initial improvement. She still cancelled sessions occasionally and seemed to regard the therapist's behavioral suggestions as simplistic. The therapist began to feel that he lacked sufficient skills and ideas, and was about to recommend a medication consult and possibly a referral to a CBT practitioner whom he had recently met at a conference. He decided to bring the case to peer supervision first, and in the process of presenting realized that his reactions to Barbara's behavior were helping him understand her experience as the family "genius" who was still not taken seriously. His reaction—the heretofore unconscious wish to flee the treatment—was exactly how she responded to narcissistic injuries in relationships.

This vignette illustrates one of the common functions of symptoms—as a nonverbal communication of some aspect of the patient's experience via projective identification (Grotstein, 1994). In fact, eating disorder symptoms commonly engender in the therapist the same feelings of helplessness and ineffectiveness as the patient has experienced (Mintz, 1992). The therapist's willingness to bear the distress associated with Barbara's projections bore fruit in the form of enhanced understanding of her inner relational world. Barbara may still have benefited from a medication consult or CBT referral, but by containing and processing his reaction first, her therapist could allow her needs, rather than his reactions, to determine the nature and timing of the interventions.

The vignette also illustrates how implementation of non-psychodynamic interventions can sometimes reflect the counter-transference difficulties of accompanying a patient on their often painful journey to healing (Zerbe, 1998). Though withholding interventions can be an equally hazardous form of enactment, therapists often end up wanting to "feed" the patient—in the form of symptom-directed interventions—too much too soon, related to the anxiety and frustration that the dangerous and highly visible symptoms arouse (Chessick, 1984). We must not forget that listening is an ancient and profound mode of healing in itself. There are few situations so urgent that there is no time to explore the patient's understanding of the worsening symptoms, and what he or she thinks might be helpful.

Discuss the Intervention with the Patient and Explore His or Her Reactions

One of the elements that makes psychodynamic therapy psychodynamic is its emphasis on the patient's unique subjectivity (Gabbard, 2004). For the therapy to remain truly psychodynamic, the patient needs to be given accurate factual information about the proposed interventions and then invited to explore his or her idiosyncratic reactions to them.

CASE 3, CONTINUED

The therapy alliance palpably improved after Barbara's therapist was able to bring some of his insights about the transference—counter-transference interactions into the treatment. It was Barbara who brought up the idea of trying fluoxetine. The therapist encouraged her to explore what it would be like to take medication for her eating disorder, and how she felt about having him prescribe it versus a referral to another psychiatrist. After spending most of the session discussing the meanings of taking medication, the therapist said, *"Now I'd like to set aside a little time to talk about the fluoxetine—what it is, how to take it, side effects, and how we'll monitor its effectiveness. That will feel different than our usual discussions, and when we're done I'll want to hear what that felt like for you, to work with me as your medication doctor."*

Any non-psychodynamic intervention that is introduced needs to be provided with skill and the expectation of efficacy. For example, a psychiatrist who prescribes medication within a psychodynamic treatment process should use the same care with informed consent and follow-up that she or he would use in a medication management case. The clinician needs to decide whether she or he has the appropriate skills or if the patient should be referred. If there is a choice between a referral or an intervention provided by the original therapist, the patient should be supported in making the choice that fits best for him or her, through a discussion of the possible impact of either alternative. If a patient is referred to another clinician for the intervention, the therapist should obtain a signed release to speak with the other clinician, and periodically follow up with both patient and clinician regarding the process and progress.

Patients at times have strong negative reactions to the therapist's recommendations; this is particularly common in working with those who have borderline personality organization. In these cases, it is helpful to look for patients' struggles with autonomy and dependency and be aware of the transference implications of their reactions. One should remind these patients that the ultimate purpose of the therapy is to restore their autonomy, which is being compromised by the eating disorder. In these challenging cases, it is most important to monitor one's counter-transference, maintain an empathic stance, give the patient ample opportunity to air his or her views, and avoid compromising the treatment frame or backing down on a recommendation if it is truly necessary (Kernberg, 1995).

Develop a Repertoire of Active Interventions

In this chapter, an underlying theme is that psychodynamic psychotherapies and active interventions can often be integrated to promote recovery, although doing so requires a thorough understanding of both approaches and skillful, empathic implementation. There is a range of active interventions that might be part of the treatment plan for eating disorder patients in dynamic or psychoanalytic psychotherapy. Among these are cognitive-behavioral techniques, educational strategies, art and expressive therapies, family interventions, and medications.

Cognitive-behavioral techniques. Although some patients may be able to recover from an eating disorder with guided CBT, probably most require more extensive interventions relying on at least some elements of psychodynamic therapy. For patients already in psychodynamic or psychoanalytic therapy, choosing when and how to introduce CBT can be difficult. Some patients may be quite resistant to CBT early in treatment. They may even have started the psychodynamic therapy process because of an earlier failure with CBT. However, during the course of psychotherapy, patients may recognize the core conflicts that have facilitated the development of the eating disorder and may try various behavioral strategies of their own without success. This may represent an optimal time for introducing cognitive-behavioral techniques into a therapy which is predominantly psychodynamic.

CASE 4

Jane had worked for several months in understanding her relationship with her mother, who had worked hard to support the family after the death of her father, but had often been unavailable to Jane. After unsuccessfully trying to cope with binge eating on her own, Jane asked her therapist for advice for the first time. This willingness to ask for help not only had many psychodynamic meanings, some of which the therapist interpreted to the patient, but it also provided the therapist with the opportunity to introduce elements of CBT into the therapy, such as addressing the distorted cognition that only certain foods are "healthy." The patient was willing to keep a diary of her food intake and trusted the therapist enough to include details of extensive binges which had previously been shameful for her. The therapist assisted the patient in devising behaviors more likely to interrupt binge eating (such as eating three meals a day and a range of foods including those that Jane particularly liked). The experience of working with the therapist in a practical definitive way on a problem was a corrective emotional experience for Jane.

Self-monitoring of thoughts, feelings, and behaviors related to the eating disorder is a key component of cognitive-behavioral therapy, and is usually initiated at the beginning of CBT treatment. Patients are typically asked to write down details about all food consumed, the time, place, context, and associated thoughts and feelings; and details about eating disorder behaviors such as binging, purging, exercising, or body checking. It is critically important that the recording be done in "real time;" i.e., at, or immediately after, the behavior occurs, and that the records be reviewed with the patient consistently (Fairburn, 2008). Self-monitoring may be a particularly useful tool for incorporation into psychodynamic psychotherapy because it can highlight behavior patterns with psychodynamic significance, as exemplified in the foregoing discussion of how Anita's conflict (Case 2) about putting jelly on her toast had its roots in her interactions with her father. In addition, self-monitoring alone is well-known to produce significant behavioral changes (Korotitsch & Nelson-Gray, 1999).

However, compliance with self-monitoring is often low. For example, members of one weekly CBT group kept records on only 42% of possible days, despite being selected for higher motivation (Nichols & Gusella, 2003). Shame, rebellion, and transference distrust or anger are just a few of the psychodynamic themes that may underlie patients' difficulties with self-monitoring. As discussed above, symptoms are expressions of patients' most intimate secrets, and they are understandably fearful of the exposure entailed by keeping and discussing detailed records. This resistance may be disguised as concerns about the time involved, fears that monitoring will worsen food preoccupations, or statements that "I tried that before, and it didn't work." When introducing self-monitoring, the clinician should address the surface concerns, but also proactively and empathically discuss the underlying issues. For example, one could say something like, "Many people find it scary or embarrassing to be this specific about their behavior. That is very understandable—we know that the eating disorder is a very important part of your life; otherwise, why would it be so hard to change? But the more we know about the truth of your experience, the better we can understand and change it."

Experts in CBT caution against using selected components of a manualized treatment apart from the entire protocol (Fairburn, 2008), arguing that the whole is greater than the parts and outcomes will not be equivalent. However judiciously incorporating elements of CBT (such as self-monitoring and regular eating) into a psychodynamic process is not the same thing. The theraphy is not labeled "CBT" by the therapist, and its theory and techniques remain predominantly psychodynamic.

Educational strategies. Many people who have eating disorders do not come willingly for treatment, or when they **do** come (often at the behest of someone else), they make it clear that they do not plan to enter into treatment or, if compelled to do so, will not cooperate. Many eating disorder patients are adept at *appearing* to cooperate with treatment when they are actually only complying with what they are asked to do. A typical example would be a young teenager with AN who is hospitalized on the pediatric unit and consumes all the calories brought to her by the nursing staff, but goes home and immediately refuses to consume adequate calories. Using the readiness for change concept, this patient is in the pre-contemplation stage. In other words, change is not even being considered. In this situation, providing information might seem irrelevant, but it may plant the seed for the patient to decide eventually to consider treatment.

A second strategy is the therapeutic use of symptoms, signs, and laboratory data. In this circumstance, the physician specifically connects the pathological behavior (e.g., purging by vomiting) to the symptoms (for example abdominal pain), signs (tenderness on palpation of the abdomen), and laboratory data (e.g. elevated pancreatic amylase and elevated lipase). The physician then explains that the patient has pancreatitis, a dangerous, potentially life-threatening condition. The patient may still only permit emergency intervention and refuse long-term treatment of the eating disorder and its underlying causes, but may remember the connections delineated by the doctor and be more willing to consider full treatment the next time there is a crisis.

A third educational example is the use of letters. After a full evaluation including interpretation of the results and recommendations for treatment, a patient with an eating disorder often does not return. Preparing a detailed letter to the patient describing all of the findings including the effects of their eating disorder on their quality of life may provide the information and impetus needed for him or her to eventually enter treatment. We have frequently been surprised to see patients several months or years later appear with our letter in their handbag saying they are now ready to take a step forward in recovering from their eating disorder.

Art Therapy. There are several situations in which art therapy can lead to very important insights, both for the patient and for members of the treatment team. Patients with eating disorders, particularly those with restricting AN, can be very conflict avoidant. Some people are able to recognize conflicts in their life via art therapy that they are otherwise unable to identify. However, this can be a two-edged sword, as illustrated by the case of Alan discussed earlier in Case 1.

Art therapy can also reveal common problems interfering with function that have not always been apparent. For example, one art therapist organized a group art project that was to be shown in the treatment center. The group chose to make a very large colorful fish that was to be hung in the central dining room of the treatment facility. Although the actual project was expected to take only a few weeks to complete, it actually took several months. The issues that became apparent were problems shared by the entire group, including: passivity (some who had agreed to participate, later wished they had not agreed); inability to cooperate in designing a feasible plan to make the fish durable; passive–aggressive behavior; and conflict-avoidance (instead of discussing disagreements about the design, some patients

FIGURE 9.1 Fish art project, a beatiful creation accomplished with great difficulty, illustrating how patients' considerable intelligence and creativity can be juxtaposed with passive-aggression, conflict-avoidance, and over-compliance. Please refer to color plate section.

simply destroyed what another had created and inserted their own creation). The fish creation still hangs in the cafeteria (Figure 9.1), and is a constant reminder to the treatment team of themes often seen in patients with eating disorders—passivity, conflict-avoidance, compliance versus cooperation, and passive–aggressive behavior traits.

As a final example, art therapy can reveal the intensity of a problem when it may not be apparent in traditional psychotherapy. Brian was a 23-year-old man with severe bulimia nervosa and alcohol dependence. Figure 9.2 ("An Eye for Alcohol") is a drawing that he did during art therapy that dramatically captures the severity of his alcohol dependence, which turned out to be more serious than his eating disorder and less responsive to treatment.

Expressive Therapies. Gestalt therapy, psychodrama, and dance and movement therapies are other examples of powerful expressive therapies that can often be very helpful. Carefully chosen at key points in treatment, they can be both diagnostic and therapeutic.

CASE 5

Alicia, a 25-year-old woman with bulimia nervosa, had been in psychodynamic treatment for several years and had improvement, but not resolution, of her eating disorder symptoms and relationship difficulties. Her psychiatrist thought that there were issues that she was not able to discuss verbally, and referred her for dance therapy. In the course of this work, particularly when any touch was involved, the patient became paranoid, suspicious, and actually ran as far from her dance therapist as she could. As she discussed this with her individual therapist, she described a series of events with her stepfather who had sexually molested her. It became apparent to the patient that her reaction in the dance group was related to the abuse, and she was able to explore it verbally and come to an emotional resolution, which eventually resulted in an improved relationship with her partner.

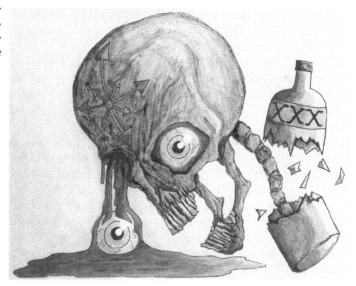

Figure 9.2 "An Eye for Alcohol," illustrating how themes previously unidentified in verbal psychotherapy may emerge in art therapy. Please refer to color plate section.

Family Interventions. Family interventions can be very powerful, but choosing how and when to proceed requires a thorough understanding of the patient. Although psychodynamic conflicts that have originated within the family are often the main themes of individual therapy, actual interventions with the family may also be helpful. With young patients who are living at home, family therapy may be crucial to recovery. The "Maudsley Method," a recently popularized treatment strategy for adolescent patients with AN, relies exclusively on family work until the patient is fully weight restored. At that point, a determination is made regarding further family or individual psychotherapy (Lock, Le Grange, Agras & Dare, 2002). Although there is often dramatic improvement in eating behaviors and weight, some family members complain that the actions taken to support weight restoration have interfered with their relationship with their daughter or son. This complaint is surprisingly similar to the concern that individual therapists have when implementing active therapies. Although the Maudsley Method has promise, developing a better psychodynamic understanding of the patient (and family) prior to implementing this kind of intervention may be more helpful in the long run and less potentially damaging to family relationships.

Medications. Timing the introduction of medication is important, for both physiological and psychodynamic reasons. It is important that the patient be ready to accept the medication and informed about its possible benefits and side effects. This can often be successfully accomplished in discussions between the therapist and patient and supplemented with educational materials. In addition, the physician prescribing the medication (who may also be the therapist) should know when the medications are likely to work. Semi-starved, poorly nourished eating disorder patients are often given medications early in treatment when they are unlikely to work. The serotonin reuptake inhibitors require adequate brain serotonin, which may not be available in patients consuming inadequate carbohydrates, and use of these medications when the patient is semi-starved is not only ineffective, but counter-therapeutic. The patient may come to believe that the medications can never be helpful, even though they might have been if given after nutritional status had improved. Delgado and his colleagues have shown that the SSRIs are ineffective when patients are on a calorie restricted diet (Delgado et al., 1990).

Cognitive behavioral techniques, educational strategies, art and expressive therapies, family interventions, and medications are just a few of the active interventions that can be incorporated into a psychodynamic psychotherapy process. The range of potential interventions is broad, and in addition to the foregoing, may include a 12-step approach, gestalt techniques, guided imagery, role playing, dialectical behavior therapy skills, body image techniques, and others. Both dynamic understanding and behavioral change can be enhanced by judicious implementation of active techniques when therapists are mindful of the theoretical and practical factors discussed above. Key considerations include respecting and deconstructing the symptoms, understanding the transference and counter-transference implications of recommending an active intervention, and carefully choosing the timing of the intervention. Also, clinicians should only offer those interventions in which they are skilled, or be prepared to refer. When possible, empirically supported interventions, such as cognitive-behavioral techniques, family therapy for adolescents, and fluoxetine for bulimia nervosa, should be preferred. However, many patients have already tried these treatments without benefit and may need alternative approaches. Finally, patients' responses to interventions should be monitored to ensure efficacy and identify previously unrecognized dynamic factors.

Monitor the patient's reactions to the intervention

Whether the therapist provides the non-psychodynamic intervention or refers the patient, she or he must monitor the patient's adherence to the recommendations, reactions to the change in the therapy process, and symptomatic improvement or lack thereof. As discussed above, patients' reactions to interventions often highlight previously unrecognized aspects of the transference or early history.

CASE 2, CONTINUED

Anita and her therapist had agreed that some family sessions would be helpful, and Anita eagerly accepted a referral list. Two weeks later, her therapist began a session by saying, "*Before we start, I'd like to take a little time to see how the family therapy has been going.*" Anita was silent, then somewhat ashamedly admitted that she had not arranged any appointments. At first she blamed her schedule, but eventually was able to relate her concerns that if she was helped by the family therapy, her therapist would feel abandoned or narcissistically injured, as her parents seemed to be when she chose not to follow in their footsteps for her career.

Monitoring needs to be an ongoing process, and each therapist will develop his or her own style for follow-up. Regardless of the style that is used, however, it is clear that close follow-up is critically important for the efficacy of behavioral interventions. For example, food diaries are powerful behavioral tools only when a therapist reviews them with the patient or returns them with comments (Herzog, Franko & Brotman, 1989).

CONCLUSION

Over time, a therapist-patient dyad will develop a dialectical process in which different non-psychodynamic intervention "puzzle pieces" are put together within a psychodynamic therapeutic frame. The interweaving of sometimes disparate theories and therapies by the clinician provides not only the real help needed by the patient but symbolically models what the patient herself must do internally. Though demanding skill and versatility, we believe that this integration is necessary in meeting the multifaceted needs of our challenging patients.

References

Almond, R. (2003). The holding function of theory. *Journal of the American Psychoanalytic Association, 51*, 131–153.

Bachar, E., Latzer, Y., Kreitler, S., & Berry, E. M. (1999). Empirical comparison of two psychological therapies: Self psychology and cognitive orientation in the treatment of anorexia and bulimia. *Journal of Psychotherapy Practice and Research, 8*, 115–128.

Bloom, C., & Kogel, L. (1994a). Tracing development: the feeding experience and the body. In C. Bloom, A. Gitter, S. Gutwill, L. Kogel, & L. Zaphiropoulos (Eds.), *Eating problems: A feminist psychoanalytic treatment model* (pp. 40–56). New York, NY: Basic Books.

Bloom, C., & Kogel, L. (1994b). Symbolic meanings of food and body. In C. Bloom, A. Gitter, S. Gutwill, L. Kogel & L. Zaphiropoulos (Eds.), *Eating problems: A feminist psychoanalytic treatment model* (pp. 57–66). New York, NY: Basic Books.

Bloom, C., Kogel, L., & Zaphiropoulos, L. (1994). Beginning the eating and body work: stance and tools. In C. Bloom, A. Gitter, S. Gutwill, L. Kogel, & L. Zaphiropoulos (Eds.), *Eating problems: A feminist psychoanalytic treatment model* (pp. 67–82). New York, NY: Basic Books.

Boris, H. N. (1984). The problem of anorexia nervosa. *International Journal of Psychoanalysis, 65*, 315–322.

Brisman, J. (1996). Psychodynamic psychotherapy and action-oriented technique: An integrated approach. In J. Werne & D. Yalom (Eds.), *Treating eating disorders.*. San Francisco: Jossey-Bass Publishers, Inc.

Bruch, H. (1982). Anorexia nervosa: theory and therapy. *Archives of General Psychiatry, 139*, 1531–1538.

Bulik, C. M., Berkman, N. D., & Brownley, K. A. (2007). Anorexia nervosa treatment: A systematic review of randomized controlled trials. *International Journal of Eating Disorders, 40*, 310–320.

Chessick, R. D. (1984). Clinical notes toward the understanding and intensive psychotherapy of adult eating disorders. *Annual of Psychoanalysis, 12*, 301–322.

Dare, C., Eisler, I., Russell, G., Treasure, J., & Dodge, L. (2001). Psychotherapies for adults with anorexia nervosa. *British Journal of Psychiatry, 178*, 216–221.

Delgado, P. L., Charney, D. S., Price, L. H., Aghajanian, G. K., Landis, H., & Heninger, G. R. (1990). Serotonin function and the mechanism of antidepressant action. Reversal of antidepressant-induced remission by rapid depletion of plasma tryptophan. *Archives of General Psychiatry, 47*, 411–418.

Dennis, A. B., & Sansone, R. A. (1990). The clinical stages of treatment for the eating disorder patient with borderline personality disorder. In C. L. Johnson (Ed.), *Psychodynamic treatment of anorexia nervosa and bulimia* (pp. 128–164). New York, NY: The Guilford Press.

Fairburn, C. G. (2008). *Cognitive behavior therapy and eating disorders*. New York, NY: The Guilford Press.

Freud, S. (1919). Lines of advance in psychoanalytic therapy. *Standard Edition, 17*, 159–168.

Gabbard, G. O. (2004). *Long-term psychodynamic psychotherapy: A basic text*. Washington, DC: American Psychiatric Publishing, Inc.

Garner, D. M., Garfinkel, P. E., & Bemis, K. M. (1982). A multidimensional psychotherapy for anorexia nervosa. *International Journal of Eating Disorders, 1*, 3–47.

Garner, D. M., Rockert, W., Davis, R., Garner, M. V., Olmsted, M. P., & Eagle, M. (1993). Comparison of cognitive-behavioral and supportive-expressive therapy for bulimia nervosa. *American Journal of Psychiatry, 150*, 37–46.

Grotstein, J. S. (1994). Projective identification and countertransference: A brief comment on their relationship. *Contemporary Psychoanalysis, 30*, 578–592.

Gutwill, S. (1994). Transference and countertransference issues: social pressures. In C. Bloom, A. Gitter, S. Gutwill, L. Kogel, & L. Zaphiropoulos (Eds.), *Eating problems: A feminist psychoanalytic treatment model* (pp. 144–171). New York, NY: Basic Books.

Gutwill, S., & Gitter, A. (1994). Eating problems and sexual abuse. In C. Bloom, A. Gitter, S. Gutwill, L. Kogel, & L. Zaphiropoulos (Eds.), *Eating problems: A feminist psychoanalytic treatment model* (pp. 205–226). New York, NY: Basic Books.

Hamburg, P. (1989). Bulimia: the construction of a symptom. *Journal of the American Academy of Psychoanalysis, 17*, 131–140.

Herzog, D. B., Franko, D. L., & Brotman, A. W. (1989). Integrating treatment for bulimia nervosa. *Journal of the American Academy of Psychoanalysis, 17*, 141–150.

Hoffman, I. Z. (1992). Some practical implications of a social-constructivist view of the psychoanalytic situation. *Psychoanalytic Dialogues, 2*, 287–304.

Hoffman, I. Z. (1994). Dialectical thinking and therapeutic action in the psychoanalytic process. *Psychoanalytic Quarterly, 63*, 187–218.

Hudson, J. I., Hiripi, E., Pope, H. G., Jr., & Kessler, R. C. (2007). The prevalence and correlates of eating disorders in the national comorbidity survey replication. *Biological Psychiatry, 61*, 348–358.

Kernberg, O. F. (1995). Technical approach to eating disorders in patients with borderline personality organization. *Annual of Psychoanalysis, 23*, 33–48.

Knekt, P., Lindfors, O., Harkanen, T., Valikoski, M., Virtala, E., Laaksonen, M. A., … Helsinki psychotherapy study group. (2008). Randomized trial on the effectiveness of long and short-term psychodynamic psychotherapy and solution-focused therapy on psychiatric symptoms during a 3-year follow-up. *Psychological Medicine, 38*, 689–703.

Korotitsch, W. J., & Nelson-Gray, R. O. (1999). An overview of self-monitoring research in assessment and treatment. *Psychological Assessment, 11*, 415–425.

Krueger, D. (1997). Food as self-object in eating disorder patients. *Psychoanalytic Review, 84*, 617–630.

Leichsenring, F., Rabung, S., & Leibing, E. (2004). The efficacy of short-term psychodynamic psychotherapy in specific psychiatric disorders: A meta-analysis. *Archives of General Psychiatry, 61*, 1208–1216.

Leichsenring, F., & Leibing, E. (2007). Psychodynamic psychotherapy: A systematic review of techniques, indications and empirical evidence. *Psychology and Psychotherapy: Theory, Research and Practice, 80*, 217–228.

Leichsenring, F., & Rabung, S. (2008). Effectiveness of long-term psychodynamic psychotherapy: A meta-analysis. *Journal of the American Medical Association, 300*, 1551–1565.

Lock, J., Le Grange, D. L., Agras, S., & Dare, C. (2002). *Treatment manual for anorexia nervosa: A family-based approach.* New York, NY: Guilford Press.

Lundgren, J. D., Danoff-Burg, S., & Anderson, D. A. (2004). Cognitive-behavioral therapy for bulimia nervosa: An empirical analysis of clinical significance. *International Journal of Eating Disorders, 35*, 262–274.

Mintz, I. L. (1992). A comparison between the analyst's view and the patient's diary. In C. P. Wilson, C. C. Hogan, & I. L. Mintz (Eds.), *Psychodynamic technique in the treatment of the eating disorders* (pp. 157–194). Northvale, NJ: Jason Aronson, Inc.

Mintz, I. L. (1985a). Psychoanalytic description: The clinical picture of anorexia nervosa and bulimia. In C. P. Wilson, C. C. Hogan, & I. L. Mintz (Eds.), *Fear of being fat: The treatment of anorexia nervosa and bulimia* (pp. 217–244). New York, NY: Jason Aronson, Inc.

Mintz, I. L. (1985b). Psychoanalytic therapy of severe anorexia. In C. P. Wilson, C. C. Hogan, & I. L. Mintz (Eds.), *Fear of being fat: The treatment of anorexia nervosa and bulimia* (pp. 83–114). New York, NY: Jason Aronson, Inc.

Mitchell, S. A., & Black, M. J. (1995). *Freud and beyond: A history of modern psychoanalytic theory.* New York, NY: Basic Books.

Nichols, S., & Gusella, J. (2003). Food for thought: will adolescent girls with eating disorders self-monitor in a CBT group? *Canadian Child and Adolescent Psychiatry Review, 12*, 37–39.

Perkins, S. J., Murphy, R., Schmidt, U., & Williams, C. (2006). Self-help and guided self-help for eating disorders. *Cochrane Database Systematic Reviews, 19*(3). CD004191

Reich, G., & Cierpka, M. (1998). Identity conflicts in bulimia nervosa. *Psychoanalytic Inquiry, 18*, 383–402.

Richards, P. S., Baldwin, B. M., Frost, H. A., Clarksly, J. B., Berrett, M. E., & Hardman, R. K. (2000). What works for treating eating disorders? Conclusions of 28 outcome reviews. *Eating Disorders, 8*, 189–206.

Rozen, D. L. (1993). Projective identification and bulimia nervosa. *Psychoanalytic Psychology, 10*, 261–273.

Searles, H. F. (1999). *Countertransference and Related Subjects: Selected Papers*. New York, NY: International University Press.

Simmons, A. M., Milnes, S. M., & Anderson, D. A. (2008). Factors affecting the utilization of empirically supported treatments for eating disorders. *Eating Disorders, 16*, 342−354.

Steinhausen, H. (2002). Outcome of anorexia nervosa in the 20th century. *American Journal of Psychiatry, 159*, 1284−1293.

Stern, S. (1986). Dynamics of clinical management in the treatment of anorexia nervosa and bulimia nervosa. *International Journal of Eating Disorders, 5*, 233−254.

Strober, M., Freeman, R., & Morell, W. (1997). The long-term course of severe anorexia nervosa in adolescents: Survival analysis of recovery, relapse, and outcome predictors over 10−15 years in a prospective study. *International Journal of Eating Disorders, 22*, 339−360.

Thompson, R. A., & Sherman, R. T. (1989). Therapist errors in treating eating disorders. *Psychotherapy, 26*, 62−68.

Thompson-Brenner, H., & Westen, D. (2005a). A naturalistic study of psychotherapy for bulimia nervosa, Part 1, comorbidity and therapeutic outcome. *Journal of Nervous and Mental Disease, 193*, 544−573.

Thompson-Brenner, H., & Westen, D. (2005b). A naturalistic study of psychotherapy for bulimia nervosa, Part 2, therapeutic interventions in the community. *Journal of Nervous and Mental Disease, 193*, 585−595.

Tobin, D. L., & Johnson, C. L. (1991). The integration of psychodynamic and behavior therapy in the treatment of eating disorders: Clinical issues versus theoretical mystique. In C. L. Johnson (Ed.), *Psychodynamic treatment of anorexia nervosa and bulimia* (pp. 374−397). New York: The Guilford Press.

Treasure, J., Todd, G., Brolly, M., Tiller, J., Nehmed, A., & Denman, F. (1995). A pilot study of a randomised trial of cognitive analytical therapy vs. educational behavioral therapy for adult anorexia nervosa. *Behaviour Research and Therapy, 33*, 363−367.

Weiss, J., & Sampson, H. (1986). *The psychoanalytic process: Theory, clinical observation, and empirical research*. New York, NY: Guilford Press.

Work Group on Eating Disorders. (2006). *Practice guideline for treatment of patients with eating disorders* (3rd ed.). Retrieved January 28, 2009 from: American Psychiatric Publishing Inc. Web site: http://www.psychiatryonline.com/content.aspx?aID=138662; doi: 10.1176/Appi.books.9780890423363.138660

Zerbe, K. J. (1993). Whose body is it anyway? Understanding and treating psychosomatic aspects of eating disorders. *Bulletin of the Menninger Clinic, 57*, 161−178.

Zerbe, K. J. (1996). Feminist psychodynamic psychotherapy of eating disorders: Theoretic integration informing clinical practice. *Psychiatric Clinics of North America, 19*, 811−827.

Zerbe, K. J. (1998). Knowable secrets: Transference and countertransference manifestations in eating disordered patients. In W. Vandereycken, & P. J. V. Beumont (Eds.), *Treating eating disorders: Ethical, legal, and personal issues* (pp. 30−55). London, UK: The Athlone Press.

Zerbe, K. J. (2001). The crucial role of psychodynamic understanding in the treatment of eating disorders. *Psychiatric Clinics of North America, 24*, 305−313.

Zerbe, K. J. (2008). *Integrated treatment of eating disorders: Beyond the body betrayed*. New York, NY: W.W. Norton & Company.

CHAPTER 10

New Pathways
Applying Acceptance and Commitment Therapy to the Treatment of Eating Disorders

Kathy Kater

In my first session with a patient who is battling body image and eating-related concerns, I routinely ask some version of this:

> *What percentage of your daily waking hours would you say you spend thinking about or taking some action related to feeling bad about your body or worrying about your weight—or engaged in some kind of thought or behavior having to do with eating, not eating, or burning calories? You know, things like worrying about how fat or thin you are, the number on the scale, wrestling with thoughts about food—planning to eat, what not to eat, actually eating or actively not eating, feeling bad about what you ate, counting calories, getting rid of calories, checking how flat your stomach feels—that sort of thing?*

In thirty years as an eating disorder (ED) specialist, the responses I have heard have nearly always fallen between 80 and 95%. This is the state of an ED mind.

While the medical consequences of an ED can be life threatening, it is the unrelenting internal dialog and the consuming behavioral rituals that briefly quiet these tortuous thoughts that cause most of the suffering. The endless stream of fear-based judgments, rules, demands, and threats can take over a mind and sometimes drive a life into the ground. It is in the face of these "eating disorder thoughts," and the difficult emotions and compulsions they provoke, that patients must triumph if they are to begin the long road to recovery. Even those who have managed to interrupt ED behaviors for weeks, months, or even years, remain vulnerable to the re-emergence of powerful mental formations that may trigger a relapse.

Given the power and irrational nature of ED thoughts, it makes sense that cognitive-behavioral therapy (CBT), an empirically supported treatment that applies rational, objective arguments to challenge, change, and reduce the frequency of illogical and harmful self-talk, has become the best supported treatment for ED. Despite this, surveys suggest that, while most ED therapists are trained in CBT, the majority do not use it as their primary approach. Instead, I, and many colleagues, employ a mix of approaches, some empirically supported and some not, in an effort to help patients create happier, healthier lives (Simmons, Milnes & Anderson, 2008; Tobin, Banker, Weisberg & Bowers, 2007).

My reason for looking beyond CBT is that I have rarely found rational arguments or objective evidence to be effective in assuaging entrenched ED thoughts, or the difficult emotions and impulses that accompany them. While it is always tempting to appeal to reason and logic to challenge and change irrational, self-destructive thinking, in my clinical experiance this just does not work. For me, the compelling methods of Acceptance and Committment Therapy (ACT, spoken as a single word) seem to have more impact. Along with other metacognitive approaches (Hayes, Follette & Linehan 2004a), ACT does not focus on rooting out or changing the content of thoughts or emotions. Instead, it aims to develop our patients' willingness to *face* these in order to *see and respond* to them in a way that is more helpful. Specifically, ACT helps patients to accept that the emergence of thoughts, feelings, and sensations is not something that is in our power to control, but *how we relate* to these mental events can be transforming (Hayes, 2004). This supports and enhances an approach that has evolved for me out of my own experience of "what works," both in my own life, and in helping patients with EDs achieve and maintain recovery.

ACT's emphasis on facing and accepting *what is not in our power to control* seems uniquely well-suited for treatment (and prevention) of body image and eating problems, particularly in today's socio-cultural context. In this environment, messages promoting a thin body ideal create and reinforce the myth that anyone can and everyone should achieve this ideal through diet and exercise. Belief in this myth can be the trigger for ED in those who are vulnerable, and is the catalyst for widespread body dissatisfaction and disordered eating in the rest of the population. Against this backdrop of widespread stigma about weight and what "should" be done about it, our patients must face the truth about what can be realistically "controlled" about the body's hunger and weight regulatory systems in general, their own body's natural predisposition in particular, and a need to abstain from "weight management" in order to recover.

Working with patients around acceptance of "what is" in order to free energy for the steps that can be taken to sustain and promote health, vitality, and self-esteem, has long been central to my clinical practice. This belief drove me to develop a universal health education approach, *The Model for Healthy Body Image and Weight*, upon which the *Healthy Body Image* curriculum is based. My model provides three guiding principles that closely parallel those found in ACT (Kater, 2005):

1. Accept what is not in our power to control about body size and shape, the biological nature of hunger, and the counterproductive effects of "dieting" (i.e., *eating for the purpose of weight loss*).
2. Understand and take action on what is in our power to choose in regard to health and wellness: eating well, physical activity, acceptance of the size and shape that results from wholesome choices, balanced attention to all aspects of personal identity, and choosing realistic role models.
3. Develop resiliency in the face of social pressures that conflict with these principles through historical perspective on today's body image attitudes, media literacy, and interpersonal support.

Given current norms, ED sufferers must not only learn to respond more healthfully in the face of difficult internal states (thoughts, feelings, sensations, and impulses), but also faulty external norms (prevalent but erroneous views about the malleability of bodies and the purpose of eating and exercise). ACT offers a new way to stand with a patient whose consuming, constricted thoughts and fears are maddeningly reinforced by the wider culture, to help them accept difficult realities, and to take the steps needed to support a life based on their most deeply held values in spite of this.

In what follows, readers should keep in mind that my experience with the application of ACT to the treatment of patients with ED is limited to an outpatient practice. If patients need more help than outpatient work provides in order to make progress, I refer them to a higher level of care where ACT interventions may or may not be appropriate.

ACCEPTANCE AND COMMITMENT THERAPY IN A NUTSHELL

ACT is one of a "third-wave" of scientifically based behavioral and cognitive therapies. It is supported by a rich body of research dating back some 25 years that documents its effectiveness with disorders of *experiential avoidance*, disorders that are driven by a compelling urge to avoid or control unwanted internal experiences—thoughts, emotions, or sensations (Hayes, Strosahl, Bunting, Twohig & Wilson, 2004; Hayes, Luoma, Bond, Masuda & Lillis, 2006). A growing body of comparative studies suggests that ACT may actually be more effective than CBT, psycho-educational approaches, and psychopharmacology in treating avoidance disorders (Hayes & Strosahl, 2005). While only one published case study documents the successful application of ACT for anorexia nervosa (AN) (Heffner, Sperry, Eifert & Detweiler 2002), ED behaviors are certainly defined by a powerful compulsion to control disturbing mental and body states (Cockell, Geller & Linden, 2002; Rasmussen-Hall, 2007).

In my experience, newly diagnosed, medically stable patients who are not in complete denial about ED thoughts and behaviors can benefit from an ACT approach, as well as patients who have struggled long enough to recognize the time, money, energy, relationships, and life experiences that have been lost to their illness. Most of my patients desperately wish to get rid of their grinding, obsessive thoughts and preoccupation with body, food, and weight, as well as their compulsion to starve, binge, purge, or exercise. Yet despite this wish to be free, their attachment to rigid, and self-limiting ED "rules" keeps them trapped, as shown by these patients' comments:

- *"I am so sick of the food rules, but I could never be happy at this weight so I cannot give them up."*
- *"I was planning to add the yogurt this week. But when it came down to it, I knew I wouldn't gain weight if I ate my cereal-bar and pretzels, and I just didn't know what would happen if I ate the yoghurt too."*
- *"I was doing really well (on vacation). And then one day there was a buffet table with these little mini-cream puffs, and without thinking I popped one in my mouth. After I swallowed I realized what I'd done. I couldn't stop thinking about the disgusting cream and dough in my stomach, and I just had to get rid of it. I grabbed some more in a napkin and ate them in the bathroom. Then I turned and threw up. I've been back at it every day since."*

It may seem counterintuitive, but ACT interventions are not intended to challenge the content of irrational or destructive thoughts and rules. Instead, progress occurs when thoughts that drive the ED are accepted for what they are—*mental constructions* spawned by associations between events that are now triggered reflexively (Hayes, Barnes-Holmes & Roche, 2001). Therapists often hear how associations between feeling bad, not eating, and in some way feeling better or "in control" can be traced to an event such as this one:

> *Right after he broke up with me he started dating the skinniest girl in our class. I felt so bad, I could hardly eat. When I started to lose weight it felt good when people complimented me. That's when I started to restrict on purpose, and the more weight I lost the better I felt. It was like the one thing I could do to feel in control.*

When events are associated like this, separate thoughts and the impulses they evoke can become *fused*, as if they were one. Thereafter, whenever any part of this fused thought is triggered by, for example, feelings of loss, rejection or anxiety, the entire fused thought arises as well, e.g., *feeling lonely, can't stand it, don't eat, feel better.* Over time any stress-inducing emotion can become associated and fused to this thought (Blackledge, 2007). In this way an ED mind may begin.

As stated above, a therapist using ACT does not respond as if a person's thoughts are "wrong," nor do we use rational arguments in an effort to convince him or her to "not think" or "not feel" as he or she does. Instead we begin by encouraging patients to examine their own experience to see what has helped and not helped in support of their most deeply held values. In doing so, patients may recognize that past efforts to stop or escape difficult mind states have not only not succeeded (or did so only momentarily), but that avoidance has caused their lives to spin ever more out of control. Patients see with their own eyes that, paradoxically, attempts to control thoughts or feelings in this way inevitably evoke more thoughts about it. Trying not to think about a tree evokes thoughts of a tree, and if you have any strong feelings about trees, these will probably come up as well. In the same way, *"Don't think about eating"* evokes thoughts about food as well as feelings and impulses associated with eating. For someone with BN, *"Stop thinking about the cream and dough in your stomach,"* is likely to evoke even more thoughts and anxiety, including *"I can't stand it!"* followed by a compulsion to purge. *"Stop feeling lonely,"* not only will *not* stop loneliness, it may also evoke anger or shame about being alone and, for someone with an ED, an impulse to escape these feelings through symptom use.

Repeated actions chosen for the purpose of avoiding difficult thoughts or feelings, which ACT calls *experiential avoidance*, actually amplify problems over time (Hayes, Wilson, Gifford, Follette & Strosahl, 1996). A patient who turns away from unpleasant internal states by using ED behaviors will enjoy transitory relief, but will reinforce the association between distress and this avoidant action in the future. I tell patients that this is like exercising a "thought-muscle": the more a person acts on impulses to avoid difficult thoughts and emotions, the stronger and more destructive these impulses become.

Early in our work, I teach patients and their families that once ED thoughts, feelings, and impulses have taken up residence in the mind, a willful decision alone cannot prevent them from arising. That said, rather than being a slave to these impulses, patients can learn to "step back" and look *at* their thoughts instead of viewing the world *from* them. Mindfulness is a 2-millennium-old method for creating this observing perspective—i.e., for watching the mind think—that is now well-substantiated (Kabat-Zinn, 2005). Mindfulness teaches that in any present moment there is a "me" that is thinking and a "me" that can "watch myself think." When patients first experience this, I often ask, "Who is watching?" Most become curious about this transcendent or *observing self.* They begin to see that it is possible to view both their self and their thoughts within a bigger context than they previously knew, creating a space in which new possibilities can arise.

ACT teaches mindfulness and cognitive defusion to help patients to watch what the mind says and does, rather than remaining a captive of its destructive and self-limiting concepts and rules. A large and growing body of research documents that the perspective from which thoughts are viewed can significantly change their impact (Begley, 2007). For example, a patient told me, *"I can't stand to feel full—it makes me feel like I will blow up!"* When fused with this thought, it seemed literally true to her. From this perspective, eating until full is akin to annihilation. After learning mindfulness she was able to observe this thought instead of being caught in it, and it became possible for her to say *"I'm having the thought that if I feel full I will explode."* In the space that opened with this new perspective, she began to feel sad that the years of eating only morsels in order to obey this thought had forced her to withdraw from training as a potential Olympic figure skater. As she faced this loss, a glimmer of a new possibility arose. Might she be willing to have her ED thoughts, and even tolerate the anxiety they evoked without acting on them, if doing so meant she could again do something that deeply mattered to her, such as becoming a skating teacher?

ACT's overall therapeutic goal is to increase *psychological flexibility*; that is, to help patients ride out the inevitable waves of emotionally provocative thoughts and impulses while committing to steps that align with what really matters most in their lives. It provides a model of psychopathology and corresponding treatment interventions (Hayes et al., 2004b). The ACT processes can be divided into two groups:

1. **The mindfulness and acceptance processes.** Helping patients to face and accept difficult internal states, to employ mindfulness of the present moment versus being caught in the future or past, to defuse from fused thoughts, and to connect with the transcendent self versus the conceptual self.
2. **The commitment and action processes.** Using mindfulness and the transcendent self to connect with chosen values, and commit to actions that align with these.

In what follows, I will briefly introduce each ACT process and offer examples of how these can be effective with ED patients. While ACT processes can be discussed separately, quite often the therapist and patient may be working on multiple interdependent and overlapping processes in a single exchange. Practitioners who want to learn more about ACT will find information about books, articles and notices of training opportunities at www.contextualpsychology.org. *Learning ACT; An Acceptance and Commitment Therapy Skills-Training Manual for Therapists* (Luoma, Hayes & Walser, 2007) is a particularly useful guide. A self-help workbook for patients with AN utilizing ACT principles is also available (Heffner & Eifert, 2004). That said, it is really impossible to "get" ACT without practicing it in your own life. ACT is challenging to patients and therapists alike because it conflicts with some of the basic tenets of Western thought, such as that *"Pain is bad,"* *"Happiness depends on feeling good ("feel-goodism"),"* and *"If you just work hard enough you can get whatever you want."* Therapists are often hooked by these beliefs in ways that are not that different from our patients'—I know I have been. It's not that we must overcome all our foibles to help others, but that we must see these for what they are. It has taken me time and practice to integrate ACT into my work. The willingness to continually ask,

"What do I do when faced with pain?" is a key aspect of this. The results have been both professionally and personally rewarding.

THE ACT PROCESSES

Willingness and Acceptance versus Experiential Avoidance

> When suffering knocks at your door and you try to say there is no seat for him, he tells you not to worry because he has brought his own stool. *(Achebe, 1967, p. 84)*

Life includes hardship and pain. What we want, or need, we may not get. What we have, we will eventually lose. There is no avoiding it. This is not negative or pessimistic, but a fundamental truth. I routinely share with patients what wise teachers have taught me: *"Pain is inevitable. Misery is not."* It is by repeatedly trying to run away from pain that we experience increased suffering. Like a Chinese finger trap, the harder we try to pull free, the more we are locked in. Whether ED patients express pain openly and directly or not, we know it is their very effort to escape their worried minds that ensnares them.

ACT proposes that language and verbal problem solving skills—one of the functions distinguishing humans from animals—sets us up to not merely have hardship and pain, but to suffer *more* because we try to run away from it. Two contrasting responses to fear show how this works:

> *With his shoe caught on the track and a train approaching, Adam felt panic. He quickly decided to sacrifice his shoe rather than be run-down. Slipping it off, he scrambled to safety. What a relief.*

Devising strategies to get out of difficult situations like Adam's has brought civilization a long way. Escape works well for solving many problems in the external world where we are dealing with hard data or tangible objects over which we have significant control or influence. Paradoxically, when this same rational strategy is used to escape internally generated problems, an added level of suffering may be created:

> *In her first day at her new school Jill felt a wave of panic as she walked into the noisy, crowded lunchroom. Quickly she decided to forego lunch rather than eat alone. Slipping out the door, she headed for the library. What a relief.*

While Jill avoided her anxiety by leaving the lunchroom, neither her hunger for food nor social connections are as expendable as Adam's shoe. Used occasionally, experiential avoidance is not a problem, but unless Jill becomes willing to face her anxiety and deal with it in another way, she may unintentionally cause herself more problems. Repeatedly skipping lunch may make it harder to concentrate in afternoon classes, be the catalyst for an after school eating binge, or, if she is genetically vulnerable, it may plant the seeds for an association between the sensation of hunger and emptiness, relief from anxiety, and a feeling of being "in control." In a culture where the thin-ideal is paramount, skipping the lunchroom scene may also provide an added "reward" if over time it results in weight loss that invites a compliment. In the meantime, Jill is learning nothing about her social anxiety, becoming dependent on avoidance as a means of coping, and is likely to remain isolated and alone in her new school.

Two Types of Pain

ACT distinguishes between two types of discomfort, "clean" and "dirty." "Clean discomfort" is the natural consequence of coming into contact with a painful event such as sickness or injury, but also situations like rejection, abandonment, or a threat to our integrity. Mary provides an example:

Growing up I was fat and was teased about it. Even though I had friends, it was a nightmare when other kids circled around me on the playground and called me names. Sometimes they pushed me down and even kicked me.

Anyone can relate to the terrible pain in this—so much so that we would consider denial of pain to be pathological. This pain is horrible, but it is primary, normal, and "clean."

In contrast, "dirty discomfort" is the *added* suffering that arises from the stories we tell ourselves to escape our unavoidable pain. Mary continues:

"I told my mother and she talked to the teacher, but it didn't help." [This is a sound and reasonable effort to eliminate a problem in the external world. Unfortunately, it didn't work in this instance.]
"I was fat and fatness is disgusting." [This is the start of a story—whether or not she was fat, if her fatness is the problem, rather than the bullies, maybe she can do something to gain control.]
"So my mother took me to Weight Watchers in the fourth grade." [This effort is chosen as a means to fix the problem, but it is based on denial of the complex contributors to weight and a false belief that weight can be reliably controlled through dieting.]
"I lost about 20 pounds, but then I gained it all back, which really made me hate myself. So then I sort of felt I deserved what I got, and I've been at it (dieting/binge eating) ever since." [Denial of the predictable results of dieting, self-blame, and shame is better than having no control over an unsafe world. If it is her fault, she can maintain the illusion that the world is safe and people will be kind when she loses weight; i.e., she can fix the problem, or be in control.]
"All I can do is keep trying to lose weight. But now I am fatter than ever and can't eat without binging—it just proves what a failure I am."

It is bad enough that this woman has suffered teasing. But it is the fatally flawed and self-perpetuating story—courageously developed to help her avoid further pain, and based on highly promoted myths about fatness, dieting, and the workability of "weight control"—that offers her no way out. This story has triggered her ED.

We cannot expect our patients to face and accept what is not in their power to control simply because we tell them it's a good idea. We can help them begin to see with their own eyes how their efforts to deny or avoid an unwanted reality through rigid adherence to eating, exercise, or other weight-related rules and behaviors have created ever more pain. Fostering a sense of what ACT calls *creative hopelessness* is an example of a method for heightening awareness of the self-perpetrating cycles of failure that naturally occur when we try again and again to control something that is not in our power to control. Metaphors like struggling to get out of quicksand, or of a hamster on a wheel, can help patients to see that it is *not* that they should try harder, but that they are trying to do something that *cannot work*. The following exchange demonstrates how fostering a sense of creative hopelessness in Mary began to undermine her belief in experiential avoidance.

Kater: *So all your life you have had this fear that people will look at you and judge you for being fat and your mind tells you have to lose weight because of this. You know I understand this is a realistic fear in our weightist world. A lot of ignorant people don't know how weight is determined and might look at you and decide you have spent your whole life eating ice cream and watching TV. But the thing I want to know is, how many years have you been trying to lose weight?*

Mary: *Since I was ten, so…eighteen years.*

Kater: *Eighteen. That's a long time. And how has this worked? Did you lose weight and stay slim until recently?*

Mary: *(Frustrated) No. I've lost weight a hundred times, but I always gain it back. I'm fatter than ever. I can't stand it. I will never be happy until I can lose weight and make it stick!*

Kater: *So you've spent more than half your life trying to lose weight and keep it off, and thinking that when this happens you can finally be happy. But in all this time, every time you've lost weight you've only gained back more. Has it ever occurred to you that maybe you're trying to make something happen that just isn't doable; that maybe trying to force bodies to be a certain size isn't what Mother Nature intended, and it just backfires?*

Mary: *So you're saying that to recover from my ED I have to accept being fat?*

Kater: *I'm not saying anything about your size, actually. What I'm asking you to consider is whether the battle you've been waging with your weight for the past 18 years is any closer to being won after all this time. Or has it just been like running on a hamster wheel—the more you have tried to lose weight, the more it has consumed your time and energy, and in the end you are not getting what you want, and suffering just as much as when you started?*

Mary: *No—I try and try! I must be doing something wrong. Oh I hate this!*

Kater: *I understand how painful this is for you Mary. But let's try to lay it all out here. And so, in terms of what you've told me is of value to you in your life, your family, your work, your spiritual growth—has the energy you have invested into fighting this battle helped or hindered you in those areas?*

Mary: *Well, hindered. A lot. I feel like a failure every day because I hate my body and all I think about is food. I get so depressed.*

Kater: *So I'd just like you to think about this. What if the very fact of trying to escape your body's natural disposition is what's trapping you—consuming your time and energy day in and out with no end in sight—while in the meantime…*

Mary: *But my doctor says I need to control my weight and I don't want to be fat. I could never accept this body!*

Kater: *Ok. I understand this is what you feel. And I'm not asking you to change how you feel or whether you want to be fat or not. What I'm asking is: how well have your efforts to reduce your body size over the past eighteen years worked? I'm also asking whether this effort is helping or hindering you from being the person you would want to hear talked about at the end of your life in your eulogy. Whether or not you want to be fat is not what I'm asking. I hear you saying that this is not what you would prefer. But it does seem that your efforts to change this have consumed years of your life while in the end the size of your body is still not in your control.*

Through this exchange Mary became willing to face this difficult reality for the first time. Instead of her failing, she was trying to do the unworkable. This was not the end of her struggle or the last time she would revisit it, but her acceptance created a space to talk about what she could do instead to develop a healthier, more successful, and therefore happier relationship with food, her body, and her life.

Cognitive Defusion versus Fusion

> Pay no attention to that man behind the curtain. *The Wizard of Oz (LeRoy & Fleming, 1939)*

While Mary became willing to begin the process of learning how to eat rather than diet, harsh judgments about her size, fears about becoming fat, and strong impulses to restrict still dominated her thoughts. ACT would say that these difficult thoughts were *fused*, and that until they are *defused*, she will remain tortured by their content. With cognitive fusion the content of words, and the events that are associated with them, become one and the same. For example, a person who has learned to fear snakes will see a snake and automatically think "danger." While "snake" and "danger" are actually two thoughts, when fused they are experienced as one literally true thought: *Snakes are dangerous*. In turn, rules about how to respond to snakes become associated, and also become fused—"*since this, then that.*" Fused rules become indisputable guides for behavior: *Snake, danger, run!* No alternative perspective seems possible.

Those suffering from ED provide endless examples of fused thinking. Our patients' fixed, rigid, and negative associations to words such as "fat," a number on a scale, or their image in a mirror are irrefutable; perception, thought, and action are one. Rules about how to avoid the inherent "danger" of these threats in turn become fused, as if following these rules is the only possible way to stay "safe." ED thoughts are thus multiple associated thoughts that have become fused, and are experienced as an indisputable fact. For example: *"fatness is disgusting—eating causes fat—you cannot eat (or must burn all calories eaten)."*

Cognitive fusion is always limiting, but it is not always problematic. When the earth was viewed as flat, people believed exploring was out of the question, but, for most people, this limited perspective did not cause trouble. On the other hand, the effects of fusion in regard to provocative or difficult emotions can range from missed opportunities for enjoyment to severe disturbances in thinking or mood. Consider these examples:

— *"You are taking a wonderful walk in a beautiful forest when you see a large snake in the path ahead. You scream and run away. Associating this path with danger, you never walk there again and never learn that what you saw was a piece of rope."*
— *"At age 12 your patient was teased about her mature figure. She learned to feel ashamed and self-conscious, and vowed to never "invite" those comments again. She began the first of many starvation diets, triggering anorexia, and later bulimia. She never learned it was her classmates' insensitivity that was wrong, not her body's size and shape."*
— *"At age 8 your patient was reading when her dad appeared, led her to the garage, committed incest and gave her a candy bar. Her mother was a depressed alcoholic, and there was no one to tell. She ate the candy bar, which gave some comfort, and pretended things were fine. By the time the violations ceased, eating sweets was the only thing she knew that could soothe her despair and loneliness, but the compulsion to eat sugar was also her greatest shame. She never learned that this shame belonged entirely to her father."*

When thoughts are fused we miss the fact that *thinking* is something we *do* that may or may not reflect the truth, and may or may not be helpful. For example, when a patient thinks, *"If I eat these potatoes my thighs will become grotesque and everyone will shun me,"* the world as colored by that thought seems so real, it is as if she is not interacting with thoughts at all. Instead of a mental construction, she is dealing with being a monster and deserving isolation. The loss of behavioral and emotional flexibility in ED sufferers is made more understandable by the concept of fusion. Why would anyone eat a potato if they believed what she does, and wouldn't everyone want a rule excluding potatoes? From the ACT perspective, this patient is responding rationally to the world as she is experiencing it, but missing the fact that she is "languaging" about it. When a 14 year-old, beautiful by any standard, girl says, *"You don't understand; at my school, looks are everything. If I don't lose ten pounds before fall I might as well not even go to high school,"* she is not dealing with a thought, but the literal end of her high school career. From this point of view, the adaptive functions of risky ED rules become understandable.

There are many ways to demonstrate "fused thinking" to patients. I sometimes take down a small painting from my office wall, asking them to hold it with their nose touching the glass. All agree that from this vantage point they can't really see it for what it is. To "know" this painting requires a little distance. In a similar way, if we are "one with our thoughts" it is hard to see them for what they are.

Since ED patients must learn to recover in spite of their fused thoughts, ACT teaches *cognitive defusion* to help them step back and see these for what they are, "just thoughts" that may or may not be helpful. Using mindfulness, patients can develop the ability to observe their thoughts and to consider them more dispassionately. The thought, *"Augh! I am so fat and disgusting,"* is very different from the thought, *"My mind is having the thought that I am fat and my mind is also having the thought that because I am fat, I am disgusting."* Instead of asking, *"What are you thinking about the sandwich you just ate?"* I might ask *"What can you notice your mind is thinking about that sandwich?"* A patient who had a rare day all to herself told me: *"I have really, really been thinking since getting up this morning that I could skip lunch and no one would know."* She had begun learning mindfulness to defuse her thoughts, and so I asked: *"What are you going to do with that thought?"* After several moments passed she said: *"I am going to say, 'Thanks mind, I know you are trying to help me, but after all these years, following your advice has led me down a path I never want to see again.' I'll go home and make a lunch to take with me."*

Those unfamiliar with mindfulness may wonder whether desperately anxious patients can realistically be able to look *at* versus *from* their thoughts. I do not mean to gloss over the deep struggle and the need for practice and repetition it may take for patients to learn to use defusion like this. That said, most patients willing to work with these methods do experience a sustainable paradigm shift regarding how they view their thoughts and feelings and the degree to which they feel controlled by them.

When patients become aware of the observing mind as distinct from the thinking mind, we can ask: *"Which part of your mind do you think has your best interest at heart?"* If the patient draws strength from spiritual or religious beliefs, I might ask: *"Which part of your mind do you think is more connected to God?"* or *"Which part of your mind is more connected to the universe?"*

The shift in the way patients relate to their thoughts differently becomes evident in comments such as:

I was walking in to the reunion, and my mind started thinking "You are so fat and stupid and ugly. You should go home." But my observing mind saw this coming, and I know that this is what my thinking mind does when I am scared. I also knew I would be sorry if I left. I've been looking forward to this reunion for months. So I went on in and my thinking mind was still yelling at me, but I just ignored it and started talking to people.

I often recommend the biographical motion picture *A Beautiful Mind* (produced by Brian Grazer and Ron Howard, Universal Studies, U.S.A., 2001) to patients because it is a remarkable demonstration of how literally true and powerful thoughts can seem to be, even when they are, in fact, delusions or gross distortions. This movie also depicts a great example of cognitive defusion. When the central character, Nobel Prize winning mathematician John Nash, observes and understands his hallucinations for what they are, he learns to just "see" them without responding to them as real, and in turn becomes free to live his life without being controlled by them. Just as our patient's minds may continue to produce ED thoughts even well along in recovery, Nash's mind continues to produce hallucinations throughout the rest of his life. In a closing scene Nash says: "*I've gotten used to ignoring them and I think, as a result, they've kind of given up on me. I think that's what it's like with all our dreams and our nightmares, Martin, we've got to keep feeding them for them to stay alive.*" Cognitive defusion can help ED patients stop feeding their thoughts.

Mindfulness as an Intervening Action versus Being Lost in the Future or Past

Meditation... helps us return to our true self. *(Nhat Hanh, 1987, p. 8)*

A thought cannot see itself. In order to work with thoughts, patients need a way to observe them with sufficient calmness to resist acting on them. This is particularly true for fused thoughts that are so intractable, anxiety provoking, and demanding. While thinking about a thought in hindsight can improve insight, responding differently the next time it arises requires learning to be aware of and observe it *as it is happening*. Mindfulness is a method for watching the mind think without judgment or action in order to see ourselves in a bigger context, connect with what is most important, and make choices that help rather than hinder our lives. A significant and growing body of research on mindfulness and neuroplasticity has documented that attention to, and awareness of, thoughts, in other words pure mental activity alone, can improve mental health (Begley, 2007). Apparently, thinking about thoughts in a new way can even lead to measurable changes in the brain's hard wiring. Given the frequent co-occurrence between obsessive-compulsive disorder (OCD) and ED, it is notable that research with OCD patients who were taught mindfulness methods, and who learned to re-label their thoughts and urges as *"an event in the mind caused by brain circuitry,"* reported after only one week that they did not feel their disease was controlling them. PET scans performed on OCD patients before and after ten weeks of mindfulness-based therapy showed that activity in the pre-frontal cortex, where OCD thoughts are triggered, had diminished dramatically (Schwartz & Begley, 2002). While I have not seen such rapid responses with my ED patients, I have found mindfulness to be very effective in reducing the power of their intrusive and obsessive thoughts and worries.

There are a variety of different models for teaching and learning mindfulness. ACT offers structured exercises for bringing attention and awareness to the present moment for the purpose of observing our thoughts for what they are, transcending our limited self-stories, and connecting with our values. I personally have learned mindfulness through the Buddhist tradition (Nhat Hanh, 1999), but I teach my patients the secular model pioneered by Kabat-Zinn (2005) in his work with stress reduction. In this model, I begin by asking patients to sit comfortably, but upright, relaxing into the chair beneath them, feeling its support rising up to securely hold them, eyes either closed or gazing without focus on a spot on the floor. I then instruct them to draw their attention to one part of the body after another, being aware of and feeling the sensations in each, noticing, but not holding on to any thoughts that might arise. With ED patients, we can expect that judging and anxiety may occur as the attention moves to different parts of the body. If this happens, they should just notice it and label it as such: "*judging, feeling anxious,*" and then return their attention to awareness of the physical sensation in that body part. I try to avoid potentially triggering words, drawing attention, for example, to their rib cage as it protects their digestive organs, rather than their tummy.

Patients then turn their attention to their breathing. "*Breathing in, know that you are breathing in. Breathing out, know that you are breathing out.*" We know that the mind will wander, because this is the nature of mind, so it is important to tell our perfectionistic patients to expect this. Thoughts will shoot into the future or the past and evaluate the present: "*this is boring, how many calories were in that muffin, I need to call Sally, that truck noise is annoying, when will I ever feel better ...?*" When they notice their attention has been taken over by thinking, they should just acknowledge this, "thinking mind, worrying mind, etc." and, without judgment, return their attention to their breath, which serves as an anchor to the present moment. They are not doing something "wrong" when their attention drifts; rather, the point of mindfulness is merely to *observe* what the mind does, including how scattered it is. Most importantly, they can begin to realize that their thoughts are not who they are, but merely events that come and go and that can be watched without requiring action.

I urge patients to practice mindfulness at home, but this is especially difficult for patients who are anxious or driven. Knowing this, I encourage them to incorporate ordinary mindfulness into everyday life. Mindful breathing while sitting at a traffic light, standing in a line, stopping to look mindfully at one leaf on a tree, feeling the sunshine on their cheek, and walking slowly in rhythm with their in and out breath are examples of non-triggering opportunities for mindfulness. I tell patients that this "ordinary" mindfulness is necessary practice for using mindfulness when they are experiencing triggering thoughts. In the following example, a patient with bulimia reported using mindfulness when she had been triggered by a bite of potato salad:

> *I was completely hooked by my thinking. My mind was screaming "You idiot! Why did you let yourself think you could eat one bite of that potato salad! You know you are as fat as a pig with no self control. Now you will have to go home and binge and purge." So I went to my car, and I was already thinking about the binge foods I would buy on the way home. But then I thought, why not try some of this mindfulness stuff before starting the car. If it doesn't help, you can still binge and*

purge. And so I did mindful breathing right there in the car for about 10 minutes, and I was surprised at how I calmed down. Even between breaths I was still planning what I would get at 7-Eleven, but in between I started to see that another part of me was really scared of meeting new people at this party. I could sort of see that even though my mind was thinking one bite of potato salad was terrible, it really isn't. I kind of think now this was just an excuse to run away. Binging and purging have always been the best way I know to do this. So I did go home, but instead of feeling disgusted with myself, I was feeling kind of sad that I have always been so shy. And I thought, if I binge and purge that would just make me feel worse. So I played the piano for a long time, and I went to bed, and I was glad.

Eating mindfully has been shown to be effective in helping people to tune into body cues of fullness and satiety (Bly, Hammond, Thomson & Bagdade, 2007), but mindfulness can also help patients choose to face, without judgment or action, rather than avoid, difficult thoughts and feelings that occur while eating. What follows is how I might lead a mindful eating exercise with a patient who was not yet willing to risk an experience of feeling full.

As you eat your lunch, look closely at each food on your plate. Where did this food come from? Where did it grow? How many elements is it made up of—water, soil nutrients, sun? Who may have planted and tended the seeds it grew from? Imagine what they might have eaten to give them energy to tend to this food? As you begin to eat, pay attention to the texture and flavor of each forkful. Think about the nutrients this food contains, how your muscles and brain need these, and why your hunger for this food is important. Picture the food as it arrives in your stomach. Allow yourself to be aware that it has to take up some space there, just as it takes up some space on your plate, so naturally your stomach will have to expand to allow for this. Just relax and notice the thoughts and feelings that arise as you allow yourself to face this reality. If you get too anxious, just return to watching your breath as it flows in and out past your nostrils. When you are calmer, notice what your mind is telling you about this food in your stomach. If you notice judgments or anxious feelings, just name these: judging thoughts—worried thoughts, and let them go. You could even say, 'Thank you mind, I know you want to protect me with these thoughts, but I am trying a different way'—and then let your thoughts glide away like leaves in a stream. Bringing your attention back to your breath, let yourself be aware that attention to the activities you carry out this afternoon will be more possible because of the energy provided by this food. To carry enough energy with you for this, your tummy must be full. If your waistband has to be a little tight for this to be possible, might you be willing to experience this if it meant you could be free from food cravings while you do your work?

As with all of aspects of ACT, it is not possible to teach mindfulness without having used it in one's own life. For treatment providers inexperienced in this method for observing the mind, guidance is offered in the books referenced above and in both non-secular and secular classes that are increasingly common in most cities and communities.

Self as Context versus Attachment to the Conceptualized Self

knowing what you know and knowing what you do not know—that is the beginning of knowledge.
(Confucius, 551–479 BCE; as cited in Li, 1999, p. 26)

ACT holds that there are two ways of understanding "self;" as the "knower" and as the "known." The known, or *conceptualized self*, arises out of fusion with the content of our experiences—out of all the many ways we learn to proclaim "*I am* ____." It incorporates all the stories we begin to believe and act on about ourselves as if this sense of self was literally "true." ACT doesn't suggest that there is anything inherently wrong with an identity based on conceptualizations, and socially it can even be helpful as we interact with others in ways that are predictable and understandable. But what works well interpersonally can also trap us in patterns that are unworkable and even destructive. If we are glued to an identity that dictates a limited perspective on possibilities, we may not be able to get what we need to be happy. For example, an ED patient whose identity was based on her membership in a doctrinaire religious group was wracked with guilt when years of self-sacrifice left her still lonely and resentful. She saw no alternative to her story, which included the thought: *"When I have sacrificed enough, I will be loved and happy."* This story offers no way out, and so her ED offered the only relief she could imagine. For ED sufferers, the conceptualized self reflects a story they have come to tell about themselves that then dictates the life they are able to live. *"I cannot eat without binging,"* but her restriction leads to binging. *"I have to weigh less than X pounds to feel acceptable,"* but X is never good enough. *"I am fat and alone and food is my only comfort,"* but relying on food limits contact with other people. There are no possible solutions within the stories that define these patients' conceptualized selves.

ACT does not try to convince people that their stories are irrational or wrong. Instead, therapists working within this model help patients to see these stories for what they are: self-limiting conceptualizations of who they are. Using mindfulness to rise above this view, the *self as context* can be distinguished. This bigger, observing self is transcendent, continuous, and stable—distinct from the thinking, feeling, impulsive, ego driven mind. The ACT literature is rich with metaphors and experiential exercises to help patients connect with self as concept, or observing self. One well-known exercise is to have a patient imagine him- or herself as the driver of a bus full of noisy, unruly passengers all of whom are yelling the worst possible insults and demands at him or her. The passengers represent all of the patient's harsh, self-limiting thoughts, ED or otherwise, such as *"you are so boring, too fat, undisciplined, lazy, too shy, a loser, impure."* The point is to help your patient see that, as the driver, he or she may be carrying around all these passengers, but they are not him or her. From this perspective your patient is free to steer the bus in the direction they choose, towards a destination that they have defined, still aware of, but not a slave to, these thoughts.

Connecting to and Committing to Action Aligned with Core Values

> The only thing greater than the power of the mind is the courage of the heart.
> *(A Beautiful Mind, produced by Brian Grazer and Ron Howard, Universal Studies, U.S.A., 2001)*

As ED clinicians, we should never be satisfied that interruption of symptoms is enough to sustain recovery. As patients find the courage to face life without using ED behaviors, we must help them connect with alternatives that truly promise a better quality of life. A central component of ACT addresses the need for patients to connect with what is of greatest importance to them and what they want their life to stand for. What lesser cause could generate a willingness to face life head on and let go of behaviors that provide such an illusion of control? Values work is the part of ACT that I find to be the most meaningful and gratifying. It helps patients re-connect to parts of themselves that have been sorely neglected in the face of the all-consuming ED. Seeing the disparity between values and how we are actually living our lives is inevitably painful, particularly for those who are spending 80 to 95% of their daily waking time consumed by ED thoughts and behaviors. At the same time, becoming aware of the disparity brings hope. At these moments, you may see a light come on in your patient's eyes as she talks about being a competitive tennis player before her BED added 100 pounds to her small frame and her knees became arthritic. You may hear a mother with anorexia who painstakingly prepares meals for her children connect with wanting to give them the benefit of actually eating meals along with them. You may see a college student, whose bulimia treatment has made him ineligible for the Peace Corps, dare to dream of what else he can do to make a difference.

Anyone who has worked with ED patients understands that voluntarily surrendering compulsive behaviors requires tremendous courage and motivation that comes from a sense of purpose bigger than themselves. ACT publications offer numerous exercises to help patients connect with their values, identify behavioral implications of commitment to them, and recognize the sacrifices necessary to live up to them as well as the cost of not doing so. A table listing many of these appears in *Learning ACT* (Luoma et al., 2007, p. 133). The intention is for patients to become aware of what they would most deeply want to be able to say about their life if ED thoughts and impulses were not controlling their actions, i.e. *"In a world where you could choose to have your life be about something, what would you choose?"* (Wilson & Murrell, 2004, p. 135). When patients feel unable to move forward on the difficult road to recovery, values work shines a compelling light on why they must.

ACT does not specify when values work should occur in the therapy process. I prefer to talk with patients about their hopes, dreams, and wishes very early on because this brings them face to face with how their ED makes the life they want to live impossible. This does bring pain to the surface, but this pain is often needed to compel movement.

If time allows, even in an initial assessment, I will ask about values.

We have been talking about your problems, but I'd like to hear about the rest of your life. What is really important to you? If your ED were not taking up so much of your time and energy, what would you be doing or want to be doing that you are not doing now? What kind of person do you most want to be? At the end of your life, what do you want to be able to say about yourself?

This set of questions is often like watering a seed that has lain on barren ground. It may wake up a part of them that the ED has suffocated.

Following is a conversation illustrating how values work helped a 38-year-old woman with AN who had relapsed after her parents died.

Kater: *Yah, those old thoughts and feelings are still there, aren't they? Once they take up residence on a little neuro-pathway in your brain (I put my hand on the left side of my forehead and hold it there), painful losses like you've experienced this year can wake them right up. And you know, those thoughts may always be there, riding around in your brain ready to harass you. The thing is, you can spend* *the rest of your life letting those thoughts control how you live, or you could even spend your life arguing with them. At the same time, if you have something more important to do, maybe you could just learn to smile and wave at them and say 'I hear you little neuro-pathway. Thank you for your opinion...' while you just keep on with what's important. Let me ask you, in the face of so much that has happened that has made you feel so sad—your parents dying, your marital problems, all the things that are not in your control, including those rude passengers in your mind telling you not to eat—what is still there that is really of value to you? I wonder deep down, what is it that you want your life to stand for, in spite of everything. What still really matters?*

Joan: (Thinks for some time.) *Well, I really want to be a good mother.*

Kater: *Ok. Yes, of course. So... what does that mean? What would that look like to be a good mother, if you were doing it the way you imagine?*

Joan: (She tells of what people said at her mother's recent funeral about how, with very few financial resources, her mother had kept her large family of 9 children closely knit.) *I'd like to think my children would feel that way about me at my funeral.* (She starts to cry.) *I don't think that's very likely to happen the way things are going.*

Kater: *Ok. I understand.* (I pause to let her cry.) *So even though you feel discouraged right now, let's say you were actually doing the kinds of things you would need to do to be the mother you'd like to be. If I could be a fly on the wall and really watch you taking the steps that align with your value of being a good mother, what would I see? What does a good mother do?*

Joan: *Well, for one thing, she sits down and eats with her kids...* (She is looking down.)

Kater: (Giving time for this to sink in.) *Joanne, look at me.* (She does, slowly.) *So in light of one of your most deeply held values, I wonder if there are any steps that you might be willing to take between now and when you come back that are in alignment with this vision of being a good mother?*

Joan: *Well... maybe I could eat something with the kids at dinner....*

Kater: *How do you think the kids would respond to that?*

Joan: *I think they'd be pretty shocked.* (laughs)

Kater: *Shocked in a good way, or something else?*

Joan: *No, in a good way.*

Kater: *And what would that be like for you?*

Joan: (She is silent for a while.) *I think I would really like to try it.*

Kater: *And what about all the noisy, anxious thoughts that will be provoked in your mind?*

Joan: *Well, they will still be there. It will be hard. I'll just have to say, 'Yah, yah, I hear you. You'll never change. But I have to do this for my kids. This is more important.'*

When we ask patients to face their pain and commit to actions that feel to them like jumping off a cliff, the foundation for this courage has to come from what really matters. While all mindfulness based therapies help patients defuse from destructive thoughts and emotions, I have found that ACT's emphasis on connecting to deeply held values, and both the recognition of and commitment to the steps needed to align with these, has added a powerful motivational piece to my work. Putting this into practice has made some form of the following question very familiar to my patients:

> *Do you think it would be possible to have this ED thought and to be terrified not to obey it, and to still go ahead and do …(a valued non-ED action for this patient) anyway—knowing that one of your deepest wishes in life is to …(a coinciding deeply valued direction)… and that to fulfill this value means you have to be able to take this step?*

CLOSING NOTE

While I have found ACT to be useful in my work with ED patients, I still use an eclectic bag of therapeutic approaches when called for. Clearly, the most important of these is the use of myself. It goes without saying that ACT processes cannot be applied without a great deal of compassion and love for our patients, understanding of their pain, and our deepest respect for the difficult road they are on. In addition to this, I add an educational component. I always teach the following basic truths: (i) there are wholesale limits to "weight management" through healthy means; (ii) size diversity is inevitable; (iii) weight loss as a goal, and dieting in particular, are intrinsically flawed methods and are likely to lead to weight gain; and (iv) regardless of size, from very fat to very thin, a healthy weight can only be determined by observing the outcome of healthy behaviors (i.e., wholesome eating and a reasonable amount of physical activity over time), never by a BMI chart, the size on a pair of jeans, or a reflection in a mirror. Our patients hear so many messages contradicting these facts every day; thus, it is essential to articulate them with authority from the start.

Even though most ED thoughts are spawned by myths that deny biological reality, using hard data to argue with them is not helpful to most patients. Rather, we, as therapists, must convey empathy for the many experiences that created these skewed perspectives. While we should have no illusion that the facts alone will diminish ED thoughts and feelings, ED sufferers need to ultimately accept their dangerous thoughts for what they are—unworkable mental constructions based on fear. Making room for this reality can be transformative.

References

Achebe, C. (1967). *Arrow of God*. New York, NY: Anchor Books (p. 84).
Begley, S. (2007). *Train your mind, change your brain*. New York, NY: Ballantine Books.
Blackledge, J. T. (2007). Disrupting verbal process: Cognitive defusion in acceptance and commitment therapy and other mindfulness-based psychotherapies. *Psychological Record, 57*, 555–576.
Bly, T., Hammond, M., Thomson, R., & Bagdade, P. (2007). Exploring the use of mindful eating training in the bariatric population. *Bariatric Times*, November–December. *Psychology Perspective* 15–17.
Cockell, S. J., Geller, S., & Linden, W. (2002). The development of a decisional balance scale for anorexia nervosa. *European Eating Disorders Review, 10*, 359–375.
Hayes, S. C. (2004). Acceptance and commitment therapy, relational frame theory, and the third wave of behavior therapy. *Behavior Therapy, 35*, 639–665.
Hayes, S. C., Barnes-Holmes, D., & Roche, B. (Eds.), (2001). *Relational frame theory: A post-Skinnerian account of human language and cognition*. New York, NY: Kluwer Academic/Plenum/Springer-Verlag.
Hayes, S. C., Follette, V. M., & Linehan, M. M. (Eds.), (2004a). *Mindful and acceptance: Expanding the cognitive-behavioral tradition*. New York, NY: Guilford Press.

Hayes, S. C., Luoma, J., Bond, F., Masuda, A., & Lillis, J. (2006). Acceptance and commitment therapy: Model, processes, and outcomes. *Behavior Research and Therapy, 44*, 1—25.

Hayes, S. C., & Strosahl, K. D. (2005). *A practical guide to acceptance and commitment therapy.* New York, NY: Springer Science and Business Media, Inc.

Hayes, S. C., Strosahl, K., Bunting, K., Twohig, M. P., & Wilson, K. G. (2004b). What is Acceptance and Commitment Therapy? In S. C. Hayes, & K. D. Strosahl (Eds.), *A practical guide to acceptance and commitment therapy* (pp. 3—29) New York, NY: Springer.

Hayes, S. C., Wilson, K. G., Gifford, E. V., Follette, V. M., & Strosahl, K. (1996). Experiential avoidance and behavioral disorders: A functional dimensional approach to diagnosis and treatment. *Journal of Consulting and Clinical Psychology, 64*, 1152—1168.

Heffner, M., Sperry, J. A., Eifert, G. H., & Detweiler, M. (2002). Acceptance and commitment therapy in the treatment of anorexia nervosa: A case example. *Cognitive and Behavioral Practice, 9*, 232—236.

Heffner, M., & Eifert, G. H. (2004). *The anorexia workbook.* Oakland, CA: New Harbinger Publications.

Kabat-Zinn, J. (2005). *Wherever you go there you are;* 10th Anniversary Edition. New York, NY: Hyperion.

Kater, K. (2005). *Healthy body image; teaching kids to eat and love their bodies too!.* Seattle: National Eating Disorders Association.

LeRoy, M. (Producer), & Fleming, V. (Director) (1939). *The Wizard of Oz* [Motion picture]. United States: Metro-Goldwyn-Mayer.

Li, D. H. (1999). *The analects of Confucius: A new-millennium translation.* Bethesda MD: Premier Publishing Company.

Luoma, J. B., Hayes, S., & Walser, R. D. (2007). *Learning ACT.* Oakland, CA: New Harbinger Publications, Inc.

Nhat Hahn, T. (1987). *Being peace.* Berkeley, CA: Parallax Press.

Nhat Hanh, T. (1999). *The miracle of mindfulness.* Boston: Beacon Press.

Rasmussen-Hall, M. S. (2007). Distress intolerance, experiential avoidance, and alexithymia: Assessing aspects of emotion dysregulation in undergraduate women with and without histories of deliberate self-harm and binge/purge behaviors. *Dissertation Abstracts International: Section B: The Sciences and Engineering, 67*(9-B), 5420.

Schwartz, J. M., & Begley, S. (2002). *The mind and the brain: Neuroplasticity and the power of mental force.* New York, NY: Regan Books.

Simmons, A. M., Milnes, S. M., & Anderson, D. A. (2008). Factors influencing the utilization of empirically supported treatments for eating disorders. *Eating Disorders, 16*, 342—354.

Tobin, D. L., Banker, J. D., Weisberg, L., & Bowers, W. (2007). I know what you did last summer (and it was not CBT): A factor analytic model of international psychotherapeutic practice in the eating disorders. *International Journal of Eating Disorders, 40*, 754—757.

Wilson, K. G., & Murrell, A. (2004). Values work in acceptance and commitment therapy: Setting a course for behavioral treatment. In S. C. Hayes, V. M. Follette, & M. M. Linehan (Eds.), *Mindfulness and acceptance: Expanding the cognitive-behavioral tradition* (pp. 120—151). New York, NY: Guilford Press.

CHAPTER

11

Outpatient Treatment of Anorexia Nervosa following Weight Restoration
Practical and Conceptual Issues

Richard A. Gordon

The treatment of patients with anorexia nervosa (AN) who are fully weight-restored allows for different priorities than the rehabilitation of those who are significantly underweight and brings new therapeutic challenges as well. Although the priority for weight gain no longer dominates the treatment process, this does not mean that the patient is "done" with that issue and no longer needs to address it. Treasure and Ward (1997) have pointed out that weight is *always* an issue in the treatment of AN. In fact, in a comparative study of patients who succeeded and those who failed at sustaining recovery after inpatient treatment (Cockell, Zaitsoff & Geller, 2004), those most likely to relapse had assumed that treatment was complete because they had restored their weight. Nonetheless, once the AN patient's weight and physical status is stabilized, the process of a more complete psychosocial recovery begins.

The term "relapse" in this chapter specifically refers to the broad range of eating disorder (ED) behaviors, thoughts, and feelings indicating a re-intensification of AN, with the most ominous warning signal for relapse being weight loss (see McGilley and Szablewski, Chapter 12, for the emerging literature on treatment outcome and the prediction of relapse). This chapter will center on the clinical issues that arise in the outpatient treatment of weight-restored patients following discharge from a hospital or residential center, focusing primarily on young adults in the college years. The existing literature is complex, but there is a consensus that vulnerability to weight relapse is highest within the first 12–18 months after discharge (Carter, Blackmore, Sutandar-Pinnock & Woodside, 2004; Strober, Freeman & Morrell, 1997). Many factors have been found to predict relapse, among them continuing body image (BI) concerns (Keel, Dorer, Franko, Jackson & Herzog, 2005), high levels of exercise at discharge (Strober et al., 1997), weight loss in the first weeks after discharge (Kaplan et al., 2009), and poor psychosocial functioning (Lowe et al., 2001). Treatment following discharge is a particularly urgent matter, best informed by a combination of research and clinical experience.

The following discussion is organized around a number of key issues that must be part of any overall treatment plan for a weight-restored patient. Until a strong evidence base suggests otherwise, clinicians must rely upon an eclectic approach, drawing upon knowledge and clinical experience about many issues and theoretical perspectives. The limited available evidence suggests that flexible and active clinical management is essential and probably more effective than more narrowly focused clinical protocols (McIntosh et al., 2005). The model proposed here includes a primary therapist working within a multi-modal, multidisciplinary clinical team and utilizing other resources where appropriate. Symptomatic issues such as weight and nutritional status, BI, and exercise are of primary concern in the early stages of treatment and recovery. However, in order for long-term recovery to be achieved, developmental, familial, and psychosocial issues ultimately need to be addressed to minimize the risk for relapse and to solidify full physical, psychosocial, and emotional recovery (Keel et al., 2005). Strategies for achieving these complex goals are discussed below using short clinical vignettes to illustrate some of the typical ambiguities and surprising outcomes of certain interventions.[1] The final section of this chapter explores a promising clinical approach, Cognitive Analytic Therapy (CAT) (Ryle & Kerr, 2002), which integrates empirically validated psychotherapy techniques with a focus on the patient's relational dynamics.

TRANSITIONING FROM INPATIENT TREATMENT

Inpatient and residential treatment programs are powerful experiences. To ensure the patient's successful transition to outpatient care, clinicians benefit from knowing the specifics of the inpatient program, and the qualities, both positive and negative, of its impact. My experience suggests that outpatient therapists need to have more information about this than they typically have. Ideally, some form of on-going contact is maintained (e.g., weekly progress reports, brief phone calls) between the primary inpatient therapist and the outpatient therapist during the patient's treatment course. At the very least, these two therapists should have a pre-discharge discussion in order to carefully coordinate the transition, and the outpatient therapist should receive a discharge summary.

Patients have a variety of feelings, ranging from highly positive to very negative, about their inpatient experiences. For some, the loss of autonomy is extremely traumatic. Conversely, others experience the environment of total immersion in a positive way, and feel uneasy with the anticipated lack of supervision when leaving the hospital. Again, it is important to know specifics about the patient's unique experience in treatment. For example, does the patient miss the total supervision over meals that occurred in the program, and does he or she feel the need for greater structure on the outside to maintain healthy eating patterns? The treatment plan, designed in collaboration with an experienced nutritionist, should reflect the patient's answer to this question. It is also important to know what aspects of the therapeutic program really "worked" for them. For example,

---

[1]Clinical vignettes are for the most part composites of stories of actual patients, to disguise identity. For this purpose, certain details that are not relevant to the story have been altered.

was the patient asked to do a specific form of journal-keeping that felt particularly powerful and might help him or her in an ongoing way? If patients feel that this or other techniques were beneficial and motivating, building them into the outpatient program is important, showing respect for the patient's experience and building a bridge between residential and outpatient care.

The shift from a round-the-clock therapeutic environment to, in some cases, once weekly sessions with a therapist is especially challenging. This is a drastic transition for a disorder as serious as AN, and the frequency is unlikely to be enough. Depending on the patient's clinical state, as well as issues of accessibility, a day hospital or intensive outpatient program may be appropriate. Patients often need this level of containment and intensity to minimize their risk for relapse. Intensive treatment, either in a program or in intensive outpatient psychotherapy, can put a great strain on financial resources, as many insurance plans will not support this level of outpatient treatment. However, the costs of relapse, physically, psychologically, and financially are also significant. For the transition to outpatient care to succeed, frequent therapeutic contact is essential.

BUILDING A TREATMENT TEAM

Patients who have been in a residential or intensive program are used to working with a treatment team. Establishing a cohesive team of outpatient clinicians is essential. In addition to the primary therapist, the three most important members are a psychopharmacologist, a physician or other primary care provider such as a physician assistant or nurse practitioner, and a nutritionist (see Tyson, Chapter 6; Croll, Chapter 8; and Levine & Levine, Chapter 7). Unless the issues are especially complicated, appointments with the medical provider and psychopharmacologist need only occur on a periodic basis. Nutritionists will typically prefer to see patients more frequently, especially in the early postdischarge stage. In addition to issues with dietary management, nutritionists can do weekly weights, communicating this information to the primary therapist as needed. Regular consultation between outpatient providers is especially critical to avoid the risks of splitting and to insure proper and timely coordination of care. For example, all treatment providers should be informed of significant changes in the patient's weight status, nutritional challenges, or medication plans.

Psychopharmacologists

Most psychiatrists who are conducting "split treatment" will be eager to have the primary therapist's insights and diagnostic evaluations of the patient's clinical state. The psychopharmacologist may also have insights and observations that will be helpful to the primary therapist. Historically, communication between psychiatrists and non-medical therapists has often been delicate. However, in the current environment, in which the use of medication is ubiquitous, the collaboration between providers is typically more collegial. As illustrated in the following vignette, non-medical clinicians, regardless of discipline, need to bear in mind that the final decisions about the use of medications is, of necessity, in the hands of the medical professional.

Catherine had done very well in an intensive residential program in the Midwest and had returned to college after a medical leave of absence. She was highly engaged in the first months of her treatment and was able to maintain her weight and normalized eating patterns. She did have some persistent issues with BI, which we were able to work on. However, as is typical of weight-restored anorexics, problems with anxiety and mood fluctuation came to the fore following her inpatient treatment. She noticed that she would go from "high" to "low" even within the space of a single therapy session. I sensed that we had uncovered an atypical mood disorder, of the rapid cycling variety. She had been discharged from the hospital on sertraline, which she continued to take and tolerate well. I suggested that the psychopharmacologist evaluate her for bipolar disorder and consider a mood-stabilizing medication. The pharmacologist agreed that there was a problem, but he did not want to treat aggressively with a medication that could very possibly induce weight gain and other side effects. The clinical decision, therefore, was to *"ride out the storm."* In fact, the patient's symptoms normalized over time and turned out to be both a product of her temperament but also of difficult situational adjustments that she was having after treatment.

Medical Providers

Medical providers with ED expertise are essential members of the outpatient treatment team (Tyson, Chapter 6). Family practitioners, adolescent medicine specialists, nurse practitioners, and physician assistants can all play an important role in the ongoing medical management of patients with AN. The following example illustrates the critical role of medical providers when patients relapse.

Sara was discharged after a 3-month hospitalization for treatment of her bulimic symptoms. Underweight prior to her hospitalization, she was discharged at a weight slightly higher than her target weight range. Sara had deeply resented her hospitalization, comparing it to a form of imprisonment. She had begun, however, to see the need to be at a more stable body weight. Sara had some particularly difficult conflicts in her family-of-origin (with whom she went back to live with before returning to school), and especially with her mother, whom she found excessively controlling and demanding. Her physical state was being monitored by her pediatrician, who was concerned with her low levels of chloride and potassium (both associated with poor food intake and vomiting).

Although Sara was making some progress in examining her interpersonal and family issues in therapy, a part of her was still invested in maintaining a low body weight. She had found it difficult to tolerate the sensation of fullness, fearing that she had eaten more food than she should have. She admitted that she was still vomiting, although it was difficult to get a straight answer about frequency. When her therapist returned from a two-week holiday, it was clear that she had lost substantial weight. Obviously, treatment of developmental and interpersonal issues could not proceed in light of her impending relapse. The physician insisted that she needed to be hospitalized due to her deteriorating electrolyte functioning. Sara's blatantly transferential description of her physician was that *"she is my mother in a white coat."* In collaboration with her therapist, instead of re-admitting her to a long-term program, the physician arranged for admission on an emergency basis to a medical ward of a local community hospital, where she was placed on IV infusions to improve her electrolyte status. This brief, non-psychiatric hospitalization was actually a turning point in Sara's treatment. She knew that she had to relinquish the fantasy of extreme thinness and accept what was probably her genetic destiny to be a full-figured woman. In fact, after the emergency hospitalization, she began to associate a more normal weight with the image of herself as a *"strong, athletic woman,"* rather than seeing herself as *"fat, just like my mother."*

Nutrition Specialists

In an *ad libitum* eating test, Sysko, Walsh, Schebendach & Wilson (2005) found that AN patients improved caloric intake after weight restoration, but were still significantly restricting compared to healthy control subjects. Many weight-restored AN patients continue to struggle with urges to restrict, dramatically raising the risk of relapse. Adding a nutritionist to the treatment team helps to address these issues.

Nutritionists who specialize in ED typically work with food diaries to obtain a realistic assessment of the patient's caloric intake. To avoid redundancy, clinicians also utilizing food journals as part of their cognitive behavioral therapy (CBT) approach should coordinate this procedure with the nutritionist. The nutritionist is often a key figure in the patient's psychological life as well, and many develop a sophisticated understanding of the dynamics of these illnesses. This creates a fruitful area for therapists and nutritionists to share and enhance their mutual clinical understanding of the patient, especially if they maintain frequent contact.

Michele was a highly motivated 18-year-old college student with AN who reached a healthy and stable weight range during her residential treatment. She connected with an experienced nutritionist at home prior to returning to school in the Northeast. Once back in school, she sought intensive, twice weekly psychotherapy. From the beginning, Michele expressed her desire to maintain telephone contact with the nutritionist from home. As this seemed important to her, I agreed, although it is generally easier to utilize the local resources for nutrition therapy.

After a number of months, the patient observed that the messages from both of us were highly consistent; we were essentially *"saying the same things, but in a different language."* This is of particular interest, as we had rather different (although not conflicting) philosophies. My approach was primarily dynamic and interpersonal, while the nutritionist emphasized spiritual recovery and motivational factors. It is quite possible, however, that these differences in therapeutic philosophy were superficial, and that the nutritionist's discourse incorporated both dynamic and interpersonal factors, while mine also emphasized motivational and spiritual factors. The gender dynamics of male therapist/female nutritionist may also have been beneficial. Michele actually commented that although the messages were essentially the same, they were being expressed alternatively in a "male" and "female" voice. There is little question that the regular and focused communications between the two professionals ensured that we remained on the same page and greatly enhanced the patient's perception of consistency.

BACK TO REALITY: FAMILY AND FRIENDS

Family Issues

The transition back to normal social and relational networks is routinely difficult, particularly for adolescents returning to family, school, and friends. Sessions with the whole family early on may be especially productive.

Research with severe mental disorders, such as schizophrenia and bipolar disorder, has established that a high degree of stressful communication within the family after discharge is an ominous predictor of relapse (Milkowitz, 2004). Although the related research for AN is limited, the impact appears to be similar (Eisler, Simic, Russell & Dare, 2007; Kyriacou, Treasure & Schmidt, 2008). In fact, because weight maintenance and normal eating remain central issues for the patient, families are especially prone to respond to their high levels of anxiety by closely monitoring their child's behavior. These issues should be discussed with the parents, prior to discharge. Parents often have many questions, and their uncertainties have been fueled to some extent by the historic divisions in the field about the issue of parental involvement (see Chapters 19 to 21 on family issues). The answers are never simple, because they depend, to a considerable degree, on the state and personality of the patient, as well as on his or her relationship with their parents. Some patients readily welcome involvement of their parents to provide some of the structure they had in the residential program. Others, as often described in the classical literature on this illness, want their parents to stay out of issues with their eating and weight, and feel that they can manage them with the help of their new treatment team. This is an issue that needs to be assessed individually.

The usefulness of family therapy for adolescents has been well established, beginning with the pioneering studies at the Maudsley Hospital in London in the 1980s (Russell, Szmuckler, Dare & Eisler, 1987). What is now known as the "Maudsley Method" has generated intense interest in the United States (Lock, Le Grange, Agras & Dare, 2001) and is considered by many to be the standard of care for adolescents with AN (Varchol & Cooper, 2009; see also Brisman, Chapter 20). Maudsley researchers have also found that, in some instances, especially when the parents are excessively critical or hostile to their child, separate sessions with the parents and the patient can be equally effective (Eisler et al., 2000). Also, the original Maudsley study cited above found that the method's effectiveness was limited to adolescents. In fact, family therapy had a minimal or worsening impact on young adult patients (i.e., over 18 years old) with regard to weight gain and a global measure of outcome. There was a trend for individual therapy to be more effective in the over-18 age group. No subsequent study has contradicted this important limitation.

Although family dynamics are still an important area for the therapist to consider in young adults, if the primary mode of treatment is going to be individual therapy, the exploration of these dynamics with the family should not preempt the goal of establishing a trusting and empathic therapeutic relationship. Often, one or both parents have their own issues that they would like to address therapeutically. In most cases, because of boundary issues, this should be carried out with another therapist. The following vignette addresses the some of the complexities of these boundary dilemmas.

Megan, a 19-year-old college sophomore who had been hospitalized for AN, had not only reached normal weight but had begun to make substantial progress in other areas as well. Despite the fact that her parents lived some sixty miles from my office, they seemed eager to meet with me, despite their awareness of Megan's need for confidentiality. Megan, however, was adamantly opposed to my meeting with her parents and she was especially averse to the idea of joint sessions. Finally, after insistent demands from her mother, she relented, under the condition that we not meet together. I met with both parents without Megan, and they indeed had many anxieties about her ability to maintain her weight and avoid relapse. They expressed concerns about other issues, including Megan's continuing social isolation. Finally, and unbeknownst to the father, the mother hesitantly and poignantly revealed that Megan had been date raped in her junior year in high school.

I had been totally unaware of this part of Megan's history, and apparently it had never surfaced during her hospitalization. I now had a problem, however. I had learned something critical that Megan had not revealed to me. In effect, a secret triangle had been set up between me, Megan, and her parents, and this was likely to cause problems later on. Fortunately, Megan herself disclosed her troubled history about two months later, and we were able to work on its consequences. I never had to disclose that her mother had told me this part of her history, but I felt uneasy about knowing it, and relieved when Megan told the story herself. Of course, there were other ways to proceed than the path that I chose. It would be conceivable to gently disclose the information provided by Megan's parents to Megan in order to facilitate her addressing the issue herself. However, one likely outcome of that communication would be to reinforce Megan's sense of not being in control of her own life and privacy. In my view, it was better to wait for her to disclose the incident when she was ready. The risk, of course, is that the disclosure may have never happened.

Secrets in therapy can have complex ethical and therapeutic ramifications challenging the therapist to titrate the obviously important emphasis on honesty with the practical ramifications of therapist disclosure of possibly disturbing communications with others (Imber-Black, 1993). In her case, I believed that allowing Megan to discuss the sexual abuse in her own time was in her best interest.

If the patient is of college-age and living away from home, the situation is somewhat different. Most parents will still be very interested in the kind of care that their child is getting, especially if they are financially responsible. Parents vary greatly in the degree to which they respect the necessary boundaries of privacy in the therapeutic situation, but the therapist must be clear about confidentiality and the need for a secure, trusting therapy relationship. The majority of parents will respect the need for confidentiality. While alleviating blame and guilt is important, this can be challenging, as most discussions take place by telephone early on in treatment, often prior to the time that the therapist has a full sense of the family dynamics.

Developmentally, college age patients living away from home may not need as much family therapy as younger patients, for whom it should be a priority. However, when the patient's permission and willingness to meet with his or her parents has been established as a precondition, it can be helpful to have occasional family sessions. This becomes a more pressing need if there is a crisis or growing threat of relapse, as seen in the following clinical vignette.

Janna was the oldest of three sisters in a family from Texas, attending a Northeastern college. She became severely anorexic during her fourth semester, forcing withdrawal from school. After intensive treatment for 2 months in a residential facility, she entered an aftercare program, returned to school the next year, and lived off-campus with several roommates. While able to maintain her weight and address many important issues in treatment, her relationships with her roommates were often turbulent. One of her roommates was actively struggling with serious depression; another roommate was a close friend with whom Janna's mother had made a sub rosa agreement to notify her if she saw any deterioration in Janna's behavior.

Janna's tense interactions with her various housemates were partially due to her new-found assertiveness (sometimes what appears to be a conflict or "troublesomeness" emanates from a patient's overcoming her excessive need to placate). Her parents came to school to deal with what appeared to be a crisis and we held a joint session. I had had little previous contact with the parents. Two issues emerged. The first was that Janna's friend may have been overreacting to the potential of Janna's "relapse" out of her own sense of personal distress in their relationship. But second, the parents very much wanted Janna to end this living situation and change her residence. Janna intensely resisted. My response was to ally with Janna, emphasizing her need to make independent and mature decisions about what she could handle. The outcome was positive despite initial tension; Janna felt validated by my support while her parents also felt more secure about Janna's treatment and reaffirmed in their role as parents.

Friends' Roles in Recovery

The preceding vignette also illustrates an issue that has been little discussed in the clinical literature. Peers who have seen a friend unravel into severe anorexia and then reappear in improved physical condition often have profound reactions. Although reassured by the physical changes, peers often remain considerably anxious about the recovering person's vulnerability. This can be productive, if the interest is primarily a constructive one. Friends can provide emotional support and opportunities for sharing of feelings, especially troubling ones. Social isolation is a typical precursor of anorexic episodes, and its continuation after treatment is obviously unfavorable for the patient's recovery. However, patients can experience their friend's concerns as intrusive and sense that they are under intense scrutiny. This perception can lead to self-consciousness, constrained interactions, and resentful or angry feelings.

BODY IMAGE, WEIGHT CONCERNS, AND EXERCISE

Although many patients show substantial improvement in BI after residential treatment (Probst, Vandereycken, Van Copponolle & Pieters, 1999), most have persistent BI concerns that include feelings of fatness and hypervigilance about not exceeding their discharge weight (see Johnston, Chapter 26). Some studies show such feelings to be highly predictive of relapse (Keel et al., 2005). Patients in partial recovery often fantasize about re-establishing their "skinny" selves, fantasies which are sometimes accompanied by brief periods of restriction. As these are normal thoughts during the recovery process, patients need a forum in which to discuss them. The therapist's attitude of acceptance towards these toxic thoughts provides a model for patients to adopt the same stance and teaches that, however much they provoke anxiety about regressing, thoughts are transient and will pass. Open discussion about negative BI allows patients to begin to see these thoughts as a part of their illness, not a part of their own character, and inimical to their well-being.

The patient's improved physical and emotional states also provide an opportunity to examine the functions and meaning of both BI concerns and impulses to restrict. With respect to the latter, it is common for patients to have a strong impulse to restrict when they become anxious about social and interpersonal concerns. Though not always possible, the therapist should make every effort to help the patient become aware of these connections, because they often represent a critical part of the dynamic that originally fueled the development of AN. Patients' ongoing journals about their thoughts and feelings can be especially helpful at these junctures.

Exercise and physical activity can play a positive role in recovery and must be addressed in outpatient treatment (Ressler et al., Chapter 24; Calogero, and Pedrotty-Stump, Chapter 25). Practices such as Yoga, Tai-Chi Chuan, or other forms of body practices that emphasize "centering" (Calogero & Pedrotty, 2004) can be especially helpful. They promote a sense of the body as experienced from within and encourage a kind of meditative attitude towards cognitive and emotional activity.

The timing and structure of exercise activities is as important as the type of movement. Recommending these activities to patients in the early stages of outpatient treatment without providing careful guidelines can be counterproductive (Calogero & Pedrotty, 2004, 2007). Nonetheless, contrary to longstanding admonitions against exercise during the weight restoration process, new evidence suggests that supervised, mindful activities can actually facilitate weight restoration and maintenance (see Calogero, and Pedrotty-Stump, Chapter 25, for a thorough discussion of this topic). Keep in mind that Strober et al.'s (1997) long-term follow-up study found that compulsive exercise is one of the most powerful predictors of relapse. Many patients know intuitively that it would be very difficult for them to return to running on a treadmill without lapsing into excess, and therapists should support their honest self-appraisals in this regard. However, it is also important for patients to gradually re-establish healthy activity levels, often by finding other avenues for movement. So while "working out" should be carefully re-introduced under supervision, regular walking, initially for prescribed periods of time, may be encouraged, with discussion about its health benefits. Even better, an emphasis on exercise that is integrated with outdoor activity in a natural context, such as gardening or hiking, is recommended. If the patient is so inclined, participation in group athletic activities that emphasize the value of exercise in a social context, can also be encouraged, again being mindful of the potential risks.

Abby's experience of her BI fluctuated considerably after residential treatment. Some days she felt that she looked fine, but on others she was tormented by the feeling of being fat. Her journals showed a correlation between these feelings and her mood. She had a conflictual relationship with a boyfriend, and following intense arguments she experienced intense fantasies of reverting to "*eating nothing*" and re-establishing waif-like status. She had made enough progress to recognize these impulses as her *"old friend anorexia,"* which she increasingly recognized as an enemy of her well-being. When she felt this way, she often increased her body checking, and would spend a certain amount of time before a mirror each morning to ensure that her abdomen remained flat. Because she spontaneously personified the anorexia in this way, I had her conduct some Gestalt-type dialogs, with her "anorexia" as the figure in the empty chair. While she found this difficult and awkward at first, as most people do, she became very engaged in these dialogs and realized the extent to which her anorexia was connected with very angry feelings and urges to *"say no."*

It took some months of treatment to get to a point where Abby was comfortable in acknowledging the connection between her anger and her ED rituals. She had also been a compulsive exerciser prior to her hospitalization and knew instinctively that it would be difficult for her to set limits for herself if she started going to the gym again. Drawn to gardening, she found that working outdoors during the summer was almost sufficient to satisfy her need for activity. During the winter, however, such activity was obviously impossible. In order to deal with her BI dissatisfaction, she joined a campus support group on BI issues, which she found helpful and to which she made many contributions. The following summer she worked for a professional gardening service. Not only did this channel her need for activity productively, but her comradeship with the regular working staff, who accepted her as an equal, was a great boost to her confidence.

INTERPERSONAL AND DYNAMIC ISSUES

Despite the need to maintain some focus on weight, food, and BI issues, the interpersonal and emotional dilemmas are core issues for long-term recovery. As noted earlier, weight restoration and improved nutritional intake allow for a shift in therapeutic priorities, with the interpersonal and emotional issues moving to the fore.

Kyra developed AN in the summer following her graduation from high school and continued to slowly lose weight over the following 2 years. After two unsuccessful attempts at outpatient treatment, both of which emphasized direct behavioral intervention, she was admitted to an inpatient program for 10 weeks, where her weight was fairly well normalized. Kyra was 10 years younger than her sister, Zoe, who had had a tumultuous adolescence, but who had achieved some success in the computer software field. Zoe was estranged from her parents, and had sworn that she would never marry or have children. The parents focused their hopes on Kyra as the child who could absolve them for being the bad or controlling parents, accusations previously made by Zoe. Kyra's anorexia had badly shaken them. She attended a local college, near home, and was an excellent science student aiming for a career in chemistry. Kyra felt that her hospitalization, while a shocking interruption to her life, had provided her with a much better ability to tolerate a higher body weight, although "being around food" was still anxiety provoking.

After 3 months of treatment, Kyra revealed that she had been pressured into having sex a number of times by a male acquaintance who had served informally as her trainer at a local gym. She very much admired the man's physical prowess and his skill as a trainer, and was afraid that she would alienate him if she refused his advances. Nevertheless, she had found the experiences painful and demeaning and bitterly regretted that she had gone along.

In addition to this secret trauma, Kyra revealed that she had felt intense anxieties around people, especially men, as far back as she could remember. These and other critical issues, including concerns about male approval, her low self-esteem, her driven and perfectionistic nature, and the way in which Kyra's conflicts were pre-figured in the lives of her parents, became the focus of treatment for well over a year. Only then was Kyra able to feel more genuinely comfortable with herself. Throughout this deepening work, Kyra continued to work with her nutritionist to help her increase her comfort level with a higher caloric intake. This nutritional stability set the foundation for the shift in therapeutic focus to the more complex interpersonal and emotional challenges.

It is difficult to prescribe specific procedures to deal with these complicated interpersonal and psychodynamic issues; the choice of interventions made by the therapist and the patient reflect the art of longer-term psychotherapy. Comprehensive, ongoing training in psychodynamic and interpersonal therapies is essential. Kathryn Zerbe's (2008) book is a particularly good resource providing a detailed discussion of longer-term psychodynamically oriented therapy with ED patients. Unlike many traditional psychodynamic approaches, Zerbe makes an effort to integrate many of the tools of a CBT approach, and incorporates cultural considerations into treatment. Zerbe's analysis of what is needed for treatment at different stages of the life cycle is valuable and instructive.

Fairburn (2008) has expanded his earlier model of CBT to address some of the mechanisms that perpetuate an overvaluation of shape and weight. The new model, CBT-E, is designed for patients across the ED diagnostic spectrum. It extends the focus on normalization of eating, weight, and shape concerns to interpersonal and emotional factors that may be sustaining the ED. The CBT-E model includes interventions for perfectionism, core low self-esteem, and interpersonal problems. A study of CBT-E for patients with AN is in progress, but has not been published at the time of writing of this chapter (Fairburn, personal communication, 2009).

Given the proven effectiveness of interpersonal therapy (IPT) for bulimia nervosa (BN) (Agras, Walsh, Fairburn, Wilson & Kraemer, 2000), and the prominence of interpersonal issues in AN, one would expect that IPT would be useful in AN. There has been little systematic work on this hypothesis to date, and the one controlled study concluded that both manualized IPT and CBT were considerably less effective in treating non-weight-restored AN outpatients than a flexible clinical management approach (McIntosh et al., 2005). I would speculate that, in their manualized forms, neither treatment addresses a broad enough spectrum of issues to be successful. For example, manualized IPT excludes a focus on behavioral issues relevant to the ED, and CBT does not sufficiently incorporate interpersonal factors. One published study did find that a broadened form of CBT, one that incorporated interpersonal factors, was superior to nutritional therapy alone for weight-restored patients (Pike, Walsh, Vitousek, Wilson & Bauer, 2003).

INTEGRATED APPROACHES

Clinicians working with ED patients need to draw from a variety of different treatment options. Cognitive Analytic Psychotherapy (CAT) (Ryle, 1990; Ryle & Kerr, 2002) effectively integrates empirically validated aspects of psychodynamic, interpersonal, and cognitive therapies into one package. In preliminary work on the application of CAT to the treatment of AN, Treasure and Ward (1997) reported its use with a severe case, and a London-based team reported on a controlled trial of CAT comparing it with family therapy, a year of focal psychoanalytic psychotherapy, and a control group of low contact, routine treatment for one year (Dare, Eisler, Russell, Treasure & Dodge, 2001). The results were only modestly successful as measured by weight gain and other improvements, but most patients in the study were below target weight. As emphasized above, under these conditions, weight issues tend to dominate, or strongly influence, the therapy. Ryle and Kerr (2002) commented that:

> Eating disorders represent the expression of inter- and intrapersonal problems through an abnormal preoccupation with weight and food. They are always associated with, but may serve to obscure, problems at the level of self processes, predominantly expressed around issues of control, submission, placation and perfectionism (p. 152).

Cognitive analytic therapy is designed to be a short-term therapy (16 sessions) for use with a wide spectrum of patients. The therapy focuses on the patient's current interpersonal dilemmas and impasses, based on the understanding that these issues originated in early interactions with family members or crucial others. In the early stages, it provides assessment tools to define the patient's essential personal issues and interpersonal impasses. The therapist's task is to formulate the crux of the patient's interpersonal dilemmas in the form of a written statement, typically in the form of a letter to the patient. The formulation is open to revision on the basis of the patient's response to it and the therapist's increasing appreciation of the patient's difficulties. This early written formulation provides a great deal of focus for the psychosocial component of therapy, and helps to keep patients motivated and connected through what is often experienced in other forms of psychodynamic treatment as a vague and open-ended process.

The therapist is also encouraged to construct diagrams depicting the patient's interpersonal dilemmas. An essential component of CAT is the notion that patients are typically trapped in vicious cycles of behavior and interactions. These recurring patterns are protective but they also serve to reinforce or worsen their anxieties; CAT focuses on changing these rigid and maladaptive attitudes towards interpersonal situations. In AN, a common pattern is to ward off feelings of anger by demanding independence or placating others, while simultaneously acting in ways that provoke others to attempt to control them further. The therapy is basically psychodynamic in that it traces the patient's self-defeating patterns to childhood experiences. But it is a far more efficient technique in that the therapist is able to arrive at a tentative formulation of the relationships between current behaviors and childhood relationships in three or four sessions.

Cognitive analytic therapy emphasizes the ways in which the patient contributes to sustaining problematic interactions ("reciprocal role relationships"). The approach also utilizes the interaction between patient and therapist as a model "in the room" for the patient's relationships in the real world, past and present. Its incorporation of an active focus on the transferential aspects of the patient–therapist relationship makes it unique among the cognitive therapies.

This integrated therapeutic model seems to have substantial potential as a technique in the treatment of weight-restored AN patients. It has demonstrated efficacy with patients with borderline personality disorder (Ryle & Kerr, 2002), but it still lacks strong empirical support for the treatment of ED. The CAT model reflects the central thesis of this chapter, namely, that a comprehensive treatment of ED requires both the management of weight and weight restoration, and a systematic approach to the associated cognitive, interpersonal, and affective features of the illness. It is this integration that makes treatment so challenging, and further research such a high priority.

ASSESSING AND ADDRESSING COMORBIDITIES

Comorbidities such as depression and anxiety often only emerge in full force once ED behaviors are interrupted and the physically toxic state of starvation is remedied. Depression and anxiety, for example, tend to be well masked or submerged in the acute anorexic state (see Levine and Levine, Chapter 7 for a review of these issues). Weight-restored patients will often be aware of a heightened sense of anxiety and moodiness, as is evident in the stories of Kyra and Michele above. At times, it is hard to distinguish between a primary mood or anxiety disorder and secondary reactions to malnutrition and disordered eating. When these symptoms are identified, medications can play a significant role, as illustrated in the following vignette.

Julie developed AN in the 11th grade during which her weight dropped from 125 to 98 pounds. She was hospitalized for a month and a half over the summer of her junior year and was discharged close to her target weight. She entered outpatient therapy immediately after her hospitalization and quickly became engaged in the treatment process. An excellent student with a high degree of perfectionism, she applied herself to her therapeutic work, writing extensive journals with enthusiasm. Her weight, however, gradually began to drop, and both the nutritionist and the therapist were concerned. During the course of one session, it suddenly became clear that Julie had had a severe episode of childhood OCD when she was around 8 years old, one that she had never disclosed to anyone, including her treatment team in the hospital. As one of her rituals, she had spent hours praying in the evening that nothing untoward happen to her parents (there was no external threat). After conferring about this and Julie's ongoing anxiety levels, the psychiatrist decided to treat her with an SSRIs, increased to a fairly high level after an initial positive response. Julie felt that the medication helped her immensely. Not only was she less anxious in general, but she was able to approach food more comfortably and re-establish her normal weight.

On the issue of mood symptoms, patients need to utilize psychotherapy to develop a clearer sense of who they are and whether some degree of social anxiety or mood fluctuation is just a part of their temperament—one they can learn to live with. Many patients who manage to embrace this position are initially ambivalent about medication. Ultimately, they usually come to accept and adapt to their own personality characteristics.

Aside from understanding anxiety and depression as medical comorbidities, we should attempt to comprehend both their dynamic significance and consequences. For example, many AN patients will experience depressive reactions to having "lost" the thrill of control they had achieved in the active phases of AN. To the extent that starvation quelled the anxieties that preceded the ED, they have a new opportunity to confront these anxieties directly, to learn what they are about, and to cope with them more effectively. In struggling with these issues, patients may have fantasies of fleeing back into comfort and protection that the starvation phase of their illness afforded them. As one said to me, *"when I think about these things [feelings of inadequacy provoked by interpersonal conflicts], I just want to starve, starve, starve."* The mere articulation of such fantasies in psychotherapy helps to establish control over the underlying impulses.

SUMMARY

The treatment of patients with AN after their weight has been restored offers the opportunity to address the fundamental issues that gave rise to the disease. These concerns tend to be obscured by malnutrition, and are of lower priority during inpatient or residential treatment. Despite the emphasis on the opportunities for thoroughly addressing psychosocial factors after weight restoration, weight and eating behaviors remain dangerous vulnerabilities for the AN patient. Body-image concerns linger; exploring their origins and meanings, both developmentally and interpersonally, can be a bridge that shifts the psychotherapy focus from body and eating to relational and developmental issues.

Ultimately, treatment for weight-restored AN patients needs to focus on the interpersonal and characterological factors that were involved in the development of the disorder. If unmodified, these factors will continue to support the maintenance of the illness (Federici & Kaplan, 2008). Although there are many interpersonal and cognitive therapies, I have argued that the CAT approach developed by Ryle (1990) has great promise.

Based on his years of experience at the Royal Free Hospital in London, Russell commented that the science and art of weight restoration in AN had been well developed, as early as 1977 (Russell, 1977). The critical question about resolving the illness, he thought, was how to help the patient after discharge. Russell felt that we knew little about what factors predict a poor outcome, and probably even less about whether outpatient treatment contributes to resolving the illness. In the intervening years, we have learned much about the course of AN and the factors that predict relapse. We know much less about the comparative effectiveness of different treatment methods.

This chapter attempts to incorporate many of the research findings about the predictors of relapse into treatment, including findings about the persistence of food restriction (Sysko et al., 2005), the predictive value of weight loss in the weeks after discharge (Kaplan et al., 2009), and the persistence of BI concerns in weight-restored patients (Keel et al., 2005). We know very little from research about the treatment of psychosocial and emotional factors, but clinical experience, as well as our patients' reports, continue to show their importance for long-term sustained recovery (Federici & Kaplan, 2008; Jarman & Walsh, 1999; Keski-Rahkonen & Tozzi, 2005; Lamoureux & Bottorff, 2005; Nordbo et al., 2008;

Surgenor, Plumridge & Horn, 2003; Tozzi, Sullivan, Fear, McKenzie & Bulik, 2003; Weaver, Wuest & Ciliska, 2005; see also McGilley and Szablewski, Chapter 12).

I have suggested the relevance of some time-limited treatment models for resolving the interpersonal and developmental issues of weight-restored anorexics. Still, much work needs to be done to put these methods on sound empirical footing. Treatment research in AN remains one of the most pressing needs of the entire field of ED (Agras et al., 2004). A few controlled trials have been conducted, with mixed results, and it is critical to undertake future research on groups of different weight status and age, for reasons enumerated in this chapter. Meanwhile, the outpatient clinician must continue to work with patients, integrating the art of psychotherapy with the emerging research. This chapter outlines some of the ingredients that make such treatment successful.

References

Agras, W., Brandt, H., Bulik, C., Dotlan-Sewell, D., Fairburn, C., Halmi, K. A., ... Wilfley, D. E. (2004). Report of the National Institutes of Health Workshop on overcoming barriers to treatment research in anorexia nervosa. *International Journal of Eating Disorders, 35,* 509–521.

Agras, W. S., Walsh, B. T., Fairburn, C. G., Wilson, G. T., & Kraemer, H. C. (2000). A multicenter comparison of cognitive-behavioral therapy and interpersonal psychotherapy for bulimia nervosa. *Archives of General Psychiatry, 57,* 459–466.

Calogero, R. M., & Pedrotty, K. N. (2004). The practice and process of healthy exercise: An investigation of the treatment of exercise abuse in women with eating disorders. *Eating Disorders: The Journal of Treatment and Prevention, 12,* 273–291.

Calogero, R. M., & Pedrotty, K. N. (2007). Daily practices for mindful exercise. In L. L'Abate, D. Embry, & M. Baggett (Eds.), *Handbook of low-cost preventive interventions for physical and mental health: Theory, research, and practice* (pp. 141–160). Amsterdam, ND: Springer-Verlag.

Carter, J. C., Blackmore, E., Sutandar-Pinnock, K., & Woodside, D. B. (2004). Relapse in anorexia nervosa: A survival analysis. *Psychological Medicine, 34,* 671–679.

Cockell, S. J., Zaitsoff, S. L., & Geller, J. (2004). Maintaining change following eating disorder treatment. *Professional Psychology: Research and Practice, 35,* 527–534.

Dare, C., Eisler, I., Russell, G., Treasure, J., & Dodge, L. (2001). Psychological therapies for adults with anorexia nervosa: Randomized controlled trial of outpatient treatments. *British Journal of Psychiatry, 178,* 216–221.

Eisler, I., Dare, C., Hodes, M., Russell, G., Dodge, E., & Le Grange, D. (2000). Family therapy for adolescent anorexia nervosa: The results of a controlled comparison of two family interventions. *Journal of Child Psychology and Psychiatry, 41*(6), 727–736.

Eisler, I., Simic, M., Russell, G. F., & Dare, C. (2007). A randomized controlled treatment trial of two forms of family therapy in adolescent anorexia nervosa: A five-year follow-up. *Journal of Child Psychology and Psychiatry, 48,* 552–560.

Fairburn, C. G. (2008). *Cognitive behavior therapy and eating disorders.* New York, NY: Guilford Press.

Federici, A., & Kaplan, A. S. (2008). The patient's account of relapse and recovery in anorexia nervosa: A qualitative study. *European Eating Disorders Review, 16,* 1–10.

Imber-Black, E. (1993). *Secrets in families and family therapy.* New York, NY: W.W. Norton.

Jarman, M., & Walsh, S. (1999). Evaluating recovery from anorexia nervosa and bulimia nervosa: Integrating lessons learned from research and clinical practice. *Clinical Psychology Review, 19,* 773–788.

Kaplan, A. S., Walsh, B. T., Olmsted, M., Attia, E., Carter, J. C., Devlin, M. J., ... Parides, M. (2009). The slippery slope: Prediction of successful weight maintenance in anorexia nervosa. *Psychological Medicine, 39,* 1037–1045.

Keel, P. K., Dorer, D. J., Franko, D. L., Jackson, S. C., & Herzog, D. B. (2005). Postremission predictors of relapse in women with eating disorders. *American Journal of Psychiatry, 162,* 2263–2268.

Keski-Rahkonen, A., & Tozzi, F. (2005). The process of recovery in eating disorder sufferers' own words: An internet-based study. *International Journal of Eating Disorders, 37,* S80–S86.

Kyriacou, O., Treasure, J., & Schmidt, U. (2008). Expressed emotion in eating disorders assessed via self-report: An examination of factors associated with expressed emotion in carers of people with anorexia nervosa in comparison to control families. *International Journal of Eating Disorders, 41*, 37–46.

Lamoureux, M., & Bottorff, J. (2005). "Becoming the real me": Recovering from anorexia nervosa. *Health Care for Women International, 26*, 170–188.

Lock, J., Le Grange, D., Agras, W. S., & Dare, C. (2001). *Treatment manual for anorexia nervosa: A family-based approach.* New York, NY: Guilford Press.

Lowe, B., Zipfel, S., Bucholz, C., Dupont, Y., Reas, D. L., & Herzog, W. (2001). Long-term outcome of anorexia nervosa in a prospective, 21-year follow-up study. *Psychological Medicine, 31*, 881–890.

McIntosh, V. W., Jordan, J., Carter, F. A., Luty, S. E., McKenzie, J. M., Bulik, C. M., … Joyce, P. R. (2005). Three psychotherapies for anorexia nervosa: A randomized, controlled trial. *American Journal of Psychiatry, 162*, 741–747.

Milkowitz, D. J. (2004). The role of family systems in severe and recurrent psychiatric disorders: A developmental psychopathology view. *Developmental Psychopathology, 16*, 667–688.

Nordbo, R., Gulliksen, K., Espeset, E., Skarderud, F., Geller, J., & Holte, A. (2008). Expanding the concept of motivation to change: The content of patients' wish to recover from anorexia nervosa. *International Journal of Eating Disorders, 41*, 635–642.

Pike, K. M., Walsh, B. T., Vitousek, K., Wilson, G. T., & Bauer, J. (2003). Cognitive-behavior therapy in the post-hospitalization treatment of anorexia nervosa. *American Journal of Psychiatry, 160*, 2046–2049.

Probst, M., Vandereycken, W., Van Copponolle, W., & Pieters, G. (1999). Body experience and eating disorders before and after treatment: A follow-up study. *European Psychiatry, 14*, 333–340.

Russell, G. F. M. (1977). General management of anorexia nervosa and difficulties in assessing the efficacy of treatment. In R. A. Vigersky (Ed.), *Anorexia nervosa* (pp. 277–290). New York, NY: Raven Press.

Russell, G. F. M., Szmuckler, G. L., Dare, C., & Eisler, M. A. (1987). An evaluation of family therapy in anorexia nervosa and bulimia nervosa. *Archives of General Psychiatry, 44*, 1047–1056.

Ryle, A., & Kerr, I. B. (2002). *Introducing cognitive analytic therapy: Principles and practice.* Hoboken, NJ: John Wiley & Sons.

Ryle, A. (1990). *Cognitive analytic therapy: Active participation in change.* Chichester, UK: John Wiley & Sons.

Strober, M., Freeman, R., & Morrell, W. (1997). The long term course of severe anorexia nervosa in adolescence: Survival analysis of recovery, relapse, and outcome predictors over 10–15 years in a prospective study. *International Journal of Eating Disorders, 22*, 339–360.

Surgenor, L., Plumridge, E., & Horn, J. (2003). "Knowing one's self" anorexic: Implications for therapeutic practice. *International Journal of Eating Disorders, 33*, 22–32.

Sysko, R., Walsh, B. T., Schebendach, J., & Wilson, G. T. (2005). Eating behavior among women with anorexia nervosa. *American Journal of Clinical Nutrition, 82*, 296–301.

Tozzi, F., Sullivan, P., Fear, J., McKenzie, J., & Bulik, C. (2003). Causes and recovery in anorexia nervosa: The patient's perspective. *International Journal of Eating Disorders, 33*, 143–154.

Treasure, J., & Ward, A. (1997). Cognitive analytical therapy in the treatment of anorexia nervosa. *Clinical Psychology and Psychotherapy, 4*, 62–71.

Varchol, L., & Cooper, H. (2009). Psychotherapy approaches for adolescents with eating disorders. *Current Opinion in Pediatrics, 21*, 457–464.

Weaver, K., Wuest, J., & Ciliska, D. (2005). Understanding women's journey of recovering from anorexia nervosa. *Qualitative Health Research, 15*, 188–206.

Zerbe, K. (2008). *Integrated treatment of eating disorders: Beyond the body betrayed.* New York, NY: W.W. Norton.

CHAPTER

12

Recipe for Recovery
Necessary Ingredients for the Client's and Clinician's Success

Beth Hartman McGilley and Jacqueline K. Szablewski

This chapter elucidates ingredients of two inextricably linked topics not well described in the clinical literature. First, despite a wide body of research now supporting the assertion that the quality of the therapeutic alliance is the best predictor of psychotherapy outcome (American Psychiatric Association, 2006), little rigorous attention has been paid to the qualities of the eating disorder (ED) therapist most conducive to a positive healing relationship. Second, even with close to 50 years of research, a comprehensive, comparable, consistent, and clinically meaningful definition of recovery has yet to be articulated and accepted in and across the ED treatment field.

To assist in bridging these gaps, we begin with a discussion of those therapist qualities associated with effective therapeutic alliance, followed by an exploration of what constitutes recovery. We hope to describe the mysterious mixture of textures, flavors, and hallmarks that indicate that the healing has indeed been done.

THE EFFECTIVE CLINICIAN'S CUPBOARD

The Importance of Alliance

Both clinical experience and scientific rigor have borne out that the quality of the therapeutic relationship is essential for successful ED treatment (American Psychiatric Association, 2006; Beresin, Gordon & Herzog, 1989; Bunnell, 2009; Burket & Schramm, 1995; Costin, 2007a). In fact, it is regarded as a better indicator of positive outcome across ED diagnoses than any specific treatment technique (Costin, 2007a). The American Psychiatric Association (2006) names the importance of a therapeutic relationship as the first principle in their psychiatric management treatment guidelines. Eating disorder practitioners attest that solid therapeutic rapport results in less attrition, fewer premature treatment terminations, and more helpful therapy. It is through the relationship that we challenge the client's reliance on their ED symptoms and engage them in developing alternative, adaptive skills in order to make effective changes in their physical, psychological, and psychosocial functioning.

Advances in neuropsychiatry and neuroscience have substantiated the relational claim, long suspected as true by practicing clinicians, and experienced as true by ED clients themselves (Beresin et al., 1989; Pettersen & Rosenvinge, 2002). For example, the discovery of mirror neurons has generated significant interest in the healing professions (Cozolino, 2002). Located in the premotor areas of the brain's frontal cortex, mirror neurons "fire in response to an observation of a highly specific relationship between an actor and some object, and also fire when the action is performed (mirrored) by the observer" (Cozolino, 2002, p. 184). The involved motor systems "in turn activate networks of emotions associated with such actions" (p. 186). In a number of ways "mirror neurons may bridge the gap between sender and receiver, helping us understand one another and enhance the possibility of empathic attunement" (p. 186). Thus, echoing early infant attachment studies, contemporary feminist thought, and the field of quantum physics, interpersonal impact is inevitable; so too, within the therapeutic relationship.

Previous chapters have addressed and explored approaches and modalities for treating ED. Effective ED clinicians must be adept in their approach and steeped in a variety of psychotherapeutic techniques. A basic body of knowledge, supervised specialty training in ED treatment interventions and techniques, clinical experience, and constant review constitute the staples of training programs for ED specialists (Andersen & Corson, 2001; Yager & Edelstein, 1987). Too often overlooked and harder to scientifically quantify are the qualities in the seasoned clinician that enliven the techniques and infuse the relationship so central to healing. Just as warmth and caring without technical skills do not suffice for sustained healing and recovery, neither do technical skills without an authentic therapeutic relationship.

Alliance Ingredients

Non-possessive warmth and unconditional positive regard. "Non-possessive warmth" (Andersen & Corson, 2001, p. 356), stands akin to the "unconditional positive regard" espoused by Carl Rogers (1961, p. 47). Informed by the tenets of feminist relational theory, it provides the base for the therapeutic connection. Elements included are:

- Basic respect for the person who sits before us
- A desire to know who that person is no matter how textbook they may seem
- Care for whom we will discover them to be in the context of their lives
- Compassionate curiosity about how and what they make meaning of in life
- Communication that is honest and direct and not placating or patronizing
- Respectful kindness honoring boundaries on both sides
- Willingness to be wrong, own mistakes, and repair therapeutic impasses
- Room for conflict and confrontation to inform and deepen the connection.

These relational conditions create breathing space for clients amidst their skepticism, self-doubt, and panic (aka resistance). They help the client to dare to stay, trust, and risk often when not wanting to and sometimes without knowing why. To the clinician, these relational conditions offer an opportunity to meet the client. Impassioned by each of these, we have a presence to offer and we begin our invitation for our clients to engage. If it is accepted, the work of recovery can proceed.

Active and worthwhile engagement. Since most clients exist within the perceived safety of the ED "atmosphere," the invitation we extend needs to be engaging enough to reach through the atmospheric resistance. Davis (2009a) speaks poignantly about "the need [at the beginning of treatment] to grab the patient's attention to wrench her away from her relationship with the ED symptoms and toward the other person [the therapist] in the room" (p. 40). He emphasizes nontraditional manners of behaving (less formal and aloof, more personal and collaborative), while being mindful of appropriate boundaries, in order, "to get the [person] thinking and wondering about you so that she starts to experience the therapist's presence" (p. 40).

In my practice, this "active and worthwhile engagement" (Davis, 2009b, p. 6) takes many forms: a comment about another NFL team if the client shows up with a Pittsburgh Steelers cap on; a compliment about the scarf or pendant a client wears to session; or an exchange about what the client's "patronas charm or dementors" would be if the client shows up reading *Harry Potter 3* (Rowling, 1999).

Such interchanges, marked by spontaneity, curiosity, and considered bits of self-disclosure, create opportunities. For a moment, the attention of the client is disarmed and moves away from the ED toward the therapist (Davis, 2009a, b). A new form of "molecular bonding," a client-allowed-crack in the resistance occurs. Somehow, through the fissures in the resistance, impact can be felt and true contact has been made.

Embodied authenticity and being real. As stated by Zerbe, perhaps "first and foremost [our] ability to be real and human [helps our client] to feel that she, too, [has] a chance to be real and human herself " (1995, p. 162). Beresin et al. (1989) describe the task of becoming real as central to ED recovery and an area in which mentoring is necessary. How we deal with a faux pas, unintended empathic lapse, or a less than graceful moment goes a long way in modeling the survivability of human imperfection, and the resiliency of relationship in the face of conflict, disappointment, frustration, and anger. These are feelings the client may have sought to avoid through the "protection" of their ED symptoms.

These real, authentic experiences in relationship, of relationship, provide exposure to a different perspective of humanness. For example, upon successful completion of a 7-year treatment course for chronic bulimia nervosa (BN), incest and abuse, a former client thanked me for my "true humanness" in our relationship. She named it as central to her "becoming a real human being again." Remarkably, this was a client whose initial grappling with the questions and meanings of humanness and perfection yielded her the insight that in her internal logic, "humanness" had defined her perpetrator, while "perfection" had attempted to separate her from abuse and the abuser. She would interpret a cookie with her lunch, an unsatisfying interaction with a co-worker, or an extra pound on the scale when she weighed herself for the fourteenth time that day, as "really messing up." In the face of such "intolerable humanness" and perceived failing by self or other, she would turn to punishing and severely denigrating behaviors. A core component of her healing involved her challenging this template in the ways she related to her body and self, as well as in thought, spirit, and action. Eventually she was able to reclaim a broader definition of humanness characterized by the alternative relational experiences she had in therapy, and slowly but surely, in the other places of her life. Loosening her grip on both the concept and practice of perfectionism made room for a healthy acceptance of the imperfections that make her authentically human and real.

In a similar vein, how we honor our own boundaries and expect others to do the same, like taking a vacation, or regularly breaking for lunch, demonstrates the permissibility of basic human needs and desires, and the process of navigating between "not abandoning self for other" and "abandoning self for fear the other will feel abandoned." How and whether we deal with or ignore a burp, hiccup, coughing fit, or tummy growl can illustrate the naturalness of body function and the vital legitimacy of having a physical self that requires both acknowledgment and care. Especially in ED treatment, the therapist's embodiment is a powerful example and teacher. "Carefully tended and appropriately nourished, the therapist's embodied experience can be … a useful tool in the efforts to help clients navigate recovery" (Costin, 2009, p. 191), teaching them to tend carefully, nourish well, and even enjoy their own embodiment.

Empathy and trust. For client and clinician alike, it can be tempting to avoid, dismiss, or numbly barrel through those life situations and challenges that threaten to expose our vulnerabilities and flaws. When finally we risk facing the challenge or developmental life task, we are also taking the tender and calculated risk of being seen, tolerated, even loved "as our worst selves" (Derenne, 2006, p. 339), in our worst light. Clinicians who have been clients, whether for ED recovery or other forms of personal growth, know both sides of this abyss. Taking that leap, believing our extension of trust will not be betrayed or belittled, is the essence of trust.

Empathy, including empathic memory of what we have learned from the client, about the client, is required to facilitate this kind of letting go. Derenne (2006) notes that remembering details, both large and small, is one of the most essential aspects of her role as a child psychiatrist and fosters her ability to connect with ED clients and their families. After many years of practice, I am still humbled by how a client can be touched and surprised by my memory of their best friend's name, the date of a particular loss they've endured, or the adjectives they've used to describe an experience in their lives. Cousin to active listening and mindful attending, empathic memory and present empathy, void of triteness, provide evidence to clients that they matter. Being listened to—heard, understood, taken in, remembered, without being intruded upon—is a reflection of the process of introjection and empathy without the loss of self, abandonment of self-identity, or annihilation of other so many clients fear.

This level of trust, then, potentiates the possibility for the client that even their "worst selves" might be tolerated without retaliation (Beresin et al., 1989; Derenne, 2006; Zerbe, 1995). Such are particularly critical moments in the therapeutic process, repeating multiple times in different forms at different stages of the therapy. Depending on the manifestation of the client's worst self, often impacted by comorbid conditions or characterological constellations, it can be more or less intense/activating for both client and clinician. As in Dante Alighieri's classic poem, *The Divine Comedia 1: Inferno* (1939), the "pilgrim," like our clients, must not be left alone in the deepest spirals of Hell. Just as Virgil, fortified by the guide Beatrice, "stays the course," so must we as clinicians, thereby deepening the therapeutic relationship and the work of healing.

Endurance and frustration tolerance. A high frustration tolerance, and the ability to endure ambiguity and the often lengthy ED recovery process, are essential qualities for the effective ED therapist (Andersen & Corson, 2001; Bunnell, 2009; Davis, 2009a, b; Derenne, 2006; Zerbe, 1995). Often permeated with high levels of anxiety and angst, ED treatment is more like a cross-country trek than a 50-yard dash. For the therapist, "staying the course" requires a stamina characterized by practiced mindfulness and compassionate curiosity toward personal experiences of counter-transference, counter-reaction, and the registering of visceral information. Without it, clinicians can run the risk of foregoing what may be clinically indicated and instead doing what might feel "easier," like end treatment prematurely, or more commonly, elect to avoid or placate rather than confront the therapeutic challenge. With "mindsight" (Siegel, 2009), informed decisions about if, how, and to what extent the therapist overtly uses visceral information, are matters of skill largely influenced by theoretical approach and timing.

Staying the course does not mean a symbiotic or parasitic joining with the client that results in collusion with the ED symptomatology, coddling or "water pouring" (Kvidera, 2007) in the fires of ED Hell. Instead, it is doing what is clinically necessary (e.g., expanding the treatment team or utilizing a higher level of care) with a mix of humanness, sound clinical judgment, honest feedback, and respectful confrontation while enduring the however-long haul of the healing journey.

Humbleness and transparency. Finding balance between humble confidence in our clinical opinions (Derenne, 2006), and a non-investment in being right can be a challenge. Narcissistic tendencies, the need to control, and competition with the client are not compatible with effective treatment (Andersen & Corson, 2001), as they replicate the very interpersonal dynamics our patients guard against through their ED symptoms (Beresin et al., 1989). Power struggles typically derail the process and certainly distract from our clients making changes or confronting their barriers to doing so.

At its heart, humbleness is the recognition that, while we may do our best to provide sagacious guidance and work in collaboration with the client, we are not in charge of the client's healing pace or decisions. This is not a cop-out, or reason to do our jobs less well. Neither is it justification for failing to improve in those areas in which we are deficient. Ultimately, it is the client's recovery, not our own, and not ours to do for them. We are neither savior, nor white knight, nor recast of another in their family or social system. Our honesty and transparency with clients on this matter demonstrates the regard we hold for them as fellow human beings, our belief in their inherent redemptive capacities, and our respect for their rights to personal power. It allows clinicians to hold the hope for healing without unrealistic expectations.

Ability to self-nurture. The daily practice of treating people with ED requires adequate self-nurturance in addition to our "base" of clinical training (Andersen and Corson, 2001; Derenne, 2006; Warren, Crowley, Olivardia & Schoen, 2009; Zerbe, 1995). Its importance is at least three-pronged: (a) to connect us with a source, singular or conglomerate, of abundance from which to receive and replenish our stores of ingredients; (b) to insulate us from chronic and debilitating burnout (Rubel, 1986); and (c) to infuse and enliven us both personally and professionally. To be of best service to our clients and to ourselves, we must contemplate and honor how, what, and by whom we are fed, as well as how we best digest, metabolize, and utilize this nourishment.

Professional self-care includes clinical supervision, peer consultation and, when appropriate, personal therapy (Andersen & Corson, 2001; Yager & Edelstein, 1987). Zerbe (1995), among others, adds to the requisites what I refer to as "convening the lineage," reading or rereading the works of forebearers in our fields of practice. This practice provides perspective, space to breathe, intellectual connection, and a means to understand our own counter-transference, improving both our clinical endurance and frustration tolerance. Developing a regular practice of continuing education, or a sense of community at professional conferences, may also serve to reconnect us to our source.

Personal self-care, of course, influences us as professionals. Healthy, satisfying relationships with friends, family, and colleagues, as well as interests and passions aside from work can remind us we need not be enveloped by the flames of our clients' ED infernos. The world is bigger and broader than what occurs within the four walls of our offices, just as our clients' worlds extend beyond the immediate treatment experience. Active engagement in our lives can keep us from a narcissistic investment in overvaluing our clients' progress as a measure of our being "good enough" as people and as clinicians.

Summary of the Clinician's Ingredients

With a connection to source, or the something-bigger, however personally defined, therapists are more able to calibrate the workings of their intuitive sensibilities. Coordinating these sensibilities with the blending of our basic clinical knowledge, accumulated experience, clinical training, and psychotherapeutic techniques, allows us to finely adjust and appropriately titrate therapeutic interventions according to the stage of treatment and the unique needs of our clients.

Spiced with courage and pinches of appropriate humor, levity, creativity, and adaptability, and folded into the therapeutic alliance, "non-possessive warmth" (Andersen & Corson, 2001, p. 356), "unconditional positive regard" (Rogers, 1961, p. 47), "active and worthwhile engagement" (Davis, 2009b, p. 6), embodied authenticity, being real, empathy, trust, endurance, high frustration tolerance, humbleness, transparency, and an ability to self nurture, heighten the likelihood of success in ED treatment. The question remains: If these are the essential ingredients for the clinician, what exactly are the fruits of our labor meant to help produce?

THE SUCCESSFUL CLIENT'S RECIPE

Product or Process: Averting Disaster in a Recipe for Recovery

Nowhere is the scientist/practitioner gap in the ED field more gaping than when it comes to answering this most elemental question: What is recovery? Fundamental to this inquiry is in whom, and by what processes, we invest the power to decide. Locating the sources of definitive authority (i.e., in the researcher, clinician, patient, and/or caregivers) determines the means by which we seek answers. Views from the ivory tower, and those from the therapy couch, provide dramatically different vantage points. The imperative to integrate these perspectives is where researchers, clinicians, and patients may ultimately find common and fertile ground.

The unfortunate divide in definitions of ED recovery parallels empirical design lines; exclusively quantitative or qualitative approaches yield vastly different results. Language is instructive in this discourse. Objective, static "outcomes" are typically the subject of quantitative research, whereas the subjective, process of "recovery" is the object of qualitative investigations. Kazdin (2009) suggests that qualitative research is a "natural way of bridging research and practice," emphasizing the need for multiple methodologies as well as their complementarity (p. 277).

A fundamental criticism of traditional quantitative ED research involves the nearly exclusive focus on the physical parameters of recovery from what are clearly biopsychosocial illnesses. Examining only the overt symptoms (e.g., restoration of weight or menses) and grouping outcomes ignores the psychosocial and spiritual dimensions of ED recovery as well as the diverse and essential personhood of the individuals subject to their torment. Emphasizing the need for outcome research to include narrative, and qualitative reports, Zerbe (2008) succinctly states that the patient is always an "n of 1" and should be "considered as a human being first, not simply as a member of a diagnostic group" (p. 289).

The remainder of this chapter will review "reports" from both sides of the empirical design divide, provide suggestions for future investigative efforts, and conclude with a synthesis of what researchers, practitioners, *and* patients bring to bear on the definition of recovery and how it is best mediated.

Outcome Literature: Coming in From the Outside

Fifty years of quantitative research devoted to ED outcomes has provided extensive data, despite failing to provide comparable, consistent, and clinically meaningful definitions of what recovery entails (Berkman, Lohr & Bulik, 2007; Couturier & Lock, 2006; Jarman & Walsh, 1999; Steinhausen, 2008; Wonderlich, Gordon, Mitchell, Crosby & Engel, 2009). Inconsistent definitions of successful outcome, as well as variations in design, measures, outcome ratings, dependent variables, populations, diagnostic categories and criteria, and the duration of follow-up, have generated an unwieldy body of literature with radical discrepancies. Indeed, given published ranges of recovery rates between 0—92% for anorexia nervosa (AN) (Steinhausen, 2002) and 13—69% for BN (Herzog et al., 1993), achieving recovery could be metaphorically construed as either a cakewalk or a death march. Additionally, these methodological inconsistencies compromise the practical interpretations and implications of the outcome data. For example, definitions of outcome applied to variable patient populations (e.g., inpatient vs. outpatient) generate different results with clinically meaningful relevance. Randomized, controlled therapeutic trials, limited mostly to tertiary care sites, are associated with high dropout rates and poorer outcomes, and their subjects may not be representative of the ED population typically seen in therapists' offices (Johnson, Lund & Yates, 2003; Steinhausen, 2002; Zerbe, 2008).

Definitions of recovery in quantitative outcome research. Despite Morgan and Russell's (1975) seminal efforts to establish and expand recovery criteria for AN to include the physical, psychological, and social aspects of functioning, subsequent research inconsistently assessed all these factors. The majority of succeeding AN outcome research relied solely on the physical parameters of weight, menses, and eating symptoms (Steinhausen, 2002). Similarly, researchers have narrowly equated a positive outcome for BN with cessation of binging and purging (Jarman & Walsh, 1999). These "outside" measures of recovery provide researchers and clinicians with static snapshots of behavioral control—momentary "product" analyses, which fail to inform us about the unfolding, multidimensional "process" of recovery.

A second common definition of recovery is simply "absence of diagnosis," meaning that the patient no longer meets the full diagnostic criteria of AN, BN or binge-eating disorder

(BED). It is not clear if these patients would otherwise meet criteria for Eating Disorders Not Otherwise Specified (EDNOS), the most frequent ED diagnosis, with comparable psychopathology to AN and BN (Fairburn et al., 2007). Moreover, patients may continue to be in psychiatric distress despite being considered subclinical or "recovered" from the gross physical and behavioral features of EDs (Jarman & Walsh, 1999). As Zerbe (2008) notes, "patients will not [achieve] life fulfillment if they still have poor social networks, feel badly about their self-image and personal well-being, lack a sense of belonging…or struggle with a lack of self-cohesion" (p. 288).

Insight is no more curative than behavioral control. What research equates with endpoints in treatment (symptom resolution), clinicians consider as starting points in recovery. From a clinical standpoint, until patients are nutritionally and physically stable, the real work cannot begin. In fact, research on AN recovery has demonstrated that *only* women who had, in addition to behavioral improvement, *also* achieved cognitive recovery, were "indistinguishable from female controls on self-report measures of body dissatisfaction…general symptomatology, endorsement of the thin ideal…drive for success, fear of failure, harm avoidance…perfectionism and self-esteem" (Bachner-Melman, Zohar & Ebstein, 2006, p. 700).

Thirdly, Steinhausen (2002) describes ratings of global outcomes (good, fair, poor) as the most common form of AN recovery classification. This nondescript conceptualization of recovery is subject to the same criticism noted above, as freedom from overt, clinical symptomatology is not equivalent to eradication of the illness. Lastly, contemporary efforts to remedy the constrictive definitions of ED recovery in quantitative research have reincorporated measures of psychological and psychosocial functioning (e.g. Noordenbos & Seubring, 2006), such as reductions in fears and preoccupations about weight and food, and improved body image. Given that weight restoration tends to occur sooner and more often than psychological improvement in AN, outcome criteria must incorporate assessment of emotional and cognitive functioning (Couturier & Lock, 2006; Jarman & Walsh, 1999; Steinhausen, 2002). Quantitative analyses of outcome are beginning to include quality of life measures (Adair et al., 2007), contributing timely and cogent insights into our understanding of recovery. Although still only "product" assessments, these efforts attempt to examine the full-bodied, robust and complex progression of recovery.

Duration as a defining factor in recovery. Empirical inconsistency in durations of asymptomatic status (from 8 weeks to 3 years) further obscures the outcome picture (Steinhausen, 2008; Von Holle et al., 2008). In general, stricter definitions requiring longer durations of both weight and psychological improvement are associated with the lowest recovery rates for both AN and BN. Investigators have recently rallied to utilize empirically derived and tested consensus definitions that incorporate the full range of ED symptomatology evaluated over a sufficient period of time (Couturier & Lock, 2006; Frank, 2005; Keel, Mitchell, Davis, Fieselman & Crow, 2000; Kordy et al., 2002; Von Holle et al., 2008). These investigations clearly distinguish remissions (briefer periods of symptom absence) from recovery (maintenance of remission for a predetermined amount of time), but the duration criteria remain critically different.

Kordy et al. (2002) suggested a 1-year minimum duration of symptom abstinence, including psychological parameters of recovery. Strober, Freeman & Morrell (1997) indicated that it took nearly 5 years for physical symptoms of adolescent AN to fully recover and another 2 years for the psychological factors to normalize. Von Holle et al. (2008) utilized a 3-year period of complete symptom abstinence to assess temporal patterns of outcome in a transdiagnostic sample. Their findings yielded sobering long-term outcomes. After 15 years, only 16% of those with AN, and 25% of those with BN, met recovery criteria. They concluded that 10 years post ED onset appears to be the critical juncture in which recovery either consolidates or the condition becomes chronic.

Summary of Outcome Definitions

Methodological variability is the one consistent factor in ED outcome literature, generating data which suggest recovery is as possible as it is improbable. Patience and perseverance, previously noted as integral ingredients in the successful ED treatment provider, appear equally essential to the patient. Quantitatively derived definitions of recovery are limited by their inherent depiction of it as a static state: a product versus a process. Additionally, these approaches locate the source of definitive authority in the researchers, creating arbitrary and inconsistent determinants of outcome, and overlooking the nuances, voice, and perspectives of the patients and their caregivers. Finally, populations conspicuously overlooked in outcome research include children, males, EDNOS, BED, late onset EDs, minorities, and primary caregivers.

Recovery Literature: Coming Out from the Inside

The shortcomings of quantitative clinical research are ubiquitous (Kazdin, 2009). The above synopsis is not meant as a wholesale indictment of research practices; indeed, data derived from quantitative inquiries provide invaluable information about group norms and variables. The relevant gap is less between science and practice as it is within scientific practices. As with most things, "the devil is in the details" and the preferential emphasis on quantitative approaches to characterize and assess ED recovery has undermined our efforts to both understand our patient's torment and improve treatment.

What has been crucially missing are inquiries that move beyond efforts to *under*stand recovery in favor of those that seek *inner*standing (Kimura, 2004) from within the canvas of our patients' lives.

> Old paint on canvas, as it ages, sometimes becomes transparent. When that happens, it is possible, in some pictures, to see the original lines: a tree will show through a woman's dress, a child makes way for a dog, a large boat is no longer on an open sea. This is called "pentimento" because the painter "repented," changed his or her mind. Perhaps it would be as well to say that the old conception, replaced by a later choice, is a way of seeing and then seeing again. *(Hellman, 1973, p. 3)*

Recovery is like Hellman's pentimento—a way of seeing and then seeing again. "Arguably, what patients know AN to be is even more important than what psychotherapists and other health professionals know" (Surgenor, Plumridge & Horn, 2003, p. 23). Professionals on both sides of the gap have failed to provide a clear, consistent, clinically meaningful definition of recovery that is process and diversity-oriented, *and* informed by those who have lived the experience. To reach innerstandings of recovery, we must get out of linear models of questioning derived from "experts" perspectives, and talk with the real experts—those who have experienced recovery.

In order to recover the definition of recovery from obscurity, we need to bring all of the relevant parties and perspectives to the proverbial table, and listen our way into the questions and answers rather than only assess answers to predetermined questions. Qualitative research is the ideal forum in which multiple voices and viewpoints can be distinguished and illuminated.

Qualitative Research: Bridging the Tower and the Trenches

Following his tenure as president of the American Psychiatric Association, Kazdin (2009) underscored the priority of improving patient care by bridging science and practice through the use of qualitative research:

> It includes an intense, detailed, and in-depth focus on individuals and their contexts….[It] can identify details of the experience; generate new theory, constructs, and measures….Traditional quantitative group research may not [be] able to reveal novel themes and processes of recovery in such an in-depth way (p. 277).

Zerbe (2008) eloquently echoed and articulated the imperative to consider ED patient's viewpoints in clinical research and practice. "Qualitative data speak to the humanity of the individual, they immerse themselves in those characteristics [patients]… include in constructing a life well lived…and with those specific skills and strengths that enable…[patients] to love…work [and] face down destructive symptoms" (p. 291). Maine (1985) was a forerunner in efforts to elucidate patients' voices in her phenomenological research regarding their understanding of the process of both their illness and recovery. Bruch's (1988) pioneering work, *Conversations with Anorexics*, and MacLeod's (1987) *The Art of Starvation* provided rich and compelling peeks into the inner sanctum of AN, while Beresin et al. (1989) offered an early empirical analysis of patients' views on recovery.

Fortunately, qualitative research regarding ED recovery is burgeoning. Some applications of these methods, such as feminist grounded theory approaches, have provided critical contributions to the recovery literature, fleshing out patients' perspectives well beyond restoration of weight and other physical parameters of improvement (Bowlby, 2008; Garrett, 1997; Jarman & Walsh, 1999; Keski-Rahkonen & Tozzi, 2005; Lamoureux & Bottorff, 2005; Noordenbos & Seubring, 2006; Nordbo et al., 2008; Serpell, Treasure, Teasdale & Sullivan, 1999; Serpell & Treasure, 2002; Surgenor et al., 2003; Tozzi, Sullivan, Fear, McKenzie & Bulik, 2003; Weaver, Wuest & Ciliska, 2005). Paralleling the outcome literature, most studies have examined the phenomenology of AN and BN, while recovery perspectives from patients with BED, EDNOS, and late onset ED, as well as males and children, have been categorically overlooked. Unlike the outcome literature, however, the findings in the qualitative studies have yielded remarkable consistency with regard to how patients define, view, and experience recovery.

The most distinct difference between quantitative and qualitative definitions of recovery is that patients clearly view recovery not as an endpoint, but as a multidimensional process in which they are simultaneously and variously making progress *and* experiencing setbacks (Pettersen & Rosenvinge, 2002). Lamoureux and Bottorff (2005) portrayed this experience as patients "inching away from anorexia," quoting one patient as saying, "*that's what characterized the struggle for me…the forward and the back*" (p. 175). This seeming incongruence is an ordinary occurrence in the clinical context: the same week a patient skipped her snacks, she risked conflict in connection, genuinely expressed and experienced disquieting feelings, and/or made a decision without seeking another's approval. As improvements within the various dimensions of the recovery process do not occur in a linear, systematic fashion, the outcome literature has been of limited use to clinicians.

In their study of the journey of recovery from BN, Peters & Fallon (1994) described three main dimensions: denial to reality; alienation to connection; and passivity to personal power. Platt (1992) also conceptualized recovery from BN as a three-stage developmental process: (a) shifting the relationship to the ED from an ego-syntonic to ego-dystonic status; (b) learning to tolerate uncomfortable physical and psychological states without ED symptoms; and (c) improving self-care and self-esteem through adaptive coping skills. Consistent with other reports (Lamoureux & Bottorff, 2005; Pettersen & Rosenvinge, 2002), coming to view their illness as the problem rather than the solution, appears to be a fundamental necessity for patients. Beyond the initial "unleashing" of denial, patterns of progression with regard to their physical, emotional, relational, spiritual, and sociopolitical well-being are profoundly personal and unique. I wrote of this in my own recovery in 1980 (McGilley, personal diary):

> If I extract myself to a different dimension, I'm aware of how tortured I am by my turbulent emotional inertia. I wonder, stoically and fearfully, if in like comparison, Sisyphus would have pursued his existential task of rolling his ill-fated rock up the hill had he seen his dilemma from afar? Faced solely with the rock, I too might persist, if only for the purpose of the struggle. But now, faced with both the rock and the weighty awareness of the "Big Picture," I find myself frantically paralyzed. In a frenzy of motionlessness. Falling with my feet stubbornly planted, gathering bruises that refuse to expel their ache. Even more frightening than the prospect of enlightenment, is that first real jolt of sentience, the piercing scream of nerves released from denial's hearty grip. Isn't there an internal gate-keeper, an emotional parachute that will ensure I don't reenter the realm of my senses free-fall and fragment into so many brittle pieces?

Across diagnoses, methods of assessment, and duration of illness, patients were unwavering in their experience that recovery entails something akin to the painter's repenting: some version of a reconciliation or reunion with one's self. Variously described as "finding me" (Weaver et al., 2005), reclaiming oneself as "good enough" (Lamoureux & Bottorff, 2005), or simply as "self-acceptance" (Pettersen & Rosenvinge, 2002), recovery demands a willingness to be "real *again*, vulnerable *again*, to the full range of human experience, all shades of gray included. Viewed from the anorexic 'shadowlands,' this invitation appears to be an absurd request, something like re-exiting the birth canal after achieving our full adult size" (McGilley, 2000). As a current patient, waxing and waning in the early stages of recovery defines it: "Recovery is a continuum of finding and [reconnecting] the part of 'you' that has been disconnected for so long, and…aligning this scared, shadowed self with a new, healthy self ready to make the transition into new life."

In one of the few qualitative analyses to include a mixed diagnostic sample, Pettersen and Rosenvinge (2002) used an open-ended interview process to assess what factors were helpful in recovery and what recovery meant to them. Patients were required to have received treatment and to have had an ED for at least 3 years. The majority of the sample had EDNOS, and the rest were nearly equally divided between AN, BN, and BED. The overarching motivation to recover was the desire for a better life. Participants' definitions of recovery were classified into seven general aspects: (a) accepting self and body; (b) ceasing to allow food to dominate life or be used to resolve problems; (c) finding a life purpose; (d) identifying and having the courage to express emotions; (e) diminishing anxiety and depression; (f) fulfilling one's potential versus conforming to other's expectations; and (g) improving social functioning. Two other findings are noteworthy. Firstly, participants noted the capacity to "function sexually and emotionally in a relationship with a stable partner" as vital to their experience of recovery (p. 68). Issues related to sexuality are rarely mentioned in the recovery literature. Secondly, referring to the mixed blessings inherent to recovery, participants emphasized "[experiencing] life as rather difficult *without* their eating disorder" (p. 68). In direct contrast to outcome literature which necessitates symptom abstinence as defining in recovery, these patients "defined themselves as recovered despite the presence of symptoms of eating disorders, anxiety, and depression" (p. 68). The authors concluded that "symptom reduction may not stand out as a goal *per se*, but rather as a means to accomplish more functional interpersonal relations, thinking, and problem solving strategies" (p. 69).

Space limitations do not allow for a full rendering of the rich and nuanced aspects of recovery unveiled by qualitative research. However, a "recipe for recovery" derived by an inpatient group I once conducted, illustrates the concordant and enduring nature of recovery phenomenology (Box 12.1).

Whether EDs are curable or chronic, and/or whether a patient is viewed as recovered or recovering is distinctly debatable among professionals and patients (Root, 1990; Schaefer, 2009). Costin (2007b), a recovered therapist, is decisive in her view that recovery is fully obtainable. Once recovered, food and weight have been put into proper perspective and

---

BOX 12.1

RECIPE FOR RECOVERY

1 strong dose of commitment
1 cup honesty
1 cup faith
½ cup openness (assertiveness)
½ cup positive attitude
¼ cup group therapy
¼ cup individual therapy
¼ cup support from family and friends
1 heaping cup of sexuality
1½ cups self-esteem and identity
1 cup moderate exercise with a good body image
1½ cups of love
1 cup of reckoning history
1½ cups sense of humor
1 cup food
Designated dose of medications

Start out with a strong dose of commitment. Mix in honesty and faith. Combine attitude with openness and add until thoroughly blended. In a separate bowl, mix group therapy, individual therapy, and family support. Beat well and *slowly* add sexuality. Peel identity down to the core and stir in self-esteem. Generously add love. Fold in food and humor, alternating with sifted history. Carefully combine the first bowl with the second. Pour into a well-greased body image. Bake at 350° as long as needed to become real and satisfied.

When cooled, top with 1 cup moderate exercise and medications as needed. Result: Healthy, happy, whole person who loves herself!

"what you weigh is not more important than who you are.... You will not compromise your health or betray your soul to look a certain way, wear a certain size or reach a certain number on a scale" (p. 164). A former patient, Claire, several years into alcohol and ED recovery, offers an alternative definition and perspective:

> Recovery...cannot be measured. It is not tangible or visible. It lies within the individual and during the process manifests itself in, although subtle, outward signs. Recovery is finding the courage to tend toward things that bring benefit to health and spirit even when in some cases the individual tends to gravitate toward destructive or harmful behaviors. It is constantly evolving and changing. Therefore, I don't think we ever truly reach a 'recovered' state. We are spiritual beings in human vessels and we won't be fully restored until we are fully spirit.

Another patient, a group member and dear friend of Claire's, whose illness and recovery unfolded in a parallel fashion, respectfully disagrees:

> Although I am on a continuum of self-improvement, I consider myself recovered. The switch from 'in recovery' to 'recovered' happened when I realized that no matter what—through illness, death, despair, and the darkness of depression—I would never go back to where I was. I still feel pain, I feel it quite often. The difference is that I have finally figured out that food or lack thereof will not ease the pain. I now cry it out, dance it out, talk it out, write it out, sleep it out, sing it out, and laugh it out. I might talk too much, move too much, cry too much and feel too much—but I will choose that 'too much' over the nothingness that the disorder gave me any day. I will never...starve it out again. It doesn't work, and that's one lesson I've learned that I just won't forget.

My model embraces both perspectives. Patients fundamentally recover from the active symptomatic aspects of their ED, what Schaefer (2009) describes as "Recovered (Period)," while remaining in a process of recovering from the underlying traumas, intrapsychic and interpersonal conflicts, emotional and temperamental vulnerabilities, and cultural stressors that co-conspired to culminate in the onset of their ED.

Spiritual issues have long been underscored in the recovery literature despite being more newly emphasized in academic forums. Somewhere within the recovery process, a spiritual shift occurs. It could be inspired by an AN patient's "first" plate of crispy fries, or by being reflected in the eyes of their beloved, but sooner or later, this dimension is tapped. Beresin et al. (1989) likened recovery to a psychological rebirth, while others have noted reconnecting with nature, finding purpose and making meaning of one's life (Garrett, 1997). In my own case, the existential leap recovery required had to do with living, loving, and losing in what had otherwise literally become an unbearable world: "Why live if loving hurts so much?" (McGilley, 2000, p. 5). For many of us, finding our place in the world, a sense of belonging, or our "connection to source—or the something bigger," is the redemptive blessing of recovery.

Two of the qualitative research contributions on ED recovery warrant further mention. The first provides a rare effort to elaborate a theory of the recovery process. Using a feminist grounded theory approach, Weaver et al. (2005) analyzed interviews of twelve women recovering from AN to discern "the central organizing process for how women recover" (p. 190). They constructed a theory of self-development, a dynamic helix, in which women move from "perilous self-soothing to informed self-care" (p. 191). In perilous self-soothing, patients wrestle with issues of identity and status in society. In this model, AN is understood as a means of providing recognition and a contrived identity at the expense of compromised health and well-being. Through improved self-awareness, self-differentiation, and self-regulation, recovering patients reach a turning point labeled "finding me" in which they gradually move towards informed self-care. This stage involves developing a sense of one's strengths and weaknesses, managing emotions, and maintaining intimate and meaningful relationships. Of greatest relevance is their comment regarding the social underpinnings of EDs and what I've referred to as the "innerstandings" of the recovery process. Emphasizing that recovery factors and their impact must be understood in context, they concluded that: "both perilous self-soothing and informed self-care arise from women's interactions within social structure and not as individual intrapsychic processes, [underscoring] the inappropriateness of relying on personality characteristics, discreet behavioral responses, and single events to evaluate AN and its recovery" (p. 202).

Finally, Jarman and Walsh (1999) were prescient in their efforts to integrate the best of what we have learned about recovery from the research/practice fields. They offered four compelling suggestions: creating a comprehensive biopsychosocial model of recovery; using both qualitative and quantitative methodologies including client's views; recognizing the limitations of different measures and methods; and connecting ED recovery research and other psychotherapy process and outcome research.

By 2009, we had achieved moderate success, at best, in applying these suggestions. Much remains to be done before we can confidently, consistently, and comprehensively evaluate and elucidate the experience of recovery from all informed perspectives.

CONCLUSIONS: BRIDGING EXPERIENCE AND EMPIRICISM

Just as the client's voice matters in the treatment process, so should it be included in efforts to define outcome. An integrated use of quantitative and qualitative research approaches would complement and expand traditionally derived empirical data. By bringing all the relevant parties to the table, "a combined methodological approach could also enable a multiple stakeholder (e.g. client, clinician, academic) perspective to be incorporated into the evaluation process" (Jarman & Walsh, 1999, p. 784). Finally, the perspective of recovered therapists is just beginning to gain serious consideration in the field, lending another gap-bridging dimension to this important inquiry (Bloomgarden, Gerstein & Moss, 2003; Bowlby, 2008; Costin, 2009; McGilley, 2000).

A change in language may also invite new perspectives. The word "integrity", which means "the state of being whole, entire, or undiminished," seems to better capture the essence of what we've been referring to as recovery. As Hillman describes (1994), integrity also has to do with a kind of wisdom, a way of "knowing together" and "accessing a more subtle kind of wisdom that depends on letting go of those old mental categories" (p. 86). If we were to conceptualize recovery as a return to a state of wholeness, and we were to go about assessing it with thinking hearts, what more could we learn about the harrowing world of EDs and the expansive world beyond its borders?

Fitting and timely, Siegel (2009) provides a compelling neurobiological basis for the concept of integration, and therapists' roles as integrators. Emphasizing the relational capacity for changing brain structures through the sharing of "information flow," Siegel argues that specific clinical interventions can literally stimulate the integrative fibers of the patient's brain. Fostering this "vertical integration" restores or improves the patient's body/mind connection, so critically impaired in the ED population. "Health," in Siegel's conceptualization, is defined as integration. "Harmony" is the subjective experience of integration. Perhaps recovery is best likened to a process of seeking harmonic healing.

In sum, bridging the research/practice gap is going to require a fundamental shift in how we approach inquiry (from asking specific questions to inviting open dialog), the degree of control we exert over variables, and the kind of consistency we expect from the answers. Certain ambiguities must be tolerated and accepted; such is the nature of both science and human healing. "Science is not about control. It is about cultivating a perpetual condition of wonder in the face of something that forever grows one step richer and subtler than our latest theory about it. It is about reverence, not mastery" (Power, 1992, p. 411). Like those we treat, we succumb to the same alluring qualities of ease and concreteness in our efforts to evaluate the hard and fluid complexities of recovery. Clinically meaningful outcome research requires contextually and collaboratively considered concepts of recovery and its fostering agents. Only then can we begin to define a true recipe for success for those suffering with ED.

References

Adair, C., Marcoux, G., Cram, B., Ewashen, C., Chafe, J., Cassin, S. E., ... Brown, K. E. (2007). Development and multi-site validation of a new condition-specific quality of life measure for eating disorders. *Health and Quality of Life Outcomes, 5*, 23.
Alighieri, D. (1939). *The divine comedy I: Inferno.* Translation/commentary by J.D. Sinclair. New York, NY: Oxford University Press.
American Psychiatric Association (2006). Practice guidelines for the treatment of patients with eating disorders (3rd ed.). *American Journal of Psychiatry* 1101–1185.
Andersen, A. E., & Corson, P. W. (2001). Characteristics of an ideal psychotherapist for eating disordered patients. *Psychiatric Clinics of North America, 24*, 351–359.
Bachner-Melman, R., Zohar, A., & Ebstein, R. (2006). An examination of cognitive versus behavioral components of recovery from anorexia nervosa. *The Journal of Nervous and Mental Disease, 194*(9), 697–703.
Beresin, E., Gordon, C., & Herzog, D. (1989). The process of recovering from anorexia. *Journal of the American Academy of Psychoanalysis, 17*, 103–130.
Berkman, N., Lohr, K., & Bulik, C. (2007). Outcomes of eating disorders: A systematic review of the literature. *International Journal of Eating Disorders, 40*, 293–309.
Bloomgarden, A., Gerstein, F., & Moss, C. (2003). The last word: A "recovered enough" therapist. *Eating Disorders: The Journal of Treatment and Prevention, 11*, 163–167.
Bowlby, C. (2008). Lessons learned from both sides of the therapy couch: A qualitative exploration of the clinical lives of recovered professionals in the field of eating disorders. *Dissertation Abstracts International, 68*, 6290.
Bruch, H. (1988). *Conversations with anorexics.* New York, NY: Aronson.
Bunnell, D. (2009). Countertransference in the psychotherapy of patients with eating disorders. In M. Maine, W. Davis, & J. Shure (Eds.), *Effective clinical practice in the treatment of eating disorders* (pp. 79–95). New York, NY: Routledge.
Burket, R. C., & Schramm, L. L. (1995). Therapists' attitudes about treating patients with eating disorders. *Southern Medical Journal, 88*(8), 813–821.
Costin, C. (2009). The embodied therapist: Perspectives on treatment, personal growth, and supervision related to body image. In M. Maine, W. Davis, & J. Shure (Eds.), *Effective clinical practice in the treatment of eating disorders* (pp. 179–192). New York, NY: Routledge.
Costin, C. (2007a). *The eating disorder source book* (3rd ed.). New York, NY: McGraw Hill.
Costin, C. (2007b). *100 questions and answers about eating disorders.* Boston, MA: Jones and Bartlett Publishers, Inc.

Couturier, J., & Lock, J. (2006). What is recovery in adolescent anorexia nervosa. *International Journal of Eating Disorders, 39*(7), 550–555.

Cozolino, L. (2002). *The Neuroscience of psychotherapy.* New York, NY: W. W. Norton.

Davis, W. N. (2009a). Individual psychotherapy for anorexia and bulimia: Making a difference. In M. Maine, W. Davis, & J. Shure (Eds.), *Effective clinical practice in the treatment of eating disorders* (pp. 35–49). New York, NY: Routledge.

Davis, W. N. (2009b, Winter). Individual psychotherapy for eating disorders: Getting a good start. *Renfrew Perspectives*, 5–7.

Derenne, J. (2006). Junior high, revisited. J. Ruskay Rabinor (Ed.) The Therapist's Voice. *Eating Disorders: The Journal of Treatment and Prevention, 14,* 335–339.

Fairburn, C., Cooper, Z., Bohn, K., O'Connor, M., Doll, H., & Palmer, R. (2007). The severity and status of eating disorder NOS: Implications for the DSM-V. *Behavior Research & Therapy, 45,* 1705–1715.

Frank, E. (2005). Describing course of illness: Does our language matter? *International Journal of Eating Disorders, 38,* 7–8.

Garrett, C. (1997). Recovery from anorexia nervosa: A sociological perspective. *International Journal of Eating Disorders, 21,* 15–34.

Hellman, L. (1973). *Pentimento.* Boston, MA: Little, Brown & Co.

Herzog, D., Sacks, N., Keller, M., Lavori, P., von Ranson, K., & Gray, H. (1993). Patterns and predictors of recovery in anorexia nervosa and bulimia nervosa. *Journal of the American Academy of Child and Adolescent Psychiatry, 32,* 835–842.

Hillman, A. (1994). *The dancing animal woman.* Norfolk, CT: Bramble Books.

Jarman, M., & Walsh, S. (1999). Evaluating recovery from anorexia nervosa and bulimia nervosa: Integrating lessons learned from research and clinical practice. *Clinical Psychology Review, 19,* 773–788.

Johnson, C., Lund, B., & Yates, W. (2003). Recovery rates for anorexia nervosa. *American Journal of Psychiatry, 160,* 798.

Kazdin, A. (2009). Bridging science and practice to improve patient care. *American Psychologist, 64,* 276–278.

Keel, P. K., Mitchell, J. E., Davis, T. L., Fieselman, S., & Crow, S. J. (2000). Impact of definitions on the description and prediction of bulimia nervosa outcome. *International Journal of Eating Disorders, 28,* 377–386.

Keski-Rahkonen, A., & Tozzi, F. (2005). The process of recovery in eating disorder sufferers' own words: An internet-based study. *International Journal of Eating Disorders, 37,* S80–S86.

Kimura, Y. (2004). *The book of balance.* New York, NY: Paraview.

Kordy, H., Palmer, R., Papezova, H., Pellet, J., Richard, M., & Treasure, J. (2002). Remission, recovery, relapse, and recurrence in eating disorders: Conceptualization and illustration of a validation strategy. *Journal of Clinical Psychology, 58,* 833–846.

Kvidera, C. (2007, March). *Dialectical behavior therapy.* Pesi Seminar: Boulder, CO.

Lamoureux, M., & Bottorff, J. (2005). "Becoming the real me": Recovering from anorexia nervosa. *Health Care for Women International, 26,* 170–188.

MacLeod, S. (1987). *The art of starvation: A story of anorexia and survival.* New York, NY: Schocken.

Maine, M. (1985). An existential exploration of the forces contributing to sustaining and ameliorating anorexia nervosa: The recovered patient's view. *Dissertation Abstracts International, 46*(6-B), 2071 (1986-52919-001).

McGilley, B. (2000). On the being and telling of the experience of anorexia: A therapist's perspective. *Renfrew Perspectives, 5*(2), 5–7.

Morgan, H., & Russell, G. (1975). Value of family background and clinical features as predictors of long-term outcome in anorexia nervosa: Four-year follow-up study of 41 patients. *Psychological Medicine, 5,* 355–371.

Noordenbos, G., & Seubring, A. (2006). Criteria for recovery from eating disorders according to patients and therapists. *Eating Disorders: The Journal of Treatment and Prevention, 14,* 41–54.

Nordbo, R., Gulliksen, K., Espeset, E., Skarderud, F., Geller, J., & Holte, A. (2008). Expanding the concept of motivation to change: The content of patients' wish to recover from anorexia nervosa. *International Journal of Eating Disorders, 41,* 635–642.

Peters, L., & Fallon, P. (1994). The journey of recovery: Dimensions of change. In P. Fallon, M. Katzman, & S. Wooley (Eds.), *Feminist perspectives on eating disorders* (pp. 339–354). New York, NY: Guilford Press.

Pettersen, G., & Rosenvinge, J. (2002). Improvement and recovery from eating disorders: A patient perspective. *Eating Disorders: The Journal of Treatment and Prevention, 10,* 61–71.

Platt, C. (1992). *Formerly chronic bulimics' perspectives of the process of recovery.* Unpublished doctoral dissertation. Berkeley: California School of Professional Psychology.

Powers, R. (1992). *The gold bug variations.* New York, NY: Harper Perennial.

Rogers, C. R. (1961). *On becoming a person: A therapist's view of psychotherapy.* Boston, MA: Houghton Mifflin Company.

Root, M. (1990). Recovery and relapse in former bulimics. *Psychotherapy, 27*(3), 397–403.

Rowling, J. K. (1999). *Harry Potter and the prisoner of Azkaban.* New York, NY: Scholastic Press.

Rubel, J. B. (1986). Burn-out and eating disorder therapists. In F. Larocca (Ed.), *Eating disorders: Effective care and treatment* (Vol. 1) (pp. 233–246). St. Louis/Tokyo: Ishiyaku EuroAmerica, Inc.

Schaefer, J. (2009). *Good-bye Ed, hello me.* New York, NY: McGraw Hill.

Serpell, L., & Treasure, J. (2002). Bulimia nervosa: Friend or foe? The pros and cons of bulimia nervosa. *The International Journal of Eating Disorders, 32*(2), 164–170.

Serpell, L., Treasure, J., Teasdale, J., & Sullivan, V. (1999). Anorexia nervosa: Friend or foe? *International Journal of Eating Disorders, 25,* 177–186.

Siegel, D. (2009, November). *Mindsight, self-regulation and attachment: Promoting health by moving from chaos and rigidity to neural integration.* Keynote presentation at the 19th Annual Renfrew Foundation Conference, Philadelphia, PA.

Steinhausen, H.-C. (2002). The outcome of anorexia nervosa in the 20th century. *American Journal of Psychiatry, 159,* 1284–1293.

Steinhausen, H.-C. (2008). Outcome of eating disorders. *Child and Adolescent Psychiatric Clinics of North America, 18,* 225–242.

Strober, M., Freeman, R., & Morrell, W. (1997). The long term course of severe anorexia nervosa in adolescents: Survival analysis of recovery, relapse and outcome predictors over 10–15 years in a prospective study. *International Journal of Eating Disorders, 22,* 339–360.

Surgenor, L., Plumridge, E., & Horn, J. (2003). "Knowing one's self" anorexic: Implications for therapeutic practice. *International Journal of Eating Disorders, 33,* 22–32.

Tozzi, F., Sullivan, P., Fear, J., McKenzie, J., & Bulik, C. (2003). Causes and recovery in anorexia nervosa: The patient's perspective. *International Journal of Eating Disorders, 33,* 143–154.

Von Holle, A., Pinheiro, A., Thornton, L., Klump, K., Berrettini, W., Brandt, H., … Bulik, C. M. (2008). Temporal patterns of recovery across eating disorder subtypes. *Australian and New Zealand Journal of Psychiatry, 42,* 108–117.

Warren, C., Crowley, M., Olivardia, R., & Schoen, A. (2009). Treating patients with eating disorders: An examination of treatment providers' experiences. *Eating Disorders: The Journal of Treatment and Prevention, 17,* 27–45.

Weaver, K., Wuest, J., & Ciliska, D. (2005). Understanding women's journey of recovering from anorexia nervosa. *Qualitative Health Research, 15,* 188–206.

Wonderlich, S., Gordon, K., Mitchell, J., Crosby, R., & Engel, S. (2009). The validity and clinical utility of binge eating disorder. *International Journal of Eating Disorders, 42,* 687–705.

Yager, J., & Edelstein, C. (1987). Training therapists to work with eating disorders patients. In P. Beumont, G. D. Burrows, & R. Casper (Eds.), *Handbook of eating disorders, Part I: Anorexia and bulimia nervosa* (pp. 379–392). New York, NY: Elsevier.

Zerbe, K. (1995). Integrating feminist and psychodynamic principles in the treatment of an eating disorder patient: Implications for using countertransference responses. *Bulletin of the Menninger Clinic, 59*(2), 160–176.

Zerbe, K. (2008). *Integrated treatment of eating disorders: Beyond the body betrayed.* New York, NY: W.W. Norton.

PART III

BRIDGING THE GAP: SPECIAL POPULATIONS

13   *Borderline Personality and Eating Disorders: A Chaotic Crossroads*   217
14   *Managing the Eating Disorder Patient with a Comorbid Substance Use Disorder*   233
15   *Comorbid Trauma and Eating Disorders*   251
16   *Healing Self-Inflicted Violence in Adolescents with Eating Disorders*   269
17   *The Weight-Bearing Years: Eating Disorders and Body Image Despair in Adult Women*   285
18   *Men with Eating Disorders*   301

CHAPTER 13

Borderline Personality and Eating Disorders: A Chaotic Crossroads

Randy A. Sansone and Lori A. Sansone

In this chapter, we discuss the complex and intriguing crossroads between eating disorders (ED) and borderline personality disorder (BPD). We begin by reviewing BPD as a clinical entity and include a description of the disorder that is based on the criteria in the Diagnostic and Statistical Manual (DSM) of Mental Disorders. We then discuss the prevalence of BPD among patients with ED (up to 28% in bulimia nervosa) and review possible diagnostic approaches. Finally, we describe the complex clinical implications when ED and BPD are comorbid, and conclude with a discussion of general treatment strategies and outcome findings.

OVERVIEW OF BPD

A DSM Description of BPD

BPD is a personality dysfunction that is classified as an Axis II disorder in the *Diagnostic and Statistical Manual of Mental Disorders* (DSM-IV-TR; American Psychiatric Association, 2000a). Within the 11 Axis II disorders in the current DSM, BPD resides in the Cluster B personality grouping, which consists of personality disorders that are characterized by dramatic, emotional, and erratic features. Within this Cluster, BPD cohabits with narcissistic, antisocial, and histrionic personality disorders.

According to the DSM-IV-TR (American Psychiatric Association, 2000a), BPD is characterized by nine clinical features. These are: (i) frantic efforts to avoid abandonment; (ii) a history of unstable and intense relationships with others; (iii) identity disturbance; (iv) impulsivity in at least two functional areas such as spending, sex, substance use, eating, or driving; (v) recurrent suicidal threats or behaviors as well as self-mutilation; (vi) affective instability with marked reactivity of mood; (vii) chronic feelings of emptiness; (viii) inappropriate and intense anger or difficulty controlling anger; and (ix) transient stress-induced paranoid ideation or severe dissociative symptoms. Because only five of the preceding nine features are required for diagnosis, there are potentially many symptomatic permutations across affected patients.

The Clinical Texture of BPD

From a clinical perspective, BPD is characterized by a distinct psychopathological texture. On the surface, patients with BPD appear to have intact social façades such that in fleeting superficial social encounters, they can appear quite normal. However, beneath this thin veneer of social adaptation resides a longstanding history of impairing *self-regulatory disturbances* and chronic *self-destructive behaviors*. This unusual and dramatic psychopathological composite—a seemingly normal exterior coupled with a chaotic interior—represents the pathognomonic clinical paradox that clinicians encounter in BPD patients.

Self-regulation difficulties. Self-regulation difficulties emerge whenever there is a disturbance in the patient's ability to regulate a core self function. If the regulatory disturbance entails eating behavior, the patient may develop anorexia nervosa (AN) or bulimia nervosa (BN), binge eating disorder (BED), or obesity. If the regulatory deficit relates to substance intake, the patient may develop alcohol or drug abuse/dependence. In the case of sexual behavior, promiscuity may emerge, and with difficulties in the regulation of pain, chronic pain syndromes may develop.

Self-destructive behavior. Self-destructive behaviors are intentional and self-damaging events that may be self-directed (e.g., hitting, cutting, burning oneself; suicide attempts) and/or provoked or elicited from the interpersonal environment (e.g., partner violence, bar fights). While self-destructive behaviors oftentimes function to regulate affect, they may also serve to elicit caring responses from others, sustain a self-destructive identity, and/or assist in fending off an impending quasi-psychotic episode.

To meet diagnostic criteria, self-regulation difficulties and self-destructive behaviors must both be consistently present in the patient's history, with symptoms typically dating back to adolescence or early adulthood. In addition, patients may report symptom substitution, in which one symptom set (e.g., binge eating) dissolves into another (e.g., substance abuse).

The Prevalence of BPD

General population. According to the DSM-IV-TR (American Psychiatric Association, 2000a), the prevalence of BPD in the community is around 2%. However, Stone (1986) believes that the rate could be as high as 10%, a figure that may include subthreshold or subclinical cases. Grant et al. (2008) reported a lifetime prevalence rate of 6%. To provide some perspective, the overall prevalence of personality disorders in the general population is around 5–10% (Ellison & Shader, 2003).

Clinical populations. The prevalence of BPD has been investigated in a number of clinical settings. For example, in an urban outpatient internal medicine setting, investigators found a prevalence rate of 6.4% (Gross et al., 2002). In resident-provider primary care clinics, which have higher loadings of indigent and low-functioning patients, we have encountered BPD rates up to 25% (Sansone, Wiederman & Sansone, 2000).

As for psychiatric settings, Quigley (2005) states that BPD is one of the most frequently encountered Axis II disorders. According to the DSM-IV-TR (American Psychiatric

Association, 2000a), the prevalence of BPD in outpatient and inpatient psychiatric populations is 10 and 20%, respectively. In more contemporary samples, we found that 22% of outpatients in a diagnostically conservative university-based clinic (i.e., residents are cautioned about premature Axis II diagnosis; Sansone, Rytwinski & Gaither, 2003) and nearly 50% of those in an urban inpatient sample (Sansone, Songer & Gaither, 2001) evidenced borderline personality traits or disorder.

Borderline personality disorder and eating disorders. Through an extensive review of the empirical literature, we found that the prevalence of BPD among patients with ED is a function of the specific DSM ED diagnosis (Sansone, Levitt & Sansone, 2005). In contrast to purely restrictive eating behavior, eating pathology characterized by impulsivity (e.g., binging, vomiting, laxative abuse) evidenced much higher rates of BPD. Specifically, the prevalence rate was 10% in patients with restricting AN; 25% in binge-eating/purging AN; 28% in BN; 12% in BED; and 14% among a sample of obese patients seeking gastric surgery (Sansone et al., 2005; Sansone, Schumacher, Wiederman & Routsong-Weichers, 2008). From a broader perspective, these data indicate that the prevalence of BPD among those with ED is 5–14 times the rate encountered in the general population. This observation conveys that in treating patients with EDs, clinicians will invariably encounter those with comorbid BPD.

BPD: A MULTI-DETERMINED DISORDER

Like many psychiatric disorders, BPD appears to be a multi-determined disorder (Bandelow et al., 2005; Paris, 2005) with contributions from genetics, repetitive trauma in childhood, parental psychopathology, family dysfunction, and a possible triggering event. Permutations of these risk factors appear to vary from patient to patient and the thresholds for symptom precipitation remain unknown.

Genetics

Recent investigations indicate that genetics meaningfully contribute to the development of BPD. Distel et al. (2008) found that genetics accounted for 42% of the variance, although what appears to be inherited is *non-specific* (Skodol et al., 2002). In other words, while BPD as a disorder is not directly inherited, core biological vulnerabilities may be genetically passed on, such as affective instability, poor impulse management, and/or dysfunctional cognitive/perceptual styles (Goodman, New & Siever, 2004). In further support of a genetic relationship, a review of the literature disclosed that, compared with controls, relatives of patients with BPD evidenced greater frequencies of impulse spectrum disorders including BPD (White, Gunderson, Zanarini & Hudson, 2003).

Early Developmental Trauma

Repetitive trauma in early development appears to be a principal risk factor for the development of BPD. The majority of empirical studies confirms a statistical association between childhood trauma and BPD in adulthood (Sansone & Sansone, 2000). Trauma types most associated with BPD include repetitive physical, sexual, and emotional abuses as well as, in some cases, the witnessing of violence. As an example, in a large sample of patients with BPD, Zanarini, Dubo, Lewis & Williams (1997) found that 85% of participants reported histories of childhood trauma.

Parental Psychopathology/Family Dysfunction

Parental psychopathology and family dysfunction also appear to contribute to the development of BPD (Sansone & Sansone, 2009; see Box 13.1). Parent and family dynamics associated with the development of BPD include neglect and a lack of empathy (Yatsko, 1996); "biparental failure" (Zanarini et al., 2000); poor relationships with parents (Norden, Klein, Donaldson, Pepper & Klein, 1995); and family interactions that are invalidating, conflictual, negative, or critical (Fruzzetti, Shenk & Hoffman, 2005). Parents have been described by offspring with BPD as negative (Gunderson & Lyoo, 1997); uncaring and over-controlling (Parker et al., 1999; Weaver & Clum, 1993; Zweig-Frank & Paris, 1991); unempathetic (Guttman & Laporte, 2000); conflictual (Allen et al., 2005); invalidating and critical (Fruzzetti et al., 2005); aversive, less nurturing, and less affectionate (Johnson, Cohen, Chen, Kasen & Brook, 2006); emotionally withholding (Zanarini, Gunderson, Marino, Schwartz & Frankenburg, 1989); over-protective (Gagnon, 1993; Torgersen & Alnaes, 1992); over-involved as well as under-involved (Allen & Farmer, 1996); and hostile (Hayashi, Suzuki & Yamamoto, 1995). These data suggest that most patients with BPD experience dysfunctional relationships with parents and develop in unstable family environments.

Triggering Events

One final risk factor for the development of BPD is the role of a triggering event. Zanarini and Frankenburg (1997) described these events as acute psychosocial stressors that appear to abruptly precipitate the onset of BPD symptomatology, although this phenomenon has received little study.

BOX 13.1

EXAMPLES OF PARENTAL DYNAMICS THAT MAY BE ASSOCIATED WITH THE DEVELOPMENT OF BORDERLINE PERSONALITY DISORDER

- Neglect and a lack of empathy
- "Biparental failure"
- Poor relationships with offspring
- Interactions that are invalidating, conflictual, negative, or critical
- Negativity
- Uncaring and over-controlling
- Unempathetic
- Conflictual relationships with offspring
- Invalidating and critical
- Aversive, less nurturing, and less affectionate
- Emotionally withholding
- Over-protective
- Over-involved as well as under-involved
- Hostile

THE DIAGNOSIS OF BPD

Because of the high rates of BPD among patients with EDs, every patient should be assessed for the disorder. Symptom assessment for the diagnosis of this Axis II disorder can be challenging in the presence of active eating pathology (Vitousek & Stumpf, 2006). Explicitly, highly symptomatic patients can appear very chaotic and borderline-like, such that accurate Axis II diagnosis may have to wait until the stabilization of the ED symptoms. Because the diagnosis of BPD is stigmatizing, clinicians need to have a reliable screening or diagnostic procedure for this complex disorder.

Clinical Diagnosis

Clinical assessment can be readily accomplished through the use of the DSM criteria, which were presented in the previous section. A briefer diagnostic alternative is the clinical adaptation of the original Diagnostic Interview for Borderlines (DIB; Kolb & Gunderson, 1980), which is shown in Box 13.2. There are only five criteria for assessment and these can be organized around the acronym, P-I-S-I-A, which enables easy recall in the clinical setting. All five criteria are required (i.e., one type of quasi-psychotic phenomenon, longstanding impulsivity, an intact social façade, chaotic interpersonal relationships, and a chronically

---

BOX 13.2

THE ADAPTED GUNDERSON CRITERIA FOR THE DIAGNOSIS OF BORDERLINE PERSONALITY DISORDER

P *Psychotic/quasi-psychotic episodes*: transient, fleeting, brief episodes that tend to emerge with stress; these may include:
- Depersonalization
- Derealization
- Dissociation
- Rage reactions
- Paranoia (the patient recognizes the illogical nature of their suspiciousness)
- Fleeting or isolated hallucinations or delusions
- Unusual reactions to drugs

I *Impulsivity*:
- *Self-regulation difficulties* (e.g., eating disorders; drug/alcohol/ prescription abuse; money management difficulties such as bankruptcies, credit card difficulties, or uncontrolled gambling; promiscuity; mood regulation difficulties; chronic pain syndromes)
- *Self-destructive behaviors* (e.g., self-mutilation such as hitting, cutting, burning, or biting oneself; suicide attempts; abusive relationships; high-risk hobbies such as parachuting or racing cars; high-risk behaviors such as frequenting dangerous bars or jogging in parks at night)

S *Social adaptation*: superficially intact social veneer; if the individual demonstrates high academic or professional performance, it is usually inconsistent and erratic.

> BOX 13.2 (cont'd)
>
> I *Interpersonal relationships*: chaotic and unsatisfying; social relationships tend to be very superficial and transient while personal relationships tend to be extremely intense, manipulative and dependent; intense fears of being alone; rage with the primary caretaker.
>
> A *Affect*: chronically dysphoric and/or labile; since adolescence, the majority of the mood experience (>80%) has been dysphoric or volatile with the predominant affects being anxiety, anger, depression, and/or emptiness.
>
> *(Kolb & Gunderson, 1980)*

dysphoric affect). For quick screening, *Impulsivity* and *Affect* seem to provide diagnostically high yields, probably because these two features are both relatively stable over time (McGlashan et al., 2005); if suggestive, the remainder of the interview is then undertaken.

Self-report screening measures. When evaluating a psychiatric phenomenon that has a high frequency in a given clinical population, a screening measure may be particularly useful. These measures function as the impetus for a more detailed clinical evaluation for BPD. Current recommendations include one of the three following self-report measures: (i) the borderline personality scale of the Personality Diagnostic Questionnaire-4 (PDQ-4; Hyler, 1994); (ii) the Self-Harm Inventory (SHI; Sansone, Wiederman & Sansone, 1998); or the (iii) McLean Screening Inventory for Borderline Personality Disorder (MSI-BPD; Zanarini et al., 2003). We have used all three measures and found them to be dependable, but over-inclusive (i.e., they tend to over-diagnose patients as borderline). Therefore, these are *truly* only screening tools. The PDQ-4 and the SHI are shown in Appendices A and B. For the PDQ-4, a score of 5 or higher is suggestive of BPD (all endorsements count as 1 point, with the last item counted as 1 point if two sub-items are endorsed). For the SHI, which demonstrates a diagnostic accuracy of 84% in comparison with the Diagnostic Interview for Borderlines (Kolb & Gunderson, 1980), a score of 5 or higher is suggestive of BPD (endorsement of 5 "yes" responses).

Other assessments for BPD. There are a number of other assessments for BPD, including the Millon Clinical Multiaxial Inventory-III (MCMI-III; Millon, Davis & Millon, 1997), which has among other scales a BPD scale. This measure for BPD and others are discussed in detail elsewhere (see Sansone & Sansone, 2007a).

IMPLICATIONS OF THE BPD DIAGNOSIS IN PATIENTS WITH EATING DISORDERS

BPD May be a Risk Factor for the Development of an Eating Disorder

Curiously, the presence of BPD may be a pre-existing risk factor for the development of an ED (Sansone & Sansone, 2007b). This potential pathway begins with childhood trauma (particularly physical, emotional, and sexual abuses), which facilitates the development of BPD. The development of BPD then results in the emergence of two key psychopathological features: (i) trauma-related disturbances in body image; and (ii) self-harm behavior. Body image disturbances then contribute to the need to "repair" one's body, a process that initially emerges in adolescence. In the presence of an actual or perceived weight disorder, efforts at body repair subsequently manifest as attempts to lose weight with attendant eating pathology. In the aftermath of the development of eating pathology, the emergence of harmful ED behaviors then complements the BPD dynamics of self-harm and self-soothing. Thus, the presence of BPD appears to heighten the risk for the emergence of an ED, particularly in the context of a weight-conscious family and/or society. Once established, BPD seems to perpetuate its own psychopathology through ED symptoms (see below).

Eating Disorder Pathology May Function as a Self-Injury Equivalent

While ED behaviors in the non-BPD patient relate predominantly to food, body, and weight issues, they tend to function as self-injury equivalents in the patient with BPD. For example, excessive exercise may not be only for weight management, but also to induce self-injury.

Comorbid BPD Requires Augmented and Lengthier Treatment

In order to produce a positive treatment outcome in the comorbid patient, the clinician will need to augment the standard ED treatment with interventions designed for BPD (American Psychiatric Association, 2000b, p. 2). Indeed, in most cases, the lengthier BPD treatment will take precedence over the ED treatment, which tends to play a secondary but important focus. In other words, ED treatment (e.g., psycho-education, cognitive-behavioral interventions) is clearly attended to, but the psychotherapy emphasis is consistently on BPD.

An exception to the preceding guideline is the low-weight, cognitively impaired patient. In such cases, treatment for weight restoration is given priority because these starved patients cannot effectively engage in sophisticated psychotherapy treatment.

Symptoms Are Worse/Outcome Less Robust in Comorbid Patients

In ED patients with comorbid BPD, overall psychiatric symptoms tend to be worse, but not necessarily ED symptoms. Yet, some studies indicate less favorable outcomes in the comorbid patient compared to those with ED symptoms only (Masjuan, Aranda & Raich, 2003; Steinhausen, 2002). However, these findings generally refer to continuing broad-spectrum psychiatric symptoms such as impulsivity and self-harm behavior and not to specific ED symptoms (Steiger & Stotland, 1996). The ED symptoms, themselves, may undergo a reasonable resolution (Grilo et al., 2003).

TREATMENT APPROACHES TO BPD

According to the *Practice Guideline for the Treatment of Patients with Borderline Personality Disorder* (American Psychiatric Association, 2001), "the primary treatment for borderline personality disorder is psychotherapy…" (p. 4). There are a number of different psychotherapy approaches to BPD and, at present, there appears to be no evidence that one treatment is superior to another (Livesley, 2005).

Systematized Treatment Approaches

De Groot, Verheul & Wim Trijsburg (2008) reviewed a number of systematized treatment approaches (i.e., formalized, structured, and/or manual-based approaches) for BPD, including schema-focused therapy, dialectical behavior therapy, cognitive analytic therapy, systems training for emotional predictability and problem solving, transference focused psychotherapy, mentalization-based treatment, and interpersonal reconstructive therapy. In their review, de Groot and colleagues (2008) point out a number of similarities between these various approaches.

According to de Groot et al. (2008), the majority of systematized approaches takes place in outpatient treatment settings and continues for at least one year, with one or two sessions per week. All of these treatment approaches incorporate individual sessions with the exception of systems training for emotional predictability and problem solving, which is undertaken entirely in group format. (Many of the remaining treatment approaches utilize group sessions, as well.) In addition, the majority of these approaches incorporate treatment contracts as well as one primary therapist for the patient.

In terms of treatment modalities, all of the preceding systematized approaches include psycho-educational and motivational techniques. In their discussion, de Groot et al. (2008) described psycho-education as any experiential component with an educational goal. The content of these psycho-educational modules typically relates to causal and maintenance factors for BPD as well as the theory underlying the treatment model. As for motivational techniques, these are described as any intervention that focuses on increasing a patient's commitment to treatment and may include but are not limited to motivational interviewing, clarification of interventions, and paradoxical interventions.

In addition to psycho-educational and motivation techniques, all of the preceding systematized treatments incorporate cognitive and interpersonal techniques. Cognitive techniques refer to any intervention that focuses on analyzing and creating insight with regard to patterns of thinking, feeling, and behaving. These patterns are then examined in the context of everyday functioning. Interpersonal techniques may be defined as any intervention that improves patients' understanding of their patterns of functioning in relationships with other people.

The majority of these systematized treatments also contain a psychodynamic component. Commonly used in psychoanalytic and psychodynamic treatments, these components address the patient's defenses as well as transference phenomena.

Finally, some of the preceding systematized approaches incorporate pure behavioral techniques as well as patient training in mindfulness (see McCallum, Chapter 23). Behavioral techniques refer to those procedures that are based on learning theory. Mindfulness, a Buddhist concept, focuses on heightening non-judgmental observation and awareness.

In conclusion, while these various systematized approaches appear somewhat structurally different at the outset, they share a variety of core psychotherapeutic strategies. This menu of treatment options suggests that one approach is not more effective than another. However, this does not exclude the possibility that one technique may be better suited than another to a specific patient.

ECLECTIC TREATMENT APPROACHES: A BRIEF EXAMPLE

The preceding systematized approaches do not exclude the use of eclectic treatments, including our own, which is described in detail elsewhere (Sansone & Sansone, 2006a). The following is a brief overview of our eclectic strategy, consisting of overlapping stages of treatment. However, it is difficult to fully describe a treatment approach in the absence of actual patient supervision.

Stage 1: Treatment entry. In the initial stage of our treatment approach, the therapist establishes a regular and reliable treatment environment. Key elements entail established and consistent office staff, location and billing procedures, stable furniture/décor, and predictable appointment times. Irregularities in the preceding office themes tend to generate mistrust and instability in patients with BPD, and may transiently derail the therapeutic relationship.

Following the development of a structured treatment environment, the therapist initiates an assessment of the patient and determines a treatment strategy. In the preceding section, we emphasized the importance of accurately diagnosing BPD, which can be effectively accomplished through the use of *DSM* criteria or the Gunderson criteria (P-I-S-I-A) (see Box 13.1). Using either, it is important for the therapist to simultaneously assess the patient's functional level, which is a clinical determination. Mid-to-high functioning patients are ideal candidates for our eclectic approach. However, low-functioning patients may be limited to supportive/maintenance treatment—a strategy that emphasizes the acquisition of basic life skills (e.g., fundamental social skills, job training, life management skills). A common pitfall in treatment determination is to direct a low functioning patient into a sophisticated psychotherapy treatment, which tends to promote regression in the patient.

For those patients who are candidates for an eclectic psychotherapy approach, we recommend a treatment-entry contract. This contract is designed, at the outset, to contain life-threatening regressive behavior during the psychotherapy treatment. All psychotherapy treatments tend to foster some level of regression in the patient. For patients who are seriously self-destructive, this normal regressive pattern could culminate in a serious suicide attempt. Therefore, the treatment-entry contract consists of the patient agreeing to *absolutely* contain high-lethal behavior in an effort to enable the psychotherapy treatment to continue. If the patient is unable to maintain this contract, they need to be diverted to a lower-risk treatment such as supportive/maintenance treatment and/or medication management.

The therapist may also elect additional contracting around other behaviors that could potentially disturb or impair the unfolding psychotherapy treatment. An example would be the proscription of substance intoxication during sessions. These adjunctive issues are highly individualized to the specific patient and may relate to some degree to the therapist's tolerance.

In addition to developing a consistent treatment environment, undertaking a reliable assessment for BPD, and establishing a treatment entry contract, we explain to the patient the intense relationship focus of the treatment. In this discussion, we emphasize the importance of the therapeutic relationship, boundaries (e.g., boundary experiences in prior relationships, boundaries in the therapeutic relationship, boundaries outside the therapeutic relationship), the importance of a good patient/therapist fit, the necessity of honesty, and the future management of potential patient-therapist conflicts.

Stage 2: The development of a working treatment relationship. During the initial phase of psychotherapy work, the therapist is entangled in a variety of tasks. These include building and sustaining a reliable and consistent therapeutic relationship, grooming healthier coping skills in the patient so as to replace longstanding self-harm behaviors, providing psycho-education, and evaluating for medication management. During this phase of treatment, we typically incorporate psychodynamic, cognitive-behavioral, and interpersonal techniques. These integrated techniques are described elsewhere (see Sansone & Sansone, 2006a).

Stage 3: Working at a deeper level with psychodynamic themes. During the third phase of treatment, deeper psychodynamic themes are tackled, including interpersonal/intrapsychic splitting as well as core self-regulation difficulties. The therapist also enhances the patient's interpersonal style of relating. In comparison, while Stage 2 is highly structured, Stage 3 tends to be more fluid and psychodynamic.

Stage 4: Treatment termination. Termination is a distinct treatment phase and is generally begun six months or more ahead of the anticipated and established end date. In other words, the termination period, itself, lasts at least several months. This duration of termination is necessary because closure of the treatment tends to result in the patient's re-experiencing of loss and abandonment. As a result, this phase provides a unique opportunity to reframe the context of loss and abandonment. Again, the details of this phase of treatment are discussed elsewhere (see Sansone & Sansone, 2006a).

Potential Treatment Quagmires

A number of potential treatment quagmires in working with patients with BPD warrant brief review. The first, as previously mentioned, is directing a low-functioning patient into a demanding psychotherapy treatment. This not only heightens the risk of patient regression, but also misuses the skills of highly trained therapists, who are relegated to staving off regressions and "putting out fires."

A second potential quagmire is assuming that all patients with BPD are candidates for therapy. A small number of such patients are high-risk in *any* treatment context. Signs of such patients include past intense and negative transferences with therapists, high levels of self-destructive impulsivity, intense reactivity to limits, pervasive anger, and a strong need to "win at any costs." In our experience, these patients solely engage in treatment to misuse the therapy as well as to engage in abusive and sadomasochistic interactions with the therapist. On rare occasion, these patients may intentfully suicide to "beat the therapist." Unfortunately, these types of patients are not candidly discussed in the BPD literature. It is essential to recognize that not all patients with BPD can benefit from treatment.

A third potential quagmire is whether to confront childhood perpetrators. Potential risks of confrontation include the revictimization of the patient through re-exposure to the perpetrator, patient retraumatization through discrediting responses from the perpetrator, the inability of the patient to contain overwhelming emotions in the aftermath of a confrontation with a resulting escalation of self-harm behavior, and/or potential damage to the therapeutic relationship in the presence of a negative outcome. An active discussion with the patient of these preceding risks is essential.

Patients who are seeking or have already achieved disability are another potential treatment quagmire. In our experience, disability unintentionally but consistently seems to impair attempts at genuine functional recovery by keeping patients unemployed. Yet, employment is a key means of providing life structure. Because employment is not achievable for the disabled patient, full stabilization is not likely to occur.

A fifth potential quagmire is comorbid substance abuse, which typically is a substantial deterrent to successful treatment (see Baker Dennis & Helfman, Chapter 14). We typically recommend that such candidates enter into substance-abuse treatment programs and maintain sobriety for a minimum of six months before entering into a demanding psychotherapy treatment.

A sixth quagmire is the potential misperception of the role and effect of medication in BPD patients. While all classes of psychotropic drugs have been explored in their treatment, most have resulted in somewhat modest and unsustained effects (see Sansone & Sansone, 2006b). Overall, the most effective medications have been low-dose antipsychotics, although in specific cases, other types of medication (e.g., antidepressants, anticonvulsants) may be temporarily useful. A reasonable response to medication is a 30% reduction in symptoms (Sansone & Sansone, 2006b). These BPD guidelines are not meant to deflect from the general beneficial effects of selective serotonin reuptake inhibitors in the treatment of BN (American Psychiatric Association, 2006, p. 32). In addition, while antipsychotics such as olanzapine have been showing some promise in low-weight patients, their use for BPD symptoms in this debilitated population has undergone very little exploration.

A final quagmire is the undertaking of a BPD treatment without any pre-determined approach. This not only facilitates patient regression but may also result in unproductive therapy hours (e.g., excessive focus on the early history) and frustration in both the patient and the therapist. Clinical experience and research have clearly demonstrated that consistency and structure are the bedrock elements of an effective BPD treatment.

OUTCOME

As we mentioned previously, treatment of the ED patient with comorbid BPD tends to result in an overall improvement in the eating pathology (Grilo et al., 2003). However, compared with non-comorbid patients, patients with BPD are likely to experience higher levels of continuing psychopathology related to non-ED symptomatology (Steiger & Stotland, 1996).

CONCLUSIONS

Patients with comorbid ED and BPD are fairly prevalent. We believe that an accurate diagnosis of BPD is essential to treatment. Patients with these comorbid disorders require treatment augmentation with a lengthier, highly consistent, and structured psychotherapy approach, whether systematized or eclectic in nature. With effective treatment, ED symptoms tend to diminish, although there may be lingering symptoms in other functional areas. These patients are truly challenging to treat but can make genuine and effective gains—given a well executed treatment opportunity.

References

Allen, D. M., Abramson, H., Whitson, S., Al-Taher, M., Morgan, S., Veneracion-Yumul, A.,...Mason, M. (2005). Perceptions of contradictory communication from parental figures by adults with borderline personality disorder: A preliminary study. *Comprehensive Psychiatry, 46*, 340–352.

Allen, D. M., & Farmer, R. G. (1996). Family relationships of adults with borderline personality disorder. *Comprehensive Psychiatry, 37*, 43–51.

American Psychiatric Association (2000a). *The diagnostic and statistical manual of mental disorders* (4th ed., text rev.). Washington, DC: Author.

American Psychiatric Association (2000b). *Practice guideline for the treatment of patients with eating disorders* (Rev. ed.). Washington, DC: Author.

American Psychiatric Association (2001). Practice guideline for the treatment of patients with borderline personality disorder. *American Journal of Psychiatry, 158*, S1–52.

American Psychiatric Association (2006). *Practice guideline for the treatment of patients with eating disorders* (3rd ed.). Washington, DC: Author.

Bandelow, B., Krause, J., Wedekind, D., Broocks, A., Hajak, G., & Ruther, E. (2005). Early traumatic life events, parental attitudes, family history, and birth risk factors in patients with borderline personality disorder and healthy controls. *Psychiatry Research, 134*, 169–179.

De Groot, E. R., Verheul, R., & Wim Trijsburg, R. (2008). An integrative perspective on psychotherapeutic treatments for borderline personality disorder. *Journal of Personality Disorders, 22*, 332–352.

Distel, M. A., Trull, T. J., Derom, C. A., Thiery, E. W., Grimmer, M. A., Martin, N. G.,...Boomsma, D. I. (2008). Heritability of borderline personality disorder features is similar across three countries. *Psychological Medicine, 38*, 1219–1229.

Ellison, J. M., & Shader, R. I. (2003). Pharmacologic treatment of personality disorders: A dimensional approach. In R. I. Shader (Ed.), *Manual of psychiatric therapeutics* (pp. 169–183). Philadelphia, PA: Lippincott, Williams, & Wilkins.

Fruzzetti, A. E., Shenk, C., & Hoffman, P. D. (2005). Family interaction and the development of borderline personality disorder: A transactional model. *Development and Psychopathology, 17*, 1007–1030.

Gagnon, P. (1993). Role of the family in the development of borderline personality disorder. *Canadian Journal of Psychiatry, 38*, 611–616.

Goodman, M., New, A., & Siever, L. (2004). Trauma, genes, and the neurobiology of personality disorders. In R. Yehuda, & B. McEwen (Eds.), *Biobehavioral stress response: Protective and damaging effects* (pp. 104–116). New York, NY: New York Academy of Sciences.

Grant, B. R., Chou, S. P., Goldstein, R. B., Huang, B., Stinson, F. S., Saha, T. D.,...Ruan, W. J. (2008). Prevalence, correlates, disability, and comorbidity of DSM-IV borderline personality disorder: Results from the Wave 2 National Epidemiologic Survey on Alcohol and Related Conditions. *Journal of Clinical Psychiatry, 69*, 533–545.

Grilo, C. M., Sanislow, C. A., Shea, M. T., Skodol, A. E., Stout, R. L., Pagano, M. E.,...McGlashan, T. H. (2003). The natural course of bulimia nervosa and eating disorder not otherwise specified is not influenced by personality disorders. *International Journal of Eating Disorders, 34*, 319–330.

Gross, R., Olfson, M., Gameroff, M., Shea, S., Feder, A., Fuentes, M., ...Weissman, M. M. (2002). Borderline personality in primary care. *Archives of Internal Medicine, 162*, 53–60.

Gunderson, J. G., & Lyoo, I. K. (1997). Family problems and relationships for adults with borderline personality disorder. *Harvard Review of Psychiatry, 4*, 272–278.

Guttman, H. A., & Laporte, L. (2000). Empathy in families of women with borderline personality disorder, anorexia nervosa, and a control group. *Family Process, 39*, 345–358.

Hayashi, N., Suzuki, R., & Yamamoto, N. (1995). Parental perceptions of borderline personality disorders in video-recorded interviews. *Psychiatry and Clinical Neurosciences, 49*, 35–37.

Hyler, S. E. (1994). *Personality Diagnostic Questionniare—4*. New York, NY: New York Psychiatric Institute.

Johnson, J. G., Cohen, P., Chen, H., Kasen, S., & Brook, J. S. (2006). Parenting behaviors associated with risk for offspring personality disorder during childhood. *Archives of General Psychiatry, 63*, 579–587.

Kolb, J. E., & Gunderson, J. G. (1980). Diagnosing borderline patients with a semi-structured interview. *Archives of General Psychiatry, 37*, 37–41.

Livesley, W. J. (2005). Principles and strategies for treating personality disorder. *Canadian Journal of Psychiatry, 50*, 442–450.

Masjuan, M. G., Aranda, F. F., & Raich, R. M. (2003). Bulimia nervosa and personality disorders: A review of the literature. *International Journal of Clinical and Health Psychology, 3*, 335–349.

McGlashan, T. H., Grilo, C. M., Sanislow, C. A., Ralevski, E., Morey, L. C., Gunderson, J. G., ... Pagano, M. (2005). Two-year prevalence and stability of individual *DSM-IV* criteria for schizotypal, borderline, avoidant, and obsessive-compulsive personality disorders: Toward a hybrid model of Axis II disorders. *American Journal of Psychiatry, 162*, 883–889.

Millon, T., Davis, R., & Millon, C. (1997). *Manual for the Millon Clinical Multiaxial Inventory-III (MCMI-III)* (2nd ed.). Minneapolis, MN: NCS Pearson, Inc.

Norden, K. A., Klein, D. N., Donaldson, S. K., Pepper, C. M., & Klein, L. M. (1995). Reports of the early home environment in DSM-III-R personality disorders. *Journal of Personality Disorders, 9*, 213–223.

Paris, J. (2005). The development of impulsivity and suicidality in borderline personality disorder. *Development and Psychopathology, 17*, 1091–1104.

Parker, G., Roy, K., Wilhelm, K., Mitchell, P., Austin, M.-P., & Hadzic-Pavlovic, D. (1999). An exploration of links between early parenting experiences and personality disorder type and disordered personality functioning. *Journal of Personal Disorders, 13*, 361–374.

Quigley, B. D. (2005). Diagnostic relapse in borderline personality: Risk and protective factors. *Dissertation Abstracts International, 65*, 3721B.

Sansone, R. A., Levitt, J. L., & Sansone, L. A. (2005). The prevalence of personality disorders among those with eating disorders. *Eating Disorders: The Journal of Treatment and Prevention, 13*, 7–21.

Sansone, R. A., Rytwinski, D., & Gaither, G. A. (2003). Borderline personality and psychotropic medication prescription in an outpatient psychiatry clinic. *Comprehensive Psychiatry, 44*, 454–458.

Sansone, R. A., & Sansone, L. A. (2000). Borderline personality disorder: The enigma. *Primary Care Reports, 6*, 219–226.

Sansone, R. A., & Sansone, L. A. (2006a). Borderline personality and eating disorders: An eclectic approach to treatment. In R. A. Sansone, & J. L. Levitt (Eds.), *Personality disorders and eating disorders. Exploring the frontier* (pp. 197–212). New York, NY: Routledge.

Sansone, R. A., & Sansone, L. A. (2006b). The use of psychotropic medications in patients with eating disorders and personality disorders. In R. A. Sansone, & J. L. Levitt (Eds.), *Personality disorders and eating disorders. Exploring the frontier* (pp. 231–244). New York, NY: Routledge.

Sansone, R. A., & Sansone, L. A. (2007a). Borderline personality: A psychiatric overview. In R. Sansone, & L. Sansone (Eds.), *Borderline personality disorder in the medical setting: Unmasking and managing the difficult patient* (pp. 3–36). New York, NY: Nova Science Publishers.

Sansone, R. A., & Sansone, L. A. (2007b). Childhood trauma, borderline personality, and eating disorders: A developmental cascade. *Eating Disorders: The Journal of Treatment and Prevention, 15*, 333–346.

Sansone, R. A., & Sansone, L. A. (2009). The families of borderline patients: The psychological environment revisited. *Psychiatry, 6*, 19–24.

Sansone, R. A., Schumacher, D., Wiederman, M. W., & Routsong-Weichers, L. (2008). The prevalence of binge eating disorder and borderline personality symptomatology among gastric surgery patients. *Eating Behaviors, 9*, 197–202.

Sansone, R. A., Songer, D. A., & Gaither, G. A. (2001). Diagnostic approaches to borderline personality and their relationship to self-harm behavior. *International Journal of Psychiatry in Clinical Practice, 5*, 273–277.

Sansone, R. A., Wiederman, M. W., & Sansone, L. A. (1998). The Self-Harm Inventory (SHI): Development of a scale for identifying self-destructive behaviors and borderline personality disorder. *Journal of Clinical Psychology, 54*, 973–983.

Sansone, R. A., Wiederman, M. W., & Sansone, L. A. (2000). Medically self-harming behavior and its relationship to borderline personality symptoms and somatic preoccupation among internal medicine patients. *Journal of Nervous and Mental Disease, 188*, 45–47.

Skodol, A. E., Siever, L. J., Livesley, W. J., Gunderson, J. G., Pfohl, B., & Widiger, T. A. (2002). The borderline diagnosis II: Biology, genetics, and clinical course. *Biological Psychiatry, 51*, 951—963.

Steiger, H., & Stotland, S. (1996). Prospective study of outcome in bulimics as a function of Axis-II comorbidity: Long-term responses on eating and psychiatric symptoms. *International Journal of Eating Disorders, 20*, 149—161.

Steinhausen, H.-C. (2002). The outcome of anorexia nervosa in the 20th century. *American Journal of Psychiatry, 159*, 1284—1293.

Stone, M. H. (1986). Borderline personality disorder. In R. Michels, & J. O. Cavenar (Eds.), *Psychiatry* (2nd ed., pp. 1—15). Philadelphia, PA: Lippincott.

Torgersen, S., & Alnaes, R. (1992). Differential perception of parental bonding in schizotypal and borderline personality disorder patients. *Comprehensive Psychiatry, 33*, 34—38.

Vitousek, K. M., & Stumpf, R. S. (2006). Difficulties in the assessment of personality traits and disorders in individuals with eating disorders. In R. A. Sansone, & J. L. Levitt (Eds.), *Personality disorders and eating disorders* (pp. 91—117). New York, NY: Routledge.

Weaver, T. L., & Clum, G. A. (1993). Early family environments and traumatic experiences associated with borderline personality disorder. *Journal of Consulting and Clinical Psychology, 61*, 1068—1075.

White, C. N., Gunderson, J. G., Zanarini, M. C., & Hudson, J. I. (2003). Family studies of borderline personality disorder: A review. *Harvard Review of Psychiatry, 11*, 8—19.

Yatsko, C. K. (1996). Etiological theories of borderline personality disorder: A comparative multivariate study. *Dissertation Abstracts International, 56*, 4628B.

Zanarini, M. C., Dubo, E. D., Lewis, R. E., & Williams, A. A. (1997). Childhood factors associated with the development of borderline personality disorder. In M. C. Zanarini (Ed.), *Role of sexual abuse in the etiology of borderline personality disorder* (pp. 29—44). Washington, DC: American Psychiatric Press.

Zanarini, M. C., & Frankenburg, F. R. (1997). Pathways to the development of borderline personality disorder. *Journal of Personality Disorders, 11*, 93—104.

Zanarini, M. C., Frankenburg, F. R., Reich, D. B., Marino, M. F., Lewis, R. E., Williams, A. A., & Khera, G. S. (2000). Biparental failure in the childhood experiences of borderline patients. *Journal of Personality Disorders, 14*, 264—273.

Zanarini, M. C., Gunderson, J. G., Marino, M. F., Schwartz, E. O., & Frankenburg, F. R. (1989). Childhood experiences of borderline patients. *Comprehensive Psychiatry, 30*, 18—25.

Zanarini, M. C., Vujanovic, A. A., Parachini, E. A., Boulanger, J. L., Frankenburg, F. R., & Hennen, J. (2003). A screening measure for BPD: The McLean Screening Instrument for Borderline Personality Disorder (MSI-BPD). *Journal of Personality Disorders, 17*, 568—573.

Zweig-Frank, H., & Paris, J. (1991). Parents' emotional neglect and overprotection according to recollections of patients with borderline personality disorder. *American Journal of Psychiatry, 148*, 648—651.

APPENDIX A

The Borderline Personality Scale of the Personality Diagnostic Questionaire—4

(Hyler, 1994)*

Instructions: The purpose of this questionnaire is for you to describe the kind of person you are. When answering the questions, think about how you have tended to feel, think, and act over the past several years. To remind you of this, you will find the statement, "Over the past several years…"

T (True) means that the statement is generally true for you.

F (False) means that the statement is generally false for you.

Even if you are not entirely sure about the answer, indicate "T" or "F" (circle one) for every question. There are no correct answers. You may take as much time as you wish.

Over the past several years

T	F	I'll go to extremes to prevent those who I love from ever leaving me.
T	F	I either love someone or hate them, with nothing in between.
T	F	I often wonder who I really am.
T	F	I have tried to hurt or kill myself.
T	F	I am a very moody person.
T	F	I feel that my life is dull and meaningless.
T	F	I have difficulty controlling my anger or temper.
T	F	When stressed, things happen. Like I get paranoid or just "black out."
T	F	I have done things on impulse (such as those below) that can get me into trouble.

Check all that apply to you:

___Spending more money than I have
___Having sex with people I hardly know
___Drinking too much
___Taking drugs
___Eating binges
___Reckless driving

*Reprinted with permission from Dr. Hyler; these items are included in: Hyler, S. E. (1994). *Personality Diagnostic Questionnaire* (fourth edition; PDQ-4). New York State Psychiatric Institute: New York. Dr. Hyler is located at New York State Psychiatric Institute, 1051 Riverside Drive, New York, NY, 10032.

APPENDIX B

The Self-Harm Inventory

(Sansone, Wiederman, & Sansone, 1998)

Instructions: Please answer the following questions by checking either, "Yes," or "No." Check "yes" *only* to those items that you have done intentionally, or *on purpose*, to hurt yourself.

Yes	No	Have you ever intentionally, or on purpose…
___	___	1. Overdosed? (If yes, number of times____)
___	___	2. Cut yourself on purpose? (If yes, number of times____)
___	___	3. Burned yourself on purpose? (If yes, number of times____)
___	___	4. Hit yourself? (If yes, number of times____)
___	___	5. Banged your head on purpose? (If yes, number of times____)
___	___	6. Abused alcohol?
___	___	7. Driven recklessly on purpose? (If yes, number of times____)
___	___	8. Scratched yourself on purpose? (If yes, number of times____)

___ ___ 9. Prevented wounds from healing?
___ ___ 10. Made medical situations worse, on purpose (e.g., skipped medication)?
___ ___ 11. Been promiscuous (i.e., had many sexual partners)? (If yes, how many?____)
___ ___ 12. Set yourself up in a relationship to be rejected?
___ ___ 13. Abused prescription medication?
___ ___ 14. Distanced yourself from God as punishment?
___ ___ 15. Engaged in emotionally abusive relationships? (If yes, number of relationships?____)
___ ___ 16. Engaged in sexually abusive relationships? (If yes, number of relationships?____)
___ ___ 17. Lost a job on purpose? (If yes, number of times____)
___ ___ 18. Attempted suicide? (If yes, number of times____)
___ ___ 19. Exercised an injury on purpose?
___ ___ 20. Tortured yourself with self-defeating thoughts?
___ ___ 21. Starved yourself to hurt yourself?
___ ___ 22. Abused laxatives to hurt yourself? (If yes, number of times____)

Have you engaged in any other self-destructive behaviors not asked about in this inventory? If so, please describe below.

© 1995: Sansone, Sansone, & Wiederman

CHAPTER 14

Managing the Eating Disorder Patient with a Comorbid Substance Use Disorder

Amy Baker Dennis and Bethany L. Helfman

The relationship between eating disorders and substance use disorders (including substance use, abuse, and dependence) is a complex and often deadly one. These disorders have the highest mortality rates across all mental health diagnoses. Harris & Barraclough (1997) performed a meta-analysis of 249 reports of mortality in those with mental disorders and found that anorexia nervosa (AN) and bulimia nervosa (BN) patients had rates of suicide that were higher than for all other psychiatric disorders and 23 times that of the general population. Among alcohol dependent women, the risk of suicide is 20 times greater than is found in the general population (Kessler, Borges & Walters, 1999). Among all eating disorders (ED), the binge purge subtype of anorexia nervosa (ANBP) is associated with the highest risk of death (Bulik et al., 2008). When ED and substance use disorders (SUD) co-occur, the risk of mortality may increase exponentially. In a study of roughly 6,000 women, Swedish researchers found that AN subjects were 19 times more likely than the general population to have died from psychoactive substance abuse (SA), primarily alcohol abuse (Papadopoulos, Ekbom, Brandy & Eskelius, 2009).

Clearly, ED and SUD are independently correlated with higher than expected rates of death. It is therefore surprising to note that despite significant comorbidity, historically, ED professionals have not integrated SA treatment into their paradigms, nor have SA professionals adequately incorporated ED treatment into their practices. Complicating matters is the fact that many randomized controlled treatment trials conducted in the ED field exclude subjects with SUDs from their studies (Gadalla & Piran, 2007). Currently, there are no empirically supported treatments for this comorbid population.

Within the ED community there is also a tendency to eschew the treatment of a patient's SUD. Instead of attempting to integrate treatment for both the ED and SUD, clinicians often refer the patient to a SUD program or specialist either prior to initiating ED treatment or concurrently. Unfortunately, many clinicians in both the ED and SUD fields are antagonistic toward the approaches of the other field. It is all too common for clinicians to simplify these cases by focusing merely on one aspect (their specialty) of the patient's presentation. Regrettably, focusing on a part of the whole can prolong a patient's suffering. Treating one disorder without the other is akin to trying to ameliorate a fever without treating the underlying infection.

This chapter explores the science/practice gaps both between and within the ED and SUD fields. What is currently known, and what areas require further research, will be highlighted. The utility of developing an integrated, individualized approach for the ED/SUD patient that incorporates evidence-based therapies (EBT) and clinical expertise from both specialties will be reviewed in an effort to improve treatment delivery and outcome. Stimulating dialogue, cross-training, and joint research efforts between the ED and SUD communities is vital. Clinicians in both fields must acquire and incorporate knowledge of the other for advancements in treatment approaches to occur.

PREVALENCE DATA

Substance Use Disorders

Based on DSM-IV (American Psychiatric Association, 1994) criteria, 3.2 million Americans abuse or are dependent on alcohol or illicit drugs. There is a widely held notion that males abuse substances at rates greater than for females. Although this is true in adults, data from the Substance Abuse and Mental Health Services Administration (2009) revealed that among adolescents ages 12 to 17, substance abuse rates were slightly lower among males than females (7.0 vs. 8.2%, respectively).

While the rate of alcohol abuse has been relatively stable, an alarming increase in the abuse of prescription drugs including Ritalin, Valium and OxyContin has occurred. These drugs are now the fourth most abused substances in America (behind marijuana, alcohol, and tobacco) with an 81% increase from 1992 to 2003 (Califano, 2007). In fact, Americans are consuming roughly 80% of the global supply of opioids, 99% of the global supply of hydrocodone, and two-thirds of the world's illegal drugs despite representing merely 4% of the world's population (Kuehn, 2007; Manchikani, 2006).

Co-occurring Substance Use Disorders and Eating Disorders

Prevalence data suggest that roughly 50% of individuals with an ED are also abusing drugs and/or alcohol, which is more than five times the abuse rates seen in the general population (The National Center on Addiction and Substance Abuse, 2003). Two meta-analyses, conducted a decade apart, examined the co-occurrence of EDs and SUDs. Holderness, Brooks-Gunn & Warren (1994), found a strong association between SUD and ANBP as well as BN, but no association between SUD and AN restrictor subtype (ANR). Familial rates of alcoholism in those with BN was 40% (Hudson, Pope, Jonas & Yurgelun-Todd, 1983) and roughly 19% of those with BN had family members with a SUD (Bulik, 1987; Kassett et al., 1989).

The second meta-analysis (Gadalla & Piran, 2007) reviewed studies of the co-occurrence of ED and alcohol use disorders (AUD) in women, and included community, clinical, and school-based samples. This study supported previous findings suggesting no association between AUD and ANR but a strong association between AUD in women with ANBP, BN, binge eating disorder (BED), and Eating Disorder Not Otherwise Specified (EDNOS). More specifically, the lowest rates of alcohol *use* were seen in restricting anorexics (0–34%). Those with ANBP, had rates of alcohol *use* that followed the general population (20–45%) but their rates of *abuse and dependence* were much higher (12–39% vs. 3–6%) than those seen in the general population.

Not only do high rates of SUD exist among women with ED, women with SUD have high rates of disordered eating (Schuckit et al., 1996). In particular, 30–40% of women with AUD (Taylor, Peveler & Hibbert, 1993) and 16.3% of women with SUD (Blinder, Blinder & Samantha, 1998) report a history of an ED as compared to roughly 3.5% of women found in the general population (Hudson, Hiripi, Pope & Kessler, 2007).

THE COMPLEX RELATIONSHIP BETWEEN EATING DISORDERS AND SUBSTANCE ABUSE

The comorbidity of ED and SUD yields a complex clinical picture that requires a thorough understanding of both conditions. A common misconception around the use and abuse of substances in restricting anorexics involves the assumption that these patients would avoid drinking alcohol not only because of the "empty calories," but the fear that intoxication might lead to "loss of control." However, animal researchers have repeatedly found that food deprivation in rats and Rhesus monkeys leads to increased self-administration of cocaine, nicotine, amphetamines, alcohol, barbiturates, phencyclidine, and opioids (Carroll, France & Meisch, 1979). Logically, it is easy to accept that alcohol use would increase during starvation (e.g., caloric replacement), but these studies demonstrated increased self-administration of virtually any drug. Furthermore, in a multicenter study exploring SUD in AN women, Root and colleagues (2010) found that 25.9% of their sample reported lifetime drug use and 19.8% met criteria for lifetime history of AUD. Rates of *use* were highest among the purging AN group while the ANBN group had the highest percentage of *drug abuse and dependence.* Despite the fact that restricting anorexics demonstrated the lowest *use* among the AN subtypes, it is interesting to note that restricting anorexics were found to *abuse* substances, most frequently marijuana and hallucinogens.

Moreover, it is widely accepted that ED diagnostic categories are fluid and exist on a continuum (Fairburn, 2008). Many BN patients report a prior history of AN. In fact, up to 50% of AN patients will develop bulimic symptoms sometime during the course of their illness (Fairburn, 2008). It is not uncommon for individuals to move between diagnositic categories during the lifetime of their ED. Clearly, the assumption that restricting anorexics do not, or will not, develop a SUD is erroneous.

Another important issue to consider is that SUD can develop before, during, or after treatment for an ED. In a study conducted by Bulik, Sullivan, Joyce & Carter (1997), 28% of participants became alcohol dependent before the onset of their BN, 38% developed their BN and alcohol dependence at the same age, and 34% developed an alcohol problem after the onset of their BN.

Clinicians should also remember that when treating adolescent or young adult ED patients, they are nowhere near passing through the "age of risk" for the development of

SUD. In fact, rates of SUD rise in older patients (Kessler et al., 2005), with one study finding 50% of BN patients diagnosed with alcoholism by age 35 (Beary, Lacey & Merry, 1986).

Finally, in a 10-year prospective follow-up study of 95 AN patients, Strober, Freeman, Bower & Kigali (1995) found that adolescents who did not have a pre-existing diagnosis of a SUD at intake frequently developed one after discharge. By completion of the follow-up study, 50% of the ANBP patients and 12% of ANR patients had developed a new comorbid SUD.

In summary, there is a high co-occurrence of SUD in all ED subgroups. Clinicians should screen every ED patient at intake for SUD, not just those that engage in binge eating and/or purging. SUD can develop at any time so patients should be monitored throughout the course of treatment.

SIMILARITIES AND DIFFERENCES BETWEEN EATING DISORDERS AND SUBSTANCE USE DISORDERS

Numerous authors have outlined the similarities between ED and SUD (Bemis, 1985; Butterfield & LeClair, 1988). Some support the notion that ED (particularly BN, ANBP, and BED) and SUD have a shared or causal etiology (Avena, 2007; Brisman & Siegel, 1984; Marrazzi & Luby, 1986), while others challenge the notion that EDs are a variant of an "addictive disorder" (Fairburn, 1995; Wilson, 1993).

Similarities

ED and SUD are often long-term illnesses fraught with repeated unsuccessful attempts to recover and frequent relapses. Both disorders are potentially life-threatening illnesses that, when left untreated, have high mortality rates. Individuals with SUD or ED tend to deny the severity of their problem. Substance abusers often underreport the frequency of use and quantities consumed just as those with AN deny restricting and low weight even in the face of malnutrition and emaciation. It is common for individuals with AN to exaggerate the amount of food consumed, minimize their exercise routines, or augment their weight through devious means. Negative physical, social, and psychological consequences associated with SUDs or EDs are often minimized or denied. Like substance abusers, individuals with AN rarely independently seek medical or psychological intervention. Most patients are referred by a concerned coach, school counselor, or doctor, or brought into treatment by a family member or friend.

Just as substance abusers crave drugs/alcohol, ED individuals (particularly BED and BN patients) describe intense food cravings and uncontrollable compulsive consuming. A preoccupation with the use of the "substance" (i.e., food, drugs, alcohol), is frequently fueled by the desire to re-experience the mood-altering effects associated with engaging in the behaviors.

Another similarity between these two disorders is the high prevalence of borderline personality disorder (BPD) found in ED patients (particularly ANBP and BN) and substance abusers (Bulik et al., 1997). Holderness et al. (1994) reported that impulsivity was the key attribute that linked SUD with BN and ANBP. This underlying personality style lends itself to global dysregulation including cognitive dysfunction, problems with affect regulation, and interpersonal difficulties (Dennis & Sansone, 1991, 1997; see also Sansone and Sansone, Chapter 13).

A growing body of research suggests the presence of genetic or neurobiological similarities between individuals with SUD and ED (Jonas, 1990). Some suggest that these two disorders may be a different expression of the same "addictive trait" (Wilson, 1991). Neurotransmitters including endogenous opioids, dopamine, serotonin, and neuropeptide Y, appear to play a role in both ED and SUD (Blinder et al., 1998).

Finally, it has long been accepted that AUD and SUD are familial, and both have been found to be highly heritable ($h^2=0.80$) with only 20% of the variability attributed to environmental influences (Hicks, Krueger, Iaconco & Patrick, 2004). Likewise, studies of monozygotic twins found that heritability estimates for AN and BN ranged from 0.54 to 0.80 (Lilenfeld, Kaye & Strober, 1997) and 0.82 for latent binge-eating (Bulik, Sullivan & Kendler, 1998). Family transmission studies have also demonstrated a 3- to 5-fold increased rate of ED in relatives of probands, as compared with relatives of non-ED controls (Strober, Lampert, Morrell, Burroughs & Jacobs, 1990). Additionally, numerous studies have found a higher than chance incidence of first-degree relatives with alcohol and drug problems in ED probands (Kassett et al., 1989; Kaye et al., 1996).

Differences

Despite the behavioral and psychological similarities between ED and SUD and their higher than expected co-occurrence rates, there has yet to be convincing evidence that ED are in fact "addictions."

Closer examination of heritability studies suggest no evidence of familial *cross-transmission* between alcohol dependence and AN or BN (Schuckit et al., 1996). Two other studies have demonstrated that although these two disorders frequently co-occur, they do not share a common transmittable familial vulnerability (Kaye et al., 1996; Kendler et al., 1995).

In several ways, ED and SUD are unique disorders that are conceptualized and treated quite differently, both from a psychological and pharmacological perspective. Substance-dependence is considered to be a chronic disease that can only be arrested, not cured. This model suggests that the individual is "powerless" over their disease and will never completely recover. Abstinence is the course of action to sustain remission. Physical, emotional, and spiritual health can be achieved through life-long abstinence, continued attendance at AA/NA meetings, and adherence to the 12 steps (Overeaters Anonymous, 1990) and traditions.

On the other hand, ED are considered to be curable mental illnesses that require aggressive psychological intervention, medical management, and in some cases, psychotropic medications. Patients must relinquish "over-control" of their eating patterns and moderate food intake (i.e., give up dietary restraint). They are encouraged to repair dysfunctional relationships and replace faulty beliefs and maladaptive coping strategies with effective life management skills. Self-help programs are not considered an essential element for ED recovery and life-long participation is not required to prevent relapse.

Some ED programs classify EDs as "addictions" and utilize an abstinence model approach to recovery. The dilemma in applying this doctrine to ED is that food, which has been used to replace the word "alcohol" in Overeaters Anonymous (OA) literature, is necessary for survival and therefore not an appropriate substance for abstinence (Overeaters Anonymous,

1990). Furthermore, no compelling human studies suggest that ED recovery can be achieved through abstinence from particular foods (i.e., white flour and sugar) as OA implies. Finally, the diagnostic criteria for addiction, which includes both "tolerance and withdrawal," do not apply to individuals with ED.

COMMON SUBSTANCES OF ABUSE

A comprehensive review of ED and SUD would be incomplete without a discussion of the atypical substances often abused by the ED patient. Inquiry into the use of these substances among ED patients will help identify problem behaviors and inform treatment planning.

Over-the-Counter Medications

The availability and use of herbal and Over-The-Counter (OTC) preparations has increased exponentially. From 1990 to 1997, the use of herbal remedies increased by 380%, with approximately 42% of Americans reportedly using some form of these therapies (Eisenberg et al., 1998). Despite this increase, the Food and Drug Administration (FDA) provides no oversight of these preparations, thus leaving issues of safety and efficacy to chance. Along with increased use of these substances has come increased misuse.

Laxatives. Of the OTC drugs, laxatives are the most commonly abused substances by ED patients (Mitchell, Specker & Edmonson, 1997). Data from Steffen and colleagues (2007) found that 67% of bulimic subjects reported using laxatives at some point in their ED. A review of 73 studies found that 15% of those with BN abuse laxatives regularly (Neims, McNeill, Giles & Todd, 1995). The importance of these data is underscored by the suggestion that ED patients who abuse laxatives may represent a more pathologically complex subgroup. In particular, laxative abusers tend to be more impulsive across multiple domains, have a history of sexual abuse, and exhibit more personality psychopathology than do non-abusers (Favaro & Santtonastaso, 1998; Johnson, Tobin & Enright, 1989).

Eating disorder patients use laxatives in an attempt to lose weight, despite the fact that they are ineffective. Treatment of laxative abuse includes prompt discontinuation of the substance. ED patients often find this intolerable due to the resultant constipation, temporary weight gain, and fluid retention that can occur. Fluid retention may last for some time, prompting patients to use more laxatives as body image distortions are exacerbated. Patients may require more frequent care during this time, as triggers for relapse are generally significant (see also Tyson, Chapter 6 for further information on medical management issues in ED).

Diuretics. Diuretics, which reduce water retention by prompting the kidneys to excrete urine, are also widely used in the ED population. Estimates suggest that 3–5% of high-school and college students use diuretics. Lifetime rates among women with BN are as high as 31% (Roerig, Mitchell & Zwaan, 2003). When used according to instructions, OTC diuretics are generally mild and safe. However, ED patients can take several times the amount recommended leading to gastrointestinal upset, nausea, and even vomiting. By contrast, the abuse of prescription diuretics (i.e., furosemide or Lasix®) can have serious medical consequences including palpitations, hypokalemia, neuropathy, and cardiac conduction defects (Mitchell et al., 1997). Diuretics may be stopped abruptly, although fluid retention, edema, and transient weight gain may occur, thus triggering urges to resume abuse.

Diet pills. Mitchell, Hatsukami, Eckert & Pyle (1985) found that 42% of ED patients used diet pills to regulate weight. Six years later, these same researchers reported a 12% further increase in diet pill use among BN patients (Mitchell, Pyle & Eckert, 1991). This increase may be due to the proliferation of new products and the ability to order these substances over the Internet. Diet pills generally contain agents that act like amphetamines on the sympathetic nervous system. Some also contain laxatives. Caffeine is commonly found in diet pill preparations and may be labeled as kola nut, guarana seed, or green tea. The daily use of 1,000 mg of caffeine has been linked with restlessness, insomnia, anxiety, tremor, and even tachycardia. With continued use, tolerance builds and when discontinued, withdrawal effects including headaches, nausea, and fatigue can occur. Diet pills may be stopped abruptly with patient education regarding potential withdrawal effects.

Syrup of Ipecac. Some ED patients also abuse emetine, which is the active ingredient in Ipecac. The intended use of Ipecac is to promote vomiting in the case of accidental poisoning, although it is no longer recommended for this purpose (American Pediatrics Association, 2003; Silber, Maine & McGilley, 2008). In a study of 851 ED outpatients, 8.8% of patients with BN reported chronic abuse (Greenfield, Mickley, Quinlan & Roloff, 1993). When looking at lifetime use, the percentage of BN patients who use Ipecac rises to 18% (Steffen, Mitchell & Roerig, 2007). Due to the toxicity and lethality of this substance, clinicians are advised to routinely screen for its use. Once detected, Ipecac use should be immediately discontinued and medical management of these patients is strongly encouraged.

Orlistat and Alli®. Orlistat and Alli® are intestinal lipase inhibitors approved for the treatment of obesity. They work by disrupting the absorption of fat in the small intestines. A few case studies have documented Orlistat abuse among patients with ED (Cochrane & Malcolm, 2002; Fernandez-Aranda et al., 2001). In particular, Orlistat has been used to promote purging by BN patients. Orlistat, now sold over-the-counter as Alli® at a reduced strength, is readily available, raising concerns of a potential increase in abuse. In addition to inhibiting the digestion of fat, these substances reduce the absorption of fat-soluble vitamins and are not benign.

Prescription Medications and Performance Enhancing Agents

Steroids. Approximately 13% of female athletes have a diagnosable ED, and of those athletes, 3.26% use anabolic steroids (Johnson, Powers & Dick, 1999). Abuse often persists to prevent significant withdrawal symptoms including mood swings, depression, fatigue, restlessness, loss of appetite, insomnia, reduced libido, and "cravings" for the drug.

Complicating matters, steroid abusers may develop additional drug dependencies in an effort to self-medicate the negative effects of steroid abuse (Cochrane & Malcolm, 2002). Anecdotal reports suggest tapering patients off steroids and treating the symptoms. Patient education regarding the effects of steroid abuse, as well as what to expect as they withdraw from the drug, is essential. During the weaning period, evaluating for the presence of depression and suicidal ideation (National Institute on Drug Abuse, 2006), as well as increased body image disturbance, is advised.

Insulin. Prevalence rates of diabetes among ED populations have been inconsistent (Jones, Lawson, Daneman, Olmsted & Rodin, 2000). A meta-analysis of diabetes and ED found that BN is more common in female patients with Type I diabetes compared with their non-diabetic peers (Nielsen, 2002). Although it is not known whether diabetes increases a patient's risk for the development of an ED, what is well known is that the co-occurrence of these disorders can negatively affect the course and treatment of an ED.

To avoid weight gain, these patients may underutilize their insulin in an effort to purge unwanted calories. Among adolescent females with Type I diabetes, 45–80% exhibit binge eating behaviors and 12–40% have deliberately induced glycosuria by reducing or omitting insulin dosages to promote weight loss (Jones et al., 2000). If insulin abuse is suspected, clinicians are advised to work closely with the patient's endocrinologist, who should be considered for inclusion on the patient's treatment team.

Ritalin® and Adderal®. There are significant rates of comorbidity between attention deficit hyperactivity disorder (ADHD) and both SUD and ED. Nearly 33% of adults with ADHD have histories of AUD and 20% have a history of SUD (Waid, LaRowe, Anton & Johnson, 2004). In a large study of adolescent girls, Biederman and colleagues (2007) found that those with ADHD were 5.6 times more likely to develop BN and 2.7 times more likely to develop AN than their non-ADHD peers. The abuse of amphetamines appears to be most common among ANBP patients (Root et al., 2010). Ritalin abuse can result in toxicity, with large doses leading to psychosis, seizures, and cardiovascular accidents.

In summary, it is imperative that ED clinicians initially and regularly assess for the abuse of alcohol and drugs, including OTC preparations, steroids, stimulants, insulin, and emetics, as SUD may arise at any point in the course of treatment.

TREATMENT

It is often difficult to assess the extent of comorbid conditions in those with a SUD. More specifically, SUD can mimic the symptoms of other mental disorders. For example, depression commonly occurs during chronic use or withdrawal from many substances, and stimulant abuse can resemble anxiety or even mania. Without a thorough assessment, including information on the history and chronology of symptoms, it can be difficult to distinguish between disorders.

As part of the assessment process, we routinely explore the adaptive function of our patients ED and SUD behaviors. Appreciating the social, psychological, physical, or interpersonal problems that these disorders manage or "solve," and identifying the secondary gains that result from engaging in these behaviors, can provide the clinician with insight into the precipitants and factors that maintain their patients' disorder. Understanding the adaptive function of these behaviors can also inform case conceptualization, team constellation, treatment goals, and the selection of appropriate levels of care, therapeutic modalities, and approach.

Empirically Supported Interventions

Empirically validated, integrated treatment approaches do not currently exist and no known clinical trials of ED treatment for SUD populations have been conducted. Gaps between what research has identified as efficacious treatment and what is actually done in clinical practice are significant. A 1998 report commissioned by the Institute of Medicine found a 17-year gap between the publication of research and its actual impact on treatment delivery in clinical settings (Lamb, Greenlick & McCarty, 1998). Moreover, a large discrepancy in funding for research exists where roughly 99% of funds are dedicated to intervention research, with only 1% available for the implementation and dissemination of research results (Fixsen, Naoom, Blase, Friedman & Wallace, 2005). To compound this problem, research data often get "lost in translation," ultimately being ineffectively interpreted or delivered in clinical settings. For example, research indicates that addiction is a chronic, relapsing disorder that can be managed, but not cured (McLellan et al., 1997). Yet in clinical practice, addiction is often treated as if it were an acute disorder requiring crisis intervention, and relapse is seen as a failure of treatment.

In light of this gap, it is not surprising that few treatment centers undertake the evaluation and treatment of these co-occurring disorders. In fact, a study by Gordon et al. (2008) of 351 publicly funded addiction treatment programs, found that while half of the programs screened for ED, only 29% admitted complex ED cases. Research also suggests that a mere 21.7% of private SA centers offer ED treatment (Roman & Johnson, 2004).

Evidence-Based Therapies

This section will focus on EBT for SUD and ED, respectively. It is intended to inform the clinician rather than provide prescriptive treatment recommendations. Patients must be treated as individuals, rather than ascribing to a "one size fits all" viewpoint. Once a thorough assessment has been conducted, the information can be utilized collaboratively with the patient to formulate a treatment plan. Components of an individualized treatment plan consider both the EBT available as well as the specific modes (individual, group, family, medical, psychiatric, and nutritional) and intensity (outpatient, intensive outpatient, day treatment, residential, or inpatient) of treatment needed for a particular individual. Other considerations include the patient's psychological issues, cognitive style and level of functioning, as well as age, gender, and social support system.

Cognitive Behavioral Therapies. Historically, the foundation of SUD treatment has been a 12-step, psychosocial, abstinence model delivered in group and residential settings. However, current emphasis is on the study and utilization of EBT. Cognitive Behavioral Therapy (CBT) is one of the few EBTs that have been found effective in the treatment of both SUD and ED. In the treatment of SUD, CBT interventions focus on enhancing self-esteem and coping skills.

Within the ED community, CBT is the treatment of choice for BN and BED (National Institute for Clinical Excellence, 2004). The focus of CBT with this population is on replacing ED behaviors such as dietary restraint, or binge eating, with normalized eating patterns. At the same time, CBT identifies the patient's dysfunctional thoughts regarding weight, shape, and appearance and replaces them with more reality-based thinking.

Interpersonal Psychotherapy. The use of Interpersonal Psychotherapy (IPT) in the treatment of SUD is limited. A clinical trial using IPT with methadone-maintained opiate addicts found significant clinical improvements, but design issues limit the generalizability of this study (Roundville, Glazer, Wilber & Weissman, 1983). Another study using IPT for depression in substance abusing female prisoners suggested that group IPT significantly decreased their depressive symptoms and improved their perceived social support, although the impact on sobriety was mixed (Johnson & Zlotnick, 2000).

Contrary to the limited data on IPT for SUD, a fair amount of data suggests that IPT is an efficacious treatment for both BN and BED. According to IPT theory, the resolution of interpersonal difficulties will result in the resolution of disordered eating. Individual IPT was found to be as effective as CBT in the treatment of BN at post-treatment and at 5-year follow-up (Fairburn, 1995). Wilfley and colleagues (2002) compared CBT group therapy to IPT group therapy for overweight individuals with BED. Both treatments were found to have intial and long-term efficacy in reducing ED behaviors and the psychiatric symptoms associated with BED.

Motivational Enhancement Therapy. Motivational enhancement therapy (MET) grew out of the need to reduce early attrition rates in SUD treatment (DiClemente & Prochaska, 1998). This model provides a framework for conceptualizing the strength of a patient's motivation for change. Research suggests that not only are motivational enhancement techniques effective; they are also efficient. Two large, multi-site trials examining treatments for alcoholism demonstrated that 3–4 sessions of MET were comparable in efficacy to 12 sessions of CBT (Project MATCH Research Group, 1997) or 8 sessions of social behavior and network therapy (United Kingdom Alcohol Treatment Trial Research Team, 2005).

Motivational enhancement therapy with ED patients is similar in its application to that seen in the SA literature. Geller and Drab (1999) have shown that ED patients' readiness for change impacts their engagement in treatment, magnitude of behavior change, and relapse risk. Additional studies are needed before MET can be considered an EBT for use with ED patients.

Couples and Family Therapy Applications. Behavioral couples and family therapy has been shown to be effective with a SA population (O'Farrell & Fals-Stewart, 2001). This approach recognizes how addiction can have a destructive impact on the family structure (Steinglass, Bennett, Wolin & Reiss, 1987). Therapy begins with a functional analysis of the patient's SA behavior and an assessment of marital and familial relationships. With regard to the marriage, interventions focus on improving communication, increasing quality time, and developing behavioral change agreements. Behavioral couples and family therapy is currently being studied for use with adult AN couples.

Family based treatment and ego-oriented individual therapy are the only two EBTs for adolescent AN (Fitzpatrick, Moye, Hoste, Lock & Le Grange, 2009; Lock, Le Grange, Agras & Dare, 2001; Robin et al., 1999). Both have been found to be effective in restoring weight and eliminating eating disorder psychopathology.

Psychopharmacological Interventions. Pharmacological treatments for SUD with FDA approval include disulfiram (Antabuse®) for alcohol dependence, methadone for heroin dependency, naltrexone (Revia®) for alcohol and opioid dependence, buprenorphine (Suboxone®) for opiate detoxification and treatment, and acamprosate (Campral®) for alcohol dependence (Johnson, Roman, Ducharme & Knudsen, 2005). Other medications that have been used to treat addiction include serotonergic and dopaminergic agents. The neurotransmitter serotonin has been implicated in the modulation of both mood and impulse. Researchers have hypothesized that low levels of serotonin in alcohol-dependent individuals may predispose them to alcohol abuse as well as maintain abuse behaviors (Kranzler & Anton, 1994). Selective serotonin reuptake inhibitors (SSRIs) such as fluoxetine (Prozac®) and sertraline (Zoloft®) may offer short-term reductions in alcohol consumption among heavy users. Sertraline and fluoxetine were also found to improve comorbid mood disorders (Cornelius et al., 1997). Ondansetron (Zofran®) and topiramate (Topomax®) are currently under investigation for alcohol dependence and stimulant abuse but larger multi-site trials are needed.

To date, the only FDA approved medication for the treatment of an ED is the SSRIs fluoxetine. Used in the treatment of BN, dosages are often higher (60 mg per day) than those that are effective for the treatment of depression (Romano, Halmi, Sarkar, Koke & Lee, 2002). See Levine and Levine, Chapter 7, for a thorough review of psychopharmacological approaches to ED treatment.

Other Interventions

Historically, 12-step models have been rooted in Alcoholics Anonymous (AA), which ascribes to the disease model of addiction. This model views addiction as a physical, emotional, and spiritual disease that can be arrested, but not cured. Recovery is viewed as a lifelong process that involves working the 12 steps of AA or NA (narcotics anonymous) and abstaining from the use of alcohol/drugs. This approach is the most common intervention utilized in the SUD field.

Dialectical behavioral therapy (DBT), a form of CBT, has also been successfully utilized in the treatment of addictions. In DBT, the focus is on affect dysregulation, and it is a treatment strategy that views SA as a maladaptive method used to manage, control, or change adverse emotional states. The goal is to replace SA behaviors with more adaptive affect regulation strategies, or as Linehan and colleagues (1999) suggest, "replacing pills with skills" (p. 282).

Dialectical behavioral therapy has also been found to be effective in the treatment of BN and BED (Safer, Telch & Chen, 2009). This approach views disordered eating as a problem of emotional dysregulation and teaches patients to more effectively cope with negative affect using mindfulness and emotion regulation skills rather than resorting to binge eating or purging.

Location of Treatment

Matching the patient with the correct level of care begins with the initial evaluation and continues throughout treatment. Levels of care typically include outpatient, intensive outpatient and partial hospitalization, residential, and medically managed inpatient hospitalization. In considering the treatment environment, it is important to evaluate the patient's medical and psychological status, their motivation for change, their social support network and their access to care. Table 14.1 provides an overview of useful treatment placement indicators.

Clinical decisions should be made in collaboration with the patient and with consideration for the probable costs, benefits, and available resources. Although it is the clinician's responsibility to determine the appropriate level of care, the active involvement of an informed patient is generally crucial to success. Ideally, patients are treated in the least restrictive environment, while maintaining their safety and optimizing their chances for a successful treatment outcome.

Integrated or Sequential Treatment

Very few treatment centers provide specialized care for the ED patient with a SUD. Unfortunately, data on the superiority of integrated treatment versus sequential treatment are limited. In an integrated approach, the same providers treat comorbidities concurrently. Sequential treatment focuses on the most acute disorder first, and is conducted with multiple

TABLE 14.1 Level of care indicators

Level of Care	Medical	Emotional and Behavioral	Readiness to Change	Environmental	Access to Care
Level I outpatient	Medically stable >85% Ideal body weight	No acute suicidal ideation	Fair to Good: Some openness and ability to change	Reasonable environmental support and structure for change	Available close to home
Level II Intensive outpatient	No serious risk for major withdrawal or seizures >80% Ideal body weight	No acute suicidal ideation	Fair: Some reluctance to change and/or limited ability to maintain change but cooperative	Some environmental support for change	Available close to home
Level III Medically or clinically monitored intensive residential	Risk for withdrawal Some level of acute or chronic medical problems <85% ideal body weight	Acute psychiatric problems Preoccupied with intrusive repetitive thoughts related to ED and/or SUD Requires supervision during and after meals	Poor to Fair: Resistance to change and/or limited ability to maintain change Somewhat cooperative within a highly structured treatment environment	Limited environmental support for change Significant family conflict, lack of structure, lack of support system; primary support is engaged in ED or SUD	May not be available close to home
Level IV Medically managed intensive inpatient hospitalization	Serious risk for major withdrawal (i.e., seizures) and/or acute or chronic medical problems: – HR <40 bpm – BP <90/60 mmHg – Electrolyte imbalance – Unstable core body temperature (i.e., <97.0° F) – Dehydration – Hepatic, renal, or cardiovascular complications <85% Ideal body weight and/or acute weight loss with food refusal	Acute or chronic psychiatric problems that could interfere with treatment Specific suicide plan with intent and plan Preoccupied with intrusive, repetitive thoughts Requires supervision during and after meals including if on TPN or nasogastric feeding	Very Poor to Poor: Resistant to change Uncooperative with treatment	Limited environmental support for change Significant family conflict, lack of structure, lack of support system; primary support is engaged in ED or SUD behaviors	May not be available close to home

Adapted and modified from American Psychiatric Association (2006) and Mee-Lee (2001).

providers or in different locations. Differences in theoretical orientations, staff training, and treatment protocols can make continuity of care difficult. Although an integrated approach holds promise for the treatment of ED and SUD, at the time of writing there are no treatment studies to suggest the best method for accomplishing this goal. However, data exist in the SA literature to suggest that, when comorbid diagnoses are treated concurrently and integrated on-site, treatment retention and outcome improve (Weisner, Mertens, Tam & Moore, 2001).

CONCLUSIONS AND FUTURE DIRECTIONS

The primary goal in this chapter was to provide a comprehensive review of the research literature on this comorbid population and translate these findings into useful treatment strategies for the practicing clinician. In the absence of EBT for patients with comorbid ED and SUD, we have extrapolated what is known individually about each disorder and attempted to provide guidelines for the selection of an appropriate treatment environment, therapeutic approach, and team composition. Although research from both disciplines holds promise for CBT as an effective initial intervention with this comorbid population, further investigation is necessary.

It appears that no formal connections link the ED and SUD professional communities. Philosophical differences and disagreements on treatment approaches, along with substantial delays in the dissemination of research findings into clinical practice, have interfered with our ability to effectively serve this comorbid group. Cross-training and research collaboration between these two fields is necessary in order to identify which EBT is most effective when applied within each level of care. Incentive programs for the adoption of EBT are on the horizon. In some states, credentialing organizations and insurance companies are moving toward requiring the use of EBT. It will no longer be viable to treat our patients based on a "treatment as usual" approach. Change, which we encourage in our patients, is also necessary in the development and delivery of programs for a more comprehensive understanding and effective treatment of these comorbid conditions.

References

American Pediatrics Association (2003). Policy statement: Poison treatment in the home. *Pediatrics, 112*(5), 1182–1185.

American Psychiatric Association (2006). *Practice guidelines for the treatment of psychiatric disorders*. Washington, DC: Author.

American Psychiatric Association (1994). *Diagnostic and statistical manual of mental disorders* (4th ed.). Washington, DC: Author.

Avena, N. (2007). Examining the addictive-like properties of binge eating using an animal model of sugar dependence. *Experimental and Clinical Psychopharmacology, 15*, 481–491.

Beary, M., Lacey, J., & Merry, J. (1986). Alcoholism and eating disorders in women of fertile age. *British Journal of Addiction, 81*, 685–689.

Bemis, K. (1985). "Abstinence" and "nonabstinence" models for the treatment of bulimia. *International Journal of Eating Disorders, 4*, 407–437.

Biederman, J., Ball, S., Monuteaus, M., Surman, L., Johnson, J., & Zetlin, S. (2007). Are girls with ADHD at risk for eating disorders? Results from a controlled, five-year prospective study. *Journal of Developmental and Behavioral Pediatrics, 28*, 302–307.

Blinder, B., Blinder, M., & Samantha, V. (1998). Eating disorders and addiction. *Psychiatric Times, 15*, 30−34.

Brisman, J., & Siegel, M. (1984). Bulimia and alcoholism: Two sides of the same coin? *Journal of Substance Abuse Treatment, 1*, 113−118.

Bulik, C. (1987). Drug and alcohol abuse by bulimic women and their families. *American Journal of Psychiatry, 144*, 1604−1606.

Bulik, C., Sullivan, P., Joyce, P., & Carter, F. (1997). Lifetime comorbidity of alcohol dependence in women with bulimia nervosa. *Addictive Behaviors, 22*, 437−446.

Bulik, C., Sullivan, P., & Kendler, K. (1998). Heritability of binge-eating and broadly defined bulimia nervosa. *Biological Psychiatry, 44*, 1210−1218.

Bulik, C., Thornton, L., Pinheiro, K., Plotnicov, K., Klump, K., Brandt, H.,…Kaye, W. H. (2008). Suicide attempts in anorexia nervosa. *Journal of Psychosomatic Medicine, 70*(3), 378−383.

Butterfield, P., & LeClair, S. (1988). Cognitive characteristics of bulimia and drug abusing women. *Addictive Behaviors, 13*, 131−138.

Califano, J. (2007). *High society: How substance abuse ravages america and what to do about It*. New York, NY: Perseus Publishing.

Carroll, M., France, C., & Meisch, R. (1979). Food deprivation increases oral and intravenous drug intake in rats. *Science, 205*, 319−321.

Cochrane, C., & Malcolm, R. (2002). Case report of abuse of orlistat. *Eating Behaviors, 3*(2), 167−169.

Cornelius, J., Salloum, I., Ehler, J., Jarrett, P., Cornelius, M., & Perel, J. (1997). Fluoxetine in depressed alcoholics: A double-blind, placebo-controlled trial. *Archives of General Psychiatry, 54*, 700−705.

Dennis, A., & Sansone, R. (1991). The clinical stages of treatment for eating disorder patients with borderline personality disorder. In C. Johnson (Ed.), *Psychodynamic treatment of anorexia nervosa and bulimia* (pp. 126−164). New York, NY: Guilford Press.

Dennis, A., & Sansone, R. (1997). Treatment of patients with personality disorders. In D. Garner, & P. Garfinkel (Eds.), *Handbook of treatment for eating disorders* (2nd ed., pp. 437−449). New York, NY: Guilford Press.

DiClemente, C., & Prochaska, J. (1998). Toward a comprehensive, transtheoretical model of change: Stages of change and addictive behaviors. In W. Miller, & N. Heather (Eds.), *Treating addictive behaviors* (2nd ed., pp. 3−24). New York, NY: Plenum.

Eisenberg, D., Davis, R., Ettner, S., Appel, S., Wilkey, S., Van Rompay, M., & Kessler, R. C. (1998). Trends in alternative medicine use in the United States, 1990−1997: Results of a follow-up national survey. *Journal of the American Medical Association, 280*, 1569−1575.

Fairburn, C. (1995). *Overcoming binge eating*. New York, NY: Guilford Press.

Fairburn, C. (2008). *Cognitive behavior therapy and eating disorders*. New York, NY: Guilford Press.

Favaro, A., & Santtonastaso, P. (1998). Impulsive and compulsive self-injurious behavior in bulimia nervosa: Prevalence and psychological correlates. *Journal of Nervous and Mental Diseases, 186*, 157−165.

Fernandez-Aranda, F., Amor, A., Jimenez-Murcia, S., Gimenez-Martinez, L., Turon-Gil, V., & Vallejo-Ruiloba, J. (2001). Bulimia nervosa and misuse of Orlistat: Two case reports. *International Journal of Eating Disorders, 30*(4), 458−461.

Fitzpatrick, K., Moye, A., Hoste, R., Lock, J., & Le Grange, D. (2009). Adolescent focused psychotherapy for anorexia nervosa. *Journal of Contemporary Psychotherapy, 10*, 9123−9127.

Fixsen, D., Naoom, S., Blase, K., Friedman, R., & Wallace, F. (2005). *Implementation research: A synthesis of the literature*. Tampa, FL: The National Implementation Research Network.

Gadalla, T., & Piran, N. (2007). Co-occurrence of eating disorders and alcohol use disorders in women: A meta analysis. *Archives of Women's Mental Health, 10*, 133−140.

Geller, J., & Drab, D. (1999). The readiness and motivation interview: A symptom specific measure of readiness for change in the eating disorders. *European Eating Disorders Review, 7*, 259−278.

Gordon, S., Johnson, J., Greenfield, S., Cohen, L., Killeen, T., & Roman, P. (2008). Assessment and treatment of co-occurring eating disorders in publicly funded addiction treatment programs. *Psychiatric Services, 59*, 1056−1059.

Greenfield, D., Mickley, D., Quinlan, D., & Roloff, P. (1993). Ipecac abuse in a sample of eating disordered outpatients. *International Journal of Eating Disorders, 13*, 411−414.

Harris, E., & Barraclough, B. (1997). Suicide as an outcome for mental disorders: A meta-analysis. *British Journal of Psychiatry, 170*, 205−228.

Hicks, B., Krueger, R., Iaconco, W., & Patrick, C. (2004). Family transmission and heritability of externalizing disorders. *Archives of General Psychiatry, 61*, 922−928.

Holderness, C., Brooks-Gunn, J., & Warren, M. (1994). Co-morbidity of eating disorders and substance abuse: Review of the literature. *International Journal of Eating Disorders, 16*(1), 1–34.

Hudson, J., Hiripi, E., Pope, H., & Kessler, R. (2007). The prevalence and correlates of eating disorders in the National Comorbidity Survey Replication. *Biological Psychiatry, 61*, 348–358.

Hudson, J., Pope, H., Jonas, J., & Yurgelun-Todd, D. (1983). Phenomenologic relationship of eating disorders to major affective disorders. *Psychiatry Research, 9*, 345–354.

Johnson, C., Powers, P., & Dick, R. (1999). Athletes and eating disorders: The National Collegiate Athletic Association Study. *International Journal of Eating Disorders, 26*, 179–188.

Johnson, J., Roman, P., Ducharme, L., & Knudsen, H. (2005). *Evidence-based care in substance abuse treatment: Comparing the public and private sectors*. Athens, GA: University of Georgia, Institute for Behavioral Research.

Johnson, C., Tobin, D., & Enright, A. (1989). Prevalence and clinical characteristics of borderline patients in an eating disordered population. *Journal of Clinical Psychiatry, 50*, 133–138.

Johnson, J., & Zlotnick, C. (2000). A pilot study of group interpersonal psychotherapy for depression in substance-abusing female prisoners. *Journal of Substance Abuse Treatment, 34*, 371–377.

Jonas, J. (1990). Do substance-abuse, including alcoholism, and bulimia covary? In L. Reid (Ed.), *Opioids, bulimia, and alcohol abuse and alcoholism* (pp. 247–258) New York, NY: Springer-Verlag.

Jones, J., Lawson, M., Daneman, D., Olmsted, M., & Rodin, G. (2000). Eating disorders in adolescent females with and without type 1 diabetes: Cross-sectional study. *British Medical Journal, 320*, 1563–1566.

Kassett, J., Gershon, E., Maxwell, M., Guroff, J., Kazuba, D., Smith, A. L., … Jimerson, D. C. (1989). Psychiatric disorders in the first-degree relatives of probands with bulimia nervosa. *American Journal of Psychiatry, 146*, 1468–1471.

Kaye, W. H., Lilenfeld, L., Plotnicov, K., Merikangas, K., Nagy, L., Strober, M., … Greeno, C. (1996). Bulimia nervosa and substance dependence: Association and family transmission. *Alcoholism: Clinical and Experimental Research, 20*, 878–881.

Kendler, K., Walters, E., Neale, M., Kessler, R., Heath, A., & Eves, L. (1995). The structure of the genetic and environmental risk factors for six major psychiatric disorders in women: Phobia, generalized anxiety disorder, panic disorder, bulimia, major depression, and alcoholism. *Archives of General Psychiatry, 52*, 374–383.

Kessler, R., Berglund, P., Demler, O., Jin, R., Merklangas, K., & Walters, E. (2005). Lifetime prevalence and age of onset distributions of DSM-IV disorders in the national comorbidity survey replication. *Archives of General Psychiatry, 62*, 593–602.

Kessler, R., Borges, G., & Walters, E. (1999). Prevalence of and risk factors for lifetime suicide attempts in the National Comorbidity Survey. *Archives of General Psychiatry, 56*, 617–626.

Kranzler, H., & Anton, R. (1994). Implications of recent neuropsychopharmacologic research for understanding the etiology and development of alcoholism. *Journal of Consulting and Clinical Psychology, 62*, 1116–1126.

Kuehn, B. (2007). Opioid prescriptions soar: Increase in legitimate use as well as abuse. *Journal of the American Medical Association, 297*, 249–251.

Lamb, S., Greenlick, M., & McCarty, D. (Eds.), (1998). *Bridging the gap between research and practice: Forging partnerships with community-based drug and alcohol treatment*. Washington, DC: Institute of Medicine, National Academy Press.

Lilenfeld, L., Kaye, W., & Strober, M. (1997). Genetics and family studies of anorexia nervosa and bulimia nervosa. In D. C. Jimerson, & W. H. Kaye (Eds.), *Bailliere's clinical psychiatry, Vol. 3. Eating disorders* (pp. 177–197). London, UK: Bailliere Tindall.

Linehan, M., Schmidt, H., III, Dimeff, L., Craft, J., Kanter, J., & Comtois, K. (1999). Dialectical behavior therapy for patients with borderline personality disorders and drug-dependence. *The American Journal of Addictions, 8*, 279–292.

Lock, J., Le Grange, D., Agras, W., & Dare, C. (2001). *Treatment manual for anorexia nervosa. A family-based approach*. New York, NY: Guilford Press.

Manchikani, L. (2006). Prescription drug abuse: What is being done to address this new drug epidemic? Testimony before the Subcommittee on Criminal Justice, Drug Policy and Human Resources. *Pain Physicians, 9*, 287–321.

Marrazzi, M., & Luby, E. (1986). An auto-addiction opioid model of chronic anorexia nervosa. *International Journal of Eating Disorders, 5*, 191–208.

McLellan, A., Woody, G., Metzger, D., McKay, J., Durrel, J., Alterman, A. I., & O'Brien, C. P. (1997). Evaluating the effectiveness of addiction treatments: Reasonable expectations, appropriate comparisons. In J. Egertson, D. Fox, & A. Leshner (Eds.), *Treating drug abusers effectively* (pp. 7–40). Malden, MA: Blackwell.

Mee-Lee, D. (2001). *ASAM PPC-2R patient placement criteria for the treatment of substance-related disorders* (2nd ed., Rev.). Chevy Chase, MD: American Society of Addiction Medicine.

Mitchell, J., Hatsukami, D., Eckert, E., & Pyle, R. (1985). Characteristics of 275 patients with bulimia. *American Journal of Psychiatry, 142*, 482—485.

Mitchell, J., Pyle, R., & Eckert, E. (1991). Diet pill usage in patients with bulimia nervosa. *International Journal of Eating Disorders, 10*, 233—237.

Mitchell, J., Specker, S., & Edmonson, K. (1997). Management of substance abuse and dependence. In D. Garner, & P. Garfinkle (Eds.), *Handbook of treatment for eating disorders* (pp. 415—423). New York, NY: Guilford Press.

National Institute for Clinical Excellence (2004). NICE. Eating disorders: Core interventions in the treatment and management of anorexia nervosa, bulimia nervosa and related eating disorders. London: NICE Clinical Guideline No. 9. Retrieved November 30, 2009, from http://www.nice.org.uk/CG009NICEguideline

National Institute on Drug Abuse. (2006). *Research report series: Anabolic steroid abuse*. Washington, DC: National Institute of Health.

Neims, D., McNeill, J., Giles, T., & Todd, F. (1995). Incidence of laxative abuse in community and bulimic populations: A descriptive review. *International Journal of Eating Disorders, 17*, 211—228.

Nielsen, S. (2002). Eating disorders in females with type I diabetes: An update of a meta-analysis. *European Eating Disorders Review, 10*(4), 241—254.

O'Farrell, T., & Fals-Stewart, W. (2001). Family involved alcoholism treatment: An update. In M. Galanter (Ed.), *Recent developments in alcoholism, Vol. 15. Services research in the era of managed care* (pp. 329—356). New York, NY: Plenum.

Overeaters Anonymous (1990). *The twelve-steps of Overeaters Anonymous*. Los Angeles: Author.

Papadopoulos, F., Ekbom, A., Brandy, L., & Eskelius, L. (2009). Excess mortality, causes of death and prognostic factors in anorexia nervosa. *The British Journal of Psychiatry, 194*, 10—17.

Project MATCH Research Group. (1997). Matching alcoholism treatment to client heterogeneity: Project MATCH post-treatment drinking outcomes. *Journal of the Study of Acoholism, 58*, 7—29.

Robin, A., Siegel, P., Moye, A., Gilroy, M., Dennis, A., & Sikand, A. (1999). A controlled comparison of family versus individual therapy for adolescents with anorexia nervosa. *Journal of the American Academy of Child and Adolescent Psychiatry, 38*, 1482—1489.

Roerig, J., Mitchell, J., & Zwaan, M. (2003). The eating disorders medicine cabinet revisited: A clinician's guide to appetite suppressants and diuretics. *International Journal of Eating Disorders, 33*, 443—457.

Roman, P., & Johnson, J. (2004). *National Treatment Center Study Summary Report: Private treatment centers*. Athens, GA: University of Georgia, Institute for Behavioral Research.

Romano, S., Halmi, K., Sarkar, N., Koke, S., & Lee, J. (2002). A placebo-controlled study of fluoxetine in continued treatment of bulimia nervosa after successful acute fluoxetine treatment. *American Journal of Psychiatry, 159*, 96—102.

Root, T., Poyastro Pinheiro, A., Thornton, L., Strober, M., Fernandez-Aranda, F., Brandt, H., ...Bulik, C. M. (2010). Substance use disorders in women with anorexia nervosa. *International Journal of Eating Disorders, 43*, 14—21.

Roundville, B., Glazer, W., Wilber, C., & Weissman, M. (1983). Short-term interpersonal psychotherapy in methadone-maintained opiate addicts. *Archives of General Psychiatry, 40*, 629—636.

Safer, D., Telch, C., & Chen, E. (2009). *Dialectical behavior therapy for binge eating and bulimia*. New York, NY: Guilford Press.

Schuckit, M., Tipp, J., Anthenelli, R., Bucholz, K., Hesselbrock, V., & Nurnberger, J. (1996). Anorexia nervosa and bulimia nervosa in alcohol-dependent men and women and their relatives. *American Journal of Psychiatry, 153*, 74—82.

Silber, T., Maine, M., McGilley, B. (2008). AED Position Statement. September 30, Over-the-counter status of Ipecac should be withdrawn. Retrieved November 30, 2009, from http://www.aedweb.org/policy/ipecac.cfm

Steffen, K., Mitchell, J., & Roerig, J. (2007). The eating disorders medicine cabinet revisited: A clinician's guide to Ipecac and laxatives. *International Journal of Eating Disorders, 40*, 360—368.

Steinglass, P., Bennett, L., Wolin, S., & Reiss, D. (1987). *The alcoholic family*. New York, NY: Basic Books.

Strober, M., Lampert, C., Morrell, W., Burroughs, J., & Jacobs, C. (1990). A controlled family study of anorexia nervosa: Evidence of familial aggregation and lack of shared transmission with affective disorders. *International Journal of Eating Disorders, 9*, 239—253.

Strober, M., Freeman, R., Bower, S., & Kigali, J. (1995). Binge eating in anorexia nervosa predicts later onset of substance use disorder: A ten-year prospective, longitudinal follow-up of 95 adolescents. *Journal of Youth and Adolescence, 25*, 519—532.

Substance Abuse and Mental Health Services Administration. Office of Applied Studies (2009). Results from the 2008 National Survey on Drug Use and Health: National Findings (NSDUH Series H-36, HHS Publication No. SMA 09-4434). Rockville, MD. Retrieved from http://www.oas.samhsa.gov/nsduh/2k8nsduh/2k8Results.cfm

Taylor, A., Peveler, R., & Hibbert, G. (1993). Eating disorders among women receiving treatment for an alcohol problem. *International Journal of Eating Disorders, 14*, 147–151.

The National Center on Addiction and Substance Abuse (2003). *Food for thought: Substance abuse and eating disorders*. New York, NY: Columbia University, The National Center on Addiction and Substance Abuse.

United Kingdom Alcohol Treatment Trial Research Team (2005). Effectiveness of treatment for alcohol problems: Findings of the Randomized UK Alcohol Treatment trial. *British Medical Journal, 331*, 541.

Waid, L., LaRowe, S., Anton, R., & Johnson, D. (2004). Attention deficit hyperactivity disorder and substance abuse. In H. Kranzler, & J. Tinsley (Eds.), *Dual diagnosis and psychiatric treatment: Substance abuse and comorbid disorders* (2nd ed., pp. 349–386). New York, NY: Marcell Dekker.

Weisner, C., Mertens, J., Tam, T., & Moore, C. (2001). Factors affecting the initiation of substance abuse treatment in managed care. *Addiction, 96*(5), 705–716.

Wilfley, D., Welch, R., Stein, R., Spurrell, E., Cohen, L., Saelens, B. E., ...Matt, G. E. (2002). A randomized comparison of group cognitive-behavioral therapy and group interpersonal psychotherapy for the treatment of overweight individuals with binge-eating disorder. *Archives of General Psychiatry, 59*, 713–721.

Wilson, G. (1991). The addiction model of eating disorders: A critical analysis. *Advanced Behavior Research and Therapy, 13*, 27–72.

Wilson, G. (1993). Binge eating and addictive disorders. In C. Fairburn, & G. Wilson (Eds.), *Binge eating: Nature assessment and treatment* (pp. 97–120). New York, NY: The Guilford Press.

CHAPTER 15

Comorbid Trauma and Eating Disorders
Treatment Considerations and Recommendations for a Vulnerable Population

Diann M. Ackard and Timothy D. Brewerton

INTRODUCTION

It feels like a typical Tuesday at work: there is a new patient, "Cortney," on the schedule. A therapist reviews her notes from a telephone call weeks ago, reminding herself that Cortney is a 23-year-old unmarried female seeking outpatient treatment for bulimia nervosa (BN). She is binge-eating and purging by self-induced vomiting and laxative abuse 3 to 5 times a week, having experienced an increase in frequency and intensity over the past 4 years. This is Cortney's first treatment; she thought she could control the symptoms on her own but they have grown stronger and more uncontrollable as time has passed. *"I can't keep this up and still do well at my job"* she stated, and so she finally sought treatment. She shared no other complaints.

Cortney presents to the office on time yet visibly anxious, twisting her hair as she speaks and darting her eyes around the therapy room as if a bird is flitting about. She recounts the history of her body image and eating concerns, dating to childhood, and the development of her binge-eating and purging behaviors in college. At first glance, the therapist believes that this might be a fairly typical eating disorder (ED) presentation. When the therapist asks Cortney if she could identify any precipitants to her ED symptoms, Cortney explains that they developed in response to stress. They help her calm down when she is wound up, and distract her when her mind goes to "dark places." With slight prompting, she expands her answer. *"Well, sometimes I think about things that are bad, things you see in the movies or hear on the news...like when someone is missing or hurt. I don't want to think about those things, but that's where my mind goes."* The therapist broadly queries if any of these types of event have happened to her, such as having been hurt or injured by someone, or having been threatened in such a way that it made her scared or tense. Cortney looks out the window, and then in a quieter voice says, *"Well, most of the girls I know had a bad date or two in college, and I did too, but mine wasn't as bad as other girls' I know. Anyway, I also get stressed thinking about my job, what my boss thinks about how I am doing, and whether or not I am in the right field, and the eating and vomiting distract me from thinking about my job all the time."* With Cortney having quickly changed the topic, and rapport not having been fully established at this first meeting, the therapist makes a note to herself to come back to the "bad date" events later.

This case description provides an example of how some patients will allude to traumatic experiences, and how clinicians can benefit from listening to covert cues of abusive events without pushing patients beyond where they are willing to go at the time. Several factors evaluated at the time of this assessment led the therapist to defer questioning about details of the trauma. First, this was the initial intake interview, a time for building a trusting rapport while gleaning enough data to make appropriate clinical formulations, and for reviewing treatment goals and expectations; this was also her first treatment. Second, Cortney had diverted the topic away from the discussion of the bad date on her own, thus giving a cue that she was not (yet) ready to talk further about details. Finally, she may have been minimizing the traumatic event as she compared it to her girlfriends' "bad dates," and deemed hers as "not as bad." With all this, the therapist felt that this topic was worth addressing in the near future as part of her treatment plan. Given the patient's dismissal of the event compared to what her friends had experienced, the first step was to provide her with education on the various types of trauma.

A WORKING CONCEPTUALIZATION OF "TRAUMA"

The Diagnostic and Statistical Manual of Mental Disorders, 4th edition (American Psychiatric Association, 1994), provides a description of the traumatic events that qualify as criteria for a diagnosis of Post-Traumatic Stress Disorder (PTSD). Specifically, DSM-IV (American Psychiatric Association, 1994) criterion A1 states:

> The essential feature of PTSD is the development of characteristic symptoms following exposure to an extreme traumatic stressor involving direct personal experience of an event that involves actual or threatened death or serious injury, or other threat to one's physical integrity, or witnessing an event that involves death, injury, or a threat to the physical integrity of another person; or learning about unexpected or violent death, serious harm, or threat of death or injury experienced by a family member or other close associate (p. 424).

However, other events that are not listed may also be traumatic, and may contribute to the complexity of a patient's clinical presentation. For example, potentially traumatic events associated with ED symptoms include bullying (Fosse & Holen, 2006; Striegel-Moore, Dohm, Pike, Wilfley & Fairburn, 2002), weight-related teasing (Haines, Neumark-Sztainer, Eisenberg & Hannan, 2006), appearance-related teasing (Sweetingham & Waller, 2008), sexual harassment (Harned, 2000; Harned & Fitzgerald, 2002), racial discrimination (Striegel-Moore et al., 2002), and generational transmission of exposure to violence (Zohar, Giladi & Givati, 2007).

For the purposes of this chapter and based on our clinical experience, we utilize the following working definition of trauma as: Any of the following past or present events that cause distress and/or dysfunction to an individual, such as physical or emotional neglect, or experience of a physical, sexual or emotional nature (including teasing, bullying, sexual harassment, racial discrimination, and witnessing violence). This definition allows clinicians to honor the significance of a broad range of traumatic events, thus validating the patient's experience.

RESEARCH ON TRAUMA AND EATING DISORDERS

Research on the role of trauma in the etiology of ED has exploded since the 1990s. A comprehensive overview of this body of work is beyond the scope of this chapter but has been extensively reviewed elsewhere (Brewerton, 2004, 2005, 2006, 2007). Brewerton (2007) summarized the significant conclusions from these studies:

1. Childhood sexual abuse (CSA) is a nonspecific risk factor for ED (Wonderlich, Brewerton, Jocic, Dansky & Abbott, 1997).
2. The spectrum of trauma linked to ED has been extended from CSA to include a variety of other forms of abuse and neglect (Allison, Grilo, Masheb & Stunkard, 2007; Fosse & Holen, 2006; Haines et al., 2006; Striegel-Moore et al., 2002; Sweetingham & Waller, 2008; Zohar et al., 2007).
3. Trauma is more common in bulimic ED than nonbulimic ED (Wonderlich et al., 1997).
4. Findings linking ED with trauma have been extended to children and adolescents with ED (Ackard, Neumark-Sztainer, Hannan, French & Story, 2001).
5. Findings linking ED with trauma have been extended to boys and men with ED (Ackard & Neumark-Sztainer, 2002).
6. Multiple episodes or forms of trauma are associated with ED (Ackard & Neumark-Sztainer, 2003).
7. Multiple types of compensatory behaviors are associated with history of trauma and PTSD (Brewerton, Dansky, O'Neil & Kilpatrick, 2009).
8. Trauma is not necessarily associated with greater ED severity (Wonderlich et al., 1997).
9. Trauma is associated with greater comorbidity (including and often mediated by PTSD) in individuals with ED (Dansky, Brewerton, O'Neil & Kilpatrick, 1997).
10. Partial or subthreshold PTSD may also be a risk factor for BN and bulimic symptoms (Brewerton, 2007).
11. The trauma and PTSD or its symptoms must be satisfactorily addressed/resolved in order to facilitate full recovery from the ED and all associated comorbidity (Holzer, Uppala, Wonderlich, Crosby & Simonich, 2008).

Clinicians can use this summary as the basis for their approach to the assessment and treatment of their patients with ED.

ASSESSMENT OF TRAUMA

Many clinicians working in the ED field are skilled at querying about overt physical and sexual trauma, especially when a patient specifically seeks treatment for CSA and physical abuse, or following a rape. However, we may inquire less often or less thoroughly about emotional and physical neglect and other traumas such as teasing, bullying, sexual harassment, racial discrimination, and witnessing violence. This oversight, or avoidance, can lead to a premature and inaccurate assumption that a patient has not suffered from a traumatic experience. In our obligation to provide the best possible clinical care, as well as to educate the public about negative responses to violence, it becomes our responsibility to help individuals understand the breadth and scope of traumatic events and their sequelae. One way to facilitate this is the use of published structured interview and self-report measures. The National Center for PTSD (www.ncptsd.va.gov) provides an extensive listing and description of structured interview and self-report assessments for adults and children pertaining to history of traumatic events, PTSD, and associated symptoms.

Clinicians are often at the front line, attending to stories that include the wide varieties of trauma. This provides an important opportunity for education about the various types of traumatic experiences. For example, a woman with presumably no history of trauma, but who presents for treatment of binge-eating disorder (BED), may describe a binge after an argument with her husband.

> *"It was a Friday night and we were getting ready to go to an anniversary party for our friends Jeanna and Paul. I was standing in front of the closet with the doors flung wide open, nearly in tears because I couldn't find anything nice to wear. My husband came into the room and screamed at me, 'What the hell is wrong with you? We've got to get going! Just throw any old potato sack over that fat ass of yours and get in the car.' I was devastated. I was already calling myself 'fat' and 'stupid' and 'weak' when he came in and confirmed that I am all of those words. I went to the living room and told him that I wasn't going. It wasn't two seconds after I heard the garage door close that I had a large spoon full of chocolate ice cream in my mouth and was rooting around the cupboard for other foods to eat."*

In this example, the precipitants to the binge episode include emotionally abusive language from her husband, and the clinician has an opportunity to both identify links to the binge event and to help her identify her husband's statements as abusive. Labeling his behavior as "abusive" can help her resist the negative internalizations of being "fat," "stupid," and "weak."

TREATING THE INDIVIDUAL WITH COMORBID TRAUMA AND EATING DISORDER

In our experience, patients with a history of traumatic events who also engage in ED behaviors present with multiple comorbid and complicated symptom constellations. There are physical symptoms such as malnutrition, bone density loss, impairment in gastrointestinal functioning, and cold intolerance. Psychological symptoms often include anxiety around issues pertaining to food and body, and emotional lability or detachment. Yet, individuals with comorbid trauma often also present with multiple layers of impaired relationship foundations such as significant distrust or fear of others, and concerns about food and body that are dually related to ED and trauma.

Kate is a high school junior who has been in cognitive-behavioral treatment (CBT) for BN for 6 months. She has a pattern of binge-eating compensated for by self-induced vomiting and dietary restriction, all contributing to a malnourished brain. Early treatment interventions (e.g., daily food logs and meal planning), helped normalize brain function by restoring adequate nutrition, yielding greater clarity in thinking and processing of thoughts, feelings, and behaviors. Once her brain was functioning more capably, she slowly disclosed, in bits and pieces of information over the course of several sessions, that she had witnessed physical and emotional abuse by her stepfather against her mother. She also revealed a past trauma of sexual victimization by a neighborhood boy. She recognized that sometimes the vomiting and restriction compensated for the binge-eating which helped to pacify mood states of depression and loneliness. More often, the purging was triggered by flashbacks from the sexual victimization in which she was forced to perform oral sex on the neighborhood boy, while the restriction served to distance or "numb" herself from further flashbacks. She believed that her body was not safe, that it existed for others' pleasure or abuse, and that she should perpetuate punishment against her body. She held an internalized belief of *"I am no good, my body is damaged, and I am damaged-goods."* Furthermore, the traumatic experiences led to a sense of danger around others. This slowed the development of rapport within the therapeutic relationship, and she frequently "tested" the relationship. The tests revolved around questions such as *"Can I trust you* [the clinician]*?"*; *"If I tell you what happened to me, will you be able to handle it?"*; *"Will you look at me differently?"*; *"Will you think that I am not worth your time?"*; or *"Will you leave me?"*

Kate's symptoms served multiple purposes. At times, her compensatory behaviors were related to fear of weight gain, whereas at others they helped to manage her trauma symptoms. Binge-eating helped her numb from intense affect, and restriction helped to keep her mind from flashing back to past traumatic experiences. Due in part to her early childhood abuses, as well as the current household tension in which she was witnessing violence, Kate had neither adequate role-models nor reliable and safe opportunities for learning and practicing healthy self-regulatory skills for managing affect. Thus, the symptoms maintained themselves, as they were helping her get through these difficulties. If she were to present for treatment of her BN without having disclosed the trauma, or without a clinician understanding the multiple "benefits" that the ED symptoms served, treatment would likely fail. In time, as both the past and current abusive experiences remained open, raw, and unmanageable, any short-term progress in symptom reduction would disappear. Consequently, it is essential to recognize the full range of presenting issues and to develop a thorough understanding of the layering of functions held by each symptom.

Treatment Approach

Brewerton (2004) has outlined a number of empirically based principles in the assessment and treatment of ED related comorbidity associated with trauma. After completing a comprehensive psychiatric evaluation and establishing the relative chronology of the onset of disorders, the clinician can begin to identify potential functional links between disorders. Treatment is a phasic endeavor in which normalization of the eating disturbance, and hence brain function, is the first and most important order of business. This initial emphasis on normalizing brain function is a necessary condition for effective psychotherapy and for the efficacy of antidepressant medications. A large body of evidence has demonstrated the profound physiological and psychological effects of semi-starvation, dieting, and other disordered eating behaviors on central nervous system function. Not until this profound dysregulation is addressed and on its way toward resolution can more in-depth treatment geared toward the resolution of trauma-related conditions proceed successfully.

It is important to prioritize the order of progression and to make sure that the cart (intensive psychotherapy) is not put before the horse (nutritional rehabilitation). A significant gap in the psychotherapy research is the absence of evidence-based psychotherapy treatments for dually diagnosed individuals with both trauma and ED. Thus, clinicians are strongly encouraged to utilize a case formulation approach (Persons, 2006) in which evidence-based treatments, grounded in theory for the specific disorders in the clinical presentation, are used and refined in active collaboration with the patient. This collaboration allows for ongoing hypothesis testing to systematically evaluate which interventions are providing the greatest assistance. Treatment becomes a matter of emphasis or focus, and this emphasis is determined by the degree of dangerousness, risk of harm to self and/or others, and/or brain—body impairment and the patient's sense of what is most effective.

For example, a case formulation for an individual presenting with both trauma and BN would include evaluation of evidence-based treatments for PTSD (Zayfert & Becker, 2007) and for BN (Fairburn et al., 1991; Le Grange, Crosby & Lock, 2008). Case formulation promotes the construction of individualized treatment plans using available evidence-based treatments, thus empowering clinicians to bridge the gap between disorders for which there is no one uniformly researched treatment, but for which evidence-based treatments may be effective at treating more than one diagnosis. Recommendations on locating information on the available practice guidelines are included in Table 15.1.

Next, the clinician and patient would evaluate together the specific symptoms that need to be addressed, and the clinician would draw on clinical interventions from evidence-based treatments to formulate the treatment plan. Regular assessment, including, but not limited to, measuring the frequency and intensity of the symptoms, allows the clinician and patient to determine if the interventions are producing the desired effect or whether they need a change in the course of treatment.

To date, and speaking very broadly, the ED treatments with the most evidence-based research supporting their effectiveness are: psycho-education, CBT, family-based therapies (such as Maudsley), interpersonal therapy (IPT), dialectic-behavior therapy (DBT), and psychopharmacology (Brewerton, 2004). Evidence-based treatments for PTSD include CBT, prolonged exposure, cognitive processing therapy (CPT), DBT, eye movement desensitization and reprocessing (EMDR), and psychopharmacology (Brewerton, 2004). After identifying the evidence-based treatments that suit the clinical presentation, the next challenge is to determine which ones to use and when.

The question on when to begin the work pertaining to the trauma will be answered by the patient. After achieving adequate brain functioning, which includes having ED symptoms under reasonable control, patients will be able to think more clearly and communicate when they anticipate being able to best manage the stress that arises during trauma treatment. The clinician and the patient must both anticipate that some of the symptoms that have *helped* the patient manage symptoms associated with the trauma may worsen during the course of thorough and intense trauma work. Regular assessment of symptom frequency and severity allows both the patient and clinician to identify worsening of symptoms and determine a threshold at which an adjustment to treatment might be warranted. The discussion around the worsening of symptoms should include a specific and collaborative plan on pausing or modifying the trauma treatment (including implementation of a safety plan that is discussed and agreed upon in advance of the commencement of treatment) in the event that other symptoms become too frequent or severe.

TABLE 15.1 Available Practice Guidelines

Eating Disorder	www.psychiatryonline.com/pracGuide/pracGuideTopic_12.aspx
	www.nimh.nih.gov/health/publications/eating-disorders/complete-index.shtml
	guidance.nice.org.uk/CG9
Post Traumatic Stress Disorder and Acute Stress Disorder	www.psychiatryonline.com/pracGuide/pracGuideTopic_11.aspx
	www.nimh.nih.gov/health/topics/post-traumatic-stress-disorder-ptsd/index.shtml
	guidance.nice.org.uk/CG26
	tfcbt.musc.edu/
	www.psychguides.com/ptsd
	www.aacap.org/galleries/PracticeParameters/PTSDT.pdf
Borderline Personality Disorder	www.psychiatryonline.com/pracGuide/pracGuideTopic_13.aspx
	guidance.nice.org.uk/CG78
Obsessive-Compulsive Disorder	www.psychiatryonline.com/pracGuide/pracGuideTopic_10.aspx
	www.nimh.nih.gov/health/topics/obsessive-compulsive-disorder-ocd/index.shtml
	guidance.nice.org.uk/CG31
	www.psychguides.com/ocd
Anxiety and Panic Disorder	www.psychiatryonline.com/pracGuide/pracGuideTopic_9.aspx
	www.nimh.nih.gov/health/topics/anxiety-disorders/index.shtml
	www.nimh.nih.gov/health/topics/panic-disorder/index.shtml
	guidance.nice.org.uk/CG22
	www.aacap.org/galleries/PracticeParameters/JAACAP_Anxiety_2007.pdf
Major Depressive Disorder	www.psychiatryonline.com/pracGuide/pracGuideTopic_7.aspx
	www.nimh.nih.gov/health/topics/depression/index.shtml
	guidance.nice.org.uk/CG23 and guidance.nice.org.uk/CG28 for children and youth
Substance Use Disorder	www.psychiatryonline.com/pracGuide/pracGuideTopic_5.aspx
	guidance.nice.org.uk/CG51
	www.aacap.org/galleries/PracticeParameters/JAACAP%20Substance%20use%202005.pdf
Dissociative Disorder	www.isst-d.org/education/treatmentguidelines-index.htm

Recommendations for the Clinician

Do not avoid trauma treatment. Patients who have experienced significant traumatic events may wish to avoid trauma treatment. This avoidance makes sense as the ED and related symptoms have served a "mental survival" purpose. For example, a binge episode may "numb" the feelings associated with trauma, and self-induced vomiting may "purge" the patient's uncomfortable and disgusting feelings about him- or herself. Dissociative experiences are another form of avoidance, as the patient "splits off" from awareness and reality and thus is able, at least temporarily, to escape intrusive thoughts and painful affects related to the trauma. While we can appreciate that each of these symptoms has a beneficial service to the patient by providing momentary respite from traumatic memories, we need to also acknowledge and reinforce the significant and life-changing benefit of helping patients develop the skills and understanding to work through (versus work around) trauma symptoms.

Continue to reassess traumatic experiences and the effectiveness of treatment. It can be very tempting, when treatment is going well, to reduce the frequency of assessment of the effectiveness of the treatment interventions, and to believe that all traumatic experiences are being addressed. However, it is not uncommon for new symptoms to emerge that were not part of the original assessment. Memories of traumatic experiences not yet disclosed may come to the patient's awareness, or the patient may be willing to impart more memories as the clinician gains their trust on the process of trauma treatment. Thus, it is vitally important that the clinician and patient revisit the initial treatment plan and modify it accordingly, adding to it new symptoms that have emerged as supplementary coping strategies while other symptoms are being treated, as well as additional traumatic experiences which the patient may be now ready to disclose.

Know your local reporting laws. No single set of laws universally applies to all patients across the globe regarding obligations to report traumatic events. However, patients can feel more at ease with a clinician who knows the local reporting laws by memory, or has them immediately available in written form. It is ethically responsible and vital to the working relationship for clinicians to be properly informed of the laws within their geographic area, noting when reporting is mandatory and what options might be available for patients in terms of securing their safety and protection.

Be aware of vicarious victimization. Vicarious victimization, also known as secondary post-traumatic stress disorder or compassion fatigue, can occur among clinicians working with patients who have a history of trauma. It occurs when the clinician experiences symptoms himself or herself that are reflective of PTSD. Symptoms may be work-related, such as fearfulness or detachment at work, forgetting facts pertaining to patients, helplessness while working with trauma patients, or dissociation or avoidance of people, places, or activities that are reminiscent of clinical work. Symptoms may also manifest physically and emotionally, such as headaches or stomach-aches, sleep disturbance, increased startle response, heightened irritability, or poorer concentration. Vicarious victimization is best prevented through regular consultation or supervision, appropriate self-care, and limiting the number of trauma patients in the caseload. If vicarious victimization does occur, clinicians need to seek additional supervision about the situation, and to consider seeking treatment as indicated.

Practice regular self-care. Treating sequelae to trauma and active ED can be rewarding, and also very challenging. In the course of trauma treatment, clinicians should develop and maintain a regular practice of self-care tailored to their own style of soothing. Optimal self-care may include personal practices such as meditation, massage, and rest, as well as appropriate nutrition and exercise. Furthermore, regular self-care also includes engaging with others outside the field of trauma, such as spending time with supportive friends and family, and decompressing by taking a walk or talking over coffee with a trusted person. Finally, it is essential to seek professional consultation and continuing education credits in areas applicable to the fields of practice, and to set boundaries on trauma-related practices, such as having days when not on-call. It is in the best interest of our patients for us to establish regular self-care practices and appropriate boundaries. A clinician who burns out and needs to leave working with this patient population, whether temporarily or permanently, may evoke unfortunate "abandonment" feelings in his or her patient and inadvertently perpetuate the patient's concerns about trust. The patient may also wonder if the traumatic experiences were "too much" for the therapist, or others, to handle.

CLINICAL VIGNETTES

The following clinical vignettes are taken from real clinical practices and illustrate many of the points noted above in terms of assessment and treatment.

PATRICIA

Patricia, a 17-year-old white single senior high school student living at home with her parents, was referred to a psychiatrist by her pediatrician for evaluation and treatment of an ED. Patricia had begun throwing up a year earlier in association with rising anxiety about her weight, which had fluctuated considerably since age 11 years. On initial evaluation she admitted to vomiting 1 to 2 times per day, stated that she had "felt fat" since age 3, and had "always" been dissatisfied with her body size and shape. She also had recently started to use over-the-counter (OTC) laxatives, diuretics, and diet pills and she had started cutting herself superficially on the arms and abdomen. She complained of depressed mood, insomnia, and trouble concentrating. Her past psychiatric history was notable for separation anxiety disorder between ages 5 and 7 following the death of several relatives within a 6-month period. She also reported having been teased during middle school about her body size and shape and about being too "emotional." She had been seen by two therapists, a psychiatrist, and her pediatrician for her ED during the 3–4 months prior to referral. She had been taking 60 mg of fluoxetine per day for several weeks but without apparent benefit. Her medical history was notable for mild microcytic anemia and childhood onset obesity. Family history was positive for alcoholism, depression, and anxiety on her father's side, and BN and obesity on her mother's side of the family. Patricia described her mother as being anxious, intrusive, and "overly controlling." Her mother's marriage to Patricia's father was her second marriage; Patricia was her first child. Patricia depicted her father as someone who was uninvolved, distant, and who frequently lost his temper. This was his third marriage, and he had one much older daughter from his first marriage. Patricia was her parents' only child together.

On initial evaluation, Patricia was 142 lbs and 5 feet 1 inches tall. Her highest weight was 169 lbs during the 10th grade, and her lowest weight was 136 lbs during the previous month. The initial diagnostic impression included: Eating Disorder Not Otherwise Specified (EDNOS; purging disorder), major depression, obsessive-compulsive personality traits, and obesity. Patricia was prescribed a course of CBT, using a treatment manual for BN. She continued to restrict, vomit several times per week, and intermittently abuse OTC diet pills. Although she grasped the CBT material, she exhibited low motivation to give up purging. She began to talk about her suppressed anger at her mother for criticizing her so much, but remained primarily angry at herself. Patricia gave herself no credit whatsoever despite her high intelligence and multiple talents, but gradually she was able to acknowledge some of her strengths.

By the fourth session, Patricia began to acknowledge some abuse, initially by a 3rd grade teacher who had been physically rough and who had yelled at her, and then by her father, who had extreme angry outbursts following a disabling accident. A diagnosis of PTSD was entertained at that point, which was later confirmed as treatment progressed. In a family session, her mother disclosed that Patricia had stopped her fluoxetine a few weeks before because "it wasn't working." Subsequently, a trial of duloxetine was initiated and her depression and anxiety significantly improved after three weeks. Her purging frequency also decreased, but she continued to restrict and make superficial cuts on her abdomen when stressed. She explained that this occurred only at home and helped to relieve inner tension and to punish herself, emphasizing that it was not about killing herself. As CBT progressed, her specific cognitions about herself were clarified (e.g., *"I'm fat, disgusting, and gross"* and *"I don't deserve to eat"*). She reported that she had received multiple messages from coaches, teachers, and fellow athletes (swimming, dancing, and track) that she had to keep her weight down *"in order to be worth anything."* These beliefs were actively challenged by repeatedly asking her questions such as: *"What is the evidence?"* or *"How do you know that's true?"*

After 4 months of treatment, Patricia had become completely abstinent from any purging for two solid weeks, and she had been eating more normally. She then reported, in the thirteenth session, a *"very disturbing experience"* when she was 10 years old of her father asking to see her completely naked body after a shower. She denied any memories of him touching her or himself. She was visibly shaken when reporting this incident and at first wasn't sure if this emerging memory was true or not. The psychiatrist, appropriately, took a neutral stance and explained that he did not know if her memories were accurate or not, but that it was important for her to learn how to calm herself using more effective coping strategies. These, he reassured her, would help her be able to "sit with" whatever was coming up for her so they could talk about it. He further explained that dissociative amnesia of traumatic events was a very real phenomenon, which had been documented scientifically, but that the reality of false memories had also been established. He suggested that her bulimic behaviors only served to numb her, did not solve any of her problems, and that with a clear and calm mind, she could learn to handle whatever had happened in her life

Several family sessions were held during the course of Patricia's treatment, two of which were attended by her father along with her mother. In the second session with her father, it became evident that he had chronic PTSD from his own childhood physical and emotional abuse. He, however, refused to pursue any treatment and, later, refused to attend any more family sessions.

During the 5th and 6th months of treatment, Patricia's periods of abstinence increased and lasted as long as three weeks before recurrence. She then reported a marked increase in purging frequency associated with the emergence of more uncomfortable feelings and vague but intense memories about her father, revealing that he had often walked around nude at home until she was age eleven. In addition, she divulged that he had masturbated the family dog to ejaculation many times since she was a small girl, even as recent as within the last three weeks, while the dog was on her lap. At age seven, he had made her touch their cat's testicles. On at least one occasion when she was around age 12, he forced her to touch their dog's penis and masturbate it to ejaculation, which horrified her. She stated that this sexual behavior with animals had occurred right in front of her mother, who just laughed about it with her father. She also reported that her father used to talk to her a lot about the penis size of men, which made her terribly uncomfortable. Recently, when her boyfriend French kissed her, she became acutely anxious, and suddenly thought how his tongue in her mouth was like having a penis her mouth. She also had an image *"pop up"* while brushing her teeth of her father "fingering" her vagina as a very young child. These last two images were particularly upsetting to her and she began to have more overt and severe PTSD symptoms, even as her ED improved. She became increasingly more anxious about having to be home alone with her father while her mother was at work at night. In order to control her hyperarousal and to suppress the intrusive imagery, she resorted to purging again regularly, stating in one of her last sessions that she consciously wanted to suppress any further memories of possible abuse by her father and the idea that he had done anything wrong. Patricia noted that she *"did not feel safe"* to stop her purging and remember the truth of what happened in her life.

When Patricia tried to tell her mother about these memories, her mother became extremely angry and upset and denied that it was even possible. Furthermore, her mother told Patricia she thought the psychiatrist may have induced false memories, and she refused to pay for any more sessions under the guise of having financial difficulties. As her PTSD symptoms worsened, Patricia's ED symptoms intensified and she became increasingly more depressed and overtly suicidal, eventually requiring acute inpatient psychiatric hospitalization. After discharge, her parents did not allow her to return to treatment and she was sent to live with her maternal aunt in another state for several months. Her aunt also denied the possibility of abuse by her father, and Patricia, who had turned eighteen, then moved 3000 miles away with her best friend, declaring that she was a lesbian.

Follow-up 2 years later revealed that she had become completely abstinent from all ED symptoms, was working and attending college, and was very happy in her relationship with her female partner. Her parents had condemned her for being homosexual and she had remained estranged from them. She indicated at follow-up that she was sure that her memories about her father walking around nude, wanting to see her naked body at age ten, and the sexual abuse with animals were true and accurate memories. She remained unsure about other memories of overt sexual abuse by her father, which had continued to haunt her. In particular, she had more recently dreamed an intensely vivid nightmare of her father vaginally penetrating her. She stated that there was no way for her to otherwise know how this felt because she had remained a heterosexual virgin. She again reiterated that she did not want to think about these things and did not want to know; it was just too upsetting and depressing.

This case illustrates the inverse relationship between ED symptom change and traumatic memory emergence associated with delayed PTSD. As Patricia's bulimic and depressive symptoms abated with a combination of CBT and antidepressant medication, very disturbing memories of CSA began to appear, along with full blown PTSD symptoms. She realized that her bulimic behaviors served to numb her anxiety and intrusive imagery and was able to learn new coping strategies. She practiced these anti-anxiety strategies effectively enough to allow her to recover from her eating and mood disorder. However, she continued to experience significant intrusive imagery in the form of occasional nightmares and trauma-related anxiety, which she avoided.

CHERYL

Cheryl, a married woman in her late forties with four school-aged children, presented for outpatient treatment of "anxiety" after her husband had caught her vomiting blood. Alarmed, he had taken her immediately to the emergency room. Results from an endoscopy revealed no medical basis for the bleeding, and Cheryl reluctantly disclosed to her husband that she had not been feeling well (*"my nerves are activated"*) and had thrown up *"a few times."* She agreed to see a psychologist for her anxiety. In the clinician's office, Cheryl spoke at length about her "nerves," how she rarely had time alone, but before having children she had enjoyed going for walks or rides on her bicycle. Now, she reported that she feels easily agitated, describing herself as a *"slave to others."* Slowly, she disclosed her "secret" that when she is really upset, the behavior that calms her nerves more than any other strategy is to go to the bathroom and throw-up.

Cheryl said that she hadn't liked her body for as long as she can remember; "My body is only good for doing things for other people." The clinician queried further, and Cheryl shared a long list, speaking slowly at first then quickening the cadence as the list lengthened, of ways in which her body has served others but not herself. *"You need to understand what my body has done over the past 10 years—my body had sex to conceive four children and then carried them around for 40 weeks, drives them to play dates and school, takes them to parks and makes sure they are entertained and safe, does the laundry, goes to the grocery store and makes meals, provides pleasure for my husband, and then collapses into bed hoping for an adequate rest which is most often disrupted by a crying child sometime in the night. Don't you see? There's no time for me in this equation! I hate my body and everything it stands for—can't everyone just understand that I'm at the end of my nerves and I need a break?"* Cheryl took several deep breaths before continuing. *"I'm just tired of everything. I want to be by myself for awhile. I've been pressured for 40 years—trying to meet everyone's demands."* The clinician asked what she meant by being pressured for 40 years, given that she hasn't been a mother or a wife for that long. Cheryl responded more quietly, "*My life has always been about other people. I didn't have a childhood—even when others were playing and having fun, I was trying to keep the peace at home.*" Cheryl continued on to describe the physical violence she had witnessed by her father to her mother, and how she, too, later became a victim of her father's violence.

Cheryl and her clinician collaborated on the treatment plan. Given the strong interplay between anxiety and purging behaviors, even in the absence of current physical abuse, the plan emphasized the development of coping skills for anxiety and stress using CBT. The plan also addressed the current symptoms of PTSD (flashbacks, nightmares, and startle responses to certain sights, smells, and noises) using Exposure and Response Prevention. Finally, regular practice of relaxation and self-care exercises, drawn from stress management and Interpersonal Therapy, was added to the treatment plan.

The first two sessions focused on the adoption of regular self-care practices such as daily mindfulness-based walks, meditation, and appropriate delegation of responsibilities to her husband and children. These practices provided applications for self-soothing strategies, and lowered Cheryl's experience of anxiety by 40% as calculated by having her estimate and log her daily anxiety experiences on a scale of 1–10. She remained prone to anxiety attacks, but decreased purging episodes by 20%, the latter determined by calculation of the frequency of purging through daily self-monitoring logs that were reviewed collaboratively by the patient and the clinician on a weekly basis. Throughout the course of treatment, regular practice of these skills, and those adopted later, were assessed for their effectiveness, and her anxiety and purging behaviors continued to wane. Sessions three through eight concentrated on the cognitions (e.g., thinking "*I can't handle this situation—I can't do this anymore—I am a failure*") and behaviors (engaging in purging behavior before trying to self-soothe) that were maintaining and exacerbating her anxiety. Cognitive restructuring to address problematic cognitive styles, and implementing delays between onset of anxiety and engaging in purging, helped her develop a sense of mastery and control. Initially, the delays were very small (10 seconds between height of anxiety and vomiting) and incrementally lengthened. After achieving success with 1-minute delays, Cheryl identified greater self-efficacy over anxiety control without the use of purging behavior.

At the ninth session, the therapist asked Cheryl if her foundation of skills seemed strong enough to begin specific work on her traumatic symptoms. Although she was reluctant to address the trauma given that it was not ongoing and that she was feeling better, Cheryl acknowledged that symptoms related to past trauma continued to occur and cause distress. Triggers to her flashbacks and re-experiences were reviewed, and "subjective units of distress" (SUD), based on an exposure hierarchy specific to CBT for PTSD (Zayfert & Becker, 2007), were established. Both *imaginal* (to memories of witnessing and experiencing physical violence) and *in vivo* (to sights, smells, and noises associated with violence) exposures were implemented and assessed systematically, beginning with the exposure experiences that generated a moderate amount of anxiety and gradually progressing to those that generated significant, heightened anxiety. Exposures were practiced in session, and at home when they were appropriate and safe for Cheryl, such as listening to an audiotape of her voice recounting details of the abuse. Progress was evaluated regularly, and movement up the hierarchy was recommended only after mastery of the current exposure was achieved and she experienced a reduction in anxiety to a lower, manageable level. Treatment progressed slowly and interventions were introduced methodically only after empirical evidence of her anxiety reductions were evident to both parties. In time, she was able to achieve abstinence from purging behavior and significantly improve management of her anxiety. Cheryl felt that she had met her treatment goals and chose to discontinue therapy after twenty-two sessions.

Cheryl's case illustrates how ED behaviors can serve a practical, albeit self-damaging, role in managing sequelae to traumatic events. Her experience also underscores how current treatment modalities can serve to dismantle the purpose of individual symptoms in a manner that allows therapeutic interventions to more specifically and strategically target the root of the symptom. Cheryl was able to make progress in self-soothing and anxiety-reduction practices in a short period of time. These practices not only helped her anxiety and the "need" for the purging behaviors, but also allowed therapeutic trauma work to proceed with greater self-efficacy than if treatment had not established that groundwork. Thus, this case emphasizes the importance of practicing and mastering self-soothing and relaxation strategies first, before the more difficult and anxiety-producing work of CBT for PTSD.

MARTINA

A social worker in a nearby town referred a 16-year-old white unmarried female junior high school student for evaluation and treatment of her ED; Martina also had a history of abuse. She had been seen by her therapist for the previous 2 years, and she had been diagnosed with AN B/P subtype. In addition, a previous psychiatrist had diagnosed her with bipolar disorder, type II, and had prescribed lamotrigine and fluoxetine. She had been treated in an inpatient ED program the previous year for 2 months and went from a low weight of 98 pounds prior to admission to 126 pounds at discharge. Weight restoration was quickly followed by resumption of her menstrual periods. During this admission, she disclosed for the first time a long history of chronic CSA by a paternal uncle beginning around age 6, which was the age when her parents separated and eventually divorced. Martina would often sleep at her paternal grandmother's home, where her paternal uncle also lived in an adjacent room. Martina's disclosure was reported by inpatient staff to child protective services, who investigated, but the patient and her family did not press charges. Her father, a fundamentalist Christian, told her that she should forgive her uncle.

In 2nd grade, after her parents divorced, Martina reported that she *"gained a ton of weight."* In the 5th grade, when both of her parents re-married within 2 months of each other, she reported decreased appetite and weight loss. In the 6th grade, she *"stopped eating on purpose,"* and she reported that she both *"liked the attention"* and *"was annoyed by it."* Her father's second wife had an ED (BN with compulsive exercising), which she said had given her the idea. Martina indicated that she had first told her step-mother about her uncle's abuse at age 9, but that her step-mother told her not to tell anyone because her father would get mad. Martina believed her and stayed silent.

Once Martina returned from inpatient hospitalization, she resumed outpatient psychotherapy with the referring social worker. As treatment progressed, Martina also disclosed to the therapist that she had been sexually abused (raped) by her step-father (mother's second husband). The therapist called child protective services who investigated the case, but it was Martina's word against his. She again did not want to press charges. However, this time her mother believed her and divorced him.

As part of a comprehensive evaluation, the psychiatrist administered a structured diagnostic interview and determined that Martina did not have bipolar disorder but met the criteria for lifetime diagnoses of AN B/P subtype, major depression, separation anxiety disorder, and PTSD. The initial focus of treatment was to get her back to a healthy weight. At the time of evaluation she was 115 pounds and 5 feet 4 inches tall, and she had been amenorrheic for several months. A goal weight range was set at 122–127 pounds and a behavioral weight gain contract requiring a weight gain of at least 1 pound per week (or risk readmission) was negotiated with the patient and her parents. She was tapered off of lamotrigine and fluoxetine was continued at a dose of 40 mg per day. In addition, a small dose of olanzapine was started to decrease anxiety, stabilize mood, and promote weight gain. She tolerated this dose well and attained her goal weight range within 7 weeks. As Martina's weight stabilized, her menstrual periods returned, but her PTSD symptoms emerged full force, including severe dissociative flashbacks, panic, anxiety, and overwhelming shame and guilt, to which she responded with intermittent cutting and restricting. During sessions she readily responded to deep breathing techniques, and she was receptive to learning new coping strategies using a CBT model.

Martina continued to process her extensive history of CSA with both her primary therapist and psychiatrist, who communicated often with each other. She began to talk more openly about the details of her experiences and their meanings to her. Despite her resistance to doing so, with lots of encouragement she eventually wrote a trauma narrative, which she used in her exposure sessions. She was taught to quantify her "subjective units of distress" or SUD using a scale of 0 to 10, 0 indicating no distress and 10 indicating the most distress she's ever had. During an exposure session, the psychiatrist would start by determining her SUD, do some deep breathing exercises with her, and then ask Martina to read aloud her trauma narrative. Periodically, he would stop her and ask for her SUD rating at that moment. If it had markedly increased, he would ask her to stop, breathe deeply, and lower it by 2–3 points before continuing on. When she had finished reading, he would thank her, praise her for her courage, and ask her to use her breathing to lower her SUD back down to where it was when they started. This series of events would be repeated at least once or twice during each session.

The psychiatrist also used cognitive strategies to help Martina identify her thoughts, assumptions, beliefs, and meanings of what had happened to her. On some occasions, Martina would develop marked depersonalization episodes in which she felt completely "numb" to what was going on. She perceived herself as *"floating above"* while watching her stepfather have sex with her. She would say, *"I'm not here right now."* Her cognitions were notable in that she believed that her uncle really loved/does love her and she could not tolerate the idea that his actions were abusive, harmful, and not loving at all. However, as time went on she began to entertain the possibility of these new ways of thinking.

One year after beginning treatment with the psychiatrist, Martina was maintaining her weight within her goal weight range, feeling much better about herself and her body, and she reported that she felt happier than any time since early childhood (prior to any abuse and parents' divorce). As Martina improved, she was surprised by how good she felt, and her trust in her therapists deepened. She continued psychotherapy using CBT with prolonged exposure and also continued to take fluoxetine 40 mg per day. Although her PTSD symptoms were certainly not gone, they had significantly decreased.

This case illustrates how reports of prior maltreatment may not emerge until well into the therapeutic process, and in particular, not until normal brain function is reestablished. Once nutritional rehabilitation was achieved, and in the context of trusting relationships, the patient was able to engage in, and benefit from, CBT with prolonged exposure, the most effective treatment for PTSD.

CONCLUSION

As this chapter explains, a "typical Tuesday at work" for clinicians who treat eating disorders is bound to include patients with significant trauma histories, as a wide variety of traumatic experiences are associated with ED symptoms and related disorders. Treatment is often challenging due to the interplay of symptoms and how they may beneficially serve the patient. A case-formulation approach, grounded in evidence-based therapies for the particular disorders in the clinical presentation, provides guidance for the clinician to assure the best treatment available. Using a case formulation approach, clinicians can sequence treatments to provide the patient with an optimal physical foundation (brain functioning) to be able to utilize the skills and techniques found to be effective in evidence-based treatments. In addition, regular assessment by both the clinician and patient of the effectiveness of the therapeutic interventions tailors the treatment to the patient's strengths and areas of highest motivation. We believe that a case-formulation approach to treating patients with ED and sequelae to traumatic experiences provides them with the highest likelihood of symptom reduction and asset enhancement.

References

Ackard, D. M., Neumark-Sztainer, D., Hannan, P. J., French, S., & Story, M. (2001). Binge and purge behavior among adolescents: Associations with sexual and physical abuse in a nationally representative sample: the Commonwealth Fund survey. *Child Abuse and Neglect, 25*, 771–785.

Ackard, D. M., & Neumark-Sztainer, D. (2002). Date violence and date rape among adolescents: Associations with disordered eating behaviors and psychological health. *Child Abuse and Neglect, 26*, 455–473.

Ackard, D. M., & Neumark-Sztainer, D. (2003). Multiple sexual victimizations among adolescent boys and girls: Prevalence and associations with eating behaviors and psychological health. *Journal of Child Sexual Abuse, 12*(1), 17–37.

Allison, K. C., Grilo, C. M., Masheb, R. M., & Stunkard, A. J. (2007). High self-reported rates of neglect and emotional abuse, by persons with binge eating disorder and night eating syndrome. *Behavior Research Therapy, 45*(12), 2974–2883.

American Psychiatric Association (1994). *Diagnostic and statistical manual of mental disorders* (4th ed.). Washington, DC: Author.

Brewerton, T. D. (2004). Eating disorders, victimization and PTSD: Principles of treatment. In T. D. Brewerton (Ed.), *Clinical handbook of eating disorders: An integrated approach* (pp. 509–545). New York, NY: Marcel Dekker.

Brewerton, T. D. (2005). Psychological trauma and eating disorders. *Review of Eating Disorders, 1*, 137–154.

Brewerton, T. D. (2006). Comorbid anxiety and depression and the role of trauma in children and adolescents with eating disorders. In T. Jaffa & B. McDermott (Eds.), *Eating disorders in children and adolescents* (pp. 158–168). Cambridge, England: Cambridge University Press.

Brewerton, T. D. (2007). Eating disorders, trauma and comorbidity: Focus on PTSD. *Eating Disorders, 15*, 285–304.

Brewerton, T. D., Dansky, B. S., O'Neil, P. M., & Kilpatrick, D. G. (2009, September 24–26). *The number of divergent purging behaviors is associated with histories of trauma, PTSD and comorbidity in a national sample of women*. Brooklyn, NY: Paper presented at the 15th Annual Eating Disorders Research Society Meeting.

Dansky, B. S., Brewerton, T. D., O'Neil, P. M., & Kilpatrick, D. G. (1997). The National Women's Study: Relationship of crime victimization and PTSD to bulimia nervosa. *International Journal of Eating Disorders, 21*, 213–228.

Fairburn, C. G., Jones, R., Peveler, R. C., Carr, S. J., Solomon, R. A., O'Connor, M. E., ... Hope, R. A. (1991). Three psychological treatments for bulimia nervosa. A comparative trial. *Archives of General Psychiatry, 48*(5), 463–469.

Fosse, G. K., & Holen, A. (2006). Childhood maltreatment in adult female psychiatric outpatients with eating disorders. *Eating Behaviors, 7*, 404–409.

Haines, J., Neumark-Sztainer, D., Eisenberg, M. E., & Hannan, P. J. (2006). Weight teasing and disordered eating behaviors in adolescents: Longitudinal findings from Project EAT (Eating Among Teens). *Pediatrics, 117*, e209–215.

Harned, M. S. (2000). Harassed bodies: An examination of the relationships among women's experiences of sexual harassment, body image, and eating disturbances. *Psychology of Women Quarterly, 24*, 336–348.

Harned, M. S., & Fitzgerald, L. F. (2002). Understanding a link between sexual harassment and eating disorder symptoms: a mediational analysis. *Journal of Consulting and Clinical Psychology, 70*(5), 1170–1181.

Holzer, S. R., Uppala, S., Wonderlich, S. A., Crosby, R. D., & Simonich, H. (2008). Mediational significance of PTSD in the relationship of sexual trauma and eating disorders. *Child Abuse and Neglect, 32*, 561–566.

Le Grange, D., Crosby, R. D., & Lock, J. (2008). Predictors and moderators of outcome in family-based treatment for adolescent bulimia nervosa. *Journal of the American Academy of Child and Adolescent Psychiatry, 47*(4), 464–470.

Persons, J. B. (2006). Case formulation-driven psychotherapy. *Clinical Psychology: Science and Practice, 13*(2), 167–170.

Striegel-Moore, R. H., Dohm, F. A., Pike, K. M., Wilfley, D. E., & Fairburn, C. G. (2002). Abuse, bullying, and discrimination as risk factors for binge eating disorder. *American Journal of Psychiatry, 159*, 1902–1907.

Sweetingham, R., & Waller, G. (2008). Childhood experiences of being bullied and teased in the eating disorders. *European Eating Disorders Review, 16*, 401–407.

Wonderlich, S. A., Brewerton, T. D., Jocic, Z., Dansky, B. S., & Abbott, D. W. (1997). The relationship of childhood sexual abuse and eating disorders: A review. *Journal of the American Academy of Child and Adolescent Psychiatry, 36*, 1107–1115.

Zayfert, C., & Becker, C. B. (2007). *Cognitive behavioral therapy for PTSD: A case formulation approach*. New York, NY: Guilford Press.

Zohar, A. H., Giladi, L., & Givati, T. (2007). Holocaust exposure and disordered eating: A study of multigenerational transmission. *European Eating Disorders Review, 15*, 50–57.

CHAPTER 16

Healing Self-Inflicted Violence in Adolescents with Eating Disorders
A Unified Treatment Approach

Kimberly Dennis and Jancey Wickstrom

Self-inflicted violence (SIV) in adolescents with eating disorders (ED) is a daunting treatment challenge requiring the integration of various clinical approaches. This chapter provides a definition of such behavior and explores its underpinnings, appreciating the factors contributed by both nature and nurture. A unified approach is proposed to treat both ED and SIV using a modified addiction model that blends Dialectical Behavior Therapy (DBT) (Linehan, 1993a, 1993b) with a 12-step model. This integrated treatment model examines the impact of SIV not only on the individual, but on peers and caregivers as well. The proposed model strives to provide a more complete therapeutic approach to best help those adolescents struggling with both ED and SIV.

DEFINITION AND SCOPE OF THE PROBLEM

"Violence" is a term that evokes strong connotations of physical force, abuse, and destruction. As suggested by Mazelis (2002), the term self-inflicted violence (SIV) will be used to describe the gamut of self injurious behaviors in which individuals engage. This characterization avoids colluding with the denial and minimization inherent to this psychiatric population, their peers, and sometimes their parents and caregivers as well. Utilizing this term more accurately describes the behavior directly and explicitly identifies the ways in which patients turn against themselves in acts of violence.

Most people imagine an adolescent cutting him or herself with a razor when they hear the term "self-injurer." The range of covert self-injurious behaviors is broad and may include: cutting, skin-picking, hair pulling, head banging, burning, inserting objects under the skin, nail biting, scratching, piercing, and tattooing, to name a few. Eating disorders can be seen as forms of SIV, as starvation, over-exercise, binging, purging, and compulsive overeating are all acts of violence on the body. On a broader spectrum, drinking, smoking, and drug abuse are included as forms of self-injurious behavior (Favaro & Santonastaso, 2000). Patients can also use other activities in ways that are violent or self-deprecating which fall under the general rubric of covert self-abuse (e.g., sexually compulsive behavior or over-working).

A major distinction between overt and covert forms of SIV is the immediacy and tangibility of the former. Favazza (1996) further distinguishes compulsive from impulsive self injury. Compulsive SIV tends to be more closely associated with obsessive-compulsive disorder and includes such behaviors as skin-picking and scratching. Impulsive self injury occurs in response to an emotional trigger, usually brought on by an interpersonal experience. The functions of SIV can include self-punishment, expression of feelings of anger, and release of overwhelming emotions (Ross, Heath & Toste, 2009). Over time, SIV consumes the patient's identity, overshadowing and drowning out her authentic personhood, much like the progression of addictions and ED.

For the purposes of this chapter, we will focus on SIV as defined by any overt act which violates social and cultural norms and results in direct physical harm to one's body in the absence of *conscious* suicidal intent. We will use the terms SIV and self injury interchangeably. Although some have made a case for self injury to be classified as its own disorder (Favazza, 1996), it is most accurately diagnosed as an Impulse-Control Disorder Not Otherwise Specified according to the DSM-IV-TR (American Psychiatric Association, 2000).

There is significant overlap between the SIV and ED populations, which may point to some common underlying etiologies. The demographics of these populations are broad, including adolescent girls and boys, adult men and women (some with onset of ED and/or SIV during adulthood), and people of a variety of educational levels, socioeconomic class, and professions (Levitt, Sansone & Cohn, 2004). Up to 50% of patients with SIV have histories of ED (Svirko & Hawton, 2007). Co-occurring self injury is also very common in patients with ED. The literature estimates that between 20 and 40% of adolescents and women with ED engage in some form of SIV (Paul, Schroeter, Dahme & Nutzinger, 2002). However, not all adolescents with ED engage in SIV outside of their eating behaviors. Of note, the research on ED with co-occurring SIV largely neglects the most common ED population—Eating Disorder Not Otherwise Specified (EDNOS). The prevalence of SIV tends to be higher in patients with purging behaviors, but not at a level of statistical significance in most studies (Cassin & von Ranson, 2005).

Patients with ED and SIV are stigmatized in both professional and lay communities. One of the functions of the "difficult to treat" labeling of SIV patients is to help treatment providers maintain distance from their transferential feelings. Labeling also blinds clinicians to the adaptive function of SIV and their identification with this population. Opening a clinician's eyes to these truths will lead to the closeness and understanding necessary for the development of an effective healing relationship. As human beings, we have all engaged in myriad and subtle forms of self-neglect. Operating from a non-judgmental stance offers patients the compassion and acceptance that they desperately need. This patient population is exquisitely sensitive to the difference between experiential knowledge and book knowledge. The gap between providers talking the talk and walking the walk may be tragically reflected in the recidivism and stigma associated with adolescents who self-injure.

UNDERPINNING: TRAUMA

Feelings, emotion, somatic memory, trauma, sensuality, and sexuality are all human experiences. Many patients with an ED and SIV have a history of childhood abuse, neglect, or interpersonal chaos in the home environment (Paul et al., 2002). Self injury later in life is strongly associated with histories of early childhood sexual abuse (Van der Kolk, Perry & Herman, 1991), but not all self-injurers have such histories. Adolescents with ED and SIV desperately need a daily means to release and regulate their seemingly unmanageable emotional experiences (Ross et al., 2009; Wildes, Ringham & Marcus, 2009). Eating disorders and self injury in part represent an avoidance response that becomes generalized to all emotions, especially for patients who have come to experience all emotions as re-traumatizing. Since adolescence is a developmental stage rife with uncertainty, identity diffusion, and intense emotions, the SIV can soon become a trusted ally and stable pillar of the adolescent's identity.

The brain areas connected to PTSD in patients are relevant to memory and affect regulation. The hippocampus plays a critical role in the formation of verbal or declarative memory. It is not uncommon for patients with PTSD to have limited declarative memory of their trauma, which is likely to be related to the neurotoxic effect on the hippocampus of high levels of stress hormones released during traumatic experiences. Over-activity in the amygdala has also been found in patients with histories of trauma and PTSD (Shin, Rauc & Pitman, 2006). The amygdala is an area of the brain involved in emotion regulation, particularly with fear. Dysfunction in emotion regulation correlates clinically with a patient's experience of having overwhelming, intense fear states (Shin et al., 2006). The impact of traumatic experiences on any areas of the brain tends to be more significant during periods of rapid development in childhood and early adolescence.

Nature

Research has revealed evidence for genetic underpinnings for ED. Multi-family linkage analysis studies have found associations of genes on specific chromosomes with anorexia nervosa (AN) and bulimia nervosa (BN) (Bacanu et al, 2005; Devlin et al., 2002; Grice et al., 2002). Other studies have examined the association of certain genetic markers with traits common to ED or SIV, such as perfectionism or impulsivity. Serotonergic deficits have been observed in studies of spinal fluid and peripheral blood in patients with impulse control disorders (including those who self-injure), in patients with AN and BN, and in those with histories of childhood abuse (Steiger et al., 2004). Dopamine reward circuitry has been studied most intensively with populations addicted to drugs and alcohol. Evidence is emerging from functional imaging studies showing similar activity of the reward pathway for patients with the so-called process addictions (including compulsive overeating and compulsive self injury) when exposed to "using cues" (Coletta et al., 2009). Patients who self-injure or purge also set up a reward circuitry in order to become dependent on the endorphin release associated with acute physical harm. As in substance addictions, the phenomenon of tolerance develops and more ED or SIV behavior is necessary to maintain the effects of achieving a state of calmness, numbness, and/or well-being.

Each individual has a biological susceptibility at the receptor level in the brain's reward center which determines the amount of relief experienced after using one type of self destructive coping mechanism compared to another. This influences the patient's unhealthy coping behavior of choice, which they turn to when faced with limited resources and overwhelming life experiences. For example, binging and purging does not "work" for everyone who attempts to use it for release, escape, or emotion regulation. Similarly, SIV doesn't leave most people feeling safe, in control, or free from pain. These types of coping mechanism develop in certain genetically and neurophysiologically vulnerable people who have a need for them—those with limited support facing uncontrollable and overwhelming circumstances.

Nurture

In addition to the biological underpinnings, SIV can have its roots in psychosocial causes (Linehan, 1993a). Learning effective emotion regulation and self-soothing skills begins in the first year of life. When developmental needs are not met, individuals learn to regulate their emotions through external measures. In the case of SIV, these measures are visceral actions taken against the body (Ross et al., 2009). This developmental disruption can be the result of many factors, including the environment in which the patient was raised. In some cases, an abusive or neglectful environment is at the base of developmental arrest. In homes with alcoholic or mentally ill parents, for example, children may not consistently receive the attention or modeling of emotion regulation that they require. The experience of abandonment, in one way, shape, or form, is common to patients with SIV.

In addition to the abusive or neglectful environments of some teenagers with emotion dysregulation, other teens may have been raised in "normal" households but were simply mismatches with the parenting styles of their caregivers. This "goodness of fit" theory has been explained in great detail elsewhere (Chess & Thomas, 1986; Thomas & Chess, 1985), and remains an important consideration when attempting to understand the causes behind the arresting of emotional development in teenagers. In situations where parents are unaware of the differences in one child's emotional needs, or cannot adjust styles to meet that child's needs, the child grows up in an invalidating, even if well-meaning, environment.

Regardless of the reason, when children are raised in environments where the acknowledgment and expression of emotion was not fostered, they do not learn how to depend on their own internal guidance to identify and properly act on feelings (Linehan, 1993a). This externalized locus of control requires teenagers to look to outside sources so as to understand how they are feeling—leading to significant problems, especially in unpredictable environments. In emotionally minimizing environments, they may need to continually express more heightened responses to get their needs met (Linehan, 1993a). Additionally, because they are unable to trust their internal responses, patients with external locus of emotional control depend on messages from others to "know" how they are feeling. This type of relationship is a set-up for patients to base their sense of self solely on what others think. Without a positive self-regard and internal locus of control, motivation for treatment and change is understandably low. Modeling these qualities to our patients can be the most effective tool for motivation enhancement.

IMPACT: PATIENT AND ECOLOGICAL SYSTEM

The impact of SIV is deep and multi-faceted. In addition to its physical and emotional toll on a patient's well-being, SIV casts a large net of pain and disruption on their entire ecological system, including family members, school peers, romantic partners, and significant others.

For the SIV patient, self injury is a trauma in and of itself. Overt SIV results in a physical wound as the body is literally under assault at the hands of its owner. The wounds may require medical attention, and, if left unattended, can become infected and result in larger, more serious health problems.

The physical trauma of SIV is, however, largely symbolic of the intangible psychological trauma lying underneath the bodily wounds. While trying to regulate emotion through SIV, patients are re-sending a message to themselves that they are worthless or deserving of physical attack. The shame and secrecy of SIV contributes to the psychological trauma; as with active ED symptoms (binging, purging, restricting), SIV is frequently done in solitude. This time spent in self-destructive seclusion contributes to their isolated state of mind as well as missed opportunities to learn to modulate emotions in the context of a healing community.

The developmental impact of SIV on teenagers is far-reaching. As with other trauma, developmental progress is halted at the time of the trauma (Trickett & McBride-Chang, 1995). Precious time is lost to self harm, leaving less of it for other aspects of teenage life. Their social development lags in regard to friendships and romantic relationships, important milestones of these years. A vicious cycle of loneliness and alienation ensues. Teens that self injure have more reluctance to form positive peer relationships and report fewer positive peer experiences than those who do not (Ross et al., 2009). Sexual development and exploration as appropriate to these years is avoided, delayed, or changed as the attack on the body can be used to dissipate feelings of sexual tension and to avoid or deny sexual feelings. In addition, the stigma associated with SIV may cause other parents to disapprove of relationships with affected teens.

Academic achievement can also be stymied due to an inability to complete work when overwhelmed with emotions, or when active symptoms or treatment needs interfere with attendance. Teachers may have little experience with SIV and may not be prepared to best support the student in their classroom. Similarly, SIV patients may experience negative impacts in their work life. Bosses may not understand what exactly is happening, and may not be prepared to support an employee with such symptoms. In short, the developmental impact of SIV can be catastrophic and reaches into every aspect of a typical teen's experience.

The family unit is rocked when a member engages in SIV. Family routines become disrupted by the SIV events, treatment, and fears of what is yet to come. As focus increases on the SIV, family dynamics can become strained, and other pressing problems neglected. Treatment can financially strain a family. Parents can become overwhelmed with anxiety about their own problems or from blaming themselves for contributing to and not being able to "fix" the SIV. Alternately, parents can experience avoidance and denial of the problem, thereby giving room for the SIV to continue unchecked. As the family focus becomes hyper-attuned to the SIV patient, other siblings may begin to feel left out.

Finally, the friends and significant others of patients who engage in SIV are also affected. In intimate relationships, the boyfriend or girlfriend may feel impotent to help. Close friends can also feel this way, and might try to avoid discussing the problem or avoid the friend entirely.

RECOVERY DEFINED

In order to define recovery from SIV, the following questions must be addressed: what is the therapeutic task? How are we authorized to help? What is our role in the process of recovery? The therapist's task is to support and facilitate the process of recovery. At its core, the healing work is about connecting on a physical, emotional, and spiritual level with our patients. Recovery is not as simple as stopping the ED or SIV behaviors (see McGilley and Szablewski, Chapter 12). Families and treatment providers may be as confused and frightened by SIV as the patient. The more others try to make a patient's behaviors stop, the more shame they swallow about their powerlessness over using SIV for comfort. In the absence of an honest acknowledgement of the benefits that SIV or ED symptoms serve, and a gentle and loving approach to addressing the roots of earlier traumatic experiences, there will be little hope for long-term recovery.

Being mindful of the benefits and costs of SIV as a coping mechanism, understanding the profound despair and pain of patients using it, and trusting in the process of recovery are the therapeutic tasks—not trying to fix, manage, and control patients or their choices. Approaching the work with ED/SIV patients as a trusted servant rather than a force of change keeps the therapist's job manageable and transmits a sense of manageability to a patient who may be starving for it. Once they have hit bottom, patients are best supported by the therapist being present and returning responsibility for recovery back over to them. Through acknowledging powerlessness over the patient's disease and choices, and focusing on what is controllable, therapists become open to seeing their true power and channeling their energy to good use. One of the most useful ways therapists serve patients is by believing in them until they can believe in themselves, caring for them until they can care about themselves, and loving them until they can love themselves (Mazelis, 2002).

While recognized recovery programs exist for other forms of addiction, those who self injure tend to have more difficulty finding support from positive group identification. An important caveat to the dearth of community support is Self Mutilator's Anonymous (SMA), whose mission statement reads:

> A fellowship of men and women who share their experience, strength, and hope with each other, that they may solve their common problem and help others to recover from physical self-mutilation...The only requirement for membership is a desire to stop mutilating oneself physically. There are no dues or fees for SMA membership. We are self-supporting through our own contributions. SMA is not allied with any sect, denomination, politics, organization or institution; does not wish to engage in any controversy; neither endorses nor opposes any causes. Our primary purpose is to stop mutilating ourselves and to help others to recover from self-mutilation. *(http://www.selfmutilatorsanonymous.org)*

SMA has begun the difficult work to become a recognized recovery and support network for those moving through their SIV recovery. Its philosophy is based on that of other 12-step programs (http://www.12step.org/), which offer ongoing support and a spiritual path to life-long recovery (see Appendix A). Despite SMA's importance and promise as a growing recovery community, currently only 20 live meetings in North America and three on-line meetings are listed on the national website.

In addition to the limited number of mutual support group meetings, those in the process of recovering from SIV confront many other obstacles. Just as the symptoms are stigmatizing, taking steps to confront and overcome SIV also carries a stigma. While alcoholism and substance abuse recovery is not shame-free by any means, it is more widely recognized since most people know active or recovering alcoholics and/or addicts. Few people know patients who compulsively engage in SIV, much less those who are open about their recoveries.

Recovery is typically a long-term process requiring intensive, specialized treatment. The tedious pace of recovery is itself good medicine for the impulsivity and urgency that otherwise characterizes the disease. Inadequate treatment, in terms of amounts or pacing, is a major cause of relapse in the ED/SIV population. Limited access to care presents further obstacles to recovery. Persisting through these obstacles, and maintaining a non-judgmental, and compassionate stance allows providers to effectively join patients and their families in their healing journeys. Patients who have successfully recovered from other addictive disorders, their treatment providers, and family may not view the SIV as a primary treatment issue to be addressed. Additionally, some patients express fatigue from recovery of other disorders and/or prioritize maintaining their recovery in, for example, their ED or alcoholism.

A UNIFIED TREATMENT APPROACH FOR PATIENTS WITH ED AND SIV

The benefits of a unified approach for this population are manifold; facilitating integration and unity are the roots of healing. Piece-meal treatment approaches with this vulnerable population risks giving them a mixed message, and, at worst, may increase their sense of disintegration and identity diffusion. A unified, experiential approach combining dialectical behavioral therapy (DBT) (Linehan, 1993a) and 12-step facilitation with their inherent philosophies of acceptance and change provides a unique way to facilitate integration and healing for this patient population. In a unified DBT recovery model, we integrate specific steps of 12-step recovery with the principles of DBT. Three core tenets of this approach are described below.

Dialectics of Step One

At the heart of a harm-reduction model is the acknowledgment of benefits of both the SIV and ED. However, there is a critical distinction between acknowledging the benefits and enabling self-injurious behavior. Some therapists will steer clear of any recognition of the benefits of SIV for fear of supporting the behavior. Linehan (1993a) found that without a balance of acceptance and change strategies, patients feel invalidated by the therapist and are less likely to engage in treatment. A clinical example illustrates this point. An older adolescent with a history of sexual trauma in childhood was in recovery from binge-purge type AN for 1 year. She began to engage in daily hair pulling and cutting in the pubic area. Although ashamed and embarrassed about it, she was not ready to stop. According to the stage of change model (Prochaska & DiClemente, 1983), she was in the contemplation stage.

The adolescent was given a homework assignment to take an inventory of the behavior (step four in a 12-step model of recovery). The assignment consisted of a written inventory of 25 positives and negatives of engaging in her SIV, and to read it at the next session with her therapist (step five). This assignment incorporates the dialectic in DBT (i.e., costs/benefits of the behaviors), as well as the concept of radical acceptance (from the therapist and within the patient). An excerpt of what she wrote follows:

> *Positives*: It feels really good; it numbs my feelings; it helps me fall asleep; it helps me avoid eating; I feel a little high during and after; it gives me a sense of control; it gives me a sense of power; it gives me a reason for my shame; it's less dangerous than binging and purging; it helps me feel less lonely.
>
> *Negatives*: Scarring; I get trapped in it; I feel shame about it; I'm abusing my body; it separates me from God; it's time-consuming; I feel depressed afterwards; I isolate during and after; I can't stop even when I feel like I want to; I feel broken afterwards.

The therapeutic pillars of this exercise included the relief associated with honesty, shame reduction, and self-disclosure. Further, it represented a practical demonstration of the delicate balance between acceptance and change strategies when working with patients who are ambivalent. The Acceptance/Change dialectic underlying DBT treatment is captured by the Serenity prayer: "God/Goddess/Higher Power, grant me the serenity to accept the things I cannot change, the courage to change the things I can, and the wisdom to know the difference." The dialectic in the serenity prayer boils down to acceptance of being powerless over the disease, while still taking responsibility for recovery by tapping into the real power inside themselves to change and heal.

Radical Acceptance

Another core principle of DBT is the concept of *Radical Acceptance* (Linehan, 1993a), which is similar to step one's principles of honesty and acceptance. Patients are supported in understanding and acknowledging that they suffer from a disease they did not ask for, and that, without help, can ruin or end their lives. Therapeutic focus is placed on what they *can* change—accepting their past and who they are in the present. Choices are identified and emphasized, the most important of which is whether or not to remain committed to recovery.

Willingness

The distinction between willingness and willfulness is another core tenet of an integrated DBT recovery approach. Willfulness is characterized by acting without regard to known adverse outcomes. It involves a lack of surrender to the "wise mind," and a rigid hanging onto old ways of thinking and behaving regardless of outcome (Linehan, 1993a). The mind of an untreated addict in distress will always take an immediate over a longer term reward, even if the latter is much bigger. During intense emotional states marked by hyperactivity of the amygdala and decreased activity in the frontal lobes, the patient loses the ability to "play the tapes forward." DBT skills, particularly mindfulness, and the use of the group or a sponsor in 12-step programs offer a means to re-engage the frontal cortex, which is essential for executive functioning, impulse inhibition, planning, and decision-making. One of the most powerful moments to witness is when a patient recognizes for the first time that a thought of, "I should cut myself," is just a thought and not an imperative. Through mindfulness, patients learn to identify thoughts as thoughts and emotions as emotions, and choose responses to these stimuli that align with their wise mind (Linehan, 1993a; Miller, Rathus & Linehan, 2007).

Willingness and open-mindedness are essential in the recovery process. Patients are encouraged to ask themselves, "Am I willing to experiment with different ways of coping?" "Am I willing to take responsibility for my choices and actions?" These qualities are captured by the second and third steps of the 12-step recovery model (see Appendix A), and the DBT skills of acting opposite to emotion, and often times opposite to the compulsion, cravings, and urges which arise in states of emotional arousal (Linehan, 1993b).

A full discussion of medication interventions used in the ED and SIV populations is beyond the scope of this chapter. Both for ED and SIV, medication is really an adjuvant treatment. Research supports only a limited role for medications in the treatment of ED and SIV (see Levine and Levine, Chapter 7).

Implementing a Unified Approach with Teens

Implementing DBT in a residential setting can be a difficult task. In outpatient DBT programs, commitment is obtained from every participant and is the basis for all further treatments. For example, when a client presents with resistance, the therapist is able to redirect the client using their commitment as an acknowledged foundation to the treatment and creating a life worth living (Linehan, 1993a). This approach allows for a greater dialectical dance of acceptance and change to occur within the therapeutic relationship. Adolescents in residential settings are not often in treatment willingly, but rather as a result of their parents' and treatment provider's consensus. Creating a "buy in" for these patients presents a challenge. Although there is no easy answer to this lack of self-motivated interest, it is possible to engage patients through flexible and determined skills lessons. The primary path to engagement in DBT skills training for adolescents is employing experiential techniques that lie within Linehan's philosophy of dialectics, validation, and non-judgmental skills coaching.

Offering two DBT sessions per week helps balance the didactic skill lessons with experiential exercises. One session is a didactic presentation teaching Linehan's (1993b) four skill areas: *Mindfulness*, *Emotion Regulation*, *Interpersonal Effectiveness*, and *Distress Tolerance*. In the second session, the focus is on learning or practicing the skills in creative ways, such as using art to solidify the week's lessons. For example, in the Emotion Regulation Handout 3, "Life of an Emotion" lesson (Linehan, 1993b, p. 137), patients learn about what happens when they experience emotions (from the prompting event, to internal and external experience, to labeling emotion name, to the aftereffects). In the initial session, Handout 3 (Linehan, 1993b, p. 137) is reviewed and group members give examples of how their emotions and reactions to emotions fit into the cycle. In the subsequent session, in groups of two or three, residents create their own version of Handout 3, using a representative piece of poster-sized work for each step of the cycle (e.g., Prompting Event, Interpretation of Event, Emotion Name). At the end of the group meeting, residents share their work with each other, explaining how that step fits in with the rest of the cycle. (See Appendix B for an example of this experiential group's artwork.) After completing this exercise, adolescents are expected to demonstrate an understanding and synthesis of the Emotion Regulation skills presented.

In addition to art projects, role-playing and skits offer adolescents experiential learning opportunities. Interpersonal effectiveness and distress tolerance skills lend themselves nicely to such techniques. Most adolescents are able to identify several real-life situations that ended negatively due to their inability to use pro-social relationship skills or effectively tolerate distress. Using these personalized examples, adolescents are encouraged to role play alternative endings using Linehan's DEAR MAN Interpersonal Effectiveness skill or various Distress Tolerance skills (e.g., ACCEPTS, IMPROVE, Self-Soothe; Linehan, 1993b). Through this interactive skill building method, group participants practice and experience how these skills apply to their particular life, so that they become more familiar and accessible when needed.

Another technique to engage the adolescent's skill building is to offer an "advanced" training group to residential patients willing to make the commitment to using and practicing mindfulness and DBT skills (Miller et al., 2007). Patients in this advanced group attend another weekly group, as well as receiving individually tailored skills training. As they display progress in their treatment, the advanced group's effectiveness becomes visible to other patients and increases their motivation to engage in both general and advanced groups.

Teaching mindfulness to teens can present another roadblock to adolescent engagement in DBT. It is highly unfamiliar for adolescents used to continual stimulation from multiple inputs to focus on a single task for any length of time. However, this obstacle is usually quickly overcome through validation and setting clear limits around participation. Once expectations are set for each group to begin with a mindfulness exercise, older group members will acquaint the newer members with the guidelines of the activities. Mixing more active mindfulness activities (playing telephone or a game of musical chairs) with the more traditional (focusing on the breath or describing an object) expands teens' means of mindful presence (Miller et al., 2007). As patients develop in their mindfulness skills and repertoire, encouraging them to lead the mindfulness practice at the beginning of the session can further engage both willing and reluctant participants.

Getting creative and employing experiential techniques while sticking to the philosophy of DBT can be enjoyable for both group members and leaders. Additionally, for those lacking initial commitment, it has the added benefit of keeping adolescents engaged and learning new skills. The combination of experiential methods, an advanced group, and group leaders' unwavering dedication to continued mindfulness practice, sets the stage for meaningful participation and skills building.

Milieu Treatment

As problems with SIV and ED will have developed within the context of a patient's family, friends, and environment, patients must transfer their new skills to these familiar relationships and situations. In outpatient programs and residential settings, family involvement in groups has proven to be a successful addition to adolescent DBT trainings and skill building (Hollander, 2008; Miller et al., 2007).

Milieu therapy is a unique element of inpatient and residential treatment. Throughout the day, residents may encounter different situations (meals, conflict with peers, upsetting group topics, etc.) that elicit emotional arousal. Treatment providers are around at all times, providing patients with opportunities to get "on the spot" skills coaching in many different contexts (Holmes, Dykstra, Williams, Diwan & River, 2003).

A challenge for milieu treatment is ensuring that active symptoms are not inadvertently reinforced. It can be triggering for a resident to see others who self-injure. Competition issues can evolve, especially for those who have used their eating behaviors or SIV to gain needed attention. Having others use those same methods can feel threatening to adolescents, who might turn back to SIV to ensure that their needs will continue to be met by the staff. Thus, it is important to have a structured milieu setting with specific approaches designed to not inadvertently reinforce SIV. For example, a careful balance must be struck between responding to any incidents of SIV without providing positive reinforcement; if the only time a patient feels that his or her emotional needs are getting met by staff is when he or she self-injures, then logically the patient will want to continue to self-injure. However, if incidents of SIV are met with matter of fact care and attention while lavish praise and attention is given to more positive coping skills, the positive skills will be quickly reinforced on the unit both for patients who self-injure and for the rest of the milieu (Holmes et al., 2003).

On the other hand, exposure to peers with similar symptoms can be a source of enormous support to suffering teens. In addition to the previously noted benefits of a 12-step program, the milieu can provide a similar type of peer understanding and help. Residents will be in varying stages of recovery, from first-time treatment seekers to those in longer-term recovery journeys. In time, they begin to trust one another, engender mutual hope and learn from the collective group experience. For example, a resident may often see in others what she cannot see in herself, and can care about others in a way that she does not yet care about herself. Likewise, her peers can do the same for her. These are potent occasions of therapeutic mirroring and reciprocity.

While the milieu setting is an excellent environment for laying the recovery groundwork, transitional planning is vital for successful long-term outcome. As patients develop and practice coping skills and the 12 steps in the milieu, situations must be continually introduced to expose them to greater approximations of the environment to which they will return (Holmes et al., 2003). It is important to keep a comfortable tension between creating confidence and familiarity in skills already learned, and presenting new learning opportunities.

IMPACT OF SIV ON TREATMENT PROVIDERS

Self-inflicted violence can be a difficult and daunting clinical presentation for many clinicians, especially when combined with a comorbid diagnosis of an ED. Staff at both residential and community treatment centers face distinct challenges when providing care to multi-impulsive adolescents. The effects of vicarious traumatization (VT) on healing providers have been well-documented in the scientific and clinical literature (Sabin-Farrell & Turpin, 2003).

Anger and frustration regarding a client's progress, or lack thereof, are common experiences of VT. These feelings can arise out of the provider's sense of hopelessness regarding the client's improvement, as well as self-doubt regarding their own effectiveness (Linehan, 1993a). Additionally, feelings of loss of control, or of not being able to fully guarantee a patient's safety, are typical in residential or community based providers (Holmes et al., 2003). The word "manipulation" gets tossed around frequently when working with the ED/SIV population, and the stigma surrounding SIV is perpetuated. Treatment providers are susceptible to projective identification and can end up carrying some of the patient's pain avoidance. To steer clear of their own pain and limitations as therapists, they may distance themselves from the SIV population. This, at its core, is counter-therapeutic.

Difficulties associated with VT underscore the importance of following the DBT framework, which stresses regular consultation with clinicians who treat similar populations (Hollander, 2008; Linehan, 1993a). Consultation agreements, one of the tenets of DBT work, provide a framework with which to approach the care both of clients and of other providers (Linehan, 1993a). Consultation groups can acknowledge and provide the support needed for working with this difficult-to-treat population. Peer support also helps clinicians recognize any of their own covert or overt SIV, and provides an avenue of accountability in those areas. Similarly, clinicians are encouraged to attend their own 12-step support groups. Al-Anon, for example, can be a very useful way to help staff learn how to take care of themselves while caring for others.

An individual mindfulness practice is an additional self care tool for therapists who treat ED/SIV patients. Mindfulness allows individuals to begin to see their own thoughts and judgments without reaction (Kabat-Zinn, 1994). This naturally extends to interactions with clients as well. Practitioners become more aware and forgiving of themselves and others, and better able to slow down their reactions to discern if they are being helpful and effective in their interactions (Shapiro & Carlson, 2009; Wilson & Dufrene, 2008).

CONCLUSIONS

Treating adolescents who struggle with both SIV and ED is an intimidating challenge for even the most experienced provider. The impact of SIV and ED reaches beyond the identified patient to her family, environment, and treatment team. The suggested unified approach that blends DBT with a 12-step model allows the patient, family, and provider to address these problems on physical, emotional, and spiritual levels, setting a framework from which integrated recovery can be achieved.

References

American Psychiatric Association (2000). *Diagnostic and statistical manual of mental disorders* (4th ed., text rev.). Washington, DC: Author.

Bacanu, S. A., Bulik, C. M., Klump, K. L., Fichter, M. M., Halmi, K. A., Keel, P., ...Devlin, B. (2005). Linkage analysis of anorexia and bulimia nervosa cohorts using selected behavioral phenotypes as quantitative traits or covariates. *American Journal of Medical Genetics Part B: Neuropsychiatric Genetics, 139B*(1), 61–68.

Cassin, S. E., & von Ranson, K. M. (2005). Personality and eating disorders: A decade in review. *Clinical Psychology Review, 25*, 895–916.

Chess, S., & Thomas, A. (1986). *Temperament in clinical practice*. New York, NY: Guilford Press.

Coletta, M., Platek, S., Mohamed, F. B., van Steenburgh, J. J., Green, D., & Lowe, M. R. (2009). Brain activation in restrained and unrestrained eaters: An fMRI study. *Journal of Abnormal Psychology, 118*(3), 598–609.

Devlin, B., Bacanu, S. A., Klump, K. L., Bulik, C. M., Fichter, M. M., Halmi, K. A., … Kaye, W. H. (2002). Linkage analysis of anorexia nervosa incorporating behavioral covariates. *Human Molecular Genetics, 11*(6), 689−696.

Favaro, A., & Santonastaso, P. (2000). Self-injurious behavior in anorexia nervosa. *Journal of Nervous and Mental Disease, 188*, 537−542.

Favazza, A. R. (1996). *Bodies under siege: Self-mutilation and body modification in culture and psychiatry* (2nd ed.). Baltimore, MD: Johns Hopkins University Press.

Grice, D. E., Halmi, K. A., Fichter, M. M., Strober, M., Woodside, D. B., Treasure, J. T., … Berrettini, W. H. (2002). Evidence for a susceptibility gene for anorexia nervosa on chromosome 1. *American Journal of Human Genetics, 70*, 787−792.

Hollander, M. (2008). *Helping teens who cut: Understanding and ending self injury.* New York, NY: Guilford Press.

Holmes, E. P., Dykstra, T. A., Williams, P., Diwan, S., & River, L. P. (2003). Functional analytic rehabilitation: A contextual behavioral approach to chronic distress. *Behavioral Analyst Today, 4*(1), 34−46.

Kabat-Zinn, J. (1994). *Wherever you go, there you are: Mindfulness meditation in everyday life.* New York, NY: Hyperion.

Levitt, J., Sansone, R., & Cohn, L. (2004). *Self-harm behavior and eating disorders: Dynamics, assessment and treatment.* New York, NY: Brunner-Routledge.

Linehan, M. M. (1993a). *Cognitive-behavioral treatment of borderline personality disorder.* New York, NY: Guilford Press.

Linehan, M. M. (1993b). *Skills training manual for treating borderline personality disorder.* New York, NY: Guilford Press.

Mazelis, R. (2002, Spring). Demystifying self-inflicted violence: Lessons learned from the past dozen years. *The Cutting Edge, 13*(49), 1−3.

Miller, A. L., Rathus, J. H., & Linehan, M. M. (2007). *Dialectical behavior therapy with suicidal adolescents.* New York, NY: Guilford Press.

Paul, T., Schroeter, K., Dahme, B., & Nutzinger, D. O. (2002). Self-injurious behavior in women with eating disorders. *American Journal of Psychiatry, 159*, 408−411.

Prochaska, J. O., & DiClemente, C. C. (1983). Stages and processes of self-change of smoking: Toward an integrative model of change. *Journal of Consulting and Clinical Psychology, 51*(3), 390−395.

Ross, S., Heath, N. L., & Toste, J. R. (2009). Non-suicidal self-injury and eating pathology in high school students. *American Journal of Orthopsychiatry, 9*(1), 83−92.

Sabin-Farrell, R., & Turpin, G. (2003). Vicarious traumatization: Implications for the mental health of health workers? *Clinical Psychology Review, 23*(3), 449−480.

Shapiro, S. L., & Carlson, L. E. (2009). *The art and science of mindfulness: Integrating into psychology and the helping professions.* Washington, DC: American Psychological Association.

Shin, L. M., Rauc, S. L., & Pitman, R. K. (2006). Amygdala, medial prefrontal cortex, and hippocampal function in PTSD. *Annals of the New York Academy of Science, 1071*, 67−79.

Steiger, H., Gauvin, L., Israël, M., Kin, N. M., Young, S. N., & Roussin, J. (2004). Serotonin function, personality-trait variations, and childhood abuse in women with bulimia-spectrum eating disorders. *Journal of Clinical Psychiatry, 65*(6), 830−837.

Svirko, E., & Hawton, K. (2007). Self-injurious behavior and eating disorders: The extent and nature of the association. *Suicide and Life-Threatening Behavior, 37*(4), 409−421.

Thomas, A., & Chess, S. (1985). The behavioral study of temperament. In J. Strelau, F. Farley, & A. Gale (Eds.), *The biological bases of personality and behavior: Vol. 1. Theories, measurement techniques and development* (pp. 213−235). Washington, DC: Hemisphere.

Trickett, P. K., & McBride-Chang, C. (1995). The developmental impact of different forms of child abuse and neglect. *Developmental Review, 15*(3), 311−337.

Van der Kolk, B. A., Perry, J. C., & Herman, J. L. (1991). Childhood origins of self-destructive behavior. *The American Journal of Psychiatry, 148*, 1665−1671.

Wildes, J. E., Ringham, R. M., & Marcus, M. D. (2009). Emotion avoidance in patients with anorexia nervosa: Initial test of a functional model. *International Journal of Eating Disorders.* Advance online publication, doi: 10.1002/eat.20730

Wilson, K. G., & Dufrene, T. (2008). *Mindfulness for two: An acceptance and commitment therapy approach to mindfulness in psychotherapy.* Oakland, CA: New Harbinger.

APPENDIX A

Twelve Steps of Self-Mutilators Anonymous

(http://www.12step.org/), as adapted by authors

Step 1   We admitted we were powerless over our self-mutilation—that our lives had become unmanageable

Step 2   Came to believe that a Power greater than ourselves could restore us to sanity

Step 3   Made a decision to turn our will and our lives over to the care of God as we understood God

Step 4   Made a searching and fearless moral inventory of ourselves

Step 5   Admitted to God, to ourselves and to another human being the exact nature of our wrongs

Step 6   Were entirely ready to have God remove all these defects of character

Step 7   Humbly asked God to remove our shortcomings

Step 8   Made a list of all persons we had harmed, and became willing to make amends to them all

Step 9   Made direct amends to such people wherever possible, except when to do so would injure them or others

Step 10   Continued to take personal inventory and when we were wrong promptly admitted it

Step 11   Sought through prayer and meditation to improve our conscious contact with God as we understood God, praying only for knowledge of God's will for us and the power to carry that out

Step 12   Having had a spiritual awakening as the result of these steps, we tried to carry this message to others, and to practice these principles in all our affairs

APPENDIX B

Representative Art from a Teenage DBT Group Regarding the Life of the Emotion

(Linehan, 1993a, Emotion Regulation Handout 3, p. 137)

Interpretation of Event. Patients created this collage to represent what might trigger an emotion. Their collage represents that both internal and external events may prompt an emotion. Through this expressive medium, the patients demonstrated understanding of Emotion Regulation Skills.

CHAPTER 17

The Weight-Bearing Years: Eating Disorders and Body Image Despair in Adult Women

Margo Maine

Although data on adult eating disorders (ED) are limited, clinicians have noticed an increase in the number of adult women seeking help for these conditions. For the first time, in 2003 a third of inpatient admissions to one prominent ED treatment center were over 30 years old (Davis, 2004, personal communication). Another treatment center reported a 400% increase in admissions of patients over 40 years of age between the years 1997 to 2007 (Cumella & Kally, 2008). Age does not immunize women from body image (BI) preoccupation and weight concerns, as has been thought in the past; in fact, disordered eating (DE) and a fear of aging go hand-in hand for many women (Lewis & Cachelin, 2001).

When women do not see older women respected and included in the dominant culture, they may begin to fear or even loathe the prospect of aging. Rarely do we see an older face, be it in film, fashion, advertising, print media, or television. Sexualized images of young, sculpted female bodies saturate everyday media. Quite simply, in Western cultures the natural process of aging disempowers women, more than it does men, and is likely to contribute to concerns about appearance, weight, and shape leading to DE and clinical ED. In this age of body technology, aging naturally is no longer the only option for women in midlife, as women are sold the myth that they can (and should) be in complete command of their bodies, even pursuing cosmetic surgeries to look younger and more attractive. Lastly, the "war on obesity" and the misinformation promulgated by the diet industry intensify many women's concerns about health at and beyond midlife, further contributing to DE practices, potentially triggering adult ED.

Today, dieting and BI concerns are normative experiences for women of all ages. For example:

- 43 million adult women in the United States are dieting to lose weight at any given time; another 26 million are dieting to maintain their weight (Gaesser, 2002)
- BI dissatisfaction in midlife has increased dramatically, more than doubling from 25% in 1972 to 56% in 1997 (Garner, 1997)

- 75% of American women, aged 25—45, report DE and BI dissatisfaction, with 10% meeting criteria for clinical EDs; 67% of non-ED women are trying to lose weight with over half of these dieters already at a normal weight (Bulik & Reba-Harrelson, 2008)
- Comparable levels of dieting and DE are found across young and elderly women (Lewis & Cachelin, 2001)
- A random sample of 1,000 Austrian women, all average BMI, desired a lower BMI; more than 80% "controlled their weight,"; more than 60% reported body dissatisfaction; 4% met the criteria for ED, and another 4% were subclinical (Mangweth-Matzek et al., 2006)
- When asked what bothered them most about their bodies, a group of women aged 61 to 92 identified weight as their greatest concern (Clarke, 2002)
- A major research project found that more than 20% of the women aged 70 and older were dieting, even though higher weight poses a very low risk for death at that age, and weight loss may actually be harmful (Berg, 2001)
- Two community studies indicate a positive relationship between a fear of aging and variables from the Eating Disorder Inventory (EDI) including a drive for thinness, body dissatisfaction, dieting, and interoceptive awareness (Gupta, 1995; Lewis & Cachelin, 2001).

The rhythmic cycles of the female body, many of which are associated with weight gain, such as premenstrual bloating, pregnancy, and the slower menopausal metabolism, are great challenges in this era of body control, glorification of thinness, and unrealistic beauty images. If a woman's power is still defined in terms of beauty and a youthful body, post-pregnancy weight gain and the pounds gained at menopause can be a source of great distress and anticipated disempowerment. The female body has never been more exposed while its natural processes, like the symptoms of menopause, are to be completely masked or controlled. Contemporary women struggle with the catastrophic consequences of ageism and weightism, suffering potentially devastating results. Unfortunately, the dominant thinking of both the public and professional sectors does not reflect current reality, nor does clinical research elucidate this issue, as few resources have been devoted to adult ED.

THE SHAPE OF ADULT EATING DISORDERS

Clinical experience suggests several different patterns in the lives of adult women with ED. Some have struggled since adolescence and have never fully escaped its grip. Others had an ED in their teens, but recovered, only to relapse later. Some have been preoccupied with food and weight for years, but were never incapacitated when younger, or they may be women who, faced with the challenges of adulthood, begin to diet for the first time in their lives. Although the popular press appears to be fascinated with the idea that women develop ED for the first time in midlife, in my clinical experience this is far less frequent. Research in this area is sparse; however, Pryor (2008) reports an adult onset in 6% of the adult women seen for ED treatment over a 2-year period. Most experienced a chronic course of illness that persisted into midlife.

Shedding some light on adult ED, Cumella & Kally (2008) retrospectively reviewed the cases of 50 women treated in their residential program whose symptoms began after age 40. On the EDI-2, this group had moderate severity of symptoms, but lower scores than younger patients on three variables: drive for thinness, bulimia, and body dissatisfaction; and higher scores on ineffectiveness, perfectionism, interpersonal distrust, and asceticism. They tended to exhibit the symptoms of restricting AN or Eating Disorders Not Otherwise Specified (EDNOS), with only 8% meeting the criteria for BN, in contrast to 40% of younger inpatients. Rates of depression and anxiety disorders were in the moderate range, similar to that of younger patients. Minnesota Multiphasic Personality Inventory scores indicated greater denial among older patients, perhaps accounting for the lower severity they reported in both their ED and comorbid conditions. The older patients indicated a higher frequency of sexual abuse (64%) than in the general population or in younger ED patients. Eighteen percent of late onset women reported substance abuse with a preference for more sedating substances such as alcohol, sedatives, marijuana, and opioids, and less stimulant use than in the younger patients. Twenty-two percent reported a history of self-harm and 28% had attempted suicide. Self-harm occurred less in this sample than in ED women in general, where the rate is as high as 45%. Cumella and Kally (2008) speculate that this may reflect generational differences in self-harm behavior. While helpful, this information is limited by being a retrospective review of patients seeking treatment in a faith-based facility. It suggests that older women with ED may have different but very serious concerns that need to take prominence in treatment and that they tend to under-report their distress.

As with young women, ED come in different shapes, sizes, and severities, from AN to BN, EDNOS, Binge Eating Disorder (BED), and subclinical or partial syndrome ED, including orthorexia, an obsessive fixation with "righteous eating," potentially leading to significant nutritional depletion, physical problems, depression, anxiety, and social isolation (Bratman, 2004). My clinical experience is consistent with the above report, suggesting that adult ED women often fall into the EDNOS or partial syndrome categories. This makes it more difficult for them to self-diagnose and seek help, as the popular self-help literature and magazine articles tend to focus on pure AN or BN via self-induced vomiting and neglect symptoms such as exercise abuse, laxatives, or chronic restricting without significant weight loss. Medical journals and opinions reflect this same bias.

The symptoms of many of my adult patients have changed over time, having initially been more purely restricting AN or more prominently BN via self-induced vomiting. Gradually, they have begun to eat more or to stop vomiting and even to reach a stable, "normal" weight, but their relationship with food is still disordered and a source of great distress. They may be engaging in ED behaviors less severely or less frequently, so may feel they no longer "have it," yet their behaviors and thoughts about food, weight, and their bodies are still distracting, obsessive, critical, and unhealthy. Such a patient might say *"but I eat now,"* while eating far less than her metabolic needs require, or might believe she is no longer purging when in fact she has substituted grueling hours of obsessive exercise or laxative abuse for vomiting.

Many common threads can be found among women of different ages suffering from significant eating and BI issues. They share personality characteristics of perfectionism, low self-esteem, and chronic feelings of inadequacy. Tending to focus on others rather than on their own needs and desires, they are often the peacemaker in the family. Ambivalent about their power and place as women, they must navigate a path in an overwhelming consumer culture that teaches them to want and to need, but not to know or meet their true wants and needs. Young or old, they are constantly bombarded by strict and unrealistic media images of beauty, by the myriad sources and marketing of body-change technology (e.g., pills, surgeries, exercise equipment, "cosmeceuticals") and the "war on obesity" fed by misinformation from the profit mongering diet industry. Regardless of the patient's age, my clinical work often reveals an intergenerational pattern of BI distress and preoccupation, restrictive dieting, and various attempts to sculpt the body to meet an unattainable ideal. Women struggling with eating and BI problems often relate that their families bred these issues, describing grandmothers and mothers who dieted chronically, exercised excessively, and even pursued cosmetic plastic surgery. Some report a similar preoccupation with weight, shape, exercise, and appearance in their fathers, grandfathers, brothers, and other male relatives, as well as highly critical comments from these important male figures about women's bodies. Kearney-Cooke (2004) traces the development of BI, demonstrating how women unconsciously internalize their family's messages about food and shape. Thus, pathogenic attitudes and behaviors surrounding BI, weight management, and food seen in women with ED are handed down from one generation to the next.

For many women, the focus on appearance and youth intensifies as their bodies age and progress through the natural stages of change that include weight gain, graying hair, and wrinkled skin. If their mothers, grandmothers, aunts, or other "role models" dieted and criticized themselves throughout their lives, they are likely to follow suit. The pressure to pursue the "perfect body" persists, creating a risk for EDs at worst and chronic dissatisfaction at best.

THE TROUBLE WITH TRANSITIONS

Much has been written about the multiple contributions to eating and BI issues in young women's lives (Levine & Smolak, 2006). We recognize the intense pressures and troubling transitions as they move from being a girl to a preteen, teen, and young adult. In fact, during the progression into adolescence and then from mid to late adolescence, girls are at the greatest risk to develop ED (Smolak & Levine, 1996). Developmental transitions create vulnerability because the normative external stressors require a major internal reorganization of the sense of self, including personality and cognitive and emotional structures. In essence, everything is in flux, both inside and out. For example, adolescence requires major changes in the female body's functioning, while, cognitively and psychosocially, it ushers in the pressure to adopt the social or cultural norms, values, and ideals, and to move away from the strict dependence on family into more engagement with the culture. The culture's relentless standards for beauty and appearance encourage unhealthy scrutiny of the body and pressures related to weight and appearance, markedly increasing the risks of ED at these developmental junctures.

Traditionally, it was thought that the field of developmental psychology stopped at early adulthood (Willis & Martin, 2005). Yet, adult life is full of transitions and stressors placing women once more at risk for DE and ED. These include: the complications of a changing, aging body; multiple role changes from school to career to marriage, mothering, divorce, empty nesting, caring for older parents; and health and mortality issues, among others. Adult women often find themselves dealing with these pressures in isolation, with little support or validation, but more and more responsibilities and burdens. Feeling intense pressure to adapt and achieve, they may have a seemingly very full life while feeling very empty, alone, and unfulfilled.

The "deadline decade," the years between 35 and 45, can be particularly disruptive (Sheehy, 1976). For some, the biological clock is ticking away, and decisions about relationships and childbearing are heightened. For others who have focused on family, their career clock may be ticking just as loudly. For the women who have tried to do it all, the what-about-me clock is ticking. To complicate their experience, women are in their sexual prime at this age, and their appetite for, and interest in, sexual relationships may be heightened, resulting in guilt and confusion. Potentially transformative, this decade is also filled with endless obligations and responsibilities, leaving women little time to reflect on the impact of all these events, emotions, and transitions.

In sum, young or old, when things are changing, feeling ambiguous, and looking uncertain, the natural human instinct is to find something stable or something to control. Contemporary culture breeds *The Body Myth*, the misguided notion that pursuing a perfect body is the answer to all the dilemmas contemporary Western women face (Maine & Kelly, 2005).

A PERFECT STORM: MIDLIFE TRANSITIONS IN AN ERA OF GLOBAL TRANSFORMATION

Cultures are constantly changing, but the current pace and intensity of global transformation is unprecedented. Transformative trends, such as sophisticated and fast growing economies, and rapid technological and market changes exert an enormous effect on the status of women, introducing a powerful global consumer culture, with unmitigated expectations about appearance and beauty as well as dramatic revisions in women's social role (Gordon, 2001). With greater access to education, increased involvement in the workplace, and the accompanying gender equity issues, women's lives and family roles are in a period of upheaval. Compounding these demands, today's Western fast food diet, and the corresponding sedentary lifestyle, has culminated in escalating rates of obesity. As a result, the gap between the body reality and the beauty ideal are increasingly in conflict, bringing up a myriad of issues, for women of all ages.

Contemporary midlife women lack female role models who have mastered the challenges of such a complex, rapidly changing culture. Their realities, especially the critical importance of appearance and body control, are substantially different from the experince of their mothers. Opportunities and expectations are endless, but these new roles constantly put women's bodies in the public eye and often in competition with younger women, or even with men. Globalized consumer culture expands the parameters for women's competition exponentially, intensified by media mavens like Martha Stewart and infinite number of Internet sites and television stations dedicated to critical domestic issues like cooking and home décor. While we may have taken women out of the kitchen, we have not taken the kitchen out of women, as they are still the primary feeders of their families. Despite the stress to look young, thin, and attractive, and the constant messages from the dieting industry, women also have to plan meals and feed their families, and do it perfectly.

The fast-paced life inherent in current postmodern culture emphasizes adaptation, achievement, and appearance, leaving little time to reflect on these new roles and expectations. Given little permission for negative feelings, and surrounded by a sociocultural environment that has labeled fat as bad, "the language of fat" is now the cover for all discomfort, anger, disappointment, and other "bad" feelings (Friedman, 1997). Unfortunately, globalization has now made the language of fat universal for women of all ages.

PREGNANCY AND PARENTING: RISK OR OPPORTUNITY?

Pregnancy, child-bearing, and parenting are stressful developmental phases, often putting a woman at increased risk for eating issues (Crow, Agras, Crosby, Halmi & Mitchell, 2008). Again, transitions into a new phase of life will challenge a woman in many ways, but pregnancy particularly puts her body in center stage. The lucky ones will experience a true joy about the miracles their bodies can achieve, and will be able to leave their ED behind. For some, pregnancy and parenting changes everything in a very positive way; for others, it is more complicated.

Dialog and education related to weight, health, and nutrition is important for all women of child-bearing age, but especially for those contemplating pregnancy or seeking fertility treatment. In fact, ED is common in women who present with fertility issues. AN occurs in approximately 4% and BN in 12% of women with infertility; these rates are triple the expected incidence. BN is associated with fewer fertility issues, but still can impact the pregnancy (Woods & Williams, 2008). Nearly 50% of cases presenting for treatment of irregular menses have EDNOS (Norre, Vandereycken & Gordts, 2001; Resch, Szendei & Haasz, 2004).

Pregnancy is an anxiety-provoking experience, but for women who have struggled with ED, the stress may be intensified due to their deficits in self-esteem, personal efficacy, and self-confidence, as well as their perfectionistic beliefs that they are never good-enough. The family of a woman with an ED history is often greatly relieved when she becomes pregnant, and may overlook or be less able to acknowledge her doubts or anxiety. Some families believe that pregnancy means that her problems are over, when, instead, pregnancy may bring new eating and BI issues. Feeling pressure to act happy and positive, the newly pregnant woman may feel deeply ashamed of these emotions and very alone.

Pregnant bodies seem to become public property, with unwanted comments and even touching from complete strangers. The normative discomfort with their bodies becomes magnified as their bodies grow and change, especially in the era of fascination with celebrity pregnancies. Shapely post-pregnancy images on magazine covers assault the self-esteem of new mothers in the supermarket checkout lane. The photos may have been altered by computer imaging, and these celebrities have far fewer responsibilities, with others taking care of their children, making their meals, and training their bodies back into shape. Media coverage of "pregorexia," describing women who gain very little during pregnancy and then instantly return to pre-pregnancy weight (CBS News, 2008), only increases the discontent that adult women have with their bodies and heightens their belief that they too can achieve this.

The research related to pregnancy and ED is limited; the few available studies have been generally retrospective, focusing on AN and BN and not on EDNOS. However, one prospective, longitudinal study (Crow et al., 2008) of women with clinical or subclinical EDs indicates that pregnancy, despite its risk for increased ED pathology, is also a potential time for positive change. Across all categories of ED, symptoms improved during pregnancy but tended to regress after the birth. Intensifying treatment during and after a pregnancy might help to maintain those improvements.

According to a review of studies of pregnancies complicated by active ED, these women are at greater risk for miscarriage, premature births, C-sections, smaller head circumferences, and lower birth weights (Woods & Williams, 2008). Pregnancy is also a risk factor for relapse, especially since women with a history of ED tend to have greater nausea/vomiting. Low weight alone (BMI less than 18.5) presents a greater risk than expected to the pregnancy (Helgstrand & Andersen, 2005). Women who become pregnant 15 years post treatment for AN still experience a higher frequency of miscarriage in the first trimester: 38% miscarried compared to 16% in the normal population. The risk normalizes as the pregnancy proceeds (Bulik et al., 1999).

Pregnancy brings up natural changes in eating and appetite, so women who have struggled with food or have restricted may begin to binge-eat, or those with pre-existing BED may continue binging (Bulik et al., 2007). Women who have chronically restricted their intake see even normal amounts of food as "a binge." Also, after such prolonged restricting, weight gain may occur faster than expected. I always warn newly pregnant patients about this and ask that they ignore charts of the expected weight gain in each month of pregnancy. Psychoeducation and collaboration with medical providers are essential to help a woman to interpret the changes in her body in a healthy way. For example, having ultrasounds earlier and even more often in the pregnancy may allow the patient to visualize her baby as the reason for the changes in her weight and body shape.

Little research is available regarding how women with a history of ED function during the transition into parenting. Koubba, Hällström & Hirschberg (2008) report that the few small, uncontrolled studies focused on feeding problems, and described moms with ED as often intrusive during feeding and play time and negative in emotional tone. Their study compared 44 first-time moms who had a history of ED prior to pregnancy with a control group from the same prenatal clinics. Of those with ED histories, 92% reported some adjustment problems versus 13% of the controls. Fifty percent had sought some sort of help for depression or other mental health issue in the first 3 months after birth. Screening for ED early in pregnancy and building in support, education, and intervention might improve the coping of these new moms and enhance the mother–child bond and baby's development.

A review of postpartum-research on ED mothers reported: three times the normal risk for postpartum depression; breastfeeding difficulties; lower than expected weight gain; maternal conflicts around normal expressions of autonomy especially related to feeding and food play; perfectionism related to parenting; and obsessional thinking regarding food, weight, and shape affecting interactions and emotional availability to the child (Woods & Williams, 2008). Another study of pregnancy and early infant development comparing women with adolescent onset AN to a normal control group, found few differences (Wentz, Gillberg, Anckarsater, Gillberg & Rastam, 2009). While no research is available on long-term developmental consequences in children born to actively eating disordered moms, general studies indicate developmental delays and lower intellectual, academic, and occupational performance (Reichman, 2005).

THE CHALLENGE OF CHILDREN

Despite the many joys it brings, parenting is a very stressful phase in adult development. Just like pregnancy, it brings both risk and opportunity when an ED is present. Many seek treatment because they want their children to be able to grow up without the burden of an ED. In fact, the desire to stop the cycle of ED and BI despair is a great motivator for many women. Jennifer's story is a great example of this, as well as of how her family history had shaped her ED. Jennifer teaches us that, even after suffering for decades, women can still recover.

JENNIFER

At age 42, decades into her ED, Jennifer sought help for the first time, motivated by concern for her 12-year-old daughter. Watching her daughter's body naturally become rounder in preparation for puberty, she recalled the pain of her adolescent body and realized that she was modeling behaviors that were not healthy. Jennifer was from a perfectionistic and body-conscious family who took her to Weight Watchers at the age of 12, concerned that she was "getting pudgy." Dieting throughout high school, Jennifer developed anorexic symptoms in college and has had some form of disordered eating since then, culminating in severe bulimia.

Seven years before seeking help, after the birth of her second child, she began purging via exercise to lose weight. When this did not restore her pre-pregnancy weight, Jennifer restricted even more, and then began vomiting almost everything she ate. As her daughter approached adolescence, she became increasingly concerned that she might pass her problems along to her, ultimately deciding to go to her PCP to ask for help. She had lost 20 pounds since her last visit a year earlier, putting her weight at the low end of normal, but significantly below her body's natural range based on her weight history, body type, and genetics. She had read about bulimia and finally was frightened enough to admit she might have this illness.

At this visit, with Jennifer feeling so desperate and confused, the nursing staff immediately praised her for her weight loss, asking for advice about dieting. Then her doctor walked in and asked, *"How does your husband like your new body?"* Jennifer was devastated and left without telling him why she was there. She was deeply depressed as she felt out of control and helpless, fearing she would be like this the rest of her life. Eventually, she found my name on the Internet.

Seven years later, with the help of outpatient therapy, nutritional counseling, and medication, Jennifer is much, much better. Physically she is stable, having regained her weight, but she still covets the thinner body of her bulimic days. Exercise is still obsessive, but much less so. Eating remains a barometer of her emotions; when facing significant stress, she restricts, but she has not purged in over 3 years. She has difficulty allowing herself to enjoy food and feels safest when she is eating rigidly. Her daughter has managed adolescence without an ED, but Jennifer had to work hard in therapy to understand her reactions to her daughter's changing body and to accept her daughter's comfort with a normal weight.

Through therapy, Jennifer has realized the many pressures she still experiences from her family of origin and has learned how to manage her reactions to their subclinical ED and appearance obsessions. She admittedly wishes she could return to her ED, but she wants to be there for her children and recognizes how her ED truly interferes with relationships, especially with her daughter. I have cautioned her to be careful about stressful periods or transition times in her life, as these place her at greater risk for behavioral relapse. Now, she comes for therapy when she hits these rough spots, but feels she can hear my voice and keep herself on track most of the time. Like many of my patients, Jennifer reports that being with a woman who does not buy into our culture's obsessions with weight and appearance is a critical and corrective experience.

CLINICAL ISSUES

Adult women are much more likely to seek help on their own, making the motivation for entering treatment quite different than that of a younger person. Often they have experienced some loss due to the ED. This may be a physical loss, such as fertility issues, miscarriages, losing teeth and encountering expensive dental bills. Or it may be a relationship loss: a failed marriage, or the transition into caring for elderly or ill parents and its attendant loss of youth. In my clinical experience, parenting appears to be the most motivating subject for women with eating and BI disorders and we can use it to the advantage of recovery. It is easy to demonstrate how their own self-care will benefit their children and to emphasize the importance of their role modeling related to weight concerns, BI, dieting, and eating, especially for female children. I frequently remind them about the directions we hear every time we board a plane: *"In case of a change in cabin pressure, an oxygen mask will drop down. You must put yours on first before turning to help your child or anyone else."* This is a repeating theme when working with my adult patients. I say *"oxygen mask,"* and they know they need to focus more on themselves.

Although they are likely to be more self-motivated, adult women may also have lived with this problem for decades, lacking skills to recognize and satisfy their true needs and hungers. Therapy can be very challenging, as they struggle to answer basic questions, like: *"What do you like to do?" "What feels good?" "What comforts you?"* These simple questions can be quite daunting to a woman who has never been encouraged to pay attention to her inner self. As she learns her own answers to these questions, she finds coping mechanisms and activities that can gradually replace the ED.

Midlife women have the same exquisite sensitivity to others that we associate with ED in vulnerable teens and young women; they easily feel shamed, shunned, and dismissed, disappearing quickly if we do not work hard to embrace them and show that we take them as seriously as we take adolescents with these problems. In groups and formal treatment programs, adult women often feel invisible and inconsequential due to the emphasis on younger patients and their developmental issues. They may also play a maternal or nurturing role to others, both to the younger patients and to the staff, as they are skilled in meeting others' needs and not their own. With their longstanding habit of sacrificing their needs to please others, they may pretend to be recovering quickly, but not truly address their underlying problems, and therefore risking serious relapse.

At times challenged to embrace the motivations that bring my adult patients to treatment, I have come to accept any reason that helps women to break their denial of how these issues are affecting the quality of their lives. For example, a woman in her early 50s sought my help as she wanted to have cosmetic plastic surgery, but knew that if she continued to vomit, the effects of the surgery would be minimal. I agreed to work with her regardless of my feelings about cosmetic surgery. We addressed her issues with aging and how these had affected her longstanding ED and BI concerns. Despite a 30-year history of daily bulimia, she achieved some degree of recovery, no longer purging, but often restricting her intake. (She still had the "face-lift.")

With such a long history of symptoms, a harm reduction model can be useful, attempting to decrease the behaviors, or "the harm," rather than to terminate them. A harm reduction perspective helps women to see progress in the small steps they can take to change their behavior and challenges the dichotomous thinking associated with ED. It also helps professionals to conceptualize what treatment can achieve, despite the severity and duration of the problems we see in those women who have suffered over decades. Some may recover quite fully, others less so, but, in most cases, treatment can notably improve the quality of their lives and decrease the risk to their health.

A feminist relational frame is especially helpful when working with adults, as it conceptualizes an ED to be a solution to a woman's dilemma. Quite simply, to recover from the problem of the ED, the dilemma must be understood and addressed. Treatment is not about taming the beast of the ED, but about understanding what the symptoms mean and how they function to address, or avoid, other issues. Jennifer's recovery required an appreciation for how ED symptoms helped her to manage long-term stressors and demanding family dynamics that she felt intensely but had never named or verbalized.

Not only does the focus of treatment surpass symptom change, but the feminist frame also explores the gender-prescribed roles in contemporary culture to see how the ED developed in response to these pressures or influences. This approach conceptualizes that sexism, oppression, and patriarchal culture are primary contributors to the problems that bring women into treatment, and acknowledges that many women have been (directly or indirectly) victimized or traumatized. Jennifer struggled with the power her father and other men had in her life and needed to find new ways to relate to men in order to feel more rooted in herself. The feminist frame also helps women to identify and value their many assets and recognizes the need women have for emotional connection and interdependence as a strength, not a weakness (Katzman, Nasser & Noordenbos, 2007; Sesan, 1994).

An essential element in the treatment of ED, psycho-education also reflects a feminist framework, as it gives the patient information that helps to guide decisions about treatment and about changing their behavior. In the traditional medical model, professionals hold the power of information. In the feminist model, the information, and, therefore, the power, is shared. Psycho-education seems to have a greater impact when working with adults. Younger patients tend to believe the rules do not apply to them, and that they are immune to ill-effects of an ED. Often, older patients have lost this adolescent omnipotence, and can better utilize information about the risks of an ED.

The Body Myth: Adult Women and the Pressure to Be Perfect, provides many resources for self-help and psycho-education, such as summarizing how Mother Nature designed women's bodies so the species will survive (Maine & Kelly, 2005). We simply cannot survive individually or collectively unless women have adequate fat stores. The biological facts are:

- Before puberty, a girl's body has about 12% body fat
- During puberty, fat cells multiply to about 17% body fat—sufficient for ovulation and menstruation
- A mature adult woman's body will have about 22% body fat—enough energy for an ovulating female to survive famine for 9 months, and long enough to bring a fetus to term
- Women gain fat first in the breasts, buttocks, hips, and thighs to protect their fertility, reproductive, and feeding organs
- In the transition through menopause, average weight gain is 8–12 pounds, and metabolism slows 15–20%

- Hormonal shifts generate an increase in the size of abdominal fat cells which produce estrogen to maintain bone density, decrease the risk for osteoporosis, and help to manage symptoms of menopause
- Moderate weight gain at midlife is associated with longer life expectancy for women
- Only 10% of women die in famines while 50% of men do (Waterhouse, 1997).

These facts demonstrate the wisdom of Mother Nature, a wisdom adult women may be finally ready to embrace. For example, a woman in her early 70s was referred to me by her physician due to a recent weight loss. Her subclinical ED behaviors had been an organizing force throughout her adult life as she faced decades of significant loss and trauma. Through psychotherapy, she found great relief in talking about these experiences and discovering how her restricted eating became a way to avoid the pain she could not face. While this provided psychic relief, she was not convinced she needed to gain weight and was distressed by the few pounds she had gained. Once we explored the biological facts listed above, she began to make peace with her body. At the next session, she proudly announced that she had always hated the little bit of fat around her middle, calling it "my spare tire," but now sees it as "my life preserver." She began to enjoy food and worry far less about her weight, her medical status improved, and she had more energy for a very full and active retiree's life. Despite decades of DE, she made substantial changes in her attitudes and behaviors related to eating.

As with younger patients, ED and BI obsessions can reflect significant family issues as well as contribute to them, and addressing them is critical. Midlife women struggling with these issues need to contextualize their symptoms within their cultural experience as women, but also within their families. Jennifer had to explore her family dynamics to begin to understand her drive to perfection and the meaning her body had assumed. With a familial emphasis on achieving and excelling and a denial of any emotions, her ED was her only coping mechanism. Understanding this and developing new resources facilitated her moving away from her symptoms. She chose to have her family of origin involved in some sessions, allowing me to see first-hand the pressures she had experienced. It is critical to allow the adult patient to decide on family involvement: the patient knows best as to whether their involvement will help or hinder.

MEDICAL ISSUES

Adults suffer from the same medical sequelae of ED as younger patients: every system in the body can be affected due to the associated nutrient deficiencies. Electrolyte imbalances create risk for cardiac arrhythmias and arrest. Endocrine dysfunction leads to menstrual irregularities, decreased bone mineral density, increased risk for osteopenia and osteoporosis, and a compromised immune system (Mehler & Andersen, 1999; Zerbe, 1999). Despite long-term medical stability and normal laboratory values, medical complications in adults with ED can occur quickly and can result in sudden death—as if one day the body decides it cannot tolerate the abuse it has experienced for decades, and gives up (Carney & Andersen, 1996; Herzog, Deter, Fiehn & Petzold, 1997).

Eating disorders also cause some unique medical issues in adult women. Depletion of fat stores exacerbates the declining estrogen level, intensifying menopausal symptoms. Dieting also causes muscle-wasting, which can reduce the metabolic rate and hasten the natural neuromuscular decline associated with aging (Holt, 2005). In the elderly, dieting is medically risky (Gupta, 1995; Lewis & Cachelin, 2001), and physical complications can occur quickly (Kay, 1987). The mortality risk associated with low weight is greater as people age (Tayback, Kumanyika & Chee, 1990) and the cognitive impairment secondary to dieting may be greater in older patients (Lewis & Cachelin, 2001).

Medical offices can be the gateway to getting help or to remaining stuck and ashamed, as Jennifer might have been if she had not searched for information on the Internet. Help came to her through that anonymous medium rather than from her trusted physician. Too often, medical providers focus on weight control without understanding the multiple genetic, psychological, and social factors that contribute to adult weight and they convey a fear of obesity rather than a health-centered approach, such as the Health at Every Size paradigm described by Burgard in Chapter 2.

OBSTACLES TO GETTING HELP

Many obstacles prevent adult women from seeking treatment for their ED. First, the shame and self-blame seen clinically in younger patients is far more intense at midlife and beyond. Adult women believe that they should know better and should have outgrown such "teenage" problems; they berate and chastise themselves, feeling that their problems are less legitimate than a younger woman's and not a worthwhile reason to seek help. With their multiple and complex roles, midlife women also have more serious everyday responsibilities, with more people to take care of, and, they fear, to disappoint, if they do start paying attention to themselves. Under constant stress, they may find the ritualized behaviors of an ED comforting and grounding, despite the long-term threat to their emotional and physical health.

The harsh reality is that adult women have had more years of denying their true appetites, hungers, and feelings, and of treating their bodies punitively; this alone may keep them from seeking help. Of course, we must also acknowledge the secondary gains that accompany BI obsessions and weight loss, due to the universal desire today to look young and avoid the loss of power associated with an aging female body.

For some, resistance to getting help is fueled by despair and a belief that they are doomed to perpetual unhappiness. For others, a sense of omnipotence develops as they continue to dodge the bullets of serious medical consequences. The fact that they have survived so far, despite their dangerous eating issues and treatment of their bodies, makes some believe that they are immune to any risks. Add these dynamics to the medical bias that eating and BI disorders are adolescent issues and it is easy to see why we have not grasped the extent of these issues in women's health.

FINAL THOUGHTS: NOTES TO, AND ABOUT, THE SELF

Most adult women with ED are competent, capable, and high-functioning in many areas of life. Seduced by this exterior, their loved ones, co-workers, and professional caregivers can ignore or underestimate the signs of ED. Treatment must challenge this façade, this "false self," but gently and sensitively, as it is frightening to be "found out" and even more disconcerting to think about being any other way.

Just like autoimmune disorders, in which the immune system attacks itself, in an ED the eating-disordered self robs the true self of confidence, self-worth, and insight. It replaces these with a relentless circular logic that devalues the core self and overvalues control over food, weight, and appearance, creating a false self. Therapy is self-to-self, requiring the therapist's healthy self to be very active, emotionally present, honest, available, and purposeful. The goals are to develop or restore basic emotional and internal awareness, to reprogram the mind and spirit in the direction of health and healing, to instill hope for change, and to ground the person within her own healthy self so the false self of the ED will seem less necessary.

With all ED patients, the therapist must be truly authentic, as discussed by McGilley and Szablewski (Chapter 12). With their well-honed radar, adults with ED read us quickly, spotting inauthenticity miles away. Consequently, therapists must be in a parallel process of growth and development, exploring and owning how we have bought into cultural standards around weight, appearance, and roles for women, and how these perspectives have affected us personally and professionally. We must walk the walk, not only talk the talk. Having treated all ages of patients suffering from ED, I feel most challenged and engaged in work with adult patients: they seem to be hungry for real women of substance. This inspires me to be my most real self. Being an alternative, authentic role model invites our patients to become more real themselves. If a woman they admire eats heartily, enjoys food everyday, and truly takes care of herself, maybe they can take that risk. Furthermore, if I can see through the false self of their ED to their true self, and convey hope and a belief in that self, they may find the strength to change. Hope and respect for her true self are the essential ingredients to successful treatment of adult ED.

BRIDGING THE RESEARCH–PRACTICE GAP

The implications of the increasing incidence of eating and BI concerns at midlife are many. In order to address this critical problem, the health care system needs:

- Real data on this problem so we can better define the range of severity and the types of ED women experience during adulthood
- Training of all health care professionals, but especially providers in primary care and obstetrics and gynecology, to screen, identify, and appropriately treat women with the full range of DE. Several simple screening questions can be added to routine medical visits. (e.g.: *"Has your weight fluctuated during your adult years?" "Are you trying to 'manage' your weight?" "If so, how?" "What did you eat yesterday?" "How much do you think or worry about weight, shape, and food?"*)
- Close collaboration between mental health and medical providers with special attention paid to women during and post child-bearing years
- An awareness of how the war on obesity, the cultural expectations of women and appearance, and attitudes toward older women, resonate in clinical providers and affect their ability to recognize and treat these issues in adult women

- Treatment options that meet the needs of adult women, as many cannot consider leaving their families and their responsibilities for any protracted period of time. Outpatient options and convenient treatment packages are critical, as is insurance reimbursement for the appropriate level and type of care
- Their partners should also have easy access to psycho-education about eating disorders and treatment as needed. Too often, the needs of the partners are ignored
- Support and education for eating disordered women in their role as mothers to create a healthy home environment and role model for their children
- Comprehensive, longitudinal research to track the most effective outreach and treatment programs for adults
- Efforts to fight for true gender equity and healthier ideals for women of all ages so that their bodies will no longer be their primary source of power
- Optimism that we can help to improve the quality of a woman's life no matter how old she is or how long she has suffered
- Emphasis on ED and related nutritional and BI problems as a major public health issue for women of all ages.

References

Berg, F. (2001). *Women afraid to eat*. Hettinger, ND: Healthy Weight Network.

Bratman, S. (2004). *Health food junkies: Orthorexia nervosa—the health food eating disorder*. New York, NY: Broadway.

Bulik, C. M., Sullivan, P. F., Fear, J. L., Pickering, A., Dawn, A., & McCullin, M. (1999). Fertility and reproduction in women with anorexia nervosa: A controlled study. *Journal of Clinical Psychiatry, 60*, 130–135.

Bulik, C. M., Von Holle, A., Hamer, R., Berg, C. K., Torgersen, L., Magnus, P., … Reichborn-Kjennerud, T. (2007). Patterns of remission, continuation and incidence of broadly defined eating disorders during pregnancy in the Norwegian Mother and Child Cohort Study (MoBa). *Psychological Medicine, 37*, 1109–1118.

Bulik, C. M., & Reba-Harrelson, L. (2008, May). *A novel collaboration between Self magazine and the University of North Carolina at Chapel Hill: Patterns and prevalence of disordered eating in a probability sample of American women ages 25–45*. Paper presented at the 2008 International Conference on Eating Disorders, Seattle, WA.

Carney, C. P., & Andersen, A. E. (1996). Eating disorders: Guide to medical evaluation and complications. *Psychiatric Clinics of North America, 19*, 657–679.

C.B.S. News (2008). *"Pregorexia" inspired by thin celebs? Moms-to-be obsessing over weight, diet, exercise so much they put baby's health in jeopardy*. Retrieved January 30, 2009 from. www.cbsnews.com/stories/2008/08/11/earlyshow/health/main4337521.shtml

Clarke, L. H. (2002). Older women's perceptions of ideal body weights: The tensions between health and appearance motivations for weight loss. *Ageing and Society, 22*, 751–773.

Crow, S. J., Agras, W. S., Crosby, R., Halmi, K., & Mitchell, J. E. (2008). Eating disorder symptoms in pregnancy: A prospective study. *International Journal of Eating Disorders, 41*(3), 277–279.

Cumella, E. J., & Kally, Z. (2008). Profile of 50 women with midlife-onset eating disorders. *Eating Disorders: The Journal of Treatment and Prevention, 16*, 193–203.

Friedman, S. S. (1997). *When girls feel fat: Helping girls through adolescence*. Vancouver, Canada: Salal Books.

Gaesser, G. (2002). *Big fat lies: The truth about weight and your health*. Carlsbad, CA: Gurze Books.

Garner, D. M. (1997). Survey Says: Body Image Poll Results. *Psychology Today*. Feb, 1997. Retrieved from. http://www.psychologytoday.com/articles/199702/survey-says-body-image-poll-results

Gordon, R. (2001). Eating disorders East and West: A culture-bound syndrome unbound. In M. Nasser, M. A. Katzman, & R. A. Gordon (Eds.), *Eating disorders and cultures in transition* (pp. 1–23). New York, NY: Taylor and Francis.

Gupta, M. A. (1995). Concerns about aging and a drive for thinness: A factor in the biopsychosocial model of eating disorders? *International Journal of Eating Disorders, 18*(4), 351–357.

Helgstrand, S., & Andersen, A. (2005). Maternal underweight and the risk of spontaneous abortion. *Acta Obstetrica et Gynecologica Scandinavica, 84*, 1197–1201.

Herzog, W., Deter, H. C., Fiehn, W., & Petzold, E. (1997). Medical findings and predictors in long-term physical outcome in anorexia nervosa: A prospective, 12-year follow-up study. *Psychological Medicine, 27*, 269–279.

Holt, M. (2005, Winter). Eating disorders in the adult population. *SCAN'S* (sports, cardiovascular and wellness nutritionists) *Pulse*, 8–9.

Katzman, M. A., Nasser, M., & Noordenbos, G. (2007). Feminist therapies. In M. Nasser, K. Baistow, & J. Treasure (Eds.), *The female body in mind: The interface between the female body and mental health* (pp. 205–213). London: Routledge.

Kay, P. A. J. (1987). Clinical aspects of geriatric eating disorders. In H. Field, & B. Domangue (Eds.), *Eating disorders throughout the lifespan* (pp. 139–146). New York, NY: Praeger.

Kearney-Cooke, A. (2004). *Change your mind, change your body.* New York, NY: Atria.

Koubba, S., Hällström, T., & Hirschberg, A. L. (2008). Early maternal adjustment in women with eating disorders. *International Journal of Eating Disorders, 41*, 405–410.

Levine, M. P., & Smolak, L. (2006). *The prevention of eating problems and eating disorders: Theory, research, and practice.* Mahwah, NJ: Lawrence Erlbaum Associates.

Lewis, D. M., & Cachelin, F. M. (2001). Body image, body dissatisfaction, and eating attitudes in midlife and elderly women. *Eating Disorders: The Journal of Treatment and Prevention, 9*, 29–39.

Maine, M., & Kelly, J. (2005). *The body myth: Adult women and the pressure to be perfect.* New York, NY: John Wiley.

Mangweth-Matzek, B., Rupp, C. I., Hausmann, A., Assmayr, K., Mariacher, E., Kemmler, G., … Biebl, W. (2006). Never too old for eating disorders or body dissatisfaction: A community study of elderly women. *International Journal of Eating Disorders, 39*(7), 583–586.

Mehler, P. S., & Andersen, A. E. (1999). *Eating disorders: A guide to medical care and complications.* Baltimore, MD: Johns Hopkins University Press.

Norre, J., Vandereycken, W., & Gordts, S. (2001). The management of eating disorders in a fertility clinic: Clinical guidelines. *Journal of Psychosomatic Obstetrics and Gynecology, 22*, 77–81.

Pryor, T. (2008). The desperate housewives syndrome: Research on midlife patients with eating disorders. *Health Within Reach Newsletter.* Denver, CO: Eating Disorders Center of Denver.

Reichman, N. E. (2005). Low birth weight and school readiness. *The Future of Children Journal, 15*, 91–116.

Resch, M., Szendei, G., & Haasz, P. (2004). Eating disorders from a gynecologic and endocrinologic view: Hormonal changes. *Fertility & Sterility, 81*, 1151–1153.

Sesan, R. (1994). Feminist treatment of eating disorders: An oxymoron? In P. Fallon, M. A. Katzman, & S. C. Wooley (Eds.), *Feminist perspectives on eating disorders* (pp. 251–271) New York, NY: Guilford Press.

Sheehy, G. (1976). *Passages: Predictable crises of adult life.* New York, NY: Dutton. 1976.

Smolak, L., & Levine, M. P. (1996). Adolescent transitions and the development of eating disorders. In L. Smolak, M. P. Levine, & R. Striegel-Moore (Eds.), *The developmental psychopathology of eating disorders* (pp. 235–258). Mahwah, NJ: Lawrence Erlbaum.

Tayback, M., Kumanyika, S., & Chee, E. (1990). Body weight as a risk factor in the elderly. *Archives of Internal Medicine, 150*, 1065–1072.

Waterhouse, D. (1997). *Like mother, like daughter: How women are influenced by their mother's relationship with food and how to break the pattern.* New York, NY: Hyperion.

Wentz, W., Gillberg, I. C., Anckarsater, H., Gillberg, C., & Rastam, M. (2009). Reproduction and offspring status 18 years after teenage-onset anorexia nervosa—A controlled community-based study. *International Journal of Eating Disorders, 42*(6), 483–491.

Willis, S. L. & Martin, M. (Eds.). (2005). *Middle adulthood: A lifespan perspective.* New York, NY: Sage.

Woods, B. K., & Williams, L. (2008). Pregnancy, fertility and eating disorders. *Remuda Review, 7*(2), 14–19.

Zerbe, K. J. (1999). *Women's mental health in primary care.* Philadelphia, PA: W.B. Saunders Company.

CHAPTER 18

Men with Eating Disorders
The Art and Science of Treatment Engagement

Douglas W. Bunnell

Most of what is known about people with eating disorders (ED) derives from the study and treatment of women. Appreciation is growing, however, for the scope and significance of ED and body image (BI) dissatisfaction in boys and men. The culture is shifting; males are the major growth category for cosmetic surgery (American Society for Aesthetic Plastic Surgery, 2001) and a quick scan of the mass media reveals countless images of highly stylized, lean, and muscular men. While gender is still a central factor in understanding ED, we seem to be moving towards gender equity in terms of BI dissatisfaction (Maine & Bunnell, 2008).

Clinicians and treatment programs around the country have reported seeing more male patients than previously and research data now validate the increased incidence (Hudson, Hiripi, Pope & Kessler, 2007). Eating disorders are different for men, different on biological, psychological, and social levels. For the most part, however, the literature suggests that treatment is essentially the same. The literature on cognitive behavioral therapy, for instance, makes scant mention of gender as a factor in treatment (see Fairburn, 2008). With the exception of a few specialized treatment centers, men struggle to locate gender informed treatment. Most outpatient clinicians see only the rare male ED patient and they have few resources or information on how to tailor their treatment approaches to best meet a man's clinical needs.

This chapter will briefly review what we know about men with ED including differences in comparison to female patients, barriers to diagnosis, etiological processes, and treatment variations. The main focus, however, will be on how to modify our psychotherapeutic approaches to best engage and motivate male patients. Male psychological development and socialization affect the psychotherapeutic process. Masculine learning styles may alter the ways in which men experience common symptom management techniques. More importantly, gender colors all relationships, including the one between therapist and patient. Conducting psychotherapy with men with ED, whether you are a male or female therapist, requires an appreciation for your own gender biases and values. This chapter concludes with an exploration of these process considerations.

AN OVERVIEW OF MALE EATING DISORDERS

Prevalence

While the conventional view is that only one in ten patients with an ED is a man, (Andersen, 1990), more recent surveys reveal higher rates of male incidence. Hudson et al. (2007) analyzed data from a large national survey of psychiatric comorbidity and reported that nearly one-quarter of diagnosable cases of eating pathology occurred in males. They argued that, due to a variety of factors limiting detection and diagnosis in males, the survey data probably underestimated the true scope of the problem. When we extend the definitions of eating pathology to include sub-threshold disordered eating (DE), the picture is even more striking: 28% of male high school students reported fasting, skipping meals, using diet pills, vomiting, or using laxatives to control their weight (Hudson et al., 2007). Another survey found that a substantial minority of men in a community sample reported ED symptoms, including overeating, body checking, and vomiting (Striegel-Moore et al., 2009). In light of this degree of DE in a nonclinical sample, researchers and clinicians need to pay attention to this issue to better determine the number of clinical cases.

Men and boys may be somewhat insulated from developing ED, due to differences in biology, temperament, vulnerability to mood and anxiety disorders, and gender socialization. A significant number of boys may experiment or flirt with DE, but only a minority will actually go on to develop full blown disorders. Still, we may be failing to detect, diagnose, and treat a substantial number of males. The diagnostic criteria, explored in more detail below, are gender biased. Men and boys are less likely to acknowledge their ED and certainly less likely to voluntarily present for treatment. Many men may feel ashamed of their symptoms because an ED is seen as a "feminine" illness. Not immune to these common perceptions, many clinicians, medical or psychiatric, are less sensitized to male eating pathology and are simply more likely to suspect these problems in female patients. Family members may also have more difficulty accepting the notion that their son, husband, brother, partner, or boyfriend has these sorts of difficulty.

CLINICAL FEATURES AND DIAGNOSIS

Many of the clinical features of anorexia nervosa (AN), bulimia nervosa (BN), Eating Disorder Not Otherwise Specified (EDNOS), and binge eating disorder (BED), a provisional diagnosis now subsumed in the EDNOS category, are similar in men and women (Andersen, 1990, 1995; Fernández-Aranda, Aitken, Badia, Gimenez, & Solano, 2004; Hay, Loukas & Philpott, 2005; Woodside et al., 2001). Comparisons of psychological functioning, including personality measures, reveal few differences. Men may show less harm avoidance, drive for thinness, and body dissatisfaction but the statistical differences may not be clinically relevant (Fernández-Aranda et al., 2004). Woodside and his colleagues (2001) concluded that male and female patients have similar psychosocial morbidities and, importantly, that ED men are significantly different from healthy men on a variety of eating and psychological variables. They reported that their results "…confirm the clinical similarities between men with eating disorders and women with eating disorders" (Woodside et al., 2001, p. 570).

There are, however, some important clinical and diagnostic differences, many of which revolve around the threshold levels for symptom frequency and severity. The boundary between overeating and binge eating, for instance, may be blurred, as overeating is more socially acceptable for men. Differentiating overeating from binge eating or BED from another form of EDNOS may also be more complicated in males. Clinicians may underdiagnose BN in men because the "binge" quantity is not large enough compared to male norms. Also, men may be somewhat less likely than women to induce vomiting in compensation for binge eating, but they are more likely, instead, to over-exercise as a form of purging. Clinicians may ignore a male patient's eating disordered exercise compulsion because rigorous exercise is more acceptable for men than it is for women.

Similarly, body weight thresholds for the diagnosis of AN in men may underestimate the severity of their weight loss. Due to their higher muscle mass, men may be at greater medical risk than women at the 85% ideal body weight criterion (Crisp & Burns, 1990). There is some additional ambiguity about diagnosis of AN in men because the amenorrhea criterion is irrelevant. Men with ED are, however, likely to have low levels of testosterone (Andersen, 1990).

Several studies (Olivardia, Pope, Mangweth & Hudson, 1995; Woodside et al., 2001) have emphasized the similarities in the core psychopathological features of ED across gender. Although the patterns of psychiatric and medical comorbidity are similar, Pope, Phillips and Olivardia (2000) described some potential and important distinctions. Whereas the drive for thinness is a critical diagnostic criterion for women with ED, it seems less powerful for men. Men are more likely to report that they are restricting, over-exercising, or binge eating and purging in pursuit of a *lean* appearance rather than a *thin* appearance. They are less likely to focus on specific weight loss goals and more likely to frame their goals in terms of leanness and muscularity (Olivardia, 2007; Pope et al., 2000). If men do not acknowledge a strong desire to be "thin," clinicians may underestimate the degree of actual eating pathology and BI dissatisfaction. Standard eating assessment instruments may fail to detect cases in males for the same reasons.

Even healthy males are increasingly worried about their bodies (Pope, Olivardia, Boroweicki & Cohane, 2001). Male BI concerns tend to focus on the upper torso, in contrast to women's concerns, which tend to cluster around the hip area. For males with ED, the preoccupation and overvaluation of BI are more shape based than weight based. Again, clinicians who are accustomed to working with females with ED may underestimate the severity of the illness when they see a male patient.

In my clinical experience, boys and men with ED have less ambivalence about regaining weight. While overly concerned with becoming too soft and undefined, men appear to be less attached to the "specialness" of thinness. As others have suggested (Andersen, 1990; Margo, 1987; Olivardia, Pope, Borowiecki & Cohane, 2004), ED may be more ego-dystonic for men than they are for women. Men are also less likely to be familiar with ED in general and, with the onset of symptoms, may be surprised and less self aware.

Physiological and Medical

Males with ED often have less fat reserves than their female counterparts. Consequently, dietary restriction, over-exercise, or chaotic binge eating and purging may drop them below critical body fat percentage levels faster than women. Crisp and Burns (1990) argued that 10%

loss of body weight in men is comparable to 15–20% loss in women. Men may, therefore, be at increased risk at a higher BMI threshold than women. Research has focused on endocrine functioning, particularly on decreased testosterone levels in men with ED. Low levels can increase the risk for osteopenia and osteoporosis (Andersen, Watson & Schlechte, 2000b; Mehler, Sabel, Watson & Andersen, 2009).

Sex and Sexual Orientation

Gay males are at a somewhat elevated risk for developing an ED but most males with ED are heterosexual (Carlat, Camargo & Herzog, 1997; Feldman & Meyer, 2007). This association is complicated, by no means well understood, and a thorough discussion of this is beyond the scope of this chapter. Certain subgroups of gay males may pursue a thin and emaciated body ideal while others may overvalue muscularity and size (Olivardia, 2007).

Decreased libido may reflect low testosterone levels and a more generalized response to weight loss and malnutrition. Weight restoration may trigger a second endocrinological puberty with a return of sexual interest and energy.

Exercise, Muscularity and Fitness

While over-exercise is often a part of ED in women, exercise, fitness, and muscularity are central factors for most men with these conditions. Men tend to see themselves as fatter than they really are and describe an ideal body with more muscle and less fat. Many male ED patients experience a profound fear of being too soft and small in contrast to simply fearing "fat." The popular press has defined "reverse anorexia" in which males compulsively pursue muscle building rather than weight loss. The professional literature describes this as muscle dysmorphia (see Olivardia, 2007). Men and boys with ED may appear to be pursuing simple fitness while a close investigation will reveal profound anxiety about smallness, softness, or "flabbiness." Often, underweight male patients struggle with weight restoration; they are able to eat enough to sustain their weight but cannot eat enough to cover their exercise energy expenditure. As noted earlier, over-exercise may be culturally valued for men and clinicians need to be wary of their own bias in this regard.

The preoccupation with bigness and muscularity leaves males particularly susceptible to anabolic steroid use. Psychologically seductive (Morgan, 2008) and readily available, these substances promise quick and reliable gains in muscle mass and many of the long-term negative effects are not immediately apparent.

Cultural and Developmental Factors

Many of the differences noted in this brief summary of the unique aspects of males with ED involve gender roles and stereotypes. Leigh Cohn, who writes and speaks about men's issues with eating and weight, notes that while women do not typically report feeling less feminine because of their ED, men who suffer are very likely to feel less masculine (Cohn, personal communication, March 2008). This important distinction underscores the significance of cultural and development factors in male eating pathology. Clinicians working with men need to be acutely aware of the nuances of male socialization and how the patient's psychological development influences symptomatology, sense of self, and reactions to psychotherapy.

The contrasts between men and women with eating issues often reflect differences in gender socialization. For many men, thinness represents weakness while it tends to be a highly desirable characteristic for most women. Morgan (2008) emphasizes that men with ED are often fearful of smallness, due to their dread of powerlessness and a fear of the feminine. The latter has a profound impact on masculine psychology. Dependence is to be avoided; self-sufficiency is the ideal. Hardening the body, denying nourishment, and shame based rules and regulations about eating can all reflect this underlying fear.

GROWING UP MALE

Differences in brain structure, biology, developmental experience, and socialization differentiate masculine psychology from feminine psychology. In the psychological literature and in the popular press, boys and men are often described as scarred by their experiences and "chronically flawed and in dire need of fixing" (Kiselica, 2006, as cited in Englar-Carlson, 2008, pp. 98–99). Clinicians, however, need to dig deeper in order to work effectively with their male patients. Men and boys do have differential rates of certain types of psychopathology, demonstrating higher rates of pervasive developmental disorder, attention disorders, oppositional and conduct disorders, obsessive compulsive disorder, substance abuse, and Tourette's syndrome (Erickson & Chambers, 2007). They are prone to externalizing disorders with this vulnerability again built on the interplay between culture and biology. Male brains differ from female brains (see Maine and Bunnell, Chapter 1), especially in the maturation of frontal cortical and temporal limbic areas (Erickson & Chambers, 2007). Men will perceive, organize, learn, and respond differently to experience than will women, particularly in areas linked to emotional and social processing.

Male socialization tends to reinforce these structural and biological differences. Male babies are expected to be more physical, more independent, more risk taking, and more active. The arc of psychological maturation also moves boys, perhaps prematurely, towards separation and independence. Early dependence on the mother often gives way to a repudiation of dependence. To identify with the same sexed parent, boys must distance themselves from this primary attachment to a female. But, as Blatt (2008) and others have argued, psychological maturation requires both separation and relatedness, describing two basic tracks of psychological development, both essential to psychological maturity. One is predominantly relational, the other predominantly individual. Earlier models of psychological development overemphasized psychological self-sufficiency as a marker of maturation and health. Feminist relational theory, along with developments in attachment and infant research, has helped to move the field towards a greater appreciation for the centrality of relatedness, but perhaps at the expense of devaluing self-definition and integration. Attempting to integrate these different dimensions, Blatt (2008) wrote:

> The development of an increasingly differentiated, integrated and mature sense of self is contingent on establishing satisfying interpersonal experience and, conversely, the development of increasingly mature and satisfying interpersonal relationships depends on the development of more mature self-definition and identity (pp. 128–129).

If mature relatedness and self-definition require each other, how does male socialization affect these two different tracks? The emphasis on self-sufficiency and autonomy can inhibit the development of relational capabilities, leaving men emotionally disconnected from other people and from their own internal experiences as well. These internalized disconnections consolidate into an enduring sense of self, or, more accurately, into a sense of self in disconnection. Men may have, due to biology and experience, less access to their inner lives and greater difficulties articulating their inner experiences (Kindlon, Thompson & Barker, 1999).

The experience of teasing and bullying is both a normative and profound experience for boys. Many men with ED, in my experience, report specific and distinct memories of having been teased and shamed about their bodies. Teasing, bullying, competition, and conflict are important forces in male socialization and are less distinct factors for most women with eating issues. Participating in the teasing relationship, from either position, can be damaging, reinforcing important lessons on vulnerability, power, and trust. For men with ED, the experience of shaming and humiliation often becomes concretized in the body. These lessons teach men to shun dependence, inhibit their ability to receive emotional support, and further complicate emotional intimacy (Levant, 2006). Fearful of dependence and wary of intimacy, the lessons of male socialization may also contribute to men's homophobia and their tendency to sexualize emotional closeness with women.

These developmental and relational challenges occur within a changing social and cultural environment. Economic uncertainty and globalization are threats to conventional male identity. Jobs are less secure, and authority rests less and less on traditional views of patriarchy. While men still enjoy an abundance of social perks for their gender, the tide is turning and men may be less secure than they were a generation ago. A profound change in media images of men has also occurred (Andersen, Cohn & Holbrook, 2000a). Today men are exposed to countless images of unattainable bodies; men may be catching up to women in this dubious category. Exposure to the gap between body ideals and body realities is increasing men's body dissatisfaction and their risk for ED (Maine & Bunnell, 2008).

Treatment

All of these biological, developmental, relational, and cultural factors come into treatment with male patients. In my experience, however, the ED usually trumps gender and even gender sensitive treatment approaches have to address eating psychopathology as the first priority. The core features for males are abnormal eating behavior, an idealization of a lean shape ideal, a fear of being too big, soft, or fat, and abnormalities in gonadotrophic functioning (Andersen et al., 2000b). Anxiety is also a common feature. These are directly comparable to the features of women with ED, with the obvious variation in the nature of the endocrine abnormalities. Treatment for men is similar to the treatment for women, and early diagnosis and intervention are critical to help stave off the long-term physical and psychological effects of an entrenched disorder.

Weight restoration and re-establishing a normal pattern of eating are the essential foundation for successful treatment and recovery. With boys and men, weight loss, or chaotic nutritional status due to binging and purging, may be even more disruptive because they encounter medical and physical trouble with less weight loss than girls and women (Crisp & Burns, 1990). As they may actually elude detection, diagnosis and referral, males may be at more acute risk than female patients at first presentation. This makes actual behavior change an even higher priority.

Males are generally less familiar with the notion of ED. Conventional etiological models, and the associated treatment approaches, may not "click" for men as male patients must learn a new language for describing their ED. Therapists, trained in standard CBT or even psychodynamic therapies, may need to become translators, or at least learn how to speak a different dialect. Male patients may not articulate a fear of fatness, a desire to lose to a particular weight level, or worry about pant size, but might express concern about the lack of six pack abdominal muscles, breasts that are too large, and a profound fear that eating "too much" will result in being soft and "mushy" rather than simply "fat." While their body dissatisfaction may be painfully intense, they may be less likely to equate the inadequacy of the body with an inadequacy in a general sense of self.

Explaining the rationale for normalizing eating and weight restoration may be more straightforward for male patients than for female patients. Sensing the problem, men, who have been socialized to fix rather than explain, may actually buy into the program with less ambivalence than many female patients. Thinness and smallness hold less magic for males and, in my experience, they seem less attached to the identity and self-definition provided by the ED. They may be less curious or interested in exploration of the relational, familial, developmental, or traumatic factors associated with their eating difficulties. But, paradoxically, this may be more open to a problem solving focus on the stabilization of their eating and less resistant to actual weight restoration.

The language of our psychotherapies, including directive techniques like CBT or DBT, places a strong emphasis on emotional, or at least cognitive, self-awareness. Identifying internal experiences is pre-requisite for learning, rehearsing, and re-internalizing coping strategies. For instance, we routinely ask a young woman to look for the pattern between her binge episodes and prior episodes of dietary restriction. Later in the process, treatment would focus on identifying other common precipitants for binge urges such as insults to her self-esteem. Psychodynamic or relational therapies look for the relational precipitants trying to detect if a relational disconnection preceded the urge to binge. These same approaches might not make intuitive sense for the average male patient. Due in part to their biology and in part to their relational and social acculturation, men may be less adept, and certainly less comfortable, with the articulation of their inner experiences and their reactivity to relational events.

Therapists need to understand that requests for their male patients to divulge internal experience may set the stage for profound feelings of shame. Not only may they be less facile with a language of internal experience; they are likely to have been taught that revealing inner experience to another is a risky move. When working with males with ED, therapists need to help them develop a language for cognitions, feelings, fantasies, urges, desires, attachment, and connection. These are not familiar words or concepts for most men, but our current therapies are heavily dependent on them.

Establishing trust and conveying respect are crucial elements of competent psychotherapy and are particularly important in working with ED. For many women, the profound ambivalence about giving up the ED can make the initial encounter with the therapist a high wire act. The therapist must simultaneously connect with the part of the patient that is reluctant to change and convey the importance of substantial and rapid changes in behavior. Male patients may be somewhat less ambivalent about change but they enter therapy with the added shame and stigma of having a "woman's illness." Andersen and co-workers (2000a) stress the importance of working in collaboration with the male patient, sensitive to his vulnerability to feeling weak and defective. Therapists need to help the patient to make his own decisions about change, steering away from direct confrontation or prescriptions in favor of a collaborative, problem solving approach.

Effective treatment of ED requires the therapist to be an active participant in the therapeutic exchange. Asking pertinent questions, even anticipating the male patient's concerns and worries, can demonstrate acceptance and understanding. Therapists should be "sitting forward" (Hoffman, 2007, p. 24) to engage in the relationship and work to convey a genuine and consistent curiosity and excitement about what is being explored. Active inquiry and expression of the therapist's own thoughts, ideas, questions, and bafflement can help the male patient learn the language of feelings and therapy. Furthermore, sitting with a therapist who can articulate feelings of conflict, ambiguity, and ambivalence will help the patient develop these same capacities, capacities that are essential to psychological maturity and for recovery from an ED.

Males benefit from talking with other men who struggle with similar issues. While many ED treatment programs treat male patients, usually only one or two men are in a program or group at any particular time. Weltzin and colleagues (2005) noted the importance of male-specific treatment opportunities, and Andersen (1990) emphasized the importance of supervised exercise and strength training. These activities can be integrated into treatment relatively early, even in patients still working on weight restoration. The shame and stigma concerns may be best addressed in single sex groups. Andersen (1990) and others also underscore the importance of addressing father–son relational dynamics and improving father–son communication. Issues about sexuality or sexual identity may be more difficult to address in mixed gender groups or milieux. Men also need a space in which to explore the scars of teasing and shame and, more generally, to sift through the cultural expectations that have sculpted their internalized sense of self and BI. Male only therapy groups can provide such a space.

While most of the intensive treatment protocols for patients with ED are certainly useful for men, we need to know more about how men actually recover. We are learning that ED have some of their roots in basic brain and cognitive mechanisms regulating learning, reward, impulse control, pleasure, and anxiety. Common cognitive features in patients with AN may explain how patients get caught up in details and how they seem to fail to grasp the "big" picture about the risks of their ED (Tchanturia, Davies & Campbell, 2007). Men's brains and women's brains may also have different mechanisms that, for the most part, result in similar functioning (De Vries & Boyle, 1998). As we learn more about men with ED, we may find that their road to recovery is significantly different than the road for women.

PSYCHOTHERAPY WITH MEN

The clinical literature on men with ED tells us very little about how to best engage men into psychotherapy. Most men with ED, if fortunate enough to find treatment, will do the bulk of their therapeutic work with an outpatient individual psychotherapist. While therapists who treat ED may have discovered how to best engage their female clients into a good working alliance, they may need to modify their approach when confronted with

CASE EXAMPLE

Bob was a 15-year-old young man who lived at home with his two parents, both highly accomplished academic professionals. An older brother was living at a therapeutic boarding school. Bob's dietary restriction started after a comment he heard in the locker room about the size of his breasts. This had been a longstanding source of shame for him but this was the first attempt to "do something." He lost weight rapidly and became medically unstable almost immediately. Hospitalized on a medical unit, he began to regain weight but was discharged prior to full weight restoration. He started individual psychotherapy and family therapy after discharge.

Our initial sessions revealed a long history of being teased about his body. Attempts at weight lifting and body building had not helped and Bob was preoccupied with his upper body. He had no interest in weight loss *per se*, but his highly restrained eating served as his defense against the possibility of becoming "soft" again. This was all playing out in the context of considerable family chaos; both the brother and Bob's mother were struggling in their own treatments. Bob felt an intense sense of duty to salvage the family. This obligation left him prone to shame and humiliation when he was unable to quell their chaos. His dietary restraint, in his mind, not only helped him defend against his upper body dissatisfaction, but also represented a statement of his own self-sufficiency and denial of need. While he was highly embarrassed about having a "girl's disease," he also confessed his fear that he was "different from other guys." He felt too sensitive, too anxious about upcoming changes in his academic program, and insufficiently competitive with other members of his tennis team. In later sessions, he began to discuss some concerns about his masculinity and was worried that he might be homosexual. He was also profoundly concerned that he was too dependent on his mother and too hungry for his father's admiration. Bob also struggled with anxiety and panic. These were often provoked by variations in his eating, exercise, or weight. His worry and anxiety were often expressed somatically; he had only a limited vocabulary for feelings, with the exception of "guilty" or "worthless."

Bob felt enormous pressure to be a man. The entire male side of his family tree was filled with professors, academic stars, and accomplished athletes. He felt humiliated by his struggle to even make the varsity tennis team and looked to me for reassurance that it was OK to have these sorts of doubts. We strategized about how to limit his sense of responsibility for his mother's struggle. We also worked with his father to limit his, largely unintentional, critical comments to Bob. He would tell me on occasions that he felt like a "whining girl" when he explored these vulnerabilities, scanning my face for expected confirmation. His emerging relationship with me contained many of these same factors including his fear of dependence, his questions about his masculinity, and his conflicted attachment to his vulnerable mother. He also felt that he had to get better and gain weight in order for me to stay loyal to him much as he felt that he had to carry the banner for his family in order to be a good son or man.

---

the occasional male patient. The following brief case example highlights some of nuances of building a therapeutic alliance with a male with an ED.

What Men Bring to Therapy

For many men, the pull for intimacy and self revelation in psychotherapy may stir something closer to dread than relief. As noted earlier, men with ED often see the disorder as a distinct, even alien, defect that needs to be eliminated. Removal becomes a project rather than an interpersonal process. Others have commented on the dread men feel when their perceived inadequacy might be exposed (Bergman & Surrey, 1997). Shame and withdrawal often follow this sense of dread, potentially complicating therapeutic engagement if not sensitively handled by the therapist.

Men are still largely defined by what they do and by how well they do it. Emphasis on work, academics, athletics, and accomplishment remain the foundation of their social sense of self. To the extent that these gender roles still rule in the early twenty-first century, they continue to have an enormous influence on how men function both interpersonally and individually (Levant, 2006; Levant & Pollack, 1995), with a deleterious impact on men's mental health (Blazina & Watkins, 1996; Cournoyer & Mahalik, 1995; Good, Dell & Mintz, 1989). Describing the covert nature of male depression, Cochran and Rabinowitz (2000) argue that stoicism and suppression of emotion, reinforced by socialization, obscure the depth of depression, and, presumably, other psychiatric symptoms. In fact, men tend to show fewer explicit signs of depression but to manifest higher behavioral risks for substance use, impulsivity, and completed suicide.

Sociocultural influences are well known factors in the etiology and maintenance of ED in women (see Maine and Bunnell, Chapter 1; and Levine and Maine, Chapter 4). It seems obvious that pervasive messaging about thinness would create a vulnerability to body dissatisfaction and eating pathology. It may seem less obvious that cultural messaging about masculinity would create similar risks in men. The masculine imperatives, or the "shoulds and musts" of masculinity, can be punishing for boys and men who do not meet the standards (Mahalik, 1999). Gender strain theory (Pleck, 1995) hypothesizes that masculine socialization can create different types of discrepancies, or strain, between masculine expectations and realities. Long-term failures to live up to masculine ideals of success, power, control, stoicism, self-sufficiency, and fearlessness can lead to an internalized sense of failure and disappointment. Even adherence to other male imperatives such as the primacy of work and the avoidance of emotional intimacy can leave men isolated and disconnected. At the extremes, masculine ideology can reinforce homophobia, distancing men even further from contact with other men.

For men with ED, the discrepancies between masculine ideals and their personal realities may be particularly discordant. For some, conflicts about sexuality, sexual identity, and emotional intimacy may be a fundamental part of the etiology of their eating difficulties. Treatment and recovery will nearly always require wrestling with questions about what it means to be a man, and particularly, what it means to be a man in connection.

What are Men like in Therapy?

The requirements for being a "good" male do not really match up well with the characteristics of a "good" patient (Betz & Fitzgerald, 1993). Most therapists are drawn to work with patients who value relationships, have a facility with emotional language, and appreciate the connections between feelings, thoughts, and actions. While female patients with ED may struggle in these same areas, they are not swimming against a tide of male socialization. That tide pushes men to be wary of closeness and makes them uncomfortable with dependence. Male patients may be even more skeptical about the value of therapy in that it threatens not only their defensive masculine autonomy but also their attachment to their ED. Many of the "resistances" noted in the clinical literature actually describe reactions that seem predictable given masculine socialization. Difficulties with emotion, fear of dependence, and a reluctance to cede control to the therapist may reflect true intrapsychic defensiveness but, for male patients, these reactions may not reflect conflict but rather a firm stance in a masculine role. Gilligan (1982) described this as a contrast between a "positional" interpersonal stance towards the therapist, in contrast to a more expectable "relational" stance.

For men, and especially those with eating concerns, the good things in therapy are likely to also be threats to their masculinity. Pollack (2001) has written extensively on the notion of defensive autonomy, addressing both the positive and negative consequences of that stance. Maximizing self-sufficiency can certainly be adaptive but can also, at the extremes, cut men off from essential sources of support. For men with AN in particular, the masculine ideal of self-sufficiency often further energizes the commitment to dietary and relational asceticism.

Scher, Stevens & Good (1987) summarized the male experience of entering psychotherapy:

> Men are restrained and constrained because of the male role. The expectations of that role cause men to be emotionally flat or repressed, imbued with a competitive spirit, fearful of intimacy, untutored in emotional responsiveness...Therapists working with men must be aware of these qualities and understand their effect on therapy, be patient, and respect the integrity of their clients (p. 29).

What Can Therapists Do?

When working with men with ED, therapists must be aware of these cultural values, as they influence a therapeutic engagement already challenged by ambivalence about letting go of the organizing power of their symptoms. Clinicians need to focus on engagement, sensitive to the resistance to establishing connection, and may need to adapt their conventional therapeutic stance. Men, for instance, may respond better to a more active and problem focused approach, at least in the initial stages. Therapists should actively ask questions, develop hypotheses, and respectfully suggest alternative words for expressing feelings. As many men enter therapy somewhat mistrustful of the process, therapists should explicitly invite male patients to discuss their perceptions of therapy. *"What does it mean to you if someone sees a therapist?"* or *"What does it say about a person if he or she sees a therapist?"* or *"Do you think it's different for men versus women?"* The answers to these sorts of question may give the therapist important information about the patient's experience of being in the room.

Therapists may find themselves less likely to call attention to the relational aspects of their interaction with male patients. Whether this reflects the therapist's own avoidance of relational intimacy or that of the patient, the avoidance is consistent with masculine "ideology." Premature focus on feelings of closeness or dependence on the therapist can be destabilizing, even frightening. The closeness in the therapeutic alliance may need to be communicated through things like joint acknowledgement of the progress made in treatment; essentially congratulations without relational detail. Men can establish closeness in the context of action and accomplishment, a sort of "action intimacy" (Englar-Carlson, 2006, p. 37). The preference for action may lend itself nicely to symptom focused and directive interventions. Some equivocal evidence suggests that men may, in fact, do better in CBT than women, although this has not been demonstrated in patients with ED (Thase et al., 1994).

A cognitive behavioral focus may, on the other hand, accentuate a disconnection from affect and emotion. Attending to affect is a critical aspect of therapeutic engagement, and research is increasingly pointing to the role of affective activation in positive therapy outcomes (Diener, Hilsenroth & Weinberger, 2007). Activating affect with male patients, particularly non-anger affects, can be challenging. Developing a language for emotional intimacy, dependence, love, and tenderness may require use of interpersonal examples outside of the therapeutic relationship. Therapists can help men develop this language by probing about feelings about their roles as fathers, sons, leaders, and husbands. Using extra-therapy relationships minimizes the risk of activation of these "non-masculine feelings" within the therapy office.

Feminist-relational approaches emphasize the role of power in relationships (see Tantillo and Sanftner, Chapter 19). Therapists working from this perspective explicitly address this issue early in treatment. This may help male patients define their work with the therapist as a collaboration rather than as a relationship in which someone with power is caring for someone who is weak or dependent.

Countertransference

Whether male or female, therapists need to be aware of their own socialized expectations of men. It is remarkably easy to lapse into a *"C'mon, be a man"* response when confronted by men in pain (Smart, 2006). Clinicians must become aware of any differences in their responses based on sex. Subtle differences such as the way you introduce yourself, the way you sit in your office, the comfort or discomfort you feel with expressed anger, or the way you manage your billing are all areas in which your selective responses to male patients may differ from reactions to females. Therapists who see only a few males are likely to be less attuned to gender countertransference predilections. ED specialists may not be as aware of the subtleties of their own emotional reactions to male patients, and may need to be more patient, perhaps less quick to make transference- and, particularly, countertransference-based comments. Therapists can, however, carefully examine their own history of relationships with men to help them anticipate predictable patient–therapist interactional patterns. If nothing else, developing this awareness of emotional responses to male patients who are complimentary, hostile, competitive, flirtatious, or rejecting will help clinicians avoid enactments that might harm the therapeutic alliance.

Special Considerations for Female Therapists

Female therapists may feel unprepared to work with men with ED. Aside from the lack of training in this specialty, many may have had little experience with male patients in general. In these cases, female therapists need to work to overcome latent gender stereotypes and to remain sensitive to issues of power, weakness, and shame. Shame and weakness are triggers for defensive autonomy, and successful engagement of men into psychotherapy usually requires a deft handling of these sensitivities. Sweet (2006) recommends that the initial interview pay explicit attention to the male patient's strengths; that is, empathic attention to sources of pain may be destabilizing while attention to successes and strengths lowers the barriers to engagement. For female therapists, male anger and aggression, which often follow experiences of shame or weakness, may be threatening. Smart (2006) recommends self analysis of issues such as: *"How do you view masculinity"* and *"Are men, at least in terms of relationships and insight, less developed than females?"* The answers will color the experience of work with men.

At some point in the therapy, many male patients will see their female therapists as objects of desire. This eroticizes the transference, masking longings for closeness, intimacy, acceptance, or understanding. These emotions risk leaving a man feeling shamed, exposed, and vulnerable. Sensitive management of this aspect of the engagement is important and therapists may benefit from consultation or peer supervision.

Female therapists should introduce the topic of gender when working with male patients with ED. Opening the door to this topic invites exploration of a long list of forces affecting male identity and relationships. Asking questions about winning, risk taking, emotional control and containment, violence, and toughness can often help the clinician and patient to develop a nuanced understanding of his masculinized identity.

Special Considerations for Male Therapists

Male therapists working with men with ED may have an initial advantage compared to their female colleagues. Patients may see them as more powerful, acceptable, instrumental, and focused on getting things done (Good & Brooks, 2005). There are disadvantages as well. For example, male therapists may have to work harder to just listen, containing their own instinct to do something and to fix the problem. This is particularly relevant in working with ED men where the pressure to fix and do can be enormous. Male therapists may also evoke instinctual distrust, competition, and resentment in their male patients.

When treating men with ED, male therapists may find that the therapeutic relationship feels different than it does with their female patients. Hirsch (2008) points out that both the male therapist and the patient may be pulled into an interaction that minimizes emotional engagement and intimacy. Mutual homosexual fears may drive both parties into safe topics, often sports. More significantly, male therapists may find that their language changes when working with male patients, with greater emphasis on doing rather than feeling. Hirsch (2008) stressed that male therapists are at risk of avoiding psychological depth with their male patients, fearful of provoking shame and humiliation in their patient but also aware of their own emotional risk in being close to another man. One toxic byproduct of this can be a mutual, if subtle and jocular, devaluation of the feminine. This may include overt comments about "women" but is more likely to involve devaluation of dependence, weakness, and vulnerability. Male therapists often have the same anxieties as their male patients about the vulnerabilities that are inherent to the masculine code.

CONCLUSIONS

More men are seeking treatment for ED and ED specialists accustomed to working with predominantly female patients will need to adapt. Appreciating the unique aspects of male eating pathology is an important first step, but it is not sufficient. Clinicians also need to learn more about what men bring to therapy and how their developmental, intrapsychic, interpersonal, and cultural experiences affect therapeutic engagement. Therapists willing to make these adaptations also need to examine their own internalized beliefs about men and masculinity. The work can be especially gratifying. For many men, psychotherapy for their ED may be their first experience of being truly heard by someone, and the experience can change their lives.

References

American Society for Aesthetic Plastic Surgery (2001). 2001 ASAPS statistics: trends—Men have over 1 million procedures, 12% of total. Retrieved September 29, 2009 from American Society of Aesthetic Plastic Surgery website: http://www.surgery.org/sites/default/files/2001stats.pdf

Andersen, A. E. (1990). *Males with eating disorders*. Philadelphia, PA: Brunner/Mazel.

Andersen, A. E. (1995). Eating disorders in males. In C. Fairburn, & K. Brownell (Eds.), *Eating disorders and obesity* (pp. 188–192). New York, NY: Guilford Press.

Andersen, A. E., Cohn, L., & Holbrook, T. (2000a). *Making weight: Men's conflicts with food, weight, shape, and appearance*. Carlsbad, CA: Gurze Books.

Andersen, A., Watson, T., & Schlechte, J. (2000b). Osteoporosis and osteopenia in men with eating disorders. *Lancet*, 355(9219), 1967–1968.

Bergman, S. J., & Surrey, J. L. (1997). The woman–man relationship: Impasses and possibilities. In J. V. Jordan (Ed.), *Women's growth in diversity: More writings from the Stone Center* (pp. 260–287). New York, NY: Guilford Press.

Betz, N. E., & Fitzgerald, L. F. (1993). Individuality and diversity: Theory and research in counseling psychology. *Annual Review of Psychology*, 44, 343–381.

Blatt, S. J. (2008). *Polarities of experience: Relatedness and self definition in personality, development, psychopathology and the therapeutic process*. Washington, DC: American Psychological Association.

Blazina, C., & Watkins, C. E., Jr., (1996). Masculine gender role conflict: Effect on college men's psychological well being, chemical substance usage, and attitudes towards help-seeking. *Journal of Counseling Psychology*, 43, 461–465.

Carlat, D., Camargo, C., & Herzog, D. (1997). Eating disorders in males: A report on 135 patients. *American Journal of Psychiatry*, 154, 1127–1132.

Cochran, S. V., & Rabinowitz, F. E. (2000). *Men and depression: Clinical and empirical perspectives*. San Diego, CA: Academic Press.

Cournoyer, R. J., & Mahalik, J. R. (1995). Cross-sectional study of gender role conflict examining college-aged and middle-aged men. *Journal of Counseling Psychology*, 42, 11–19.

Crisp, A., & Burns, T. (1990). Primary anorexia nervosa in the male and female: A comparison of clinical features and prognosis. In A. Andersen (Ed.), *Males with eating disorders* (pp. 77–99). Philadelphia, PA: Brunner/Mazel.

Diener, M. J., Hilsenroth, M. J., & Weinberger, J. (2007). Therapist affect focus and patient outcomes in psychodynamic psychotherapy: A meta-analysis. *American Journal of Psychiatry*, 164, 936–941.

De Vries, G. J., & Boyle, P. A. (1998). Double duty for sex differences in the brain. *Behavioral Brain Research*, 92, 205–213.

Englar-Carlson, M. (2006). Masculine norms and the therapy process. In M. Englar-Carlson, & M. Stevens (Eds.), *In the room with men: A casebook of therapeutic change* (pp. 13–47). Washington, DC: American Psychological Association.

Englar-Carlson, M. (2008). Counseling with men. In C. M. Ellis, & J. Carlson (Eds.), *Cross-cultural awareness and social justice in counseling* (pp. 98–99). New York, NY: Routledge.

Erickson, C., & Chambers, R. (2007). Adolescence: Neurodevelopment and behavioral impulsivity. In J. Grant, & M. Potenza (Eds.), *Textbook of men's mental health* (pp. 23–46). Arlington, VA: American Psychiatric Publishing.

Fairburn, C. G. (2008). *Cognitive behavior therapy and eating disorders*. New York, NY: The Guilford Press.

Feldman, M., & Meyer, I. (2007). Eating disorders in diverse lesbian, gay, and bisexual populations. *International Journal of Eating Disorders*, 40, 218–226.

Fernandez-Aranda, F., Aitken, A., Badia, A., Gimenez, L., & Solano, D. (2004). Personality and psychopathological traits of males with an eating disorder. *European Eating Disorders Review, 12,* 367–374.

Gilligan, C. (1982). *In a different voice: Psychological theory and women's development.* Cambridge, MA: Harvard University Press.

Good, G. E., & Brooks, G. R. (2005). Introduction. In G. E. Good, & G. R. Brooks (Eds.), *The new handbook of psychotherapy and counseling with men* (pp. 1–13). San Francisco, CA: Jossey Bass.

Good, G. E., Dell, D. M., & Mintz, L. B. (1989). Male role and gender role conflict: Relations to help seeking in men. *Journal of Counseling Psychology, 36,* 295–300.

Hay, P. J., Loukas, A., & Philpott, H. (2005). Prevalence and characteristics of men with eating disorders in primary care: How do they compare to women and what features may aid in identification? *Primary Care and Community Psychiatry, 10,* 1–6.

Hirsch, I. (2008). *Coasting in the countertransference: Conflicts of self-interest between analyst and patient.* New York, NY: The Analytic Press.

Hoffman, I. (2007, Summer). Excerpts from Irwin Hoffman's keynote address: Therapeutic passion in the countertransference. *Psychoanalytic-Psychologist,* 23–27.

Hudson, J., Hiripi, E., Pope, H., & Kessler, R. (2007). The prevalence and correlates of eating disorders in the national comorbidity survey replication. *Biological Psychiatry, 61,* 348–358.

Kindlon, D. J., Thompson, M., & Barker, T. (1999). *Raising Cain: Protecting the emotional life of boys.* New York, NY: Ballantine Books.

Levant, R. (2006). Foreword. In M. Englar-Carlson, & M. Stevens (Eds.), *In the room with men: A casebook of therapeutic change* (pp. xv–xx). Washington, DC: American Psychological Association.

Levant, R. F., & Pollack, W. S. (Eds.), (1995). *A new psychology of men.* New York, NY: Basic Books.

Mahalik, J. R. (1999). Men's gender role socialization: Effect on presenting problems and experiences in psychotherapy. *Progress: Family Systems Research and Therapy, 3,* 13–18.

Maine, M., & Bunnell, D. W. (2008). How do the principles of the feminist relational model apply to the treatment of men with eating disorders and related issues? *Eating Disorders: The Journal of Treatment & Prevention, 16,* 187–192.

Margo, J. (1987). Anorexia nervosa in males: A comparison with female patients. *British Journal of Psychiatry, 151,* 80–83.

Mehler, P. S., Sabel, A. L., Watson, T., & Andersen, A. (2009). High risk of osteoporosis in male patients with eating disorders. *International Journal of Eating Disorders, 41,* 666–672.

Morgan, J. F. (2008). *The invisible man: A self help guide for men with eating disorders, compulsive exercise and bigorexia.* East Sussex, UK: Routledge.

Olivardia, R. (2007). Body image and muscularity. In J. E. Grant, & M. N. Potenza (Eds.), *Textbook of men's mental health* (pp. 307–324). Washington, DC: American Psychiatric Publishing.

Olivardia, R., Pope, H., Borowiecki, J., & Cohane, G. (2004). Biceps and body image: The relationship between muscularity and self-esteem, depression, and eating disorder symptoms. *Psychology of Men and Masculinity, 5,* 112–120.

Olivardia, R., Pope, H. G., Mangweth, B., & Hudson, J. (1995). Eating disorders in college men. *American Journal of Psychiatry, 152,* 1279–1285.

Pleck, J. H. (1995). The gender role strain paradigm: An update. In R. F. Levant, & W. S. Pollack (Eds.), *A new psychology of men* (pp. 11–32). New York, NY: Basic Books.

Pollack, W. S. (2001). "Masked men": New psychoanalytically oriented treatment models for adult and young adult men. In G. R. Brooks, & G. E. Good (Eds.), *The new handbook of psychotherapy and counseling with men: A comprehensive guide to settings, problems, and treatment approaches.* (Vol. 2) (pp. 527–543). San Francisco, CA: Jossey-Bass.

Pope, H., Phillips, K., & Olivardia, R. (2000). *The Adonis complex: The secret crisis of male body obsession.* New York, NY: Free Press.

Pope, H. G., Olivardia, R., Boroweicki, J., & Cohane, G. H. (2001). The growing commercial value of the male body: A longitudinal survey of advertising in women's magazines. *Psychotherapy and Psychosomatics, 70,* 189–192.

Scher, M., Stevens, M., & Good, G. (Eds.), (1987). *Handbook of counseling and psychotherapy with men.* Newbury Park, CA: Sage.

Smart, R. (2006). A man with a "woman's problem": Male gender and eating disorders. In M. Englar-Carlson, & M. A. Stevens (Eds.), *In the room with men: A casebook of therapeutic change* (pp. 319–338). Washington, DC: American Psychological Association.

Sweet, H. (2006). Finding the person behind the persona: Engaging men as a female therapist. In M. Englar-Carlson, & M. A. Stevens (Eds.), *In the Room With Men: A casebook of therapeutic change* (pp. 69–90). Washington, DC: American Psychological Association.

Striegel-Moore, R., Rosselli, F., Perrin, N., DeBar, L., Wilson, G., May, A., & Kraemer, H. C. (2009). Gender difference in the prevalence of eating disorder symptoms. *International Journal of Eating Disorders, 42*, 471–474.

Tchanturia, K., Davies, H., & Campbell, I. (2007). Cognitive remediation for patients with anorexia nervosa: Preliminary findings. *Annals of General Psychiatry, 14*, 1–6.

Thase, M., Reynolds, C., Frank, E., Simons, A., McGeary, J., Fasiczka, A. L.,… Kupfer, D. J. (1994). Do depressed men and women respond similarly to cognitive behavioral therapy? *American Journal of Psychiatry, 151*, 500–505.

Weltzin, T., Weisensel, N., Franczyk, D., Burnett, K., Klitz, C., & Bean, P. (2005). Eating disorders in men: Update. *The Journal of Men's Health and Gender, 2*, 186–193.

Woodside, D. B., Garfinkel, P. E., Lin, E., Goering, P., Kaplan, A. S., Goldbloom, D. S., & Kennedy, S. H. (2001). Comparisons of men with full or partial eating disorders, men without eating disorders, and women with eating disorders in the community. *American Journal of Psychiatry, 158*, 570–574.

PART IV

BRIDGING THE GAP: FAMILY ISSUES

19 *Mutuality and Motivation in the Treatment of Eating Disorders* 319
20 *When Helping Hurts* 335
21 *The Most Painful Gaps* 349

CHAPTER

19

Mutuality and Motivation in the Treatment of Eating Disorders
Connecting with Patients and Families for Change

Mary Tantillo and Jennifer Sanftner

"You know, I have been watching you during the group, and I can tell that this is not just a job for you. It is a labor of love." **(A father's comments after Multifamily Therapy Group)**

Engaging patients with eating disorders (ED) and their families in treatment is strongly related to the therapist's ability to enhance motivation for change while promoting a sense of mutual empathy and empowerment. Our ability to intervene early and continuously, alleviate suffering, and conduct treatment outcome research is hampered by a high treatment drop-out rate. We desperately need innovative models for building and sustaining motivation for change. This chapter presents an integrated relational/motivational (R/M) approach that fosters mutual connection with patients and families and increases their motivation and readiness for change. Following a brief discussion of the theoretical and empirical support for this approach, we address therapist and patient/family challenges to developing mutuality and motivation and present several clinical vignettes to illustrate various R/M strategies and the therapist stance within this approach.

AN INTEGRATED RELATIONAL/MOTIVATIONAL APPROACH

Theoretical and Empirical Roots

The integrated R/M approach described in this chapter is grounded in three main bodies of literature: Stages of Change Theory (SOCT), Motivational Interviewing (MI), and Relational-Cultural Theory (RCT). Understanding the basics of these models builds a foundation for the application of the integrated R/M approach presented in this chapter.

Stages of Change Theory and Motivational Interviewing

SOCT and MI comprise a modified cognitive-behavioral approach that combines evaluation of the patient's Stage of Change (SOC) with a process shown to increase motivation for change and readiness for treatment (Miller & Rollnick, 2002; Prochaska, Norcross & DiClemente, 1994). Identifying a patient's SOC can help therapists match interventions specific to the patient's readiness for change. Stages of Change include: (i) Precontemplation (denial of illness); (ii) Contemplation (ambivalence regarding change); (iii) Preparation for action (intending to make behavioral change within 30 days); (iv) Action (modifying behaviors or environment to promote change); (v) Maintenance (experiencing behavior change for over 6 months); and (vi) Termination (ED no longer presents any temptation, and confidence regarding coping without fear of relapse) (Prochaska et al., 1994). Table 19.1 summarizes the Processes of Change active in various stages and corresponding therapist interventions.

Motivational interviewing is a person-centered and collaborative approach for helping patients resolve ambivalence and move through the Stages of Change (Miller & Rollnick, 2002). In MI the therapist assumes that the resources for change reside within the patient and that motivation for change increases as the therapist helps patients verbalize their own perspectives, goals, and values, and how these things contrast with the impact of the ED (Miller & Rollnick, 2002). Motivation for change and resistance are not seen as patient traits but as products of the interpersonal interaction. Resistance is seen as a signal that the therapist

TABLE 19.1 Processes of Change and Therapist Interventions

Process of Change	ED Therapist Interventions
Consciousness-raising	Help increase knowledge regarding self, ED, and recovery process. Teach functional analysis of behavior (i.e., identify antecedents and consequences to behavior).
Helping relationships	Encourage/assist with development of open and trusting relationships in therapy and the community to promote recovery.
Social liberation	Explore and practice new alternatives in the community that promote change efforts, e.g., social activism.
Emotional arousal	Assist to experience/express feelings related to negative consequences created by the ED.
Self-reevaluation	Assist with assessment of how the person thinks and feels about her/himself in regard to the ED. Create discrepancy by examining goals/values in relation to ED and practice decisional balance (i.e., identify pros/cons related to behavior).
Commitment	Strengthen beliefs that the person can change. Create a realistic plan of action and involve supports.
Countering	Assist to identify and practice alternative coping/self care strategies.
Environmental control	Help to assess and restructure environment to promote recovery.
Reward	Assist with identifying rewards from self and others to celebrate changes

Note. Consciousness-Raising, Helping Relationships and Social Liberation are most helpful in Precontemplation, while the same processes of change along with Emotional Arousal and Self-Reevaluation are helpful in Contemplation (see Prochaska et al., 1994).

is assuming greater motivation or readiness for change than what is actually experienced by the patient. The goal of the therapist is to empathically engage with patients, accurately judge their SOC, and use motivational strategies to maintain their forward movement in recovery.

Stages of Change Theory, Motivational Interviewing, and Eating Disorders

Assessments for motivation for change and readiness have been specifically developed for patients with ED. These scales predict the decision to engage in intensive treatment, symptom change at end of treatment and follow-up, and the likelihood of drop out and relapse (Geller, Cockell & Drab, 2001; Geller, Drab-Hudson, Whisenhunt & Srikameswaran, 2004; Rieger & Touyz, 2006). MI based approaches have been shown to increase motivation for change (Feld, Woodside, Kaplan, Olmsted & Carter, 2001) and reduce symptoms of ED (Cassin, von Ranson, Heng, Brar & Wojtowicz, 2008; Treasure et al., 1999). Thus, research exists to support the application of SOCT/MI to the treatment of ED, although additional research is needed to determine the circumstances under which these approaches will be most beneficial (Miller & Rollnick, 2002; Wilson & Schlam, 2004).

Relational-Cultural Theory and Research

The Integrated R/M approach is rooted in the SOCT and MI principles. Relational-Cultural Theory, with its emphasis on how the relational experience of mutual empathy and empowerment fosters motivation and change, is the heart, or art, of this integrated approach. Traditional psychoanalytic and psychodynamic theories view separation—individuation and self-sufficiency as the markers of psychological maturity (Mahler, Pine & Bergman, 1975). According to RCT, psychological development occurs through *"relational differentiation and elaboration rather than through disengagement and separation"* (Jordan, Kaplan, Miller, Stiver, & Surrey, 1991, p. 87). RCT asserts that mutual empathy and empowerment occur when individuals in the relationship experience "perceived mutuality" (PM) (Genero, Miller, Surrey & Baldwin, 1992; Miller & Stiver, 1997; Tantillo, 2004, 2006). PM involves a bi-directional movement of thoughts, feelings, and actions (Genero, Miller, Surrey & Baldwin 1992), an attunement and responsiveness to the subjective experience of the other that embraces both similarities *and* differences. PM honors space for each person in the relationship, while also honoring the integrity of the connection with the other. The markers of psychological development are *differentiation within connection* and *interdependence*. PM in relationships leads to increased knowledge of oneself and others, improved self-worth and validation, increased zest and vitality, increased ability to act on behalf of oneself and others, and a heightened desire for connection (Miller & Stiver, 1997).

Relationships lacking PM are marked by a pattern of disconnections, i.e., psychological experiences of rupture that occur when one is prevented from engaging in a mutually empathic and mutually empowering interaction with another individual (Miller & Stiver, 1997). Disconnections can range from minor disconnections, e.g., changing the subject when someone is talking, to serious and persistent disconnections such as chronic invalidation, non-responsiveness, neglect, and sexual, physical, and emotional abuse. While occasional disconnections are common to all relationships and can be transformed into growth-fostering moments in life and therapy, chronic and severe disconnections are quite damaging to personal and therapeutic relationships. Disconnections without opportunity for relational repair are particularly damaging. For example, a repeated lack of validation for the patient's and family's experiences and a lack of empathy about where they are in the process of change, are examples of disconnections that obstruct motivation toward change, psychological growth, and recovery (Tantillo, Nappa Bitter & Adams, 2001).

RCT has a growing base of empirical data to support its application to the psychotherapy of patients with ED (Sanftner, 2004; Tantillo & Sanftner, 2003, in press). Several measures of PM and other characteristics of growth-fostering relationships have been developed (Genero et al., 1992; Liang et al., 2002; Tantillo & Sanftner, in press). In addition, our own research suggests that low levels of PM with close friends and romantic partners (Flis et al., 2004; Sanftner, Sippel, Cameron, & Taggart, 2002; Sanftner, Cameron, Walter, & Ohaco, 2003; Sanftner & Tantillo, 2001), as well as with mothers and fathers (Sanftner et al., 2002; Sanftner, Tantillo & Seidlitz 2004; Tantillo, Sanftner, Noyes & Zippler, 2003), are associated with higher levels of ED symptomatology and body image dissatisfaction in men and women (Sanftner, Ryan & Pierce, 2009). An individual's level of PM with her mother and father may be a better predictor of ED attitudes/behaviors than more traditional family variables, such as parental attachment, perceived criticism, and overinvolvement (Sanftner et al., 2006; Tantillo & Sanftner, 2006). PM with parents may predict treatment response, including improvement in several major symptom categories (e.g., binge eating and vomiting); for example, a pilot study found RCT based treatment to be as effective as cognitive behavioral treatment in reducing symptoms of bulimia (Tantillo & Sanftner, 2003).

In a preliminary pilot study, Sanftner and Tantillo (2001) reported that low PM with mothers was associated with the Precontemplation Stage, and high PM with mothers was associated with the Action Stage. These findings offer preliminary evidence that movement along the Stages of Change is associated with increased PM, but randomized controlled studies are required to further examine the etiological link between PM and eating pathology and treatment effectiveness.

Relational-Cultural Theory and Eating Disorders: The Patient

Despite the paucity of empirical data, a substantial amount of theoretical and clinical support suggests that women's PM in relationships influences the etiology, treatment, and recovery from psychiatric disorders such as ED and depression (Gilligan, 1991; Jordan et al., 1991; Miller & Stiver, 1997; Steiner-Adair, 1991; Surrey, 1984; Tantillo, 2000, 2006). RCT conceptualizes ED as "diseases of disconnection" that emerge from, and are perpetuated by, various biopsychosocial risk factors, including a serotonergic disturbance, relationship losses, and a societal value on appearance, especially thinness (Tantillo, 2006; Tantillo & Sanftner, in press). The interaction of these biopsychosocial risk factors contributes to a psychological/relational experience marked by disconnection from one's authentic internal experience (thoughts, feelings, and needs), dissatisfaction with and disconnection from one's body, and disconnection from others, creating risk for an ED. This experience may obstruct the development of PM in relationships with others before onset of illness. In addition to predisposing the individual to ED, disconnections can also precipitate and/or perpetuate them. Once the ED has developed, disconnections and decreased PM can trigger further symptoms. For example, starvation adversely affects the patient's ability to process information. Thus, she can misinterpret and negatively distort information, leaving her feeling more alone and vulnerable to coping through symptom use.

RCT emphasizes that relationships characterized by higher levels of PM will survive and, in fact, often strengthen after disconnection. PM facilitates relational growth as relationships move cyclically from connection to disconnection to re-connection (Miller & Stiver, 1997; Siegel & Hartzell, 2003). Therefore, a central task for the therapist in promoting motivation and engagement in treatment is to develop PM in relationships with their patients and families. As PM develops in the therapeutic relationship, the patient begins to identify the connections between his or her relationship with self (genuine thoughts, feelings, and needs), with others, and with his or her ED. For example, the patient may see that urges to restrict food or binge/purge allow him or her to avoid or regulate difficult feelings including anger, loss of control, or neediness. The patient may also start to recognize how these experiences affect relationships with family members.

Relational-Cultural Theory and Eating Disorders: The Family

In RCT, family stress and illness can increase experiences of low self-worth, disempowerment, inability to tolerate difference, tension, feeling "locked up or locked out" in relationship, self-doubt, and increased isolation. These disconnections often lead to the family problems long identified as part of the presentation of ED (Bruch, 1973; Gull, 1874; Minuchin et al., 1975; Morgan & Russell, 1975). The concept of families being viewed as important resources for treatment and not solely responsible for the development and maintenance of ED is relatively new (Bryant-Waugh & Lask, 2004; Dare & Eisler, 1997; Le Grange & Lock, 2007; Lock, LeGrange, Agras & Dare, 2002; Lock & Le Grange, 2005; Tantillo, 2006; Treasure, Smith & Crane, 2007). Similarly, RCT does not presume that a severe pattern of family disconnection predates the onset of the ED. However, disconnections are apt to exist after onset. These may include parental shame, guilt, fear, high expressed emotion, and conflict at meal times. Many families struggle to understand the illness and do not know how to be helpful to their loved ones. Parents often worry that whatever they say or do may worsen symptoms or will make them lose any connection they have with their son or daughter. As the patient's self-knowledge about his or her own internal experiences plummets because of starvation, fatigue, and cognitive limitations, the patient is less able to communicate these experiences or to see his or her impact on others, further inhibiting PM.

Repeated disconnections, characterized by the inability to embrace difference in a relationship (e.g., different thoughts, feelings, needs) and a lack of relational repair, erode the PM required for growth-fostering connections. These disconnections lead to the development or reinforcement of (often unconscious) negative and distorted relational images and meanings for the patient (e.g., *"No one values what I think or feel"* because *"I am unlovable and defective"*). Parents may come to believe, for instance, that *"My daughter must think I am a failure because I am inadequate and never know how to help."* Siblings may come to believe that *"I don't matter to my brother because he is selfish and refuses to see how he is affecting me."* These negative relational messages prevent patient and family members from accessing large portions of their authentic experience (e.g., the needs and feelings they feel too emotionally vulnerable to identify). They are then prone to use their own strategies for disconnection (to deal with underlying emotional pain/conflict), including more ED symptoms, avoidance, isolation, or aggressive interactions.

This same pattern of disconnection can play out between therapist and patient or between therapist and the patient's family. The patient, family members, and therapist can all feel the strain, fatigue, and disconnection created by the ED. These disconnections can also occur in a parallel process ranging from the patient's internal experience to relationships on the treatment team. Disconnections within the treatment team are often metaphors for the disconnections experienced by the patient within him- or herself, and between the patient and other family members. An RCT-informed approach to recovery requires that patients and families build PM with one another and with the therapist to decrease and/or repair disconnections, thereby fostering motivation, growth and healing (Miller & Stiver, 1997; Tantillo, 2000, 2004, 2006; Tantillo et al., 2001; Tantillo & Sanftner, 2003; Walker & Rosen, 2004).

THE NEED FOR AN INTEGRATED RELATIONAL/MOTIVATIONAL

Approach to Patients and Families

Recovery from ED involves not only a change in self-efficacy (a focus of MI), but a change in self-definition and relational patterns (Tantillo et al., 2001). RCT puts PM at the center of the understanding of the etiology of and recovery from ED. Therapy from this perspective emphasizes that patient, family, *and* therapist grow through their work by naming difference and disconnection and by engaging in repair. In order to do this work, the therapist must not only be expert in matching motivational strategies with the patient's present SOC (see Table 19.1), but must also deliberately work to create PM in relationships with patient/family. Box 19.1 gives a brief summary of Therapist Strategies to Increase Mutuality and Motivation. The RCT therapist must bring more of him or herself into the work. The therapist recognizes his or her own impact on the patient/family, but, most importantly, he or she allows the family to influence him or her. The therapist conveys how the family has moved him or her in the work.

RCT also enriches SOCT and MI because it provides a gender-informed way to conduct motivational work. RCT posits that PM is important for male and female psychological development, resistance to illness, motivation for change, and recovery from ED (Miller & Stiver, 1997; Steiner-Adair, 1991; Surrey, 1984; Tantillo, 2004). While PM may be more helpful for women because they have been socialized to grow in and through connection with others (Chodorow, 1978; Gilligan, 1991; Miller & Stiver, 1997), an increasing body of literature suggests that disconnections are damaging for men as well (Bergman, 1991; Levant & Pollack, 1995; Pollack, 1998; Real, 1997; see also Bunnell, Chapter 18).

AN INTEGRATED RELATIONAL/MOTIVATIONAL APPROACH FOR EATING DISORDERS

General Principles and Strategies

The integrated R/M treatment approach described below is especially important at the start of treatment when patients experience denial and ambivalence (in Precontemplation

BOX 19.1

THERAPIST STRATEGIES TO INCREASE MUTUALITY AND MOTIVATION

Relationally reframe the illness and recovery. Educate regarding biopsychosocial risk factors that create a vulnerability to disconnection from self and others and possible development/maintenance of ED.

- Externalize the illness from the patient and family.
- Acknowledge the burden/strain incurred by the illness.
- Normalize instead of pathologize patient and family experiences.
- Use validation and avoid shame and blame.
- Be empathic and warm while being deliberate and consistent.
- Be real, genuine, and emotionally present and responsive.
- Build and empower the "We"—put connection at the heart of the relationship.
- Teach and model that lapses and mistakes are opportunities for learning.
- Provide anticipatory guidance (regarding development, the illness, recovery).
- Acknowledge and apologize for errors.
- Convey humility, use well-timed humor, and foster collaboration.
- Build a new culture of shared meanings through learning the language and values of family and patient related to illness, recovery, and relationships.
- Judiciously use therapist self-disclosure to convey how patient and family have moved you in the session.
- Teach and model flexibility, openness to change and difference, and the ability to tolerate uncertainty, unpredictability and ambiguity.
- Name all or nothing thinking and its contributions to disconnections.
- Name relational dilemmas and let patients/family members see how you are thinking and feeling, while helping them name their own experience.
- Help patient and family name points of tension and disconnection and one's responses to these events.
- Teach/model the importance of not assuming what others think/feel. Check it out.
- Honor difference *and* the integrity of connections with one another.
- Emphasize that all relationships naturally move from connection to disconnection and back and that there are no perfect relationships. The best ones come from continual hard work and repair.
- Be aware of your own strategies for disconnection, e.g., intellectualization, etc.
- Encourage opportunities for connection that don't include the ED.
- Remind patient and family that they are more than the ED. Identify their strengths.
- Help patient/family members identify family values and goals and how the ED helps or hinders them from living according to these things.
- Help patient and family identify supports with whom they experience mutuality.
- Engage in your own self-care and with personal and professional supports with whom you experience mutuality.

and Contemplation Stages of Change). It can be used, however, throughout treatment, particularly at times when it seems that patient or family are experiencing a lapse or relapse and across treatment modalities including individual (Tantillo, 2004), group, family, or Multi-Family Group Therapy (MFTG) (Tantillo, 2006). Also, in this R/M approach, "family members" refer not only to biologic family members but to all those closest to the patient and comprise her or his "community for healing" or recovery network.

While the description of interventions in Table 19.1 is linear, therapists must always be prepared to circle back and use earlier Processes of Change if they sense disconnection occurring in relationships between the patient and family, patient and therapist, or patient/family and therapist. In a group setting, this would also include disconnections that occur between group members and between the group and the therapist. Interventions listed in Box 19.1 enable therapists to prevent or to identify and repair disconnections quicker and more accurately and, thereby, promote continued engagement, motivation, and change.

The following vignette demonstrates how the therapist returned to an earlier Process of Change in response to a patient's relapse. This patient had moved into the Action SOC but slipped back into Precontemplation and Contemplation SOC. Using the Process of Change, Consciousness-Raising, and corresponding intervention, Functional Analysis of Behavior, this example describes the work with the patient's father. The therapist started by externalizing the illness, identifying the disconnections created by the ED, and by promoting PM in the relationship between the father, his wife, and the therapist.

Dad: *(directed to the therapist) But how do you know this weight loss is because of the ED? Maybe she does not really have an ED? People who run track can lose some weight.*

Mom: *You still are having trouble believing our daughter is sick. What will it take to get you to understand this? Of course the weight loss is about the ED.*

Therapist: *First I want to remind us that the ED is at work right now trying to disconnect us from one another by creating doubt and conflict among us. It has tried this before. Second, I know you both love your daughter very much and want what is best for her. Dad, it is hard to look at your daughter and see her as sick, especially when it seems her track time gets better with her weight loss and she says she is fine. You want her to be healthy, to feel good about herself, and to be successful. It's hard to separate her from the ED and the effect it has on her, especially after you all have worked so hard to restore her health.*

Dad: *I just don't understand how suddenly, it's back. I thought we were done with it.*

Mom: *I keep telling you, it is not like having a cold or something that goes away. It sticks around and waits for her to feel vulnerable and then gets the best of her.*

Therapist: *Dad, remember how we discussed that transitions are really rough for your daughter. Well, she is interviewing for college now, competing for scholarships, and there are many things going on. (The therapist reminds him about previous transitions that were difficult and the patient's perfectionism, dichotomous thoughts, and her increased symptoms at these times.) Mom is right about her increased vulnerability. But tell us what you are feeling right now.*

Dad: *I feel helpless because I can't stop this thing from hurting her. And she can't control it either. I am tired of it. I just want it to stop. It is dragging all of us down.*

Mom: *(to Dad) I know you feel down and out of control with all this because you feel you have to fix it. But that is not your job. This relapse is disappointing, but we have to keep going.*

Therapist: *Dad, is mom right about how you are feeling?*

Dad: *Yes. But it's hard not to feel that way. It is so disappointing. She's supposed to be getting ready for college. I wanted her to feel good about all this.*

Therapist: *It is disappointing, and I feel frustrated for all of you because I know how hard you have all worked. But maybe your daughter is somehow telling us that she is fearful about this transition. It's good for us to slow things down and help her examine what is going on. Being able to do this is part of recovery and life. Dad, let's talk about what you need from mom and me to help you remember that it is not your job to fix this. We will figure out a way to stick together on this like we have before. The ED would like us to feel helpless and alone, but we can't let it take advantage of us.*

Within the integrated R/M approach, each dynamic in the therapeutic relationship with patients and families is understood as an effort to create/maintain connection or to move out of connection. Also, the therapist defines the ED as the primary agent creating disconnection in the hope that the patient, family members, and therapist will remain divided from one another. R/M work generally begins with Consciousness-Raising and Education regarding the SOC model and the recovery process, reflecting elements of the Precontemplation stage. The therapist conveys acceptance of each individual's present SOC and emphasizes that "Action" (behavioral change) is only one stage on the road toward sustainable change. Much work must be accomplished in the earlier stages for successful recovery. The therapist should emphasize that the ED is a "disease of disconnection" and identify the biopsychosocial risk factors that predispose, precipitate, and perpetuate disconnections and contribute to development and maintenance of ED. One common feature, the all-or-nothing thinking that accompanies the ED, is particularly important to address, empowering everyone in a relationship to notice it in interactions with one another. While the therapist seeks to understand the patient's and family's world (learning and using their language and meanings), he or she also integrates a common relational language and meanings that can be used by all individuals in a relationship. Motivation and connection grow with this shared language as everyone works to disable the disease by placing it outside the patient/family/therapist relationship.

Next, the therapist engages in Processes of Change associated with the Contemplation SOC. In this stage, the focus is on the development of patient and family relational skills. This work involves cultivation of knowledge regarding the self in relationships with others while introducing the concept of Decisional Balance, expanding the Functional Analysis of Behavior, and promoting Emotional Arousal, Discrepancy, and Self-Reevaluation (see Table 19.1.) For example, the therapist encourages the patient and family to identify the behavioral sequence leading up to ED behaviors, the pros and cons related to the ED, and whatever disconnection the illness creates in their relationships. This is done within the context of promoting PM by teaching and modeling how to name and withstand difference while being connected. For example, if the patient says that she cannot tell her parents what she really feels about recovery, the therapist will initially use reflection and reframing (a fundamental aspect of MI) of the patient's statement to decrease any associated shame/blame that could increase disconnection. She may say, *"It sounds like it is difficult to speak about what you really feel with your folks and you end up protecting them from what you really feel."* The therapist may ask the patient how she feels when this occurs or ask her about her worst fear if she were to genuinely disclose to them. The therapist will also ask if there is a specific example the patient can share that would demonstrate her relational dilemma and would encourage her to identify what happened just before she could not disclose to her parents, and what occurred after she withheld information. The therapist might then turn to the parents and state, *"Your daughter has certain assumptions about how you may be thinking and feeling in this scenario. These assumptions may be different or similar to what you would actually feel or think."* This demonstrates how they can remain connected in the face of difference, tension, and disconnection.

The therapist might then return to the patient and ask what it is like hearing her parents' feedback. If this scenario occurred in MFTG, the therapist would suggest that the patient and family are not the only ones who have experienced a dilemma like this and would ask other parents and patients to discuss how they felt as the patient and family described their situation, as they may feel freer to name certain feelings that are at play. This intervention promotes PM because everyone in the relationship shares their experiences. It also promotes Emotional Arousal; the patient deeply "feels", as opposed to just cognitively "understands," the impact of the ED in his or her life and the lives of family members (Prochaska et al., 1994). Patient and family members hear and feel the support of others who have also experienced similar disconnections, emotions, and consequences created by the illness.

The therapist might also ask what other patients and families would have felt in this scenario and follow-up by asking, *"Is it similar or different from what they have heard here?"* This helps to normalize either possibility. The therapist would explicitly state how being able to embrace and work through difference in connection is a hallmark of recovery and how this eliminates the ED as a ready solution for dealing with the disconnections it also causes. The therapist could end with a Decisional Balance discussion that identifies the pros and cons of protecting one's parents from genuine thoughts/feelings (Prochaska et al., 1994). Circling back, the therapist can ask the patient whether *"protecting your parents from your true feelings/thoughts* (based on the pros and cons analysis) *helps you feel more or less connected to yourself and others?"*

Developing discrepancy between the ED and the patient's life goals and values is another tool for enhancing connection. For instance, in the prior vignette, the therapist could also ask the patient whether her choices about communicating with her parents move her toward or away from her life goals and values. This line of exploration promotes Self-Reevaluation, by helping her evaluate the present negative consequences of ED behaviors and to develop a forward looking assessment of a new potential self, i.e., a healthier and changed view of herself (Prochaska et al., 1994; See also Kater, Chapter 10). The therapist might also help the patient track differences in her symptom use between occasions in which she could be open with her parents versus times in which she maintained secrecy. This intervention helps her define the connections between her relationships with her illness and her relationships with her parents and herself.

In an integrated R/M approach, the therapist also spends time helping patient and family evaluate the growth-fostering potential of their relational environments at home, school, and work. The therapist can help them to develop ways to support "the we" or sense of mutual connection with one another (Shem & Surrey, 1998) by encouraging them to:

1. routinely find time to be together in safe and comforting places (e.g., family room, favorite place in the park or vacation spot);
2. identify the high risk times that threaten their sense of "we" (e.g., transition times from school or work to home, or visiting less supportive extended family or friends); and
3. find ways to connect with one another that have nothing to do with the ED (e.g., including a family movie/board game night and action projects like volunteering together).

A Relational/Motivational Approach: Therapist Stance

The R/M approach with ED patients and families requires the therapist to sit with a degree of openness, flexibility, and emotional vulnerability that is not required in more traditional psychodynamic psychotherapy. The best way to eliminate shame and share vulnerability is to demystify the therapist role and therapy process by leading by example (Tantillo, 2004). As Lawrence-Lightfoot (2000) states, "Making oneself vulnerable signals trust and respect, as does receiving and honoring the vulnerability of others" (p. 93). In this integrated approach, the therapist fosters PM and motivation for change by sharing how he or she is influenced by the patient and family members. He or she does this through the use of the real relationship with patient and family, and through authentic and judicious self-disclosure (Jordan et al., 1991; Miller & Stiver, 1997; Tantillo, 2004). These are always utilized with an ongoing awareness of the potential transference relationships among therapist and patient/family. The therapist can also be authentic, i.e., trying to represent oneself more fully in the relationship (Miller et al., 1999), without self-disclosing. For example the therapist can use verbal interventions such as validation and reflection or a non-verbal stance that emphasizes attentiveness and emotional presence in the moment (Tantillo, 2004). However, one of the most powerful ways to promote PM is for the therapist to communicate how he or she has been moved by patient and family in the here and now of the work. This directly models openness to the influence of others and tells the patient and family that what they say and do does matter. They empathize with the therapist's response, try it on for size and, in doing so, become more aware and empathic about their own internal experiences and how these influence one another. During this kind of interchange, the therapist can also acknowledge his or her own limitations or strategies for disconnection, either in relation to something occurring in the therapy or to a past experience, as in the following vignette.

During the second session of MFTG, the group therapist described again how the ED is separate from the patient, and group members shared how important and challenging that is to remember. She noticed this week that Peter (a 16-year-old with anorexia nervosa [AN]) was rolling his eyes toward the ceiling when she described it this way and remembered he had done the same thing last week. This time she asks for his feedback.

Therapist: *Peter, I'm hearing from many folks in group how they see the ED in the way I have described it, but it is also important to know if some of us might see it in a different way. Do you see it in the same or in a different way? I'm getting the feeling it might be different.*
Peter: *It doesn't matter* (he looks down).
Therapist: *What you say does matter though. Remember how we discussed the importance of being able to discuss our differences and our similarities in group? You may be thinking about something very important that can help all of us here.*
Peter: *Well, I felt the same way last week too. And I didn't want to come back and hear the same thing again. It's the way you talk about the ED, like it's outside or separate from me. It's how my parents talk about it too. They don't get it* (he looks over at them).
Therapist: *Do you feel you can say a little more about what we don't get because it sounds like all of us have upset you in the same way?*
Peter: *You don't understand how the ED is not separate from me. It **is** me. When you say it is separate from me, it's, it's, I don't know. It doesn't matter.*
Therapist: *You do matter, Peter. It's what? Can you tell what you are feeling right now?*
Peter: *It's just so annoying. Everyone keeps talking about the ED instead of me. I am the one restricting. The ED isn't making me do anything. They don't get it.*
Therapist: *So you are pretty annoyed with us because it's like we don't see you? We keep talking about the ED and its effect on you, instead of talking with you about what you are feeling or doing. Like you are invisible?*
Peter: *Yes, and you don't understand that by doing what you say, sending the ED away, means I have to sacrifice myself. They don't understand this.*
Therapist: *I can see why you would be so annoyed with us now and what a terrible position this puts you in. Basically you have to stop being you to let go of the ED. That also sounds pretty scary. It's hard to think about how you could be you without restricting food and staying thin.*
Peter: *Yes, exactly. I have to give in and be someone else to be who they want me to be.*
Therapist: *I want you to know that I feel badly that you had to feel like this since last week. I noticed you roll your eyes last session when I mentioned how the ED was separate from the person. Part of me wanted to ask you what was going on to better understand your thoughts, but the other part of me didn't want to put you on the spot and make you feel uncomfortable during the first group session. I am sorry about that. I am thinking now that I should have asked you then. I feel sad that you have felt this way since last week. I'm sorry. I missed the boat on that one. It's important for you to be able to say when*

you feel annoyed or feel differently about something. You end up sacrificing even more of you when you have to keep it in. (The therapist turns to Peter's parents to ask if they knew how annoyed and misunderstood he felt about this issue and has them share what they felt as he spoke. After they acknowledge that they also had no idea he felt this way and their own frustration and sadness about not communicating well with one another, the therapist encourages other parents and patients to describe times when they also struggled with similar situations.)

In this vignette the therapist models naming and understanding disconnection without blame or shame and begins to restore mutual connection. She does so by joining with the parents as an adult who annoys Peter, while aligning herself with Peter about his annoyance and feelings of invisibility. This kind of relational movement in connection with the patient and family eventually helps them to respond similarly in relationships with the therapist, with each other, and with others (Miller and Stiver, 1997; Tantillo, 2004, 2006; Tantillo et al., 2001; Walker & Rosen, 2004). In contrast to MI, RCT specifically embraces the notion that mutually empathic and empowering relationships also enhance and transform the therapist. As the therapist "stretches" to match or understand the patient's and family's experience, he or she is also moved by them in the change process (Jordan, 1991).

CONCLUSION: IT TAKES AT LEAST TWO

Treasure and colleagues (2007) emphasize the following six Cs when teaching family members how to support their loved one with an ED: calmness, communication, compassion, cooperation, consistency, and coaching. Smith (2007) encourages professional carers to do the same when interacting with patients and families in order to *"build constructive conversations"* (p. 133). It takes at least two partners, and in the case of recovery, three partners, i.e., patient, family members, and the therapist, to create strong connections and constructive conversations that can lead to engagement in treatment, motivation for change, and continued growth in recovery. Specifically, therapists need to be not only technical experts in terms of ED, SOCT, and MI, but also able to bring more of who they are into the working relationship and to clearly demonstrate to patient and family that their feedback is understood and valued, and most importantly, that it has somehow moved them.

Helping ED patients and their families move from Precontemplation toward Action is challenging because either or both parties may not be ready for change. Also, the ED does whatever it can to fuel conflict, high expressed emotion, or isolation among patient, family members, and therapist in order to perpetuate itself and remain connected to the patient. These challenges, along with the therapist's own misattunements, fatigue, intense feelings, and inaccurate judgments regarding readiness for change, complicate connections for patients and families looking for change. Therefore, a main goal of the therapist is to establish a mutually engaging and motivating stance while honoring the many differences (e.g., of opinions, feelings, and attitudes) and working through the disconnections that inevitably arise on the road toward change. It is the therapist's ability to model and teach the value of this mutually empathic and empowering stance that strengthens engagement, increases motivation for change, and fosters ongoing collaboration in treatment.

References

Bergman, S. J. (1991). Men's psychological development: A relational perspective. *Work in Progress, No. 48*. Wellesley, MA: Stone Center.

Bruch, H. (1973). *Eating disorders, obesity, and the person within*. Houston, TX: Basic Books.

Bryant-Waugh, R., & Lask, B. (2004). *Eating disorders: A parents' guide*. New York, NY: Brunner-Routledge.

Cassin, S. E., von Ranson, K. M., Heng, K., Brar, J., & Wojtowicz, A. E. (2008). Adapted motivational interviewing for women with binge eating disorder: A randomized controlled trial. *Psychology of Addictive Behaviors, 22*(3), 417–425.

Chodorow, N. (1978). *The reproduction of mothering*. Berkeley, CA: University of California Press.

Dare, C., & Eisler, I. (1997). Family therapy for anorexia nervosa. In D. M. Garner & P. E. Garfinkel (Eds.), *Handbook of treatment for eating disorders* (2nd ed.) (pp. 307–324). New York, NY: The Guilford Press.

Feld, R., Woodside, D. B., Kaplan, A. S., Olmsted, M. P., & Carter, J. C. (2001). Pretreatment motivational enhancement therapy for eating disorders: A pilot study. *International Journal for Eating Disorders, 29*(4), 393–400.

Flis, K., Allen, T. D., Schlosky, K., Walter, C., Walters, A. M., Zippler, E. S., ... Sanftner, J. L. (2004, April). *The association between relational functioning and eating disorder behaviors in women*. Poster presented at the annual meeting of the Eastern Psychological Association, Washington, DC.

Geller, J., Cockell, S. J., & Drab, D. (2001). Assessing readiness for change in anorexia nervosa: The psychometric properties of the readiness and motivation interview. *Psychological Assessment, 13*, 189–198.

Geller, J., Drab-Hudson, D., Whisenhunt, B. L., & Srikameswaran, S. (2004). Readiness to change dietary restriction predicts short and long term outcomes in the eating disorders. *Eating Disorders: The Journal of Treatment and Prevention, 12*, 209–224.

Genero, N. P., Miller, J. B., Surrey, J., & Baldwin, L. M. (1992). Measuring perceived mutuality in close relationships: Validation of the mutual psychological development questionnaire. *Journal of Family Psychology, 6*, 36–48.

Gilligan, C. (1991). Women's psychological development: Implications for psychotherapy. In C. Gilligan, A. G. Rogers, & D. L. Tolman (Eds.), *Women, girls & psychotherapy: Reframing resistance* (pp. 5–31). New York, NY: Harrington Park Press.

Gull, W. W. (1874). Anorexia nervosa (apepsia, hysteria, anorexia hysterica). *Transactions of the Clinical Society of London, 7*, 222–228.

Jordan, J. V. (1991). The meaning of mutuality. In J. V. Jordan, A. G. Kaplan, J. B. Miller, I. P. Stiver, & J. Surrey (Eds.), *Women's growth in connection: Writings from the stone center* (pp. 81–96). New York, NY: The Guilford Press.

Jordan, J. V., Kaplan, A. G., Miller, J. B., Stiver, I. P., & Surrey, J. (Eds.), (1991). *Women's growth in connection: Writings from the Stone Center*. New York, NY: The Guilford Press.

Lawrence-Lightfoot, S. (2000). *Respect: An exploration*. New York, NY: Perseus.

Le Grange, D., & Lock, J. (2007). *Treating bulimia in adolescents: A family-based approach*. New York, NY: The Guilford Press.

Levant, R. S., & Pollack, W. S. (1995). *A new psychology of men*. New York, NY: Basic Books.

Liang, B., Tracy, A., Taylor, C. A., Williams, L. M., Jordan, J. V., & Miller, J. B. (2002). The Relational Health Indices: A study of women's relationships. *Psychology of Women Quarterly, 26*(1), 25–35.

Lock, J., LeGrange, D., Agras, W. S., & Dare, C. (2002). *Treatment manual for anorexia nervosa: A family-based approach*. New York, NY: The Guilford Press.

Lock, J., & Le Grange, D. (2005). *Help your teenager beat an eating disorder*. New York, NY: The Guilford Press.

Mahler, M. S., Pine, F., & Bergman, A. (1975). *The psychological birth of the human infant*. New York, NY: Basic Books.

Miller, J. B., Jordan, J. V., Stiver, I. P., Walker, M., Surrey, J., & Eldridge, N. S. (1999). Therapists' authenticity. *Work in progress, No. 82*. Wellesley, MA: Stone Center Working Paper Series.

Miller, J. B., & Stiver, I. P. (1997). *The healing connection: How women form relationships in therapy and in life*. Boston, MA: Beacon Press.

Miller, W. R., & Rollnick, S. (2002). *Motivational interviewing: Preparing people for change.* New York, NY: The Guilford Press.

Minuchin, S., Baker, L., Rosman, B. L., Liebman, R., Milman, L., & Todd, T. C. (1975). A conceptual model of psychosomatic illness in children. *Archives of General Psychiatry, 32,* 1031–1038.

Morgan, H. G., & Russell, G. F. M. (1975). Value of family background and clinical features as predictors of long-term outcome in anorexia nervosa. A four year follow-up study of 41 patients. *Psychological Medicine, 5,* 355–371.

Pollack, W. (1998). *Real boys.* New York, NY: Random House.

Prochaska, J. O., Norcross, J. C., & DiClemente, C. C. (1994). *Changing for good.* New York, NY: William Morrow & Co., Inc.

Real, T. (1997). *I don't want to talk about it.* New York, NY: Scribner.

Rieger, E., & Touyz, S. (2006). An investigation of the factorial structure of motivation to recover in anorexia nervosa using the Anorexia Nervosa Stages of Change Questionnaire. *European Eating Disorders Review, 14,* 269–275.

Sanftner, J. L., & Tantillo, M. (2001, June). *A relational/motivational group treatment approach for eating disorders.* Paper presented at the annual Research Network Forum of the Jean Baker Miller Training Institute, Boston, Massachusetts.

Sanftner, J. L., Sippel, J. A., Cameron, R. P., & Taggart, J. M. (2002, November). The association between interpersonal functioning, eating disorder behavior, and ethnicity in college undergraduates. Poster presented at the annual meeting of the Eating Disorders Research Society, Charleston, South Carolina.

Sanftner, J. L., Cameron, R. P., Walter, C. P., & Ohaco, G. N. (2003, September). The relative importance of difference individuals and the general community in predicting eating disorder symptoms. Poster presented at the annual meeting of the Eating Disorders Research Society, Ravello, Italy.

Sanftner, J. L., Tantillo, M., & Seidlitz, L. (2004). A pilot investigation of the relation of perceived mutuality to eating disorders in women. *Women and Health, 9*(1), 85–100.

Sanftner, J. L., Cameron, R. P., Tantillo, M., Heigel, C. P., Martin, D. M., Sippel-Silowash, J., & Taggart, J. M. (2006). Mutuality as an aspect of family functioning in predicting eating disorder symptoms in college women. *Journal of College Student Psychotherapy, 21*(2), 41–66.

Sanftner, J. L., Ryan, W. J., & Pierce, P. (2009). Application of a relational model to understanding body image in college women and men. *Journal of College Student Psychotherapy, 23*(4), 262–280.

Shem, S., & Surrey, J. (1998). *We have to talk: Healing dialogues between men and women.* New York, NY: Basic Books.

Siegel, D. J., & Hartzell, M. (2003). *Parenting from the inside out.* New York, NY: Penguin Putnam, Inc.

Smith, G. (2007). *Families, carers, and professionals: Building constructive conversations.* West Sussex, England: John Riley & Sons.

Steiner-Adair, C. (1991). New maps of development, new models of therapy: The psychology of women and the treatment of eating disorders. In C. Johnson (Ed.), *Psychodynamic treatment of anorexia nervosa and bulimia* (pp. 225–241). New York, NY: Guilford Press.

Surrey, J. (1984). *Eating patterns as a reflection of women's development. Work in progress, No. 83–06.* Wellesley, MA: Stone Center.

Tantillo, M. (2000). Short-term relational group therapy for women with bulimia nervosa. *Eating Disorders: The Journal of Treatment and Prevention, 8,* 99–121.

Tantillo, M. (2004). The therapist's use of self-disclosure in a Relational Therapy Approach for eating disorders. *Eating Disorders, 12,* 51–73.

Tantillo, M. (2006). A relational approach to eating disorders multifamily therapy group: Moving from difference and disconnection to mutual connection. *Families, Systems, and Health, 24*(1), 82–102.

Tantillo, M., Nappa Bitter, C., & Adams, B. (2001). Enhancing readiness for eating disorder treatment: A relational/motivational group model for change. *Eating Disorders: Journal of Treatment and Prevention, 9,* 203–216.

Tantillo, M., & Sanftner, J. L. (2003). The relationship between perceived mutuality and bulimic symptoms, depression, and therapeutic change in group. *Eating Behaviors, 3,* 349–364.

Tantillo, M., Sanftner, J. L., Noyes, C., & Zippler, E. (2003, June). The relationship between perceived mutuality and eating disorder symptoms for women beginning outpatient treatment. Poster presented at the annual meeting of the Eating Disorders Research Society, Ravello, Italy.

Tantillo, M., & Sanftner, J. L. (2006, June). The impact of disconnections for individuals with eating disorders: Evaluating the level of perceived mutuality in relationships with parents using the Connection–Disconnection Scale. Paper presented at the International Conference on Eating Disorders, Barcelona, Spain.

Tantillo, M., & Sanftner, J. L. (in press). Measuring perceived mutuality in women with eating disorders: The development of the Connection–Disconnection Scale.

Treasure, J., Katzman, M., Schmidt, U., Troop, N., Todd, G., & de Silva, P. (1999). Engagement and outcome in the treatment of bulimia nervosa: First phase of a sequential design comparing motivational enhancement therapy and cognitive behavioural therapy. *Behaviour Research and Therapy, 37*, 405–418.

Treasure, J., Smith, G., & Crane, A. (2007). *Skills-based learning for caring for a loved one with an eating disorder.* London, England: Routledge.

Walker, M., & Rosen, W. B. (2004). *How connections heal.* New York, NY: The Guilford Press.

Wilson, G. T., & Schlam, T. R. (2004). The transtheoretical model and motivational interviewing in the treatment of eating and weight disorders. *Clinical Psychology Review, 24*(3), 361–378.

CHAPTER 20

When Helping Hurts
The Role of the Family and Significant Others in the Treatment of Eating Disorders

Judith Brisman

Not knowing when the dawn will come I open every door. **(Emily Dickinson, 1960)**

Controversy abounds with regard to the role of the family and significant others in the treatment of eating disorders (ED). The most well-documented research indicates that conjoint family-based treatment (FBT), in which parents are directly involved in re-feeding, is most beneficial for teenagers or children with anorexia nervosa (AN) (Le Grange, 2005). However, other research and clinical experiences contradict these findings and allow for the possibility of more varied means of involving the family. This chapter hopes to bridge the gap between extensive research pointing to the need for direct parental intervention and alternate research and clinical experience reminding us that one size does not fit all.

Considering the evolution of the family's role in ED treatment, this chapter will explore research and treatment modalities that inform therapeutic decision-making by primary caregivers. How do we heed the recommendations of research investigations while at the same time recognizing that the research itself may narrow viable possibilities for effective engagement of family and significant others? Researchers and clinicians alike have been dedicated to finding answers in hope of providing patients, parents, and significant others a path to recovery. The answers, however, have sometimes been as problematic as the questions.

DAWN: WHERE WE BEGAN

Early investigations in the arena of ED looked directly to the family as both the source of trouble and the ballast of cure. As early as the 1800s, intra-familial conflict was seen as an integral catalyst to AN (Brumberg, 1989). The linking of family dysfunction to symptomatic development endured over time, with specific family characteristics and traits seen as precursors to particular disorders with food. For example, the anorexic family was found to be more rigid in its organization, more avoidant of conflict, and more interdependent, while the bulimic family displayed marked disorganization, chaos, and hostility (Humphrey, 1988). Poor or insecure attachment within the ED family was also seen as causal in symptomatic development (Ward, Ramsay, Turnbull, Benedettini & Treasure, 2000).

In particular, the work of Minuchin, Rosman & Baker (1978) and Selvini-Palazzoli (1978) set the stage for considering family dysfunction as the cause of ED. Using different strategies, both Minuchin and Selvini-Palazzoli considered that patterns of family interaction, such as over-protectiveness, rigidity, conflict avoidance, and triangulation (pulling a third person into an issue involving two other people), aided in the development and perpetuation of disordered eating, particularly AN. The anorexic behavior was seen as a thwarted attempt to maintain family connections and to stabilize emotional expression. Indeed, a success rate of 86% was reported with regard to the anorexic behavior when families became aware of and changed problematic family patterns (Minuchin et al., 1978).

A focus on systemic interactions and a consideration of boundaries, communication, and emotional expression within the family were typical of the early ED literature for families (e.g., Siegel, Brisman & Weinshel, 1989). At that time, the goal for symptomatic recovery was specifically to engage parents in ways that could facilitate interactional and emotional change in the family. While Siegel, Brisman & Weinshel (1997, 2009) changed the message significantly in later editions of their book, initially the family was urged to step aside from the actual battles over weight and eating.

Clinicians began to recommend parental disengagement, focusing on developmental problems of separation in response to repetitive familial patterns of interaction that seem to keep everyone stuck in the symptoms. Nina and her family illustrate this problem. Over the course of her freshman year, Nina lost 40 pounds and was transformed from a self-described bulky but quite effervescent freshman to a hauntingly ethereal 5'7" 108 lb anorexic. During the early fragile attempts to work on an outpatient basis, a family meeting was called. The session quickly devolved into a discussion of what the daughter could eat at the upcoming Easter dinner, with each parent offering a litany of well-meaning suggestions:

> You don't have to eat Grandma's lamb. Why not just eat vegetables? How about we bring in our own piece of chicken? What if you sit in the other room for dinner? I can tell them you're not feeling well. I can tell them you developed an allergy to meat.

"Why don't you tell them all to go to hell!" yelled the overwhelmed teen as she stormed out the door. These were well-intentioned, frightened parents who worried that their daughter would plummet into even worse self-destruction in the face of the upcoming large meal. The parents' urgent efforts to rescue were met with equally defiant dismissals by their daughter. When the daughter finally rejoined her parents in my office, I asked her what she wanted to do at the holiday dinner. Almost predictably, she said *"I don't know,"* and the tsunami-like force of the parents trying to help started all over again.

With a systems model of disengagement, every time the person with the ED implicitly or explicitly calls out for help from significant others, the work is to keep them out of the actual struggle regarding food and weight. Inevitably, food will be left untouched, vomit will remain on toilet seats, and the question "Do I look fat?" will be asked repetitively. The work of the family is to give their daughter the space and the time to listen to herself—not to others—and to decide what she does or does not want to eat. Rather than have the parents directly engage with the food behaviors, the task is to encourage them to set and enforce real consequences if their child is not responsible for her own self-care. Within the embrace of supportive others, the patient is enlisted to source and enliven her own recovery potential.

WHAT WE THOUGHT WE KNEW: SYSTEMS THEORY AND DISENGAGEMENT FROM THE BATTLES WITH FOOD

With systems theory as a base, the encouragement of an adolescent's self-care does *not* mean that there is a lack of extensive support; nor does it mean that parents are shunted to the side (Siegel et al., 2009). From this perspective, a treatment team, along with the family and patient, actively establishes food plans, behavioral interventions, and day-to-day health related behaviors (i.e., amount of exercise, weigh-ins, therapy meetings) that allow for the resolution of the ED. Clear contracts and consequences are set up at each step of the way. For example, if a certain amount of weight is lost or not gained, a dance class might be dropped, or, in more urgent situations, hospitalization might be required. Parents and significant others are critically involved as agents of support, not control. Their role is to develop an environment in which the patient can take responsibility for healthy eating.

For example, with a 16-year-old who binges, the parents' role might be to stock the house with healthy food options, to support treatment financially, and to open pathways for better communication. But it would not be to set limits, punishments, or rewards for eating behaviors or weight loss. Or, for a 15-year-old with AN, a team would establish meal plans and a weekly weight gain goal. The parents' goal would be to firmly support the goals and consequences established with the team.

Interestingly, in these kinds of situation, clinical experience repeatedly shows that many patients are able to change their eating habits or, in the case of AN, keep weight within a pound of the established goal. Yet, actual research regarding the success of family systems interventions is sparse at best. Minuchin (1978) was one of the few practitioners to systematically record the results of this kind of family intervention. However, his research contained numerous methodological problems including no comparison treatments and no independent researchers (the researchers were the clinicians involved), and the studies varied in length of follow-up from several months to several years (Asen, 2002). Other studies (Herscovici & Bay, 1996; Martin, 1985) have replicated positive findings regarding family systems interventions, but results have been obscured when both individual and inpatient treatments were used while studying the use of family therapy.

Thus, empirical validation of the effectiveness of family systems interventions in ED treatment is lacking, illustrating a significant gap between research and clinical practice, and further compromising the urgent needs of ED patients and their families.

WHAT WE HAD TO LEARN: PARENTAL NEGLECT

The goal of disengagement was to work with parents in ways that could facilitate change in the family and ultimately with the ED patient. But the clear message was that family and friends needed to step back from actual interventions with food. Unfortunately, disengagement came to mean that parents should be separated from the child's treatment. When most abused, this approach resulted in pathologizing of the family. Parents were thought to be harmfully intrusive; their concerns and ideas about treatment were considered ineffective, even damaging. Parents were unwittingly shunted to the sidelines of treatment and blamed for the problems facing their child (Graap et al., 2008; Halvorsen & Heyerdahl, 2007; Honey et al., 2008; Vandereycken & Louwies, 2005).

As a result, this direction of treatment began to be questioned. Researchers and clinicians faced with severe AN raised an eyebrow at the implication that these patients, lying in their potential death beds, could attend to any developmental task or self-experience other than learning how to eat. To some, the thinking about parents with ED children was beginning to parallel the much outdated thinking regarding the parents of autistic and schizophrenic children (i.e., that if only the parents did something differently, the problems would not have arisen). This stimulated the questioning of researchers at the Maudsley Hospital in London, who, in the late 1980s, took these issues to task (Russell, Szmukler, Dare & Eisler, 1987).

NEW DIRECTIONS: FAMILY-BASED TREATMENT AND PARENTAL INTERVENTION WITH FOOD

The Maudsley approach, as this FBT came to be known, focused predominantly on research with anorexic teenagers (see the section below on empirical evidence for FBT). Treatment involved overtly guiding the family to restore normal eating behaviors in their daughter or son. Conjoint family sessions, in which parents and siblings were seen in sessions with their ill family member, became the centerpiece of treatment. The focus was not merely on the management of food, but on the establishment of parental authority in a family unit where structure had usually all but fallen apart in the face of the ED crisis. Contrary to abiding therapeutic perspectives on family dysfunction, in this treatment model, the question of etiology was eschewed. The thinking was that problems *other than* the ED developed *as a result* of the ED, not the other way around.

Three phases define the FBT goals and remain the cornerstones of the therapeutic trajectory (see Lock, Le Grange, Agras & Dare, 2001, for a thorough discussion of this treatment approach). Phase I consists of 10–15 sessions, with 85% weight restoration as a goal. Parents are urged to take complete control over the eating and weight of their child, under the guidance and care of trained professionals. Siblings and significant others are enlisted as sources of support, distraction, care, and encouragement. The assumption is that the adolescent is not ready to make healthy choices on his or her own and is not ready to eat the food on the meal plan without support.

In this FBT modality, the family must be available for ongoing treatment and support with their child's eating. One family sat with their 14-year-old daughter every meal over the course of a seemingly endless summer. The TV was put on as a distraction, the mother fed the daughter, the father supported the mother, and the sister rubbed the back of this teenager as, bite by bite, she ultimately put on the needed 8 pounds so that she could stay out of a hospital.

In general, consequences are established for noncompliance. For example, not eating might initially mean that a teen cannot participate in a longed for school trip. However, the consequence may quickly escalate such that immediate hospitalization will be initiated if progress is not made or the teen's condition deteriorates. Increasing empirical evidence demonstrates that children of families adopting this approach effectively restore weight and overall health (Le Grange, 2005, 2008).

Phase II of FBT is initiated when 85% of normal weight is maintained. At this point, parents are encouraged to help their child to take more control over eating once again. Parental feeding is reduced, and ego-building techniques and interpersonal exploration are a part of the treatment, encouraging self care until weight is restored to 95% or above of ideal weight.

The last phase of FBT, Phase III, is initiated once the symptoms are predominantly resolved and the patient is eating normally. Achieving age-appropriate autonomy becomes the patient's treatment focus. Goals include supporting the independence of the adolescent, establishing healthy parent–child boundaries, and assisting parents in the pursuit of their own individual and marital needs.

EMPIRICAL SUPPORT FOR FAMILY-BASED TREATMENT

Research comparing FBT with individual psychodynamic treatment for recent onset AN in teens found that conjoint family therapy was significantly more effective in preventing relapse. Initial investigations of FBT reported recovery rates as high as 90%, with symptom-free behavior at 5 years post hospitalization (Russell et al., 1987; Eisler et al., 1997, 2000). Further FBT research revealed that 70% of patients reached a healthy weight by the end of treatment and a majority of patients started or resumed menstruation (Le Grange, 2005). At 5 years post treatment, 75–90% of patients were fully recovered and no more than 10–15% remained seriously ill.

These compelling FBT research findings precipitated a seismic shift in the ED field, substantiating the positive effects of parental intervention with AN, and transferring the focus of treatment from therapist control to parental control. By insisting that parents were not to blame for their child's disorder, and that they were the best source of support and help in the recovery process, families were finally recognized as pivotal agents of change. For many practitioners, parental involvement in patient's food management came to be seen as the rule, not the exception. Indeed, FBT research indicated that facilitating an early therapeutic family alliance enhanced the teen's treatment potential and decreased their dropout rate (Pereira, Lock & Oggins, 2006).

While the lion's share of FBT research has focused on work with anorexic patients, applications for family interventions with bulimic adolescents have also been assessed (Le Grange, 2008). Family-based treatment was compared to a manual-based supportive psychotherapy in which underlying problems were explored, goals set, and therapeutic discussions were directed by the patient. Results indicated that FBT was more effective with bulimic teens with lower levels of ED pathology. For all other teens, however, the rates of remission proved to be similar. With regard to binge eating disorder (BED) or eating disorder not otherwise specified (EDNOS), research has yet to be systematically initiated.

LIMITATIONS OF EMPIRICAL RESEARCH

With the research protocols substantiating FBT, this approach has begun to be adopted with lightening speed by clinicians and health care facilities throughout the United States and Great Britain (Le Grange, 2005). A closer examination of their clinical success, however, necessitates further questioning as opposed to definitive answers. For example, early success rates may well be due to the fact that FBT research was generally conducted with two groups: young patients immediately after their ED onset (8 months or less into the duration of the illness); and patients already well on their way to recovery after treatment was in progress. With this limited group, those early in the illness process and those already responding, treatment is likely to yield favorable results (Fairburn, 2005; Vitousek & Gray, 2005).

Perhaps of most significance is the fact that the Maudsley researchers are the only clinicians steadily assessing their means of intervention. Indeed, as other research has begun to be initiated, the extreme rate of success reported by the Maudsley researchers has been contradicted. One study (Robin, Seigel, Moye & Tice, 1994) found that a somewhat varied version of FBT was only slightly more effective than ego-oriented psychotherapy for adolescent patients. Ball and Mitchell (2004) found that Maudsley FBT was not more effective than cognitive behavioral therapy in a mixed sample of teens and young adults. Both treatment interventions were equivalent in symptom abatement in patients.

The results of the early FBT studies are further limited by the fact that they were single studies using very few participants (8–10 patients) and were inappropriately extrapolated across patient groups of differing ages, duration, and severity (Wilson, Grilo & Vitousek, 2007). Thus, the patients who did well in these studies were those who did not drop out of treatment, were more motivated, and had families who were dedicated to the treatment. It is unclear how much the family's motivation and interest played a part in the recovery rates. Further research is needed to establish the specific beneficial effects of FBT for anorexic adolescents as "it is not clear that the effects of [FBT] are due to its involvement of the family or indeed to any property of the treatment. The changes could simply reflect the good prognosis of AN in adolescence" (Fairburn, 2005, p. S29).

Geist, Heineman, Stephens, Davis & Katzman (2000) compared the effects of two family-oriented treatments (FBT and family group psycho-education) on girls with AN in a randomized inpatient trial. At 4 months, significant improvement in weight was noted in both groups, with no significant difference between the two groups. This study was noncontrolled, so it is difficult to determine the specific results of the treatments in the context of other treatments received by the patients. However, the study suggests that FBT and family psycho-education may be equally helpful with respect to weight gain in the course of inpatient treatment with AN teens.

Rather than comparing FBT to other modalities, the current emphasis in FBT research is on fine-tuning the approach through examination of different formats used *within* the FBT model (amount of time needed for interventions, etc.). For example, some preliminary steps have been taken to develop intensive programs for adolescent anorexics who are not responding to FBT. Multiple-family day treatment is being developed in which families explore how the ED and the interactional patterns in the family have become entangled, and how this entanglement has made it difficult for the family to get back on track with its developmental course (Dare & Eisler, 2000; Scholz & Asen, 2001). This approach allows for more of a convergence of the standard re-feeding techniques with that of a traditional family psychotherapy. In this regard, families are encouraged to find their own solutions to helping and are not just bound to a manual-based approach. Another direction, behavioral systems family therapy (Robin, Seigel, Moye & Tice, 1999), expands the more traditional Maudsley model such that cognitions and problems in the family structure are examined while parents are still in charge of weight restoration.

With regard to the treatment of bulimia nervosa (BN), as noted, the initial reports of FBT led to promising results (Le Grange, 2008). However, a keen look at the research again raises questions. For example, in this study, FBT was compared to a supportive interpersonal therapy that did not directly employ interventions regarding the symptoms. Treatment of BN should include interventions that directly focus on changing the eating behavior while also allowing for a therapeutic exploration of the psychological issues involved (Brisman, 1992; Davis, 1991; Goodsitt, 1983). It is, thus, unclear in this study whether the positive findings associated with FBT were due to the familial involvement or with the more specific focus on the eating behaviors.

The Maudsley researchers have noted from the start that this approach is not for every family. When there is a significant level of criticism, hostility, or maternal emotion in the family (for example, if, during an initial assessment, a mother makes three or more critical comments), direct intervention by the family is not recommended (Le Grange, Eisler, Dare & Hodes, 1992). This limitation, however, in and of itself, provokes challenging questions. Can we really evaluate maternal emotion or parental hostility by overt critical comments? Often, hostility and emotion are covered over by the parents' genuine attempts to be caring and helpful. It may take months of work to understand the fury behind an eye movement or tightened lip.

Research also indicates that, for patients with severe obsessive–compulsive disorder or those who do not come from intact families, a longer version of FBT is needed (Le Grange, 2005). In general, this population has been extremely hard to assess because the teens drop out of the ongoing treatment studies (Halmi et al., 2005).

Not only does the research literature question the unilateral veracity of FBT, but clinical observations, unsupported by specific research, also indicate that this approach is not universally effective or applicable. Indeed, some parents reported that FBT provided within a structured inpatient facility was not helpful when their child was still in a pre-contemplative stage of recovery—or was just not eating at all (Pereira et al., 2006). For other families, despite extraordinary efforts and ongoing therapeutic support, FBT just does not work. Many thoughtful, determined, and caring parents have decided that the initially inspired and motivated attempts at re-feeding have only resulted in maddening frustration and anger. As one mother put it: "The Maudsley therapists don't blame the parents—but do you know what a failure it feels like not to be able to get your daughter to eat? The implicit message is that I really am the one to blame." In general, posing the question of whether evidence based treatment of AN is possible, Fairburn (2005) concludes, "The answer must be 'barely', a disquieting conclusion given the seriousness of the disorder" (p. S29).

WHAT DO WE DO NOW? ASSESSING TREATMENT DIRECTIONS

Despite the inconclusive reports regarding the effectiveness of FBT, attention to the role of parents in treatment is now unswerving. Recommendations from the American Psychiatric Association (2006) and the National Institute for Health and Clinical Excellence (NICE, 2004) stress the need for parental involvement when a child or teenager is in need of treatment for an ED. In particular, family therapy should especially be considered for adolescents who still live at home or older patients with ongoing conflicted interactions with parents (as well as with older patients with marital discord). Interventions may include sharing of information, advice on behavioral management, and the facilitation of communication. For women with ED who are mothers, parenting help and interventions aimed at assessing and, if necessary, aiding their children should be included.

These umbrella guidelines still stop short of recommending whether parents should intervene with, or disengage from, the actual issues with food. The effect research has had on this pivotal question cannot be stressed strongly enough. In the current therapeutic climate, decisions about clinical directions are now significantly influenced by economics and the financial viability of insurance companies and heath care facilities. Marketing alone can determine what kind of treatment is recommended to a family. Thus, the extensive and prolific expanse of the Maudsley-based FBT research has already had a sweeping effect on treatment recommendations, with hospitals throughout the USA changing clinical policy to mandate Maudsley-based parental intervention as part of the treatment program (Jones, 2007). The culture, the insurance companies, and certainly clinicians and families alike are relieved to find answers, and to have a clearly documented path to follow.

It is nonetheless unclear whether standard answers are always in our best interest. Over-reliance on the "truth" can obscure discovery. Alternately, discovery often upends previous truths. Where does clinical wisdom stand in our modern fact- and data-based world? Most clinicians are not researchers, nor do they routinely systematize what they do. Yet, a myriad of clinicians who have spent years in the trenches, with the most complicated of patients, know that there are many possible paths to recovery. Clinical judgments are based on experience, history with other patients, and accumulated knowledge of what has and has not worked. Thus, research findings do not consistently inform or parallel the clinical process. As a result, the serious question facing clinicians and families alike is whether we disregard what we have learned from experience if these results are not validated by research statistics. Perhaps, in order to assess treatment intervention models, one needs to begin, not by knowing what to do, but by *not* knowing what to do next. Questions, not answers, must follow.

FACTORS INFORMING TREATMENT RECOMMENDATIONS

Age and Developmental Capacity

One of the first questions to ask when determining the role of the family concerns the age and developmental capacity of the patient involved. What specific factors in the development of an ED teen, for example, allow one family intervention to be more effective than another?

Without clear guidance from research protocols, clinical knowledge guides and reminds clinicians that a teen's developing sense of self and independence are fragile processes in the best of circumstances. When that teen has AN, while most signs point to developmental crisis and abatement, clinicians must make sure that any initial seeds of self-functioning are not trampled in the process. As a result, if the ED patient has any capacity to self-soothe, to be reflective about his or her disorder and/or emotional experiences, or if he or she is motivated on their own to change, parental re-feeding may need to be questioned.

Age is also of importance. Intervention with a 19-year-old, for example, is likely going to look very different from that of a 13-year-old. With the older teen, the goal may be to allow for more separation and independence than one would with the patient who is 6 years younger. No matter what the age, when determining the role of the parents, clinicians are encouraged to assess the young patient's capacity to be reflective and thoughtful about his or her own experience, the motivation to be separate from the parents, and the hint of an interest in their own life, apart from any obsession with weight.

Another question relates to the developmental capacities and goals of the family as a whole. In one family, the mother was a 45-year-old, highly successful, well-functioning woman who chronically restricted her food intake. Whenever the mother was anxious about anything, most particularly if things in general were not "in control," she focused on her daughter's body and food intake. The daughter was a binge eater who had, in recent years, gained a significant amount of weight. While the daughter was only 13, the work involved helping the mother and daughter disengage from the struggles around food. Both needed support in distinguishing their own boundaries so that each could better identify, monitor, and address their own needs and emotions.

Family Interactions that May Interfere with Re-feeding Interventions

Assessment of parental involvement also involves the consideration of issues involving rights, responsibilities, privacy, and communication in the family. The Maudsley researchers have assessed that overt hostility may preclude the possibility of effective parental re-feeding (Le Grange et al., 1992). However, other factors also interfere with effective parental intervention. Clinical wisdom and experience with complicated family situations warrant attention to issues involving boundaries, capacity for communication, and respect for individual rights and responsibilities of each family member (Siegel et al., 2009). When there are significant problems with regard to communication and mutual respect, regardless of whether these issues precede the ED or are its consequence, the potential for effective parental intervention with food needs to be closely questioned. More research and clinical exploration is needed to assess what factors, other than overt hostility, enhance or impede the process of parental re-feeding.

Previous Treatment Approaches

Ultimately, when considering family treatment interventions, clinicians must ask what has already been tried and how it has worked. Treatment recommendations should take into account complications, gaps, or problems with regard to previous clinical interventions. A fine tuning of direction may be needed or, at times, a completely new approach may need to be tried.

In one family, the parents had spent a full year actively involved in a re-feeding regime with their 14-year-old AN daughter. The parents sat through every meal with their teen and ultimately supported her in gaining a much needed 10 pounds. Now at a safe weight of 110 lbs and 16 years old, the daughter's menstrual cycle resumed and she was able to avoid more intensive care. However, months after the parents slowly retreated from their daily involvement with the meals, the daughter's weight plummeted. More work was needed to help the daughter slowly wean herself from parental care. But at this stage of treatment, the parents had had it. They no longer felt they could sit at their daughter's side for 2 hours at a time making sure she ate. They feared their own resentment and anger, and they now felt that the feeding intervention was best left to professionals. The daughter agreed to meet with a team involving a psychologist, nutritionist, psychiatrist, and physician who closely monitored food plans, weight, and other arenas in which the teenager had difficulty moving forward. The family was seen together in family treatment—not to monitor the food—but to do everything possible to support the daughter in her own efforts to keep to the plan. The parents were still present for support and help, but it was now up to their 16-year-old to decide whether she would eat or not. It was up to her to decide what she wanted to do to avoid residential care.

WHEN THE PATIENT IS NOT A CHILD: SPOUSES, PARENTS, SIBLINGS, AND OTHERS AS THE EATING-DISORDERED PATIENT

While current research efforts focus predominantly on the role of parental intervention with young anorexics, ED have now become so entrenched in our culture that we are currently seeing the manifestations of their generational effects. Children, spouses, and siblings of those struggling with AN, BN, or EDNOS are now coming to us for help.

Research involving family therapy with adults is sorely lacking. Russell et al. (1987) found that individual therapy tended to be superior to family therapy by the end of active treatment but that this difference disappeared at follow-up. Dare, Eisler, Russell, Treasure & Dodge (2001) concluded that, in the treatment of adult anorexics, specific treatments, including family therapy, were superior to generic treatments and treatments rendered by inexperienced clinicians. However, family therapy was not significantly more effective than other types of individual therapy.

Research with couples demonstrates the potential fallout in relationships when an ED is in the picture. Couples in which one person struggles with an ED tend to censor negative communication, but they also do not provide positive messages to one another. This pattern results in communication that may not be overtly destructive but is somewhat bland and unrewarding (Van Den Broucke, Vandereycken & Vertommen, 1994). A deadening complacency develops in which couples can co-exist without really knowing one another. Food becomes refuge, subterfuge, and a much welcomed haven for the embrace of the unspoken.

This kind of unspoken emotional fallout is no doubt apparent in other configurations of relationships in which an ED is present. For example, the impact of an ED on siblings is well documented. Siblings have described intense and conflicted emotions about their sister or brother, and feelings that the ED is a pervasive phenomenon in their lives (Garley & Johnson, 1994). Family therapy was found to be helpful with regard to discussing the context of the illness, the disruption of intra- and extra-familial relationships, the special status awarded the symptomatic sister, and finding a means to cope with the illness. Interestingly, the NICE guidelines (2004) consider that involving siblings in family treatment helps the siblings more than the person with the ED. While family therapy allows for fears to be calmed and questions answered, the addition of siblings adds little in terms of the actual outcome of treatment.

With regard to BN, family therapy had beneficial effects in one large case series of adults (Schwartz, Barett & Saba, 1985). Otherwise, sufficient research is lacking to provide specific recommendations regarding the role or effectiveness of family-oriented treatment for adults with BN, BED, and EDNOS.

WHAT WE KNOW, WHAT WE DO NOT KNOW...AND MAYBE WHAT WE SHOULD NOT KNOW

What we now know is that, in families afflicted by an ED, parents matter critically, both in terms of preventative care and in terms of treatment of the ED itself. In addition to the research regarding FBT, a broad scope of research findings reflect the importance of parental involvement in general in the treatment of children and adolescents with ED (Bryant-Waugh, Turner, Jones & Gamble, 2007; Honey et al., 2008; Zucker, Marcus & Bulik, 2006).

We know that educating parents about the signs and symptoms of ED also facilitates treatment because early identification and intervention are the most significant factors in positive treatment outcome (Nilsson & Hagglof, 2005; Steinhausen, 2002).

In addition, we know that families can play a part in developing a setting that best allows for health and recovery. For example, parental support, a decrease in talk about weight and bodies, a supportive home food environment, and modeling of healthy eating habits and physical activity all provide the preventative base of knowledge and care that is needed to ground children in positive relationships to their body, weight, and food (Loth, Neumark-Sztainer & Croll, 2009).

However, we also know that parents have been criticized and marginalized, and that information aimed at encouraging health and prevention can unwittingly be used to assume parental blame if indeed an ED develops. These are dangerous and complicated waters we tread, at once struggling to allow for a direction of care without implying blame if things go awry.

Emerging out of this chaotic terrain is the much grounding position paper established by the Academy of Eating Disorders (AED) (Le Grange, Lock, Loeb & Nicholls, 2010). The position paper asserts that, while family factors can play a role in the genesis and maintenance of ED, current knowledge refutes the idea that they are either the exclusive or even the primary mechanisms that underlie risk. The AED stands firmly against any etiologic model of ED in which family influences are seen as the primary cause, and condemns generalizing statements that imply that families are to blame for their child's illness.

Clinicians and researchers are left to further clarify when and how to most effectively involve families and significant others in the ED treatment arena. "Further exploration in the form of randomized controlled trials to establish the true significance of the role of the family in AN treatment is sorely needed" (Le Grange, 2005, p. 144). Certainly, with regard to other EDs, such as BN, EDNOS, and BED, there is just too little documented research to conclude anything definitive at this point in terms of what kind of family intervention is best.

Not knowing hardly leaves ED professionals directionless. There is an urgent need to establish an inviolable role for parents and significant others with regard to the treatment and prevention of ED. Certainly, in keeping with the "oxygen mask" theory of care demonstrated on airlines, parents also need to make sure that they have support in dealing with their own struggles so they can better care for their children.

In that spirit, clinicians also need their own version of oxygen masks. Training in ED, particularly with regard to family treatment, is sorely lacking in all clinical fields. Clinicians are in desperate need of assessment tools that determine what is effective (and why) with every family and patient who seeks treatment.

Research is needed to document the benefits of traditional family therapy and to better understand the benefits of FBT. Pressing questions include: How do we integrate developmental capacities, age, and prior clinical experience when investigating the efficacy of parental re-feeding programs? What characteristics of the family picture should be considered (other than hostility) when recommending or discouraging parental intervention with food? And what pre-treatment programs might be employed to better actualize the parental role both in the re-feeding treatments and in structural family treatment modalities?

Lastly, perhaps the most important question is: How do we proceed when research varies in results and has yet to consider a multitude of critical questions? As clinicians and researchers, we stand as models for our patients, finely balancing the safety of knowledge with the stalemate of curiosity when we know too much. Research tells us where to look but may unwittingly obscure alternative directions. In this culture of documentation and facts, how do we ask our patients and families what they are hungry for, without definitively answering for them? How do we ask this of ourselves? This is the ongoing, but ever hope-filled challenge of our field.

References

American Psychiatric Association (2006, June). Practice guideline for the treatment of patients with eating disorders (3rd ed.). *American Journal of Psychiatry* 1101–1185.

Asen, E. (2002). Outcome research in family therapy. *Advances in Psychiatric Treatment, 8,* 230–238.

Ball, J., & Mitchell, P. (2004). A randomized controlled study of cognitive behavior therapy and behavioral family therapy for anorexic patients. *Eating Disorders: Journal of Treatment and Prevention, 12,* 303–314.

Brisman, J. (1992). Bulimia in the late adolescent: An analytic perspective to a behavioral problem. In J. O'Brien, D. Pilowsky, & O. Lewis (Eds.), *Psychotherapies with children and adolescents: Adapting the psychoanalytic process* (pp. 171–208). Washington, DC: American Psychiatric Press.

Brumberg, J. J. (1989). *Fasting girls: The history of anorexia nervosa.* New York, NY: Plume.

Bryant-Waugh, R., Turner, H., Jones, C., & Gamble, C. (2007). Developing a parenting skills-and-support intervention for mothers with eating disorders and pre-school children. Piloting a group intervention. *European Eating Disorders Review, 15*(6), 439–448.

Dare, C., & Eisler, I. (2000). A multi-family group day treatment program for adolescent eating disorders. *European Eating Disorders Review, 8,* 4–18.

Dare, C., Eisler, I., Russell, G., Treasure, J., & Dodge, L. (2001). Psychological therapies for adults with anorexia nervosa: Randomized controlled trial of out-patient treatments. *British Journal of Psychiatry, 178,* 216–221.

Davis, W. (1991). Reflections on boundaries in psychotherapeutic relationship. In C. Johnson (Ed.), *Psychodynamic treatment of anorexia and bulimia* (pp. 68–85). New York, NY: Guilford Press.

Dickinson, E. (1960). In T. H. Johnson (Ed.), *The complete poems of Emily Dickinson.* Boston, MA: Little, Brown and Co.

Eisler, I., Dare, C., Hodes, M., Russell, G., Dodge, E., & Le Grange, D. (2000). Family therapy for adolescent anorexia nervosa: The results of a controlled comparison of two family interventions. *Journal of Child Psychology and Psychiatry and Allied Disciplines, 41*, 727–736.

Eisler, I., Dare, C., Russell, G., Szmukler, G. I., Le Grange, D., & Dodge, E. (1997). Family and individual therapy in anorexia nervosa: A 5-year follow-up. *Archives of General Psychiatry, 54*, 1025–1030.

Fairburn, C. G. (2005). Evidence-based treatment of anorexia nervosa. *International Journal of Eating Disorders, 37*, S26–S30.

Garley, D., & Johnson, B. (1994). Siblings and eating disorders: A phenomenological perspective. *Journal of Psychiatric Mental Health Nursing, 1*(3), 157–164.

Geist, R., Heineman, M., Stephens, D., Davis, R., & Katzman, D. K. (2000). Comparison of family therapy and family group psycho-education in adolescents with anorexia nervosa. *Canadian Journal of Psychiatry, 45*, 173–178.

Goodsitt, A. (1983). Self regulatory disturbances in eating disorders. *International Journal of Eating Disorders, 2*(3), 51–60.

Graap, H., Bleich, S., Herbst, F., Scherzinger, C., Trostmann, S., Wancata, J., & de Zwaan, M. (2008). The needs of carers: A comparison between eating disorders and schizophrenia. *Social Psychiatry and Psychiatric Epidemiology, 43*(10), 800–807.

Halmi, K. A., Agras, W. S., Crow, S., Mitchell, J., Wilson, G. T., Bryson, S. W., & Kraemer, H. C. (2005). Predictors of treatment acceptance and completion in AN: Implications for future study designs. *Archives of General Psychiatry, 62*, 776–781.

Halvorsen, I., & Heyerdahl, S. (2007). Treatment perception in adolescent onset anorexia nervosa: Retrospective views of patients and parents. *The International Journal of Eating Disorders, 40*(7), 629–639.

Herscovici, C. R., & Bay, L. (1996). Favorable outcome for anorexia nervosa patients treated in Argentina with a family approach. *Eating Disorders: Journal of Treatment and Prevention, 4*, 59–66.

Honey, A., Boughtwood, D., Clarke, S., Halse, C., Kohn, M., & Madden, S. (2008). Support for parents of children with anorexia: What parents want. *Eating Disorders: Journal of Treatment and Prevention, 16*(1), 40–51.

Humphrey, L. (1988). Relationships within subtypes of anorexic, bulimic and normal families. *Journal of American Academy of Child & Adolescent Psychiatry, 27*, 544–551.

Jones, J. (2007, November). *Getting them in—or out—the door: Working with parents whether they want to or not.* Workshop presented at the 17th Annual Renfrew Foundation Conference, Philadelphia, PA.

Le Grange, D. (2005). The Maudsley family-based treatment for adolescent anorexia nervosa. *World Psychiatry, 4*(3), 142–146.

Le Grange, D. (2008). Predicting treatment outcome for teens with bulimia nervosa. *Journal of American Child and Adolescent Psychiatry, 47*, 464–469.

Le Grange, D., Eisler, I., Dare, C., & Hodes, M. (1992). Family criticism and self-starvation: A study of expressed emotion. *Journal of Family Therapy, 14*, 177–192.

Le Grange, D., Lock, J., Loeb, K., & Nicholls, D. (2010). Academy for eating disorders position paper: The role of the family in eating disorders. *International Journal of Eating Disorders, 43*(1), 1–5.

Lock, J., Le Grange, D., Agras, W. S., & Dare, C. (2001). *Treatment manual for anorexia nervosa: A family-based approach.* New York, NY: Guilford Press.

Loth, K. A., Neumark-Sztainer, D., & Croll, J. K. (2009). Informing family approaches to eating disorder prevention: Perspectives of those who have been there. *International Journal of Eating Disorders, 42*, 146–152.

Martin, F. E. (1985). The treatment and outcome of anorexia nervosa in adolescents: A prospective study and five year follow-up. *Journal of Psychiatric Research, 19*, 509–514.

Minuchin, S., Rosman, B. L., & Baker, L. (1978). *Psychosomatic families: Anorexia nervosa in context.* Cambridge, UK: Harvard University Press.

NICE (2004). *Eating disorders: Core interventions in the treatment and management of anorexia nervosa, bulimia nervosa and related eating disorders: Clinical Guideline 9.* London, UK: British Psychological Society.

Nilsson, K., & Hagglof, B. (2005). Long-term follow-up of adolescent anorexia nervosa in northern Sweden. *European Eating Disorders Review, 13*, 89–100.

Pereira, M., Lock, J., & Oggins, J. (2006). Role of therapeutic alliance in family therapy for adolescent anorexia nervosa. *International Journal of Eating Disorders, 39*, 677–684.

Robin, A. L., Seigel, P. T., Moye, A. W., & Tice, S. (1994). Family therapy versus individual therapy for adolescent females with anorexia nervosa. *Developmental and Behavioral Pediatrics, 15*, 111–116.

Robin, A. L., Seigel, P. T., Moye, A. W., & Tice, S. (1999). A controlled comparison of family versus individual therapy for adolescents with anorexia nervosa. *Journal of the American Academy of Child and Adolescent Psychiatry, 38*, 1482–1489.

Russell, G. F., Szmukler, G. I., Dare, C., & Eisler, I. (1987). An evaluation of family therapy in anorexia nervosa and bulimia nervosa. *Archives of General Psychiatry, 44*, 1047–1056.

Schwartz, R. C., Barett, M. J., & Saba, G. (1985). *Family therapy for bulimia*. New York, NY: Guilford Press.

Scholz, M., & Asen, K. (2001). Multiple family therapy with eating disordered adolescents. *European Eating Disorders Review, 9*, 33–42.

Selvini-Palazzoli, M. (1978). *Self-starvation: From individual to family therapy in the treatment of anorexia nervosa*. New York, NY: Jason Aronson.

Siegel, M., Brisman, J., & Weinshel, M. (1989). *Surviving an eating disorder: Strategies for family and friends*. New York, NY: Harper Collins.

Siegel, M., Brisman, J., & Weinshel, M. (1997). *Surviving an eating disorder: Strategies for family and friends* (Rev. ed.). New York, NY: Harper Collins.

Siegel, M., Brisman, J., & Weinshel, M. (2009). *Surviving an eating disorder: Strategies for family and friends* (Rev. ed.). New York, NY: Harper Collins.

Steinhausen, H. (2002). The outcome of anorexia nervosa in the 20th century. *American Journal of Psychiatry, 159*, 1284–1293.

Van Den Broucke, S., Vandereycken, W., & Vertommen, H. (1994). Psychological distress in husbands of eating disordered patients. *American Journal of Orthopsychiatry, 64*, 270–279.

Vandereycken, W., & Louwies, I. (2005). "Parents for parents": A self-help project for and by parents of eating disorder patients. *Eating Disorders: Journal of Treatment and Prevention, 13*(4), 413–417.

Vitousek, K. M., & Gray, J. A. (2005). Eating disorders. In G. O. Gabbard, J. S. Beck, & J. Holmes (Eds.), *Oxford textbook of psychotherapy* (pp. 177–202). Oxford, UK: Oxford University Press.

Ward, A., Ramsay, R., Turnbull, S., Benedettini, M., & Treasure, J. (2000). Attachment patterns in eating disorders: Past and present. *International Journal of Eating Disorders, 28*, 370–376.

Wilson, G. T., Grilo, C., & Vitousek, K. (2007). Psychological treatment of eating disorders. *American Psychologist, 62*(3), 199–216.

Zucker, N. L., Marcus, M., & Bulik, C. (2006). A group parent-training program: A novel approach for eating disorder management. *Eating and Weight Disorders, 11*(2), 78–83.

CHAPTER
21

The Most Painful Gaps
Family Perspectives on the Treatment of Eating Disorders

Robbie Munn, Doris and Tom Smeltzer, and Kitty Westin

INTRODUCTION

It is easy for clinicians, especially if working individually, to forget how the family is affected by their child's eating disorder (ED). This chapter is meant to help us remember.

Kitty Westin tells the story of her daughter, Anna, revealing the tragic gaps that can still exist between patients, families, clinicians, and third-party payers. Robbie Munn's description of her family's experience with an ED highlights the implications of multigenerational risk factors and the pressures families face to become expert partners in their child's treatment. Tom and Doris Smeltzer lost their daughter Andrea to an ED. They conclude this chapter by enumerating thirteen messages that, "in a perfect world," would have been shared with them during their daughter's illness.

In the not-too-distant past, a "parentectomy" was seen as standard care for patients with ED. That perspective has waned over the years and families are increasingly seen as key partners in comprehensive treatment. Family perspectives offer clinicians important strands of information about how to refine their clinical practices. Research, evidenced-based practice guidelines, and our own clinical experience are all important sources of information, but truly sophisticated clinical work requires the unique perspectives of our patients' families. These three stories are poignant, and ultimately hopeful, reminders of how much is at stake in our work with people with ED.

IN MEMORY OF ANNA SELINA WESTIN: NOVEMBER 27, 1978–FEBRUARY 17, 2000

Kitty Westin, Minneapolis, MN

How does a parent cope with the death of a child and how do you plan a funeral for a 21-year-old who was once vibrant and full of life? Is it possible to recover from the horror and deal with the guilt and anger when your child dies from a treatable illness? On 17 February, 2000, my life changed forever and I was forced to consider these questions. On that day, my precious daughter Anna died from anorexia nervosa (AN). I hope that telling our family's story will help others avoid the mistakes and fill in the "gaps" that led to Anna's unnecessary death. Learning to steer clear of them could go a long way in saving lives and avoiding the heartbreak our family experienced the day Anna died.

Part One: Anna Is Welcomed into the World

"IT'S A GIRL! Congratulations!" It was Anna Selina Westin's birthday. I had just given birth to a "perfect" baby girl and I was thrilled. Anna had a little rose bud mouth and bright blue eyes, she had a full head of curly blonde hair, and it appeared that she was born with the ability to charm just about everyone. I remember looking at my husband with tears of joy streaming down my face and seeing that he was also crying. As I gazed into my infant daughter's eyes, I saw a future full of promise and possibility. I had no doubt that this beautiful child was destined for a life filled with love and happiness. How could this not be true? Anna was born into a family who cherished her and we were prepared to do everything in our power to give her the opportunities she deserved.

I loved being Anna's mother and having the ability to comfort her when she was sick or injured with a gentle kiss, a kind word, and a mom's healing touch. I felt confident in that role, having learned skills from my own mother and grandmother, both of whom were wise and loving women. My mother taught me many things about parenting, including the highly effective treatment called "kiss and make it all better." A kiss and a colorful cartoon decorated bandage would fix just about anything. Knowing that my love and attention were, more often than not, the best medicine, gave me a sense of satisfaction and accomplishment. I trusted my intuition to know when Anna needed my care, and when she needed to see a doctor, and I knew the limits of "kiss and make it all better." There was every reason to believe that Anna would be successful and achieve anything she set her mind to. She had a loving and supportive family and she had vision, drive, and a wonderful sense of humor. She cared deeply for others and gave of herself freely. Anna's thoughtfulness and compassion are legendary and her family and friends still talk about her ability to make them feel like they were the most important person in the world.

Part Two: Anna Is Diagnosed with Anorexia Nervosa; August 2000

Fast forward 16 years: *"Your daughter has anorexia and will die if she is not immediately hospitalized."* I was horrified, and to this day I doubt that there are many words that could cause me more terror than *"your child has an ED."* I remember sitting in an office at a local hospital the day Anna was diagnosed with this potentially fatal illness and feeling like my breath was being sucked out of me. Tears were streaming down my face. My "perfect" baby was gravely ill. I looked at my husband and noticed that he looked terrified and had tears in his eyes. I looked into Anna's eyes and she was staring blankly into space. At that moment I knew we were in for a fight with a formidable foe. And, within minutes, we encountered the first of many of the gaps and obstacles to Anna's recovery.

Anna was first diagnosed with AN in 1995 when she was 16 years old. When I look back, I recognize certain longstanding traits and behaviors that could have been early warning signs of risk for developing an ED. She was a precocious child who loved adventure, learning, and discovering everything the world had to offer. A perfectionist who needed to do everything "right," even as a young child, Anna was extremely hard on herself when she made a mistake, easily dissatisfied with herself, and very intolerant of her perceived imperfections. But, she was more often happy than not and seemed fully engaged in life.

Anna went on her first diet when she was 15 years old. This proved to be a fatal decision. I remember the gradual progression from limiting a few foods to the elimination of almost everything from her diet. The day Anna announced that she was a vegetarian was a surprise, but I was not overly concerned (although I was confused). When I asked her about her decision, she assured me that she just wanted to be healthy and that she was aware that she would need to learn how to eat a well balanced, nutritious diet. Naïve at the time, I was satisfied with her explanation. It did not take long for her restriction to move from meat, to sugar, to fats, to carbohydrates, and eventually to most foods. In only a few weeks this restricting behavior moved from disordered eating to a full blown ED. What were the gaps and how could things have been different? Who should have noticed what was happening and who should have taken action sooner? Is there someone to blame? These are the questions I have repeatedly asked myself. I wonder if the outcome would have been different if the people closest to Anna had been more observant and knowledgeable about ED. If health care professionals, school personnel, her family, and friends had known what to watch-out for, could we have changed the course of history? I still wonder if the lack of public awareness about eating disorders created gaps that contributed to her death.

Once I noticed that Anna was losing weight and that her personality was changing, I made an appointment for her with our family doctor. Dr. K. had known Anna all her life and, after a physical exam, she told me that she suspected Anna was suffering from AN. She recommended a full evaluation at a specialized ED clinic. This was another encounter with a gap. Anna waited several months for an appointment. She lost considerably more weight in the interim and her ED became more entrenched. If Dr. K. had suspected that Anna was suffering from cancer and had referred her to a specialist, I am sure she would have been seen within days, not weeks or months. Delays could have cost Anna her life.

My husband and I accompanied her to the hospital for her evaluation. We wanted to be available to answer questions and we hoped to better understand the disease that was taking our daughter away. The team recommended that Anna begin outpatient treatment and my husband and I were told that our job was to bring Anna to her appointments and that they would take it from there. We did not get any advice or guidance about nutrition, no family therapy was offered, and we were sent home with almost no information about the illness. We left the hospital confused and frightened, not knowing where to look for answers. I remember that it seemed odd to me that we (her loving and caring parents) were not invited to fully participate and assist in her care. In fact, we were basically told to back off, and to let the professionals do what they do best. I was inexperienced and did not know enough about the treatment of ED to insist that we be included as part of Anna's treatment team. I did feel some relief because I believed Anna was in the right place to get the specialized care she needed. I did not know then that this was just the beginning of a hellish 5 year journey into uncharted territory.

Unlike so many other serious illnesses like diabetes, heart disease, or cancer, we had little idea where to turn for information, education, or guidance. We could see that Anna was extremely ill, and we were terrified. My heart was breaking. I felt like I had let Anna down, that I had failed as her mother because I couldn't protect her. My self-confidence tumbled as I watched my daughter struggle and fight for her life. When I look back, I am appalled at the lack of resources for parents. At that time it was hard to find accurate, up-to-date information that was designed to help us understand the disease that had consumed our daughter. I felt alone, confused and, for the first time as a mother, helpless and powerless.

Part Three: Anna Recovers

Anna seemed to recover fully after she completed her first round of treatment and I felt thankful that we had our daughter back. During the next 4 years she routinely reported feeling healthy and happy. At times she complained of feeling depressed and anxious, but she seldom complained of ED symptoms. Between the ages of 17 and 20, we were confident that Anna had beaten her ED and that she was on her way to fulfilling her dreams.

Anna graduated from high school and went off to college and, once again, I felt like a competent mother and thought that we no longer had to worry about an ED stealing our daughter. How could I have known that the average time between diagnosis and recovery from an ED is 7 years? I am still angry that nobody involved in my daughter's care told me that the relapse rate for ED is very high and that we should remain vigilant for several years. What a shock it was to discover that AN had crept back into Anna's life sometime during her sophomore year of college. By the time Anna returned from college in the summer of 1999, she was gravely ill. Because Anna had responded well to her first treatment, we were not overly concerned, and had no reason to doubt that she would fully recover. However, like most patients with ED, Anna was at times resistant, angry, and in denial that she needed treatment. Her father and I had lost the little bit of power we had because Anna was now a 20-year-old adult. I understood that Anna had the right to make her own decisions about her care, but it was absurd to think that she could make rational, informed decisions in her very ill state. She was starving herself to death and was not able to make good decisions regarding her health. I begged her team to include me and her father in her care and to view us as an important part of her recovery. We were willing and able to help and we were ready to listen and learn. After all, we knew Anna better than anyone and we were certain that we had something to contribute.

Anna's ED was taking its toll on our entire family. My husband and I argued about our role and what we should or could be doing to help Anna; our anger and fear was often directed towards each other. Our younger daughters struggled to understand the disease and found it very difficult to accept that their older sister could possibly die. We all walked on eggshells, fearing that, if we did or said the wrong thing, we would upset Anna and make things worse. Anna's treatment included both individual and family therapy, and we were always willing to participate. We looked forward to our sessions because we desperately wanted to understand our part in this puzzling disease and we hoped to be given some direction about our role. I recall asking Anna's therapist to teach us strategies and to give us suggestions on how to be supportive without being overbearing. We felt scared, worried, and helpless. The reality was that, during our sessions, we were often told what not to do, but not what to do. It was extremely frustrating to hear what we were doing wrong, while getting little direction on how to do it "right." On one occasion, I asked the dietician to imagine herself in a similar situation. Did she really think she could resist being the "food police" when food was the life-saving medicine? I think she would have found it almost impossible to ignore the skipped meals, the rituals around food, and the over-exercising. I argued that she would have been unable to resist screaming "STOP" if it was her daughter running on the treadmill and she knew her heart could give out at any moment.

Part Four: The Most Painful Gap: The Beginning of the End

Anna was admitted to a hospital-based ED treatment program in September 1999 and, in spite of periodic miscommunication and frustration, she appeared to be doing well. Anna wrote about the painful battle against AN and how hard she was fighting. Without the support of professionals, she told us, she knew she would not survive and she put her trust in the people who had been trained to help her fight the disease.

I know Anna put her heart and soul into the battle against AN. There were times when she was angry and/or resistant, sometimes refusing to fully participate in treatment. But people who treat patients with ED should expect this. What Anna and we (her parents) needed most during these trying times was extra support and compassion. In fact, what she got were threats that she would be discharged from the program if she did not fully cooperate, and admonishment that she was not working hard enough.

Anna spent the last 6 months of her life in and out of the different levels of care including inpatient, partial hospitalization, and outpatient treatment. In December 1999, Anna was in the outpatient program, but asked to be readmitted to the partial program. In her own words, she needed to "jump-start" her program which had become increasingly difficult for her to follow. She felt unable to control her symptom use in the outpatient program. She requested a meeting with her team to discuss options and she invited my husband and me to attend. We expected Anna's individual therapist, doctor, dietician, and the director of the partial program to participate. When we arrived for the meeting, we were surprised to learn that her doctor and therapist would not be attending. This turned out to be a grave mistake that contributed to Anna's death less than 2 months later.

Anna was clearly motivated and sincere in her desire to participate in the program. She left the meeting with an assignment to write her goals and fax them to the partial program. If she completed this task, she would begin the program the following day. She worked on her goals all day and asked if my husband and I would review them before she sent them. She had created a list of goals that included following her meal plan, developing strategies to help her fight her ED, and reducing symptom use. We were proud of her. She had taken her assignment seriously and we were hopeful that her commitment would lead her toward recovery.

We faxed the document to the partial program in the evening believing that the therapist would see them in the morning and review them before the program stated. Anna arrived at the hospital as planned but was told to go home because they had not received her goals in time. Anna was devastated and embarrassed. When she returned home unexpectedly, she tearfully explained the situation to me as best she could. For the first time, she expressed her fear that she was hopeless and that she would never recover from the ED. I called the program asking for someone to explain the problem to me. Late in the afternoon, Anna's individual therapist called and told her that she had been discharged from the program and that she would need to seek treatment elsewhere. Anna was speechless and handed the phone to me. When I asked Anna's therapist what she had said to her, she reiterated that Anna had been discharged from the program. The treatment team had met and decided that Anna was not engaged, was not working hard enough, and had not made enough progress to stay. She cited Anna's failure to complete the goal setting assignment. She had not submitted the requested goals "on time" in order to return to the partial program. I was horrified and reminded the therapist that she was not present at the meeting and that I was. I was stunned when the therapist insisted that, in spite of my objection, Anna was no longer welcome to return to the program.

Anna felt helpless. If the specialized program was giving up on her, she was ready to give up on herself. I suspected that Anna was contemplating suicide, and I was in a panic knowing that she felt abandoned by the only people who had been able to help her. It still stops my heart when I remember the next few weeks and what Anna experienced. I begged the program to reconsider their decision to discharge Anna when she needed it most. After much discussion, they agreed to let her participate in the outpatient program, clearly an inadequate level of care given her declining state of health.

In my opinion, Anna began to die on the day she was discharged from the ED program, abandoned by the very same "team" who had assured her they could and would help her recover: December 22, 1999. This certainly was the most painful gap in Anna's treatment. How could a treatment program that understands the complexity and tenacity of ED tell a patient that she was failing treatment because she was not working hard enough? How can this be interpreted as the best option for a dying patient? Why did Anna's professional caregivers give up on my daughter when she so desperately wanted to be free from the disease that was killing her?

Anna's faith that treatment could help her recover had eroded and she was resistant to continuing. We were well aware that she needed more care and we were considering a long-term residential program. However, before we were able to find a program, Anna lost all hope and committed suicide on February 17, 2000. She was just 21 years old and she had her whole life ahead of her, but she could not see her way to recovery from her ED.

Part Five: We're in This Together—How Families, Treatment Professionals, and Insurance Companies Can and Should Work Together

The story you just read is tragic. It was difficult to write and is painful to read. Families must learn as much as they can about ED so they can advocate for their daughters and sons. Treatment professionals must understand that families have much to offer and can be an important resource when treated with respect and invited to participate as part of the "team." ED programs must find ways to help parents and families get the information they desire and should offer support, education, and advice whenever asked. And third party payers need to support and pay for the medically necessary treatment that is recommended by professionals and tailored to the needs of the patient and family.

The most painful gap in the treatment of ED is the lack of trust by some professionals that parents can participate effectively in treatment. Parents need and want direction; we need to be treated as part of the team and consulted when appropriate. We deserve to be treated respectfully and valued as knowledgeable partners in the care of our loved one. If we work together, we can prevent the repetition of these mistakes and help fill in the gaps in treatment that contributed to Anna's death.

* * * * * * * *

THROUGH THE LOOKING GLASS

Robbie Munn, Charlottesville, VA

Hope is not a plan.
— a father speaking of his daughter who died from the long-term effects of her decades-long ED.

People make fun of eating disorders until one comes into their lives.
Then they no longer think it is funny.
— another father speaking of his experiences with his daughter who survived her ED.

During the summer of 1996, our family of four tilted on its axis when our daughter, Genny, barely 13, developed severe depression and AN. Her struggles and roller-coaster recovery encompassed the next decade, robbing her of her entire adolescence and much of her early adulthood. Within a six-week period that summer, she would go from the bright, school-loving, risk-aversive daughter and older sister to a gaunt, black-clad, scantily dressed girl, unrecognizable but still fiercely loved. She would take up smoking, drinking, and drugs and choose new "friends" who were equally lost, angry, and not timid about letting the world know. Instead of being welcomed into friends' homes, parents feared and shunned her. Eventually, we, her parents, would also be avoided. Genny obsessed for years about hurting herself and did so in various ways. She seriously contemplated suicide many times. Sometimes she told someone; sometimes she didn't.

Genny's grades and behavior became poor and unpredictable enough that she passed through five different schools, some for only weeks, before finally graduating high school a year late. While having taken the SATs in fifth grade as part of a gifted program, Genny would attend three colleges before managing to graduate, another year late, due to her ongoing emotional and physical instability. Her father and I urged her to finish at that point only because we knew she would obsessively regret dropping out. She was just beginning, at age 22, to be able to hold onto her progress and to a semblance of normal life. She needed desperately to know that she could succeed at something.

Recovery from her AN was ragged, rough, and seemingly random. When she was able to hold onto it, Genny's recovery progressed at a glacial pace, while real life predictably presented new complexities and challenges. It was many years after Genny first got sick before I dared to believe again that she had a chance to lead a full life. For the earliest years of her illness, I feared she might not survive; as time went on, I feared that the rest of us might not either.

In the 1990s, few families and/or experts believed that an ED could be the driving force behind such a protracted and chaotic recovery. Fifteen years later, little has changed. Even more unfortunate is that too many clinicians, medical and psychiatric, still do not understand the power and destructiveness of ED. Clinical experts still fall critically short in their assessment and treatment. This was and remains unacceptable.

Our daughter survived her ED, and all that fueled and accompanied it, but at a tremendous cost. That long, dark decade affected each of us in our family as well, mostly for the better, but again in raw and unenviable ways. In research labs around the world, much is now being learned about the biological and genetic aspects of ED. The better hospital and residential treatment programs analyze their long-term results and re-structure themselves accordingly. Yet what is being taught—or not taught—at medical schools and graduate psychology programs determines what the front-line responders will "see" when they first encounter someone with an ED. These gaps lead to ongoing delays in appropriate diagnoses and long-term treatment.

Society's obsession with thinness repeatedly manages to trump reason, science, and fact. People's love–hate relationship with ED, and their own distorted view of the importance of thinness, feed the denial that lives are being derailed and lost. If one were to describe any other illness in their family to another person, the predictable, appropriate reaction is sympathetic and empathetic. Even if the illness is esoteric and previously unknown, common sense tells one that it would be better if that disease never crossed their family's threshold. In contrast, if one talks of a loved one's ED, the response is usually one of regret over the sufferer's stupidity or lack of willpower. Ironically, the reaction also often includes a variety of comments such as *"Oh, but I wish I had one, too. Just a little one."*

Welcome to the Rabbit Hole

One Journey

What remains remarkable to me is how early the signs of Genny's potential AN emerged and how easily they might again be missed in another child today. Genny had always been bright, precociously so. She spoke and read early and readily grasped abstract concepts. However, even as a young child she had more difficulty than normal fitting in easily and playing with other children. At each developmental juncture, she seemed to lack age-appropriate social skills. Within known circumstances and in smaller groups, she functioned well and often rose quickly to the top through her verbal and cognitive skills. Throw her into the pool of life, however, and she seriously struggled from age 3 on. Like many educated, observant, and loving parents, we assumed this would improve with time.

By the time Genny reached middle school, the discrepancy between her scholastic and intuitive skills and her social and emotional maturation reached a flashpoint. She was desperately unhappy, strangely immature in certain ways, and truly did not seem to understand how her reactions were isolating her further from her peers. The year before Genny became ill, her father and I separated. There is no doubt that the heightened anxiety around difficult family issues played a pivotal role in the timing of her symptoms; however, I now believe that "something" would have happened regardless. Genny had inherited a predisposition toward certain psychiatric illnesses given our family's DNA and history.

EDs, we now know, often run in families, just as do cancer, alcoholism, and learning disorders. If one has a close relative who suffers from an ED, one is at higher risk. The genetic contributions to ED were not understood until recently and are still not widely known today. My mother, born in 1910, suffered from AN, along with a host of other digestive and psychiatric symptoms, most of her life. Yet, she was never diagnosed, which was not uncommon for that era, before her death in 1979. My mother's often bizarre perfectionism, increasingly apparent during my youth, was called by other names. It was not until Genny became ill that I learned how many different ways AN could be packaged.

Genny often reminded me so much of my mother that I would be stunned into silence by the similarities. My mother died before I was married; the two most important women in my life, my mother and daughter, never met. Still, they navigated life in frighteningly similar ways. Both, at their high points, were intelligent, diversely creative women with real determination, and persistence. Both could be witty and truly engaging. In darker times, their chronically high anxiety and obsessive-compulsive disorder (OCD) tendencies could no longer be contained and seriously interfered with daily life. They each suffered from horrible self-esteem. Both equated fatness with being lazy, slothful, and undeserving, and each viewed restrictive eating as a solution to most of life's problems. They had trouble initiating and maintaining friendships and kept people away with their unpredictable outbursts and rages. At the worst times, they each had trouble leaving their houses: my mother to take care of me, her youngest child, and my daughter to go anywhere typical for her age.

From the beginning of Genny's rapid onset of AN, we, her parents, received conflicting and confusing information, advice, and directives from experts. I was fortunate to locate a wonderful, skilled, caring therapist, who became Genny's lifeline for years to follow. The therapist immediately realized what was happening and accurately diagnosed her. During Genny's first, brief emergency hospitalization, she was evaluated and prescribed Paxil by a psychiatrist, Dr. P. Traumatized by the events leading to the hospitalization, the stay itself, and her reaction to the Paxil, she lost 20 pounds in 6 weeks from her already slim pubescent frame. Over that summer, Dr. P. repeatedly refused to re-evaluate Genny for an ED and told Genny that I was *"overreacting and hysterical."* His thinking was archaic; he believed that Genny would have to "work it out" in therapy. I distinctly remember confronting him and saying that if she continued at this rate of overall deterioration, I was not sure she would be alive to do so. Genny began to engage in risk-taking behaviors we never could have anticipated earlier. Her ED had moved in.

Over the summer, Genny began to write notes and poems about suicide. Again, Dr. P. dismissed them. When she wrote a suicide note on her bedroom wall, in black permanent marker and five feet high, I finally admitted that I was no longer sure we could keep her alive. I finally removed Genny from Dr. P's care, against others' advice. This discrepancy between the psychotherapist's and the psychiatrist's overall assessment of Genny was stunning. Unfortunately, Dr. P's evaluation "counted" more than her therapist's, even though it was grossly incorrect. This is a common scenario even today; families feel forced to make agonizing choices between professionals who may be unwilling to listen to one another or to reevaluate their patient's status. By the time Genny was evaluated by a new psychiatrist, the diagnosis of AN was no longer in question. She was so depleted physically and emotionally that her attempts to return to school that fall were unsuccessful. Her therapist and new psychiatrist agreed that Genny required residential treatment. After being asked to leave school on a medical leave, she barely fought it.

At the time, few ED treatment programs admitted 13-year-olds. It literally took over 100 phone calls across the country before we were able to locate three programs in which we had any faith. Each was several hundred to a thousand miles away. Both then and today, many families would be unable or unwilling to send their desperately ill daughter to an unknown facility a thousand miles away to be treated by strangers. To do so, one has to be completely desperate as a parent. We were.

It took three plane rides to reach our destination. My greatest fear was that I would find the program so poorly run and therapeutically inept that I would not be able to let her stay. If that happened, I did not know what we would do next; I could easily see her attempting suicide out of emotional exhaustion and despair. The first time I broke down that day was from sheer relief at the competence of the staff and the warmth of her welcome to the patient community. When I finally arrived at my hotel, it was past midnight, 20 hours since I had left home the day before to bring her to treatment. I called her father and began sobbing in earnest.

This first, two and a half month stay at an ED program did little more than stabilize Genny medically. As much as they tried to provide her with the tools and skills to be able to return home and to school, she was highly resistant to the nuances of what she was being taught. After returning home, she was more vulnerable emotionally than she had been. She quickly returned to anorexic behaviors but hid them better. Over time she began to binge and purge, as attempts to attend school and have "normal friends and a life" repeatedly failed. She began smoking pot more regularly. She drank. She repeatedly denied it all.

Neither her father nor I could understand how AN alone could account for this downward spiral, but we wanted to believe her. She was seeing her therapist and attending a weekly group meeting, but crises happened more frequently. She was hurting herself more often and more deeply; she deliberately dehydrated, and she had to wear a heart monitor. She grew frailer, more frantic, more unreasonable about everything, and began to refuse to stay with her father. She was failing school again and I, once more, felt like I was on a 24-hour watch every day. I was exhausted.

Finally, during an icy, frigid weekend, Genny made a number of incredibly poor and dangerous decisions. I stopped denying what was in front of me and searched in specific places for information. I did not normally "violate her privacy." All the rules had changed, however, and I was now at war for her life. That night after dinner, I did not give Genny permission to leave the table. Every night she had been purging while I thought she was doing homework. She was also purging whatever little breakfast she had consumed while walking down the drive to the school bus. At school she was skipping lunch but ingesting unknown drugs. I told her what I had pieced together. I called her therapist and said that I no longer thought Genny was safe in any way; she agreed. I called Genny's father, and we began making plans to re-admit her. This time Genny had some say in the choice of her hospitalization; she requested that it include access to AA and NA groups. However, she was unaware of two critical decisions we had made. If she suddenly backed out, we were re-hospitalizing her regardless; we had to before she died accidentally or on purpose.

In addition, we were asking the program up front to *"keep her as long as was needed "* to try to identify or rule out all neurological and psychiatric problems. First they had to stabilize her medically, wean her from daily excess laxative use, and heal her bleeding throat. Genny was there 5 long months. She told me later that our commitment to hospitalize her saved her life.

That was the beginning of Genny's real recovery. The years that followed remained difficult and unpredictable. Genny did well, and she did horribly. She continued to hurt herself and was consumed by obsessions. In 10th grade, she was finally in a school committed to see her through, but she was still struggling with life and with eating on a daily basis. I resolved that if Genny needed to be re-hospitalized yet again, I would not make the unforgivable mistake of waiting so long. I knew better now.

The common course of recovery from AN can take up to 7 years. Data suggest that as young people move out of adolescence and into their twenties, certain neurological and cognitive maturational processes occur which in themselves encourage recovery from ED and other psychological issues (C. Johnson, personal communication, September 18, 2008). As Genny was unsteadily making it through high school and college, I felt like my goal was to do everything I could to keep her from dying prematurely. Time would be on her side if she let it be. I no longer had to understand everything; I just had to urge her to continue to hold on. She would then have the chance to do the rest.

Concluding Thoughts

I am aware of how fortunate our family has been. We were surrounded and supported by family and friends who cared about what we were each going through. We were able to find compassionate and competent clinicians and a residential program that did not fail her. Since becoming involved in the larger field of ED, I have met with scores of parents whose stories vary widely and end differently. Each story is unique, and yet there are eerily common themes in all. I can look into another family's eyes, say that I understand, and know that I do. The clinical field has matured in significant ways, no longer excluding families and blaming them. Families are being encouraged to become "part of the team" and to learn as much as they can. They are viewed as assets versus liabilities.

There are holes in the fabric everywhere. People continue to be diagnosed improperly; thousands of them run out of insurance coverage for out-patient services long before they have mastered critical coping skills. Daily, many more are refused hospitalization because they are not yet "sick enough." Standardization of treatment, whether on an out- or in-patient basis, has yet to be formalized. The process of credentialing both programs and individual clinicians remains in its infancy. We still lack accepted protocols for various modalities and therapies on an outpatient basis before seeking hospitalization. At the same time, ED have become a "hot" and marketable specialty practice. Some clinicians claim expertise after a weekend of training. Programs promote themselves and report their respective 85 or 95% "success rates" without factually informing prospective families about what that means. Distraught, anxious, frightened families believe these pronouncements, unknowingly perpetuating a vicious cycle of unsuccessful treatments, relapses, and detoured lives.

ED can be successfully identified and treated. Our daughter is but one illustration of that. Following graduation from college, Genny attended 2 years of post-graduate training in the field she loves. As of today, she lives a few hours away from home, working and supporting herself, surrounded by people who care about her. She does not consider herself "cured" of AN but nor is she consumed by this as she once was. She is thriving.

* * * * * * * *

OUR "IN A PERFECT WORLD" DESIRES

Doris and Tom Smeltzer, Napa, CA

"You must give me five minutes, Officer, to tell my wife what you have just told me." My husband's measured words continued, *"I will then call you back to answer your questions."* My right hand tightened on Tom's strong shoulder. I stood to the left of his high-backed desk chair. I could not see his face, but my heart knew what he was about to tell me. He replaced the receiver in its cradle and swiveled the chair gently toward me. The room was filled with my husband's broad six-foot frame as he rose to face me; tears streamed into his closely cropped salt-and-pepper beard. His voice cracked with emotion as he relayed the message that I knew was coming and yet, for the previous tortured eight hours, had prayed would not be true. *"They've found the dead body of a young girl in the home where Andrea was house sitting."* (Smeltzer & Smeltzer, 2006, p. 15)

Our book, *Andrea's Voice: Silenced by Bulimia,* begins with the ending of our daughter's life. This was intentional. Ultimately we wanted the book's message to be one of hope—death does not need to be the outcome of an ED—but we wanted to be certain readers understood that an ED is a serious illness that *can* kill. Andrea began seeing a therapist, physician, and nutritionist within 2 weeks of the first time she made herself throw up. Tragically, she died a mere 13 months later when an electrolyte imbalance caused her heart to stop beating while she slept. No one had ever told us that an ED is a deadly illness.

We would like to share with you thirteen messages that, in a perfect world, would be offered by professionals to every parent of a child diagnosed with an ED. These messages might have been difficult for us to hear, as we now realize how deeply steeped in denial we were. Andrea, the younger of our two daughters, was a vibrant, talented, spirited, and determined young woman. We felt certain bulimia nervosa (BN) was a not-too-serious, short-term anomaly in her life. Chipping away at our denial would have taken a tremendous amount of patience, a lot of modeling, and continual repetition of pertinent information. But knowing what we know now, we wish someone had tried.

We must state here that we blame no one for our daughter's death. We are completely cognizant of the fact that even had our denial lifted and all of our wished-for knowledge and skills become a reality, the outcome may still have been the same. The professionals working with Andrea did the best they could with what they knew at the time, as did we. We offer our perspective only in the hope that, as time goes on, practitioners and parents can and will do better. It is our belief that guidance and information of this sort would have radically changed our understanding of our daughter's illness, and our approach to loving her through it.

1. Iterate to parents, as many times as necessary, the deadliness of their child's illness.

Although we have already raised this concern, it merits repeating. Yes, it is a difficult message to convey and an even more difficult message to hear. But how else are parents to know the urgency of making the illness the highest priority? Andrea told us that people live with these illnesses for many years without major adverse affects. Those treating her assured us that, because we had gotten on it immediately, Andrea would most definitely heal. Knowing that BN could kill our daughter would have given us an important tool to help us counter her insistence that she was fine.

We needed to hear precisely how damaging her behaviors were. We needed to be told that, each day the behaviors continued, they actually became further entrenched in her brain's "do- over-and-over-again" pathways. We needed to know that an ED could be a terminal diagnosis: If it does not kill the body, it most definitely kills the soul. If practitioners convey these messages to parents, then parents will be better able to respond to the illness with the seriousness that it deserves.

2. Help parents realize that, while they did not cause the ED, they can make changes to ensure their actions do not help sustain the illness.

Due to the brevity of Andrea's illness we had a very steep learning curve. Although we were barely scratching the surface of understanding when she died, we know that no one was to blame for her death, and, in particular, that *we* did not cause our daughter to develop BN. ED are complex illnesses with a multitude of contributing factors.

That said, we now realize how much we had bought into our culture's idealization of a thin body, and how we conveyed that to our daughter in subtle yet powerful ways. If professionals can help parents recognize how ubiquitous those faulty ideals are, and explore how they might unknowingly convey them to their children, those parents might be better able to begin defusing the power of an unhealthy body image.

3. Convey to parents that there is no need for guilt or shame.

Practitioners need to normalize the feelings of guilt and shame as common human responses. At the same time, they must work to explore the ways these feelings can be a distraction, or a way to avoid dealing with other responses to the illness that are uncomfortable or difficult to express. Doris struggled with this more than did Tom. Andrea spent a great deal of time attempting to assure her mother that she was not the source of the ED, and Doris' self-blame caused her, at an unconscious level, to avoid having other more necessary conversations with Andrea.

4. Be sure that parents know that recovery and healing can be a very long process.

Two years ago, our older daughter Jocelyn spent many months in intensive care on life support, completely paralyzed with a rare variant of the autoimmune syndrome Guillain-Barré. From the outset, every health professional we came into contact with warned us that healing from this syndrome would *not* be a sprint—we needed to have the mind-set of running a very long marathon and be sure to pace ourselves accordingly. Indeed, it was nearly a year before Jocelyn was released from the hospital, and a few more months before she could walk. Even now, she is still in leg braces.

We would not have thought to ask Jocelyn to heal more quickly, and yet our initial attitude with Andrea was "just stop." We actually thought she could heal over one summer and were surprised and disappointed in her when she did not. It would have been helpful to us if the marathon metaphor we were given for Jocelyn had been used with Andrea.

Parents need to hear that there are no quick fixes or magic bullets and that it can take as long to heal from an ED as it takes to develop one.

5. Guide parents in recognizing why secrecy and silence are not helpful responses.

We allowed Andrea to talk us into keeping her illness a secret. *She* would determine with whom and when to share. We were unaware that this secrecy isolated Andrea, permitting the ED to flourish unchecked. It also cut us off from the resources and support of others.

Professionals need to explain to parents that silence is a request the illness makes—it does not come from the child's healthy self. Families need modeling and counseling on how to talk to others openly about this topic, how to inform their child that they will not keep her secret, and how to effectively respond to her reaction, whatever it may be. We were shocked at the number of friends, family, and colleagues who "came out" about their own or a loved one's ED at Andrea's memorial service. Had we not been silent, maybe others would have come out sooner, alleviating the stigma we felt and allowing us to see that we were not alone.

6. Practitioners must encourage family and friends to take care of themselves.

Even in the short time Andrea suffered, we experienced exhaustion, more from the stress of worry and not knowing how to respond and help our daughter than anything else. We needed guidance in how to develop strong self-care skills so that the ED, although real and of primary concern, did not take over our lives. As the ED developed, Andrea systematically disconnected from nearly everything in her life, especially from herself and her own needs. Hence, it was particularly important for us to model self-care.

The treatment team needs to remind parents to take time to ensure they are eating well, exercising, getting enough sleep, and scheduling regular date nights with each other. Parents will be better able to see the value of these actions if they are presented as ways to model self-care for their child.

7. Practitioners need to emphasize that recovery must take precedence over everything else.

Our daughter's health and healing needed to be our foremost concern. Initially, we erroneously thought Andrea could simply continue to live her very full life (college, work, volunteering) while she attempted to heal. Eventually, we did encourage her to cut back a bit by dropping a few classes, deferring volunteer work for a while, and reducing her work schedule, but we marvel today at that attitude back then. We would not have asked Andrea to continue these activities if she had suffered with a deadly cancer, nor would we want her to ask that of herself.

8. Professionals need to let parents know that dieting is dangerous.

We were unaware of the dangers of dieting. Parents need to be told not only of the potential dieting has to trigger ED behaviors but of the importance to model a non-dieting approach to wellness. An environment where parents are counting calories and restricting intake makes healing far more difficult. Practitioners can suggest changing these behaviors as a tangible testament to a parent's willingness to "do their own work" to aid in their child's healing.

9. Help parents recognize their own, shared characteristics common to ED.

Parents need to hear of the importance of exploring characteristics in themselves that mirror their child's. If they are guided to examine their own perfectionism, their tendency toward dichotomous thinking, or their habit of putting the care of others before their own self-care, it may help their child to do the same. As above, their willingness to do their own work will be a valuable model as well as a powerful statement of their love and support.

10. Parents need to know the severe impact—cognitive and physical—of the illness and the covert nature of the symptoms.

We did not realize that our daughter's thinking, in areas beyond the distorted thoughts of the ED, was also challenged. Up until the day she died, Andrea appeared to us to be a healthy, rational, and responsible being. Although her heart was starving, she never physically appeared starved. We were as duped by the ED as were those who were treating her.

Parents need to be educated on the effects of starvation, binging, and purging on the brain and body, especially the ways ED behaviors compromise rational thought.

11. Educate parents about normal adolescent and physical development.

Although we are both educators, we needed to be reminded of the developmental and physical stages of adolescence. A simple fact that we were somehow ignorant of—and that would have been helpful to us long before Andrea developed BN—was that females must increase body fat by 120% for menstruation to begin (Frisch, 1991). This significant increase of body fat during puberty is a natural biological process in the female body—a fact that appears obvious to us today, but when we first heard it after Andrea's death it was an "ah-hah" moment. While Andrea was alive we pictured healing as a straightforward, linear endeavor. We were surprised to learn of all the stops and starts and movement forward and back within the change process. It is truly no wonder that Andrea could not heal over one summer!

12. Parents need to see that returning to the old "normal" is not the goal.

At the time, we did not realize that "normal" had not been working for quite a while, that we needed instead to allow ourselves to grieve the family of times past, and the past experience of, and relationship with, Andrea. Grieving would have allowed us to focus on the now and helped us see how as a family we could work together to create a future that included more helpful, productive ways of being in relationship with each other while embracing a new vision of what normal looked like for us. When practitioners help us grieve what used to be, it allows us to open up to new possibilities.

13. Teach parents how to express their feelings in productive ways.

Since Andrea's death we have discovered the value of therapy, specifically, for us, emotionally focused couple's therapy. It has helped us come to recognize the habitual patterns of communicating thoughts and feelings that cause us to disconnect from each other, and ways that we can stay connected and authentic during intense interactions. Given the high cost of treatment, suggesting that parents add their own therapy (as a couple or individually) to the treatment protocol may seem like one too many expenses, but these are skills that will serve the entire family. It is important for practitioners to lead parents in that direction.

Although the lessons we have shared here may not provide much new information for those educated in the field of ED, they may very well be new for clients and their families. In a perfect world, these lessons would be integrated into a comprehensive treatment protocol. But adding just a few could make a huge difference in attitudes and outcomes for patients and parents. They would have made an enormous difference to us.

As is our practice at the end of each presentation, we would like to close with Andrea's words. The following is the last stanza from a poem she wrote shortly before her death:

> Look carefully
> judge kindly
> read under and between lines
> The Journey is never so clear as the Destination
> and the telling is more confusing still. — *Andrea Smeltzer, 19*

* * * * * * * *

CONCLUSION

These three powerful stories remind us of the real impact of the gaps in our understanding and treatment options for patients with ED. Faced with our own limitations, and the lack of certainty about what we can and cannot promise to our patients and families, we, as clinicians, may be at risk for tuning out the terror and helplessness of our patients' loved ones. Clinicians working with patients and families with ED have to face these limitations because it will help them appreciate and respect the experience of being caught up in the ED storm. Perhaps some of the rules and regulations of treatment programs exist more to contain clinician anxiety than they do to actually meet the needs of patients and families.

Stigma matters. There is still a chasm, not a mere gap, between what is known about ED and what most people still believe. This lingering perception that ED are not serious and lethal disorders still clouds, perhaps unconsciously, our own clinical judgment but it also underpins the lack of resources allocated to these issues. Where is the research? Where are the clinical training programs? Why does it take so long for research findings to make it into general clinical practice? These are not merely academic and professional questions. The answers touch the lives of millions across multiple generations.

Clinicians need to encourage their patients and families to become active partners in collaborative treatment. Help your patient's loved ones by encouraging them to educate themselves about ED. This is not only about providing access to resources, it is also about a spirit and attitude of partnership. You need to invite their questions and to validate and respect their experience and "gut" knowledge of what is happening.

Our field has come a long way in recent years in welcoming families back into the treatment of patients with ED. Yet, as these three stories emphasize, we still have a long way to go.

References

Frisch, R. E. (1991). Body weight, body fat, and ovulation. *Trends in Endocrinology and Metabolism, 2,* 191–197.

Smeltzer, D., & Smeltzer, A. (2006). *Andrea's voice silenced by bulimia: Her story and her mother's journey through grief toward understanding.* Carlsbad, CA: Gürze Books.

PART V

BRIDGING THE GAP: MIND, BODY, AND SPIRIT

22   The Role of Spirituality in Eating Disorder Treatment and Recovery   367
23   The Case for Integrating Mindfulness in the Treatment of Eating Disorders   387
24   The Use of Holistic Methods to Integrate the Shattered Self   405
25   Incorporating Exercise into Eating Disorder Treatment and Recovery   425
26   Body Talk   443

CHAPTER 22

The Role of Spirituality in Eating Disorder Treatment and Recovery

Michael E. Berrett, Randy K. Hardman, and P. Scott Richards

Religion and spirituality are important to the majority of people in the United States. Approximately 95% of Americans profess a belief in God, 65% are members of a church, and 60% say that religion is very important in their lives (Gallup, 2003). Many also practice a non-religious spiritual lifestyle which is important to them. Given this, it is likely that religion or personal spirituality is important to the majority of eating disorder (ED) patients.

The widespread public interest in spirituality highlights the need for health care professionals to be aware of and sensitive to the spiritual values and needs of their patients. The ethical guidelines of many organizations in the social sciences and health care professions now recognize religion and spirituality as types of diversity that professionals need to understand and respect (e.g., American Psychological Association, 2002), although this has not been an easy task for many professionals. Because of the alienation that has existed historically between the health care professions and religion, the spiritual concerns of patients have long been neglected in training programs and continuing education offerings (Koenig, McCullough & Larson, 2001; Richards & Bergin, 2000). Fortunately, this situation is changing.

Hundreds of research studies on spirituality and health have now been published which provide evidence that religious and spiritual influences may be beneficial in human health and healing (Benson, 1996; Koenig, 1998; Koenig et al., 2001; Plante & Sherman, 2001). These empirical findings have encouraged changes in health care practice and training. Over half of U.S. medical schools now offer courses in spirituality and healing (Puchalski, Larson & Lu, 2000). Increasingly, graduate training programs in psychology are giving attention to religious and spiritual aspects of diversity (Brawer, Handal, Fabricatore, Roberts & Wajda-Johnston, 2002; Puchalski & Larson, 1998). Growing numbers of continuing education workshops concerning spiritual issues in medical and psychological treatment are being offered, and many professional books have been published about spirituality, health, psychotherapy, and medicine (e.g., Benson, 1996; Koenig, 1998; Koenig et al., 2001; Plante & Sherman, 2001; Richards & Bergin, 2005).

More and more medical and psychological practitioners now believe that treatment may be more successful if patients' spiritual issues are addressed sensitively and capably, along with their other concerns (Benson, 1996; Koenig et al., 2001; Puchalski et al., 2000; Richards & Bergin, 2000; Sperry & Shafranske, 2005). Spiritual perspectives and interventions have been integrated with many mainstream therapeutic traditions and treatment modalities and have been used with a variety of clinical issues and patient populations, including addictions, anxiety disorders, eating disorders, dissociative disorders, trauma victims, antisocial and psychopathic personality disorders, and postpartum depression (Benson, 1996; Plante & Sherman, 2001; Richards & Bergin, 2005; Sperry & Shafranske, 2005). Although much research work remains to be done, some empirical evidence indicates that spiritual approaches are as effective, and sometimes more effective, than secular ones, particularly with religiously devout clients (McCullough, 1999; Smith, Bartz & Richards, 2007; Worthington & Sandage, 2001). Other studies suggest that non-religious psychotherapists can effectively incorporate spiritual interventions into psychotherapy (e.g., Propst, Ostrom, Watkins, Dean & Mashburn, 1992).

Despite the widespread professional interest in spirituality that now exists, the relationship between spirituality and ED has rarely been addressed in scholarly work. A systematic review of the *International Journal of Eating Disorders* found that only 0.8% ($N = 8$) of over 1033 empirical studies published from 1993 to 2004 included religion or spirituality as a variable. Only 2.2% ($N = 4$) of 186 empirical studies published from 1999 to 2004 in *Eating Disorders: The Journal of Treatment and Prevention*, included religion or spirituality as a variable (Richards, Hardman & Berrett, 2007). Few have been published in these journals since then.

The lack of empirical attention to religion and spirituality is also reflected in the *Practice Guideline for the Treatment of Patients with Eating Disorders*, published by the American Psychiatric Association (2006). This important publication mentions the possible role of religious and spiritual issues in etiology or treatment on only a few brief occasions. First, it points out that "patients' religious and cultural practices must be considered and discussed to limit patient rationalizations for restricted eating" (pp. 41–42). Second, it acknowledges the potential beneficial role that 12-step groups can play, but cautions that their "effectiveness and potential adverse effects... have not been systematically studied " (pp. 47, 56). Third, it mentions "12-step models" as one alternative form of psychosocial therapy that needs further research investigation (p. 88). No mention is made of other spiritual treatment approaches that have been described in the literature (e.g., Benson, 1996; Sperry & Shafranske, 2005), or of the potential role faith and spirituality may play in facilitating ED treatment and recovery (Richards, Hardman & Berrett, 2007).

The lack of research attention given to the topic of spirituality and ED is all the more striking when one considers that the integration of spirituality into clinical treatment appears to be a growing trend in the ED field. We conducted an exhaustive Internet search in order to determine how many contemporary ED treatment programs advertise on the world-wide-web that they address issues of spirituality. Of the 150 ED treatment programs we located in this search, 67 of them (44.7%) indicated that they address religious or spiritual issues in some way (Richards & Susov, 2009). Thus, it appears that sizable numbers of treatment centers and practitioners in the ED field regard spirituality as important in treatment, but researchers have given it relatively little attention to date. Clearly, there is a wide gulf between practice and research in regard to the subject of spirituality and ED.

This chapter attempts to help bridge the gap that currently exists by sharing information about how spiritual perspectives and interventions can be integrated ethically and effectively into ED treatment in order to promote patients' healing and recovery. Spirituality is only one component of rigorous clinical treatment, but it is vital. This chapter describes research findings about the role of spirituality in recovery from ED; common spiritual issues in ED patients; process guidelines for integrating spirituality into treatment, including strategies for establishing a spiritually safe environment, assessing patients' spiritual frameworks, and implementing spiritual interventions during treatment; six spiritual pathways towards recovery from an ED; and suggestions for research and training.

FAITH AND SPIRITUALITY AS RESOURCES IN TREATMENT AND RECOVERY

A detailed review of theory and research that addresses the relationship of religion, spirituality, and eating disorders is beyond the scope of this chapter. We have done this elsewhere and at that time offered several tentative conclusions (Richards et al., 2007):

1. Religious teachings about asceticism and fasting may contribute to the development of anorexia nervosa (AN) in some women.
2. Patriarchal or male dominated cultures may contribute to women's feelings of powerlessness and AN may represent an attempt by some women to exert control in their lives in such cultures.
3. The development of bulimia nervosa (BN) may be associated with a decline in religious devoutness in some women.
4. Many women who have ED believe that their personal faith and spirituality helped in their treatment and recovery (pp. 208–209).

This chapter briefly discusses only those empirical studies providing evidence that women with ED, and those who have recovered, believe that faith and spirituality were helpful in their treatment and recovery.

Mitchell, Erlander, Pyle & Fletcher (1990) reported that, in a follow-up study of patients with BN, "the single most common write-in answer as to what factors have been helpful in their recovery had to do with religion in the form of faith, pastoral counseling, or prayer" (p. 589). To further explore these issues, they surveyed 50 women with BN, finding that 88% believed in God and approximately 60% prayed or worshipped privately at least several times a month. They concluded that "religious issues are important for many patients with bulimia nervosa. Therapists should consider discussing each patient's religious needs as part of their intervention and be willing to refer to a member of the clergy when appropriate" (pp. 592–593).

Hsu, Crisp & Callender (1992) did follow-up interviews of six patients who had recovered from AN to find out what the patients believed had helped them recover. One patient indicated that her religious beliefs, including prayer, church attendance, and faith in God had helped in her recovery. Hsu et al. (1992) acknowledged that the influence of religion on recovery was "an area that we did not inquire about at all in our interview(s)" and that it was thus "unclear whether it played a part in the recovery of others" (p. 348).

Hall and Cohn (1992) asked former patients (366 women and six men) what activities had aided their recovery from BN and other forms of problem eating. Fifty-nine percent of the respondents said that "spiritual pursuits" had been instrumental and 35% said that a spiritually oriented 12-step program was helpful to them. Rorty, Yager & Rossotto (1993) interviewed 40 women in recovery from BN to find out what had promoted their recovery. Many of the women (25 to 40%) reported that the spiritual aspects of 12-step programs (e.g., Overeaters Anonymous) were useful and others mentioned other forms of spiritual guidance.

Garrett (1996) interviewed 32 anorexics about their self-starvation and recovery. Participants "regarded anorexia nervosa as a distorted form of spirituality." For many of them, recovery involved rediscovery of the self (p. 1493). Garrett explained that "participants claimed that recovery requires an experience of something (a material reality and/or an energy) beyond the self. They named it 'spirituality' or referred to it as 'love,' as God or as Nature" (p. 1493).

Smith, Richards & Hardman (2003) quantitatively examined the correlation between growth in spiritual well-being and other positive treatment outcomes in a sample of 251 women who received inpatient treatment for their ED. Improvements in spiritual well-being were positively correlated with improvements in attitudes about eating and body shape, as well as psychological functioning.

Marsden, Karagianni & Morgan (2007) qualitatively interviewed 11 patients at St. George's inpatient ED unit in London, United Kingdom, who indicated that their religion was important to them, in order to examine relationships between ED, religion, and treatment. They found that patients often understood their ED in religious ways and that spirituality enhanced the motivation and improved treatment adherence. They also concluded that for "patients with strong religious faith, spiritual practice is helpful in recovery, and spiritual maturation goes hand in hand with positive psychological changes" (Marsden et al., 2007, p. 11).

A survey of 36 women who had successfully completed an ED inpatient program asked open-ended questions about how their spirituality promoted recovery (Richards et al., 2008). The patients reported that their spirituality gave them purpose and meaning, expanded their sense of identity and worth, helped them experience feelings of forgiveness towards self and others, and improved their relationships with God, family, and others, sustaining them during their most difficult challenges.

COMMON SPIRITUAL ISSUES OF EATING DISORDER PATIENTS

Our clinical experience has also convinced us that religious and spiritual issues are significant in both ED etiology and recovery (Richards et al., 2007). Providing treatment to hundreds of women in our specialized treatment center we have noticed that deep spiritual struggles are a major impediment in many women's recovery process. Many sufferers have previously felt a connection to God or a Higher Power and a degree of personal spirituality. Some have participated in religious practices, but they have lost these connections through the course of their ED.

Almost all women with ED lose touch with their sense of spiritual identity and worth. They lose the ability to see and affirm the various aspects of their identity, as well as the ability to see and experience their positive characteristics, such as goodness, kindness, honesty, and compassion. In the later stages of the illness they may see themselves exclusively *as* an ED, or as the expression of an ED (Richards et al., 2007).

Women with ED often feel unworthy, unlovable, and incapable. They lose their ability to feel connected to family and friends. As their relationships with God and with loved ones deteriorate, they rely ever more exclusively on their ED as their way of coping with pain and problems (Richards et al., 2007). They have difficulty paying attention to spiritual feelings because negative thoughts and false beliefs consume all of their energy and time.

Patients often have false beliefs about the ED providing solutions to all of their problems. For example, a patient may erroneously believe that her ED will help her effectively communicate her pain and suffering, or that it will compensate and atone for her past mistakes (Hardman, Berrett & Richards, 2003). Such false and unhealthy beliefs contribute to patients' feelings of alienation from God and other spiritual supports such as the supportive influence of other people.

ED patients may also struggle with a number of other spiritual issues, including: (i) negative images of God and religious leaders; (ii) fear of abandonment by God; (iii) guilt and shame about sexuality; (iv) reduced capacity to love and serve; and (v) dishonesty (Richards et al., 1997). We have written about these issues in other publications (e.g., Richards et al., 2007).

PROCESS GUIDELINES FOR A MULTIDIMENSIONAL SPIRITUAL FRAMEWORK

Our treatment approach is grounded in current research findings and accepted clinical guidelines (American Psychiatric Association, 2006) and includes contemporary state-of-the-art approaches, such as medical, pharmacological, nutritional, cognitive-behavioral, individual, family, group, experiential, and recreational therapeutic interventions (Richards et al., 2007). Into this multidimensional approach, we integrate a non-denominational spiritual emphasis suitable for patients from a wide variety of religious and spiritual backgrounds (Richards & Bergin, 2005). We encourage patients to explore their own spiritual beliefs and to draw upon those beliefs (Richards et al., 2007). Although this approach is most suitable for patients and therapists who have a theistic spiritual orientation, many aspects can be applied in culturally sensitive ways with patients who believe in Eastern, humanistic, or other forms of spirituality. We encourage therapists to respect and work within the belief systems of their patients, whatever they may be.

Given the lack of consensus in the professional literature concerning the definitions of religion and spirituality (Zinnbauer, Pargament & Scott, 1999), it is best to use a broad definition of *spirituality* in order to give patients the opportunity to figure out and declare what spirituality means to them. Religious or spiritual activities and practices such as reading scriptures, attending worship services, yoga, meditation, or prayer, are all potentially important and helpful, but spirituality is much more than these practices. Spirituality also includes experiences such as feeling compassion for someone, expressing love to others, accepting love, being able to feel hope, receiving inspiration, feeling enlightened, being honest and congruent, feeling gratitude, and feeling a sense of life's meaning and purpose.

Establish a Spiritually Safe Therapeutic Environment

The inclusion of questions about religious and spiritual interests at intake opens the door to spirituality. This conveys that it is permissible and appropriate to explore spiritual issues should patients so desire and that their spiritual beliefs may be a potential resource for their treatment. Clinicians must communicate interest and respect when patients self-disclose information about their religious tradition and spiritual beliefs and use spiritual resources and interventions that are in harmony with the patients' beliefs. They must also keep in mind potential ethical pitfalls associated with incorporating spirituality into treatment, such as: engaging in dual relationships (religious and professional); displacing or usurping religious authority; imposing religious values on patients; and violating work setting (church-state) boundaries (Richards & Bergin, 2005).

Conduct a Religious and Spiritual Assessment

When patients are considered for ED treatment, a thorough assessment of their functioning should be conducted, including their physical, nutritional, psychological, social, and spiritual functioning. The goal is to gain a clear understanding of each patient's current spiritual framework so that staff can work within that belief system in a sensitive and respectful manner. Information about patients' spirituality can be gathered through written intake questionnaires, clinical interviews, and standardized measures (Richards et al., 2007). As spirituality cannot be adequately assessed through a list of questions, a well-crafted and clinically-mature interview process is needed where patients are invited to share more about their true thoughts, feelings, and beliefs about spirituality and its role in their ED and in their lives.

Patients are typically relieved that spirituality is something they can talk openly about and explore in their treatment. If they are uncomfortable or unwilling to discuss spiritual issues, this should be respected and treatment can proceed without a spiritual focus. Communicating respect and tolerance for patients' religious and spiritual beliefs is an essential part of the assessment process. Information that patients disclose about their religion and spirituality is often experienced as sacred; trivializing it in any way can be extremely painful and undermines the therapeutic relationship.

Conducting a religious and spiritual assessment uncovers and clarifies the possibilities, hopes, and desires about what patients truly want, how they want their spiritual life to be, and how they might want to incorporate spirituality into their treatment and recovery. Once the patients' strengths and their personal resources are identified, treatment can tap into them. While religious beliefs and background may contribute to their conflicts, impasses, and difficulties, it also helps to understand the positive, the uplifting, and the strengthening aspects of patients' spirituality and religiosity that may assist them in recovery.

Implement Spiritual Interventions in Treatment

Spiritual interventions in treatment can begin after a full assessment of the patients' psychological functioning, spiritual background and beliefs, and attitudes about exploring spiritual issues during treatment. This facilitates working within their value frameworks, so the interventions used are in harmony with their beliefs. Spiritual interventions are probably more risky and less effective when patients are young (children and young adolescents), severely psychologically disturbed, anti-religious or non-religious, or spiritually immature, or view spirituality as irrelevant to their presenting problems (Richards & Bergin, 2005). With these cautions in mind, faith and spirituality are often powerful resources for change.

Spiritual interventions can be integrated into virtually every treatment modality, including individual, group, and family therapy and in all treatment settings: inpatient, residential, and outpatient. We offer a 12-step group that has been adapted to the spiritual needs of female ED patients, and which affirms their worth, autonomy, and personal power in their relationships with family, religious leaders, and their Higher Power (Richards et al., 2007). We also offer a spiritual exploration group and provide patients with opportunities to attend worship services, seek spiritual counsel and direction from their religious leaders, engage in acts of altruistic service in the community, and participate in private spiritual practices such as yoga, meditation, prayer, spiritual journaling, and private worship. We have described these interventions in more detail elsewhere (Richards et al., 2007).

SIX SPIRITUAL PATHWAYS TO RECOVERY FROM AN EATING DISORDER

This section examines six spiritual pathways to recovery. For each of these pathways, we describe practical clinical guidelines and interventions.

Listening To and Following the Heart (Box 22.1)

Listening to and following the heart is the foundation of all spiritual pathways to recovery and healing and is a common thread throughout each of the other five spiritual pathways. The heart is a metaphor for patients' eternal spiritual identity (Richards & Bergin, 2005; Richards et al., 2007), with a language both universal and ecumenical in nature. For thousands of years, people from diverse cultures and religions have considered the heart as a source of emotion, courage, wisdom, and spirituality (Childre & McCraty, 2001). Helping patients get in touch with, and affirm their spiritual worth and goodness through the metaphor of the heart, is non-threatening and well-accepted and enables them to discover or rediscover their sense of identity as spiritual beings (Richards et al. 2007).

According to McCraty, Bradley & Tomasino (2004−2005), a considerable amount of contemporary scientific evidence supports the idea that the heart is itself a sensory organ with its own nervous system that processes information, learns, remembers, and makes decisions independent of the brain. The heart's flow of signals to the brain actually influences the higher brain centers' activities such as perception, cognition, and emotional processing. While an intricate neural communication network links it to the brain and body, the heart also uses electromagnetic field interactions to communicate information to both. Growing evidence also suggests that the heart plays an important role in spiritual experiences, intuitive ways of knowing, and healthy physical and emotional functioning (e.g., Childre & McCraty, 2001; McCraty, Atkinson & Tamasino, 2001; McCraty, Atkinson & Bradley, 2004a,b).

Patients with ED often acknowledge a lost or broken heart. As the ED develops, they become disconnected from their heart and may even fear that the shame and self-contempt that they feel is coming from the heart. When patients are attempting to control their mind through obsessive preoccupations, ruminations, and distractions, their mind becomes complicated. The heart, on the other hand, is uncomplicated. The heart says "yes" or "no." The complicated ruminations of the mind take patients away from the quiet simplicity of

> **BOX 22.1**
>
> ## LISTENING TO AND FOLLOWING THE HEART: CLINICAL GUIDELINES AND INTERVENTIONS
>
> 1. Adapt the language and discussions of "listening to the heart" to the spiritual framework of the individual patient. Listening to the heart is labeled and expressed by patients in different ways including the following: sensitivity, sensibility, impressions, intuitions, God's voice, the spirit, the inner voice, the true self, the source of wisdom, the soul.
> 2. Directly teach patients the concept of listening to the heart. Teach that there are three significant aspects of the self to listen to and learn from: thoughts, feelings, and heart. Teach them that the heart can be the most powerful resource, and that, even though the heart is more subtle and quiet, it is essential and uplifting in leading patients towards hope, healing, and recovery.
> 3. Ask religious patients to look for God's hand, providence, or miracles in their lives on a daily basis and to write those understandings down in their journal for regular review.
> 4. Ask patients to start a "The Writings of my Heart" journal, writing down first impressions, spiritual promptings, and personal intuitions which come to them in quiet moments. Then have them take this raw list, gathered over weeks or months, and compile or rewrite them in one place so that they can have a place to return to often for a clear reminder of how they want to live and be (Richards et al., 2007).
> 5. Encourage patients to take "solo" or "quiet" times where the time, place, and environment are free of interruptions and external distractions, and 2where they can find solace and solitude and access to the heart.

the heart. Fear does not speak to the heart. We encourage our patients to look beyond their fears to find out what truly speaks to their hearts.

To become aware of their own hearts as a source of information and a resource for self-direction, leads patients to the center of both physical and spiritual well-being. We speak of patients' hearts as the inner core of their being, their deepest place for desires and emotions, and their receptor for spiritual impressions. The heart is more than feelings and it is more than thoughts. It is deeper and quieter. It does not speak in sentences, but rather it conveys an impression, a sense of knowing, and an understanding (Richards et al., 2007).

The experiences that speak to a person's heart are often the simple but profound expressions of what is best in all of us and in life. Tenderness, compassion, understanding in the face of difficulty, kindness, generosity, love, forgiveness, and mercy speak to the heart. The truth speaks to the heart. The key is to help patients begin to look for the things in their lives that speak to their own hearts. Learning to trust the heart is not a quick fix, but rather, an ongoing development of inner understanding and a sense of purpose and direction (Richards et al., 2007).

Learning a Language of Spirituality (Box 22.2)

As therapists come to understand patients' spiritual language, they can join, respect, support, challenge, and nurture patients in their beliefs in a way that facilitates recovery. As patients develop and understand their own spiritual language, it strengthens their sense of spiritual identity and stability. And, as their spiritual language is understood and accepted by others, the opportunity for spiritual intervention and support begins (Richards et al., 2007).

In learning the language of patients' spirituality, clinicians must acknowledge that the patients are the teachers and they are the students. Patients help therapists understand their views of spirituality, what is most important to them, and what has become lost or disconnected during the development of the ED. In encouraging this relationship of sharing and understanding, therapists create a spiritual space and safe environment in which patients can do their work. Just as giving permission to patients to express emotion creates safety in psychotherapy, so giving permission to discuss spiritual matters creates safety in the therapeutic relationship.

To develop spiritual literacy, our treatment center provides patients with a spirituality workbook that explores 11 universal spiritual themes: faith, transcending suffering, spiritual self-worth, meaning and purpose, responsibility, forgiveness, gratitude, belonging, congruence, love, and spiritual harmony (Richards, Hardman & Berrett, 2000). While these themes are common and easily talked about with others, as patients travel the recovery path, they take on individual significance and deepen the meaning of their spiritual experiences.

A language of spirituality can begin to replace the negative ED voice, which includes all or nothing thinking and rigid dichotomies (e.g., fat-thin, good-bad, success-failure) that fuel perfectionism and hamper spirituality. The language of spirituality is abundant in that it is

---

BOX 22.2

LEARNING A LANGUAGE OF SPIRITUALITY: CLINICAL GUIDELINES AND INTERVENTIONS

1. Assess and understand the spiritual framework and language of each patient.
2. Give patients handouts with quotes from various and differing spiritual thoughts from well accepted and wise people from diverse spiritual traditions. Ask them to read the quotes and identify which quotes they related to, resonated with, or were touched by. Then ask patients to discuss—either with the therapist or within a group—how that quote was meaningful to them (Richards et al., 2007). This can help patients realize and articulate their own spiritual beliefs.
3. Give patients an invitation, permission, and time to talk in-session about any spiritual thoughts, feelings, or experiences which they might have had during the previous week. This reinforces safety to talk about the spiritual, and talking about it openly helps patients begin to develop and feel comfortable about their own language of spirituality.

respectful, enlarging, uplifting, and tied to a purpose that is bigger than self. The patients' language of spirituality gives a vehicle for sharing, which often strengthens their deeply held beliefs and becomes an anchor in difficult times as deeply held beliefs find expression.

The language of spirituality is the language of the heart, and is affirming, reassuring, kind, and hopeful. It edifies and builds patients up towards understanding who they really are, and who they can become. As patients become more aware of their spirituality and better at sensing those things that speak to their heart, they begin to connect to positive understandings of self that greatly challenge and contradict the negative messages of the ED.

Mindfulness and Spiritual Mindedness (Box 22.3)

Mindfulness is being aware of feelings, thoughts, and the heart, each of which can be consciously attended to. It is an awareness of both internal mental and emotional processes and of external circumstances. Through mindfulness, patients can learn to recognize, label, and express emotions, and to accept their emotional feelings as helpful information.

Spiritual mindedness is connecting to the heart, paying attention to what speaks to the heart, and hearing and knowing the quiet answers of the heart. It is living in the present moment, rather than living in the conflicts, confusion, negative thinking, and over-analysis of the ruminating mind. Spiritual mindedness is about recognition of the truth in the moment. When truth is listened to and followed it becomes a powerful change agent.

As the center of spiritual mindedness, the heart is a pathway or conduit to spiritual experience, connecting patients to their priorities, principles, purpose, and deepest desires that lead them to life changing decisions, daily confidence, and peace of mind. The loud confusion of the negative mind masks the quiet and steady impressions of the heart. Spiritual mindedness can help patients to step outside of and away from the repetitive and negative

---

BOX 22.3

MINDFULNESS AND SPIRITUAL MINDEDNESS: CLINICAL GUIDELINES AND INTERVENTIONS

1. Help patients understand the reality of and the differences between thoughts, feelings, and the heart. All three exist, and all three are important. Listening to and following the heart is the primary activity connected to spiritual mindedness.
2. Help patients understand that pondering, reflecting, meditating, and praying are all practices that can promote spiritual mindedness through the inner works of solitude.
3. Ask patients to write impressions of their heart versus writing their thoughts and feelings. Then discuss in session and help the patient learn to differentiate the difference between the spiritual (heart) and the psychological (thoughts and feelings).
4. Ask patients to look for evidence of providence, divine influence, or miracles in their lives and to write in their journal about it each day.

messages that are a part of the obsessive thinking of an ED, listen to their heart, live in the here and now, and act according to such impressions.

Those struggling in an ED tend to hold the past, and their behaviors, thoughts, and feelings about the past, against themselves. In spiritual mindfulness, a patient's heart, or higher power, speaks to them in the present, generating hope and self-forgiveness and conveying that where they are going is more important than where they have been.

In spiritual mindedness, goals and priorities are of a spiritual nature. Spiritual mindedness, therefore, is a pathway of growth and development where patients nurture, listen to, focus on, and take care of their spiritual nature. Faith and hope are also a part of spiritual mindedness. Looking for and finding evidence of divine influence, small miracles, or providence in their lives helps to develop spiritual mindedness.

Humility is also part of spiritual mindedness. In humility patients recognize that there are influences in the world greater than theirs. Humility allows potential for openness and a connection to the love, light, influence, and goodness which comes from a higher source. For some patients that source is God. For others it might be other people or love.

Principled Living (Box 22.4)

Principled living is a common thread of spirituality in the broad community of humanity. Individual spirituality includes clarification of which principles are critical to live by, and

---

BOX 22.4

PRINCIPLED LIVING: CLINICAL GUIDELINES AND INTERVENTIONS

1. Have regular discussions during therapy sessions with patients about courage and bravery in living in harmony with their heartfelt convictions.
2. Make self-correction and learning from mistakes without self-judgment a consistent theme of progress in the process of eating disorder recovery.
3. Help patients make promises and commitments to themselves, others, and God, which will empower their heartfelt decisions. Support the lessons learned, efforts made, and progress gained in these choices to be true to their own hearts.
4. Help patients work actively on only one or two principles at a time in their striving for principled and congruent living. Pacing is important in building confidence and hope, and to avoid overwhelming them.
5. Help patients explore for themselves the significance of the statement, "What will it mean for me to be true to my heart?" as they address the challenges and dilemmas of recovery, relationships, and life.
6. Initiate the 24-hour honesty rules with patients as follows: (1) "If you break your commitment to yourself or to me then you have to let me know within 24 hours." (2) "If you find yourself falling short of 100% honesty at any moment then you have to correct that by being fully honest with yourself and me within 24 hours."

efforts to live a life congruent with those principles. Integrity comes from a consistent pattern of living true to self-chosen beliefs and values. In principled living, patients follow their hearts, and act on these internal prompts even in times when they feel overwhelmed, undeserving, or fearful.

Once patients recognize what their heart is saying to them, they can start to make promises and commitments to themselves, God, and others and to act upon their heartfelt understandings and values. This important step is critical for successfully following the heart. Therapists can help patients make this step conscious and deliberate.

Commitments can be followed by making necessary sacrifices and embarking on the hard work of change. This can lead to the fruition of the deeply held desire or intention. In being true to their heart, in being consistent, and following through on impressions of the heart, patients can move forward with increased confidence which comes from knowing how they will act in whatever circumstance comes their way.

An important phase of recovery is when patients begin to trust that they can live congruent with what their heart is telling them. They begin to put their faith in heartfelt truths and desires ahead of the ED behavior as they begin to have a sense of choice that was not previously available to them. Patients experience a measure of healing when they are congruent with themselves. The commitment to self to keep true to their deepest selves, if honored, generates an internal strength that leads to dignity and self-respect. Dignity of self is expressed by increased self-confidence, faith, and determination to live with real purpose and meaning.

When patients' words and actions are congruent with their desires and intentions of the heart, therapists can frame or label this as courage. This includes the courage to admit weaknesses and mistakes, and the courage to self-correct. Congruence between the internal and the external self can become a powerful way of being (Rogers, 1961).

Principled living leads to living a life with integrity and brings peace of mind, but it does not mean perfect or perfectionistic living. When teaching patients about principled living, we explain to them the idea of honesty and of self-correction without judgment. In a moment of incongruence to a chosen principle, honesty, learning from the mistake, and self-correction can all happen within a context of love, kindness, and forgiveness. Mistakes are made, yet mistakes are not who patients are, just as the ED illness is not who they are. If patients strive to live according to the spiritual principles they have embraced, and self-correct when they fall short, they will progress much in their recovery.

Principled living is an important process for patients to regain or renew their spiritual identity which has been lost to the ED. Reconnecting with and strengthening spiritual identity cannot happen through self-criticism and self-punishment. Patients learn to bypass the negative mind, to recognize that their hearts are good even if they make mistakes, and to understand that the enemy is the negative beliefs and pursuits of the ED and not themselves.

Giving and Receiving Good Gifts of Love (Box 22.5)

To reconnect with their hearts, ED patients need to once again let love in and let love out. Therapy explores how patients refuse love and what they can do to begin to open up their hearts and to receive love from others. It also examines how they can show their love for someone else and increase the expression of their loving and heartfelt desires through

> **BOX 22.5**
>
> **GIVING AND RECEIVING GOOD GIFTS OF LOVE: CLINICAL GUIDELINES AND INTERVENTIONS**
>
> 1. Teach patients to directly and clearly ask for the gifts and support they need from God and significant others. Help them realize that teaching others what they need is an act of kindness to themselves and to their loved ones who want to love them well.
> 2. Teach patients to show love by giving consistent attentiveness to self through affirming, reassuring, and kind messages to self. Self-kindness opens the heart to receive loving kindness from God and others.
> 3. Helping patients to understand the reasons for their refusal of love can also help them open their hearts to receive the love available in their life. Constant self-judgment, self-criticism, and self-blame are roadblocks to love and need to be actively addressed in therapy.
> 4. Help patients see their own moments of giving good gifts of love to others and to recognize the good gifts of love that are coming into their lives from others, including God. Help them understand and begin to accept that good gifts received and given are real and genuine gifts of the heart.
> 5. Use experiential interventions and activities in therapy to amplify the gifts and power of love. For example, in a group setting, therapists can ask one patient to sit on a chair in the middle of the group. Ask other patients to take turns kneeling in front of the chosen patient and express how they feel about him or her, including their feelings of love for them and any other positive thoughts or feelings they have. With each expression of love, the group members are asked to hand the patient a white tissue so the love is visible. The patient gathers and holds the tissues (Richards et al., 2007). Or, ask group members to express their love and admiration and to share positive things they have experienced and noticed about each other. Ask patients to deliver their messages (their gifts) face-to-face, looking each other in the eye. Ask the patient receiving the feedback to say after each shared message, *"Thank you. I accept your gift to me."*

both word and action. The key in this process is to label love as an expression of the heart and experience love from others as receiving a gift from the heart of another. By doing so, patients can begin to increase awareness of how much love there is around them and how much love they have to share.

We seek to help patients who believe in God to understand that God's acceptance is not encumbered by the damaging, external conditions or expectations so often found in human relationships (e.g., based on performance and appearance). Godly acceptance can bring peace, comfort, and hope rather than anxiety and pressure to please others. When patients feel this confirming acceptance, they feel encouraged to become kinder and more accepting of themselves. Acceptance and love are worth the risk when patients develop honest and real connections with God and other people. By developing an inner sense of the love available to them from God and significant others, patients can more easily experience themselves as a whole person with much to offer, and not "just a body."

Compassionate service promotes spiritual mindedness. Not only does service require an understanding of the needs of another, but it also requires some willingness to let go of self-preoccupation. Service can open up the heart to others through love. Many ED patients give little of themselves to others out of fear that their gift of self is defective and unworthy of giving. They also often reject the gifts of others, feeling undeserving of the gifts.

When patients give and receive the gift of love, they also bring into their lives the gift of hope. In this way, faith, hope, and love are interconnected. If they take the step of letting love into their lives, and the step of truly giving the love which they have to give, then hope will return into their lives. Hope comes from knowing that they possess the faith and the inner qualities that can help them change their lives, and that God and others have both the desire and the capability to give them the sustaining support they need in the difficult work of recovery.

ED patients typically make themselves the exception to love: they believe that love is available for everyone but themselves. This belief is a roadblock. Self-judgment and self-contempt are also roadblocks to receiving and giving love. Understanding that their self-judgments keep them from feeling loved, helps patients to realize that it is not who they inherently are that prevents them from feeling loved, but rather, it is their inaccurate beliefs and false pursuits that have led to the restriction of love in their life.

Therapists can be loving with their patients by: (i) keeping proper physical, sexual, and emotional boundaries; (ii) modeling for their patients self-love and self-kindness; (iii) expressing belief in patients' ability to grow and progress; (iv) sharing unconditional acceptance; (v) letting patients learn in word and deed that they truly care about them; and (vi) learning to ask patients better questions and listen carefully to their answers. As patients receive affirming messages of love from their therapist, they begin to be affirming, reassuring, and kind to themselves as an expression of love to themselves. Love is a gift, whether it comes from God, oneself, or from significant others. The heart connects to love. Love is what heals the lonely, the empty, and the shameful sense of self.

Holding Up a Therapeutic Mirror That Reflects Spiritual Identity (Box 22.6)

Holding up the mirror that reflects spiritual identity encompasses all of the steps in listening and following the heart, since the heart is the literal and symbolic representation of a person's spiritual nature. Spiritual things come from the heart, and they go to the heart. The heart is a major receptacle of and a primary vehicle for spirituality.

Therapists can often see patients better that they can see themselves and they can give feedback to help patients affirm their spiritual worth and identity. "Holding up the mirror" is to say, "*Look. Here is how I see you*" or "*This is something I am noticing about you.*" As therapists love, respect, listen to, and give value to the things that patients say, they are holding up the therapeutic mirror to help patients see their worth and goodness.

During the development of an ED, patients endure a cycle which takes them through stages of attempts at self-betterment, obsessive ruminations, and addictive and compulsive behavior. Eventually they reach a point in the illness where, in their minds, they *become* the disorder. Treatment can help patients again separate from the illness so that they can have

> **BOX 22.6**
>
> **HOLDING UP A THERAPEUTIC MIRROR WHICH REFLECTS SPIRITUAL IDENTITY: CLINICAL GUIDELINES AND INTERVENTIONS**
>
> 1. When holding up the mirror for patients, point out their successes, their strengths, their good intentions and motives, successes with giving and receiving love, and their spiritual qualities and gifts.
> 2. Metaphors and stories, parables and allegories, often speak deeply to the soul. Where possible, use symbols, ceremonies, metaphors, and stories to help patients see who they are and to help them understand the depth of their goodness and their spirituality.
> 3. Ask patients to tell stories of people or heroes in their life that have spiritual qualities that they respect. Help them see the similarities between them and the hero in their story.
> 4. Help patients see a new reflection of themselves in the messages that they receive from their therapist, significant others, and a higher power that they have discounted, ignored, or missed in the past. Support them in this expanded view of themselves and encourage their self-acceptance of these new and important truths.

a more truthful perspective of their identity and a sense of power over the illness. This approach is well described in the book, *Life Without ED: How One Woman Declared Independence from her Eating Disorder and How You Can Too* (Schaefer, 2004).

The separation of the illness from the person allows patients to understand that much of their behavior is a part of an illness, and not the result of personal deficiency or flawed willpower. An important theme during treatment is, *"You are not your illness and you are not simply your behaviors, thoughts, or feelings."* As patients begin to understand the developmental versus the willful process of the illness, they begin to have understanding and even compassion for their painful journey into the ED. With a sense of hope and self-empowerment, they can begin to heal from the shame they have about the ED and take personal responsibility for their choices in the present which have power to change self-defeating patterns (Richards et al., 2007).

Shame, self-judgment, and self-contempt create a false identity. But as therapists hold up the mirror for patients and help them see love in them and in their lives, then healing can take place. They gain confidence that if they listen to their hearts, they can give and receive love even while they are not completely recovered because love is part of who they are. It is part of their identity.

Many ED patients have a difficult time seeing the good in themselves. By pointing out the good and the strength in patients, therapists can be a therapeutic mirror to the qualities of which patients are truly blind. When therapists witness patients' faithfulness, hard work, endurance, resilience, humility, congruence, and kindness, they can point these qualities out in affirming ways.

Receiving the reflected witness of their goodness through feedback, compliments, kindness, and love from others helps patients' spiritual identity grow in positive ways. When patients see who they really are, they see that they are separate from their ED. When patients live in harmony with their hearts, their spiritual identity is manifest in their countenance and in the way they live moment by moment. They feel at peace and lose the need for the ED illness.

RECOMMENDATIONS FOR RESEARCH

In our view, the American Psychiatric Association's (2006) recommendations concerning future research directions about ED are incomplete because they almost completely ignore the need for more research about religion, spirituality, and ED. We previously proposed some general research questions about religion and spirituality and ED that we hope researchers and scholars will investigate (Richards et al., 2007, pp. 210–211). Here we list several of these research questions most relevant for clinical practice:

1. In what ways do women who are recovering from ED believe that faith and spirituality are helpful in treatment and recovery?
2. How do helping professionals go about implementing spiritual perspectives and interventions in the treatment of ED?
3. What percentage of ED patients would prefer to receive treatment from programs, or from helping professionals, who incorporate spirituality into treatment?
4. Does using spiritual interventions in the treatment of ED enhance the effectiveness of treatment?
5. Does patients' faith and spirituality help prevent relapse and promote long-term recovery from ED?
6. How can spiritual interventions be integrated ethically and effectively into medical and clinical treatment of ED?

Many other research questions may be of interest concerning the topics of religion, spirituality, and ED. We hope that, during the next decade, scholarship in this neglected domain will receive more attention from scholars and researchers in social sciences and health care.

RECOMMENDATIONS FOR TRAINING

It would be ideal if all ED professionals could respond affirmatively to the following four "spiritual competency questions" (see Gonsiorek, Richards, Pargament, & McMinn, 2009, pp. 389–390):

1. Do I have the ability to create a spiritually safe and affirming therapeutic environment for my patients?
2. Do I have the ability to conduct an effective religious and spiritual assessment of my patients?
3. Do I have the ability to use or encourage religious and spiritual interventions, if indicated, in order to help patients access the resources of their faith and spirituality during treatment and recovery?

4. Do I have the ability to effectively consult and collaborate with clergy and other pastoral professionals (e.g., chaplains) when it might benefit my patients?

Much has been written about each of these competency areas in other sources (e.g., Aten & Leach, 2008; Gonsiorek, Richards, Pargament & McMinn, 2009; McMinn, Aikins & Lish, 2003; Plante, 2009; Richards & Bergin, 2000, 2005; Sperry & Shafranske, 2005). We encourage ED professionals who have not yet received adequate training in religious and spiritual aspects of diversity to seek more education and understanding in this domain. Taking courses, reading books and articles, and/or watching videos about the world religions, the psychology of religion, and spiritual approaches for counseling and psychotherapy can be valuable. Growing numbers of graduate courses and continuing education workshops about these topics are available. It is also valuable to consult with and seek supervision from professionals who have already developed expertise in spirituality and ED.

CONCLUSION

The overall spiritual goals of our approach to working with patients with ED is to help them learn to: (i) listen again to their hearts; (ii) actively engage in spiritual pathways of recovery; (iii) overcome their ED and related illnesses; (iv) reconnect with a higher power, such as God, with significant others, and with themselves in meaningful relationships; and (v) regain a sense of spiritual worth and identity which brings confidence and peace (Richards et al., 2007). We encourage therapists to be open to the power of the spirit, the power of light, the power of miracles, the power of the deepest beliefs, the power of principled living, and the power of spiritual influences in the lives of their patients. As therapists honor and attend to patients' spirituality it can promote a sense of safety and trust in the therapeutic relationship. When spiritual influences touch the lives of our patients, and when the heart begins to heal and change, that change is likely to be a major catalyst in recovery with the potential to endure throughout patients' lives.

References

American Psychological Association (2002). *Ethical principles of psychologists and code of conduct.* Washington, DC. Accessed at http://www. apa.org/ethics
American Psychiatric Association (2006). *Practice guideline for the treatment of patients with eating disorders* (3rd ed.). Arlington, VA.
Aten, J. D., & Leach, M. M. (2008). *Spirituality and the therapeutic process: A comprehensive resource from intake to Termination.* Washington, DC: American Psychological Association.
Benson, H. (1996). *Timeless healing: The power and biology of belief.* New York, NY: Scribner.
Brawer, P. A., Handal, P. J., Fabricatore, A. N., Roberts, R., & Wajda-Johnston, V. A. (2002). Training and education in religion/spirituality within American Psychological Association-accredited clinical psychology programs. *Professional Psychology: Research and Practice, 33,* 203—206.
Childre, D., & McCraty, R. (2001). Psychophysiological correlates of spiritual experience. *Biofeedback, 29*(4), 13—17.
Gallup, G. H. (2003, January 7). Public gives organized religion its lowest rating. *The Gallup Poll Tuesday Briefing.* The Gallup Organization, Princeton, NJ: pp. 1—2.
Garrett, C. J. (1996). Recovery from anorexia nervosa: A Durkheimian interpretation. *Social Science Medicine, 43*(10), 1489—1506.

Gonsiorek, J. C., Richards, P. S., Pargament, K. I., & McMinn, M. R. (2009). Ethical challenges and opportunities at the edge: Incorporating spirituality and religion into psychotherapy. *Professional Psychology: Research and Practice, 40*, 385–395.

Hall, L., & Cohn, L. (1992). *Bulimia: A guide to recovery.* Carlsbad, CA: Gurze Books.

Hsu, L. K., Crisp, A. H., & Callender, J. S. (1992). Recovery in anorexia nervosa—The patient's perspective. *International Journal of Eating Disorders, 11*, 341–350.

Hardman, R. K., Berrett, M. E., & Richards, P. S. (2003). Spirituality and ten false pursuits of eating disorders: Implications for counselors. *Counseling and Values, 48*, 67–78.

Koenig, H. G. (1998). *Handbook of religion and mental health.* San Diego, CA: Academic Press.

Koenig, H. G., McCullough, M. E., & Larson, D. B. (2001). *Handbook of religion and health.* New York, NY: Oxford University Press.

Marsden, P., Karagianni, E., & Morgan, J. F. (2007). Spirituality and clinical care in eating disorders: A qualitative study. *International Journal of Eating Disorders, 40*(1), 7–12.

McCraty, R., Atkinson, M., & Tomasino, D. (2001). *Science of the heart: Exploring the role of the heart in human performance.* HeartMath Research Center. Boulder Creek, CA: Institute of HeartMath. Publication No. 01–001

McCraty, R., Atkinson, M., & Bradley, R. T. (2004a). Electrophysiological evidence of intuition: Part 1. The surprising role of the heart. *Journal of Alternative and Complementary Medicine, 10*(1), 133–143.

McCraty, R., Atkinson, M., & Bradley, R. T. (2004b). Electrophysiological evidence of intuition: Part 2. A system-wide process. *Journal of Alternative and Complementary Medicine, 10*(2), 325–336.

McCraty, R., Bradley, R. T., & Tomasino, D. (2004–2005). The resonant heart. *Shift: At the Frontiers of Consciousness, 5*, 15–19.

Mitchell, J. E., Erlander, Revd., M., Pyle, R. L., & Fletcher, L. A. (1990). Eating disorders, religious practices and pastoral counseling. *International Journal of Eating Disorders, 9*, 589–593.

McCullough, M. E. (1999). Research on religion-accommodative counseling: Review and meta-analysis. *Journal of Counseling Psychology, 46*, 92–98.

McMinn, M. R., Aikins, D. C., & Lish, R. A. (2003). Basic and advanced competence in collaborating with clergy. *Professional Psychology: Research and Practice, 34*, 197–202.

Plante, T. G. (2009). *Spiritual practices in psychotherapy: Thirteen tools for enhancing psychological health.* Washington, DC: American Psychological Association.

Plante, T. G., & Sherman, A. C. (Eds.), (2001). *Faith and health: Psychological perspectives.* New York, NY: The Guilford Press.

Propst, L. R., Ostrom, R., Watkins, P., Dean, T., & Mashburn, D. (1992). Comparative efficacy of religious and non religious cognitive-behavioral therapy for the treatment of clinical depression in religious individuals. *Journal of Consulting and Clinical Psychology, 60*, 94–103.

Puchalski, C. M., & Larson, D. B. (1998). Developing curricula in spirituality and medicine. *Academic Medicine, 73*, 970–974.

Puchalski, C. M., Larson, D. B., & Lu, F. G. (2000). Spirituality courses in psychiatry residency programs. *Psychiatric Annals, 30*(8), 543–548.

Richards, P. S., Hardman, R. K., Frost, H. A., Berrett, M. E., Clark-Sly, J. B., & Anderson, D. K. (1997). Spiritual issues and interventions in the treatment of patients with eating disorders. *Eating Disorders: The Journal of Treatment and Prevention, 5*, 261–279.

Richards, P. S., & Bergin, A. E. (Eds.), (2000). *Handbook of psychotherapy and religious diversity.* Washington, DC: American Psychological Association.

Richards, P. S., & Bergin, A. E. (2005). *A spiritual strategy for counseling and psychotherapy* (2nd ed.). Washington, DC: American Psychological Association.

Richards, P. S., Hardman, R. K., & Berrett, M. E. (2000). *Spiritual renewal: A journal of faith and healing.* Orem, Utah: Center for Change.

Richards, P. S., Hardman, R. K., & Berrett, M. E. (2007). *Spiritual approaches in the treatment of women with eating disorders.* Washington, DC: American Psychological Association.

Richards, P. S., O'Grady, K. A., Berrett, M. E., Hardman, R. K., Bartz, J. D., et al. (2008, May 16). Exploring the role of spirituality in treatment and recovery from eating disorders: A qualitative survey study. Paper presented at the Academy of Eating Disorders International Conference on Eating Disorders, "Bridging science and practice: Prospects and challenges," Seattle, Washington.

Richards, P. S., & Susov, S. (2009). *The role of spirituality in eating disorder treatment programs: An Internet survey of contemporary clinical practice*. Manuscript in preparation. Brigham Young University, Provo, Utah.

Rogers, C. R. (1961). *On becoming a person*. Boston: Houghton Mifflin.

Rorty, M., Yager, J., & Rossotto, E. (1993). Why and how do women recover from bulimia nervosa? The subjective appraisals of forty women recovered for a year or more. *International Journal of Eating Disorders, 14*, 249–260.

Schaefer, J. (2004). *Life without ED: How one woman declared independence from her eating disorder and how you can*. New York, NY: McGraw Hill.

Smith, T. B., Bartz, J. D., & Richards, P. S. (2007). Outcomes of religious and spiritual adaptations to psychotherapy: a meta-analytic review. *Psychotherapy Research, 17*, 643–655.

Smith, F. T., Richards, P. S., & Hardman, R. K. (2003). Intrinsic religiosity and spiritual well-being as predictors of treatment outcome among women with eating disorders. *Eating Disorders: The Journal of Treatment and Prevention, 11*, 15–26.

Sperry, L., & Shafranske, E. P. (Eds.), (2005). *Spiritually oriented psychotherapy*. Washington, DC: American Psychological Association.

Worthington, E. L., Jr., & Sandage, S. J. (2001). Religion and spirituality. *Psychotherapy, 38*, 473–478.

Zinnbauer, B. J., Pargament, K. I., & Scott, A. B. (1999). The emerging meanings of religiousness and spirituality: Problems and prospects. *Journal of Personality, 67*, 889–919.

CHAPTER 23

The Case for Integrating Mindfulness in the Treatment of Eating Disorders

Kimberli McCallum

Predictable and tenacious concerns about self-regulation, dissociation, and critical self-evaluation complicate the psychotherapy and recovery process of those suffering from eating disorders (ED). Overwhelmed by shame, patients avoid paying attention to their underlying problems and emotions. Thinking they deserve to suffer, they respond to their pain with contempt and disgust and cling to beliefs that they are flawed and therefore unlovable, identifying with their body or with their illness. Without an ED, they fear they will not have an identity. Flooded with negative emotions, they find it difficult to make use of supports or process corrective information. Restrictive eating, control of body weight, purging, and compulsive exercise become the main strategies for self-regulation. Thinking becomes trapped in loops of negative self-evaluation. Self-acceptance is, in the patient's view, contingent on achievement and maintenance of an ideal body shape and weight. In reality, the desired ideal image is unsustainable and dangerous.

These predictable self states and mindsets become more the norm as the ED progresses. Traditional psychotherapy techniques lose power as the therapist is held at a distance. Thinking and behavior become increasingly rigid, and desire becomes stifled, dreaded, and forbidden. As the individual struggles with more severe symptoms, these take their toll. Decreased available energy, compromised blood flow, electrolyte abnormalities, and hormonal derangement lead to increasing biologic dis-regulation. Body cues are no longer accurate.

In order to maintain a connection with the person struggling with an ED, therapists attempt to create an environment of safety, respect, and trust. Yet, as long as the mind and body are deprived of necessary energy and water, the patient will remain dis-regulated. The limbic system, the amygdala in particular, is involved in emotional processing and memory. Activation of the amygdala is implicated in the "fight or flight" response. Anxiety disorders are believed to result in part due to anomalies in amygdala function. When anxiety and shame over-activate the amygdala, cognitive processing is inhibited. As long as the sufferer maintains a focus on control rather than on acceptance, on self-hate rather than love, the ED will thrive. To recover, nutritional balance must be restored and patients must learn skills to increase self-regulation, over-ride cognitive dysfunction, develop resources to accept discomfort associated with changes in the body, and practice accepting themselves without clinging to judgments. In short, patients must become mindful to recover.

Infusing mindfulness into psychotherapy focuses the work with this goal in mind. Mindfulness is neither easy nor is it equivalent to relaxation. In fact, most people initially find mindfulness practice quite difficult, needing both encouragement and regular practice. Patients with ED, or severe anxiety and mood disorders, benefit when a skilled clinician can integrate mindfulness into treatment. The more mindful the therapist, the more access the therapist will have to empathy and understanding of the patients' dis-regulated mind states. This chapter introduces a model for the inclusion of mindfulness in the treatment of ED.

WHAT IS MINDFULNESS?

Mindfulness practice is nothing new; in fact, it springs from ancient Buddhist philosophy. Its intent is simply to increase wellbeing. Pain and suffering are understood as inevitable, compassion the natural outcome of an openness allowing the individual to become aware of suffering. Kabat-Zinn (2003) defines mindfulness as "awareness that emerges through paying attention in a particular way: on purpose, in the present moment and non-judgmentally to the unfolding of experience moment by moment" (p. 145). The emphasis is on clearly seeing and accepting things as they are, without trying to change them. Stable states of mindfulness are sometimes referred to as being in "wise mind," when awareness of all thoughts, impulses and sensations are welcomed with equanimity. Concepts of self are experienced as fluid rather than fixed. These concepts about identity shift from notions of moral status and social identity, stories describing our past and present, self viewed as image, to ideas about personality traits, desires, and vulnerabilities.

Mindful states can be contrasted with being on "autopilot," or in other inattentive states including preoccupation with past or future concerns. A resource which optimizes self-regulation, mindfulness allows increased freedom from automatic thinking and promotes wellbeing. While it is not possible to remain in a mindful state at all times, the degree to which a person is able to return to the mindful state voluntarily should be correlated with symptom reduction and less negative emotion.

Kornfield (2008) describes several domains of mindful awareness. Attention may be focused on the body: movement, impulses, and sensations; on emotions; on desire and appetites; on thoughts, beliefs, fantasies or stories; and on the state of consciousness itself. Mindful states welcome awareness of each of these elements without judgment. The goal is neither to cling to nor avoid states, but to allow or witness; specifically to notice how they deepen, shift, and transform. The intention is to become better able to witness one's experience with compassion for the inevitable pain life brings. Mindfulness is not a way to escape feeling; instead, it is a way to get better at feeling, to live fully, and to cultivate states of wellbeing.

How does mindfulness work? Four key steps for mindful transformation can be taught with the acronym RAIN (Kornfield, 2008). The first step is *Recognition*. When we are out of balance, stuck or avoiding in our lives, we must be willing to see it. Recognition of pain, however, may come with resistance. With recognition, we let go of denial. Thus, the second step, *Acceptance*, is an active courageous step. Acceptance is not the same as complacency, but instead an attitude of openness and non-judgment. Acceptance is necessary for full investigation and exploration of states of body, thoughts, emotions, and consciousness. The third step, *Investigation*, is an exploration of the flow of thoughts, images, emotions, and sensations as they pass through experience. This step can lead to what Buddhists call "seeing the waterfall." The intensity and complexity of this flow of mind on body states and impulses can be overwhelming. Practice helps keep the focus on witnessing and not clinging to any particular state. This final step is *Non-identification*. Non-identification allows a natural flow and deepening of experience. States of compassion follow naturally from awareness of the suffering associated with painful thoughts, emotions, and judgments; states of appreciation and contentment are associated with love and connection.

Investigators have linked states of mindfulness with states of coherence in brain and heart function (Siegel, 2007). In states of coherence, the brain is able to focus attention appropriately without too much interference or noise. Mind and body are integrated, resulting in stability and flexibility of the autonomic nervous system. Blood pressure is in balance with the body's needs. Pulse is regular and appropriately responsive. Breathing is deep. Emotions, thoughts, and sensations flow through consciousness freely. In a state of wellbeing, a person is able to experience emotion, sensation, and perception fully without getting stuck or disregulated. In addition, the capacity for mindfulness is a core construct of the character trait, self-directedness. According to Cloninger (2004), self-directedness correlates with mental health and better outcomes in those who struggle with mental illnesses.

In many ways, mindfulness is a spiritual practice, building on values of acceptance and connection. Although mindfulness principles are most explicitly developed in Buddhist philosophy, mindfulness does not necessarily challenge or conflict with any particular religious teachings. Instead, anyone can become more confident and transcendent by developing a capacity to act from a wise/calm self, for this promotes states of coherence, characterized by self-acceptance, compassion, gratitude, creative impulses, curiosity, calm, or heightened awareness. By contrast, states of incoherent brain and body function are experienced as chaotic (impulsive) or rigid (compulsive) states. Overwhelmed by intense emotions, individuals engage in critical self-evaluation and judgment and are likely to feel disconnected, dissociated, distracted, stagnant, hopeless, and stuck.

In *The Mindful Brain*, Siegel (2007) describes mental health as the ability to generate stable states of coherent thought. Consciousness is energized and behavior is flexible, adaptive to the environment. These states are consciously experienced as the flow of states of wellbeing. He proposes a helpful image of wellbeing experienced as the flow of a river bounded by two banks: "One bank is that of rigidity, the other, chaos." Clinging to or avoiding thoughts, emotions, and beliefs will wash us up on the banks. When in rigid self states, we are preoccupied, self-critical, and focused on control. When in chaotic states, we are distracted, dissociated, and impulsive. Learning and practicing mindful reflection will help us to spend more time in the flow of the river, accepting the inevitability of visiting the banks.

Without mindfulness skills, change can overwhelm us. We risk getting caught in details. ED are associated with neuropsychological dysfunction of just this type. Mindfulness helps to cultivate the universal dimension, to think of the big picture without clinging to or avoiding details of feelings, thoughts, or beliefs. Mindfulness practice has lofty goals: to grow in compassion and acceptance; to increase flexibility and the capacity for self-regulation; not just to gain pleasure and avoid pain. States of wellbeing allow a deeper appreciation of and participation in pleasurable experiences and an acknowledgement that pain is inevitable. In mindfulness-based therapy, the intention is to notice and discourage ways in which we automatically add to our pain by clinging, avoidance, and identification.

MINDFULNESS AND THE BRAIN

Increased mindfulness may be a key mechanism for how psychotherapy works. Research suggests that mindfulness practice may increase the coherence of brain function, increasing positive emotion and reducing anxiety (Davidson, 2003). These changes are associated with changes in brain activation demonstrated by both EEG and fMRI. Schwartz (2002) described the effects of a mindfulness-based cognitive behavior therapy (CBT) intervention on the brain activity of patients with an obsessive-compulsive disorder (OCD). In patients with OCD, the brain regions involved in error detection, self-judgment, and processing rewards and punishments were over-active at baseline. This activity fell dramatically with mindfulness-based CBT treatment. Increased neural firing in the prefrontal cortex is linked to states of gratitude, contentment, and compassion reported by subjects practicing mindfulness meditation. Activity of this brain region quiets the limbic brain and is experienced, in the mind, as a quieting of worry and rumination and, in the body, as reduced tension and regulation of arousal.

MINDFULNESS-BASED THERAPIES

Was Freud influenced by Buddha? Perhaps, since mindfulness is nothing new to psychoanalytic or psychodynamic psychotherapy. In fact, the core psychoanalytic technique, free association, encourages compassionate, non-judgmental reflection on mind—body self states, cultivating curiosity about thoughts (concerns, ideas, wishes, and impulses), body sensations, and emotions. Epstein (2006), Kornfield (2008), and Brach (2003) have modified traditional psychoanalytic technique to focus interpretation more directly in order to enhance mindfulness.

Sensory motor therapy (Ogden, Minton & Pain, 2006) uses mindfulness-based techniques to increase awareness and acceptance of body sensations and impulses. This technique encourages an increased awareness of the body, particularly the level of arousal, changes in voice, breath, movements, and posture associated with states of wellbeing. A language for describing body states emerges as the therapist notices changes in these. Interventions teach mindfulness, focusing the client on the present experience of core organizers: thoughts, emotions, movement, sensation, and the five sense perceptions with a specific goal of enhancing self-modulation. Mobilization is associated with states of compassion, acceptance, exploration, and self-care. The therapist notices, celebrates, and builds on emerging skills as clients become more able to listen and respond to the body, gradually shifting to a state of wellbeing. These skills are necessary to reduce the dis-regulation associated with very painful memories or emotions. It follows that these techniques may be particularly useful in those who have suffered sexual trauma or who have extreme physical dis-regulation due to an ED.

Investigators have used mindfulness techniques to enhance traditional CBT by focusing on building tolerance to cues and triggers and changing the relationship to thoughts rather than changing thoughts themselves. Schwartz (2002) suggests a focused CBT-based intervention in which therapists teach patients suffering with OCD to re-label, re-attribute, refocus, and revalue troubling thoughts, images and impulses. Rituals lose power as the individual stops clinging to them as a way of neutralizing anxiety.

Dialectical behavioral therapy (DBT) and acceptance commitment therapies (ACT) are examples of mindfulness interventions embedded in psychotherapy technique. Both explicitly build on mindfulness as a core therapeutic concept. DBT (Linehan, 1993) teaches mindful meditations and encourages "mindful reflection" on emotions. Especially helpful in the treatment of borderline personality disorder, it improves self-efficacy and distress tolerance. ACT (Hayes, 1999) integrates mindfulness principles teaching acceptance, diffusion, contact with the present moment, and transcendence of self, reducing the avoidance and distress associated with anxiety disorders. Mindfulness-based stress reduction (MBSR) and mindfulness-based cognitive therapy (MBCT) have been shown to be effective in reducing symptoms in anxiety disorders (Kabat-Zinn et al., 1992) and in OCD (Schwartz 2002) and to reduce relapse rate of depression (Baer, Fischer & Huss, 2005; Segal, Williams & Teasdale, 2002; Teasdale, 1999).

Several studies integrate mindfulness in the treatment of ED, with the best evidence to date in the treatment of binge eating disorder (BED). Research (Kristeller, 2003; Kristeller & Hallett, 1999) indicates some promising effects of a behavioral treatment and mindfulness-based eating disorder training. This treatment integrates elements from mindfulness-based stress reduction MBSR and CBT, including guided eating meditation and body meditations, while Baer et al. (2005) showed some benefits with MBCT. Safer, Telch & Agras (2001) found some improvement when utilizing DBT in treating bulimia. Eifert and Forsyth (2005), studying the benefits of ACT for patients with AN, reported favorable results.

BRINGING THE BRAIN INTO TREATMENT

Exciting advances in our understanding of the brain are beginning to illuminate the biologic basis for consciousness, self-regulation, self-concept, and wellness. More importantly, understanding shifts in brain function that underlie recovery should help us to develop more effective and efficient psychotherapy strategies to reduce suffering.

For patients, knowing about the brain can demystify experience. Therapists can help the patient understand that brain function is modular with the mind constantly integrating and interpreting inputs from many interconnected circuits to make sense of experience. Left brain cortical circuits are involved in interpreting experience and generating stories to explain the meaning of the information it receives. A sense of self is experienced partly as a series of linked experiential memories, a narrative story of oneself through time, the self viewed as an image, and the links assigned between personality traits, experience, and stored memories. These cortical pathways will assign meaning even when the information is incomplete or distorted! This explains the experience of someone struggling with an ED. Sensations and sense perception may grab attention, trigger a memory, or evoke an emotional reaction. Then, out-of-balance circuits may mis-assign meaning and generate false stories to explain these shifts, attributing any disappointment, difficult emotion, or pain to issues such as weight, shape, or food.

Limbic (ventral) brain circuits are important for generating emotional response to stimuli. Cognitive (dorsal) circuits are involved in selective attention, planning, and effortful regulation of emotion. Imbalances in these circuits have been implicated in several mental illnesses including anxiety disorders and mood disorders (Phillips, Drevets & Rauch, 2003). Brain activation in these regions is also abnormal in ED. Research suggests that individuals with anorexia nervosa (AN) have trouble modulating emotional responses and over-activate brain processes concerned with planning and consequences (Kaye, Fudge & Paulus, 2009). Marsh et al. (2009) showed that the ventral circuit-based self-regulatory processes are impaired in patients with bulimia nervosa (BN). Balance of these circuits may lead to symptom reduction and the sensation of wellbeing. Therapies in the future should be tested for their ability to shift the brain towards more balanced, coherent function.

Therapists introduce patients to their brain by explaining that there are at least three major ways that the brain processes experience: thoughts (neo-cortex); emotions (limbic system); and sensations (brain stem). As patients develop an increased ability to reflect their own brain process, the therapist might reflect that traps of thoughts are natural and understandable, but can be overwhelming, especially when the patient has trauma or states of starvation. The therapist can focus the therapy by setting goals around self-awareness and self-regulation. Mindfulness allows the patient to recognize their traps of thought and behavior and to develop new ways of coping. To many, the concept of mindfulness can seem lofty, intimidating and out of reach, so clinicians should warn that it will take time and practice to develop a mindful stance.

BEGINNING THERAPY WITH AN EMPHASIS ON MINDFULNESS

Mindfulness-based values and skills can provide a focus and power in any type of therapy. The therapist begins by setting goals to increase wellbeing, freedom, and flexibility, and to improve decision-making and the quality of relationships. In general, these goals are readily accepted. On the other hand, precisely because patients are so attached to stories of unworthiness and habits of self-reproach, they may initially reject the core principles and values of mindfulness (self-compassion, non-judgment, and acceptance). For those struggling with ED, a lack of compassion for oneself is the norm. Furthermore, poor nutrition causes greater imbalance in brain and body regulation, intensifying states of psychic pain and disease, making it more difficult to achieve states of wellbeing. When patients do pay attention to their experiences, it is with the goal of controlling or eliminating distress or imperfections, creating the "perfect storm'" for suffering. Simply put, they are addicted to un-mindful strategies of self-modulation.

Beginning the practice of mindful exercises and meditation can be difficult, triggering disappointment and self-critical states. Kornfield (2008) and Epstein (2006) describe simple modifications to traditional psychoanalytic technique, with the therapist exploring the awareness, frequency, and transformation of states. I begin to infuse mindfulness directly into psychotherapy by: expressing curiosity about internal states; naming and examining complex emotions, sensations, thoughts, and impulses; and identifying parts of behavior which are associated with clinging or avoiding. Borrowing from CBT techniques, therapists can recognize and re-label maladaptive states associated with ED, naming typical experiences such as body checking, food rituals, feeling fat, overvaluing a thin body ideal, body disgust, impulses to purge, and relentless self-criticism. Some of our patients call these obsessive thoughts "ED HEAD." Therapists facilitate dis-identification by: reframing, "that's not you, it's your ED head"; reattributing, "you are having these thoughts because of ED or because you are stuck in gear;" and refocusing and suggesting, "you must be active. It's time for courage! Do a different behavior or face a fear." Practicing mindful eating, meditation, and breathing can help the individual develop internal resources. Goals are set to try to see the big picture and follow core values rather than clinging to the pursuit of thinness, avoiding discomfort and fears.

Beginning mindful reflection with patients quickly ushers in familiar states of confusion, states of hypo-arousal, judging mind, and comparing mind. Initially, the work is typically slow, painstaking, and best delivered individually, rather than in a group. With starvation increasing their dis-regulation, patients are preoccupied with fullness and lack confidence in the face of fearful states. The therapist can verbalize the distress and pain the mind causes and facilitate non-identification by gentle reminders of the temporary, partial nature of feelings, suggesting future release. At all times, clinicians should model compassion and acceptance. Practice of a focused compassion meditation with the patient, as described in Box 23.1, may demonstrate how self-acceptance will naturally emerge.

Patients with ED struggle to avoid strong emotions, lacking confidence in their ability to self-regulate. Emotions are connected with thoughts, perceptions, and body processes which become organized into "self states." Emotions code intensity, modify attention, and organize our experience. In an attempt to avoid fear, anger, and shame about their bodies and desires, clients may mobilize self-destructive impulses, such as purging or cutting. The therapist notices suffering caused by rigid identifications and clinging to the sick role. Mindfulness encourages exploring rather than avoiding such emotions. The idea that possession of the ideal body and perceived control of the body will bring happiness and interpersonal success

---

BOX 23.1

COMPASSION MEDITATION

Sit, breathing softly. Feel your body. Treasure your life.

Notice ways you guard yourself.

When you are ready, bring to mind to someone you dearly love.

Picture them and feel your natural caring for them. Be aware of their measure of suffering.

Recite: may you be held in compassion, may your pain be eased, and may you be at peace.

Now, turn compassion towards yourself.

Recite: May I be held in compassion, may my pain be eased, may I be held in peace.

From Kornfield (2008).

should also be discussed and challenged. Mindfulness values suggest, instead, that contentment will arise from self-acceptance, letting go of the judging and comparing mind.

The therapist may use mindful reflection to explore emotional states, witnessing but not clinging to a story. In truth, a stance of mindfulness will do what the ED attempts and fails, as it increases a sense of control and acceptance, brings courage and compassion, and allows one to be open to attachments with others. This approach will help the client identify chaotic (impulsive) or rigid (compulsive) self states, refocusing attention when stuck in states of disease.

States of distraction can be particularly intense when the body is malnourished. Sufferers feel uncomfortable: full, bloated, constipated, cramping, struggling with disgust associated with regurgitation. Low blood pressure and dehydration cause dizziness and weakness. Hunger, body temperature, energy, sex drive, sleep are all dis-regulated. Other predictable states of dis-ease include the judging mind (critical self-evaluation), the comparing mind, stagnant (hopeless, stuck) states, dissociated states (lack of concern for body), states of shame and disgust (body image, feeling fat), and states of contraction of consciousness. "Here and now" internal mind and body "noise" associated with the established ED will increase suffering and inevitably impact the therapeutic work if not addressed directly. To aid the re-feeding process, the therapist must help the patient embrace the goal of facing rather than avoiding discomfort and fullness. The therapist should have a clear idea about what is required for compassionate care: Weight gain? Rest? Changing eating habits? Active efforts to work toward recovery are affirmed by focusing on a mindful process such as: facing a fear, food rule, or ritual response; letting go of identification; taking steps toward increased flexibility; and accepting a healthy ideal.

SKILLS FOR WELLBEING

Wellbeing can be understood as the functional ability to direct life, deal with frustrations, marshal resources, create and care for self and others, and cope with conflict. In many senses, mindfulness is necessary to achieve wellbeing. Thus, all effective therapy techniques will increase mindfulness.

Most states of mental illness, especially psychosis, depression, and ED, are characterized by persistent negative, distorted, or intrusive *thoughts*, specifically thoughts that undermine or interrupt wellbeing. When thoughts are in conflict with each other they can lead to distress; when linked to strong negative emotions they can be overwhelming and disrupt coherence of the brain and autonomic nervous system. Mindfulness-based therapy and meditation techniques encourage a stance of being open to thoughts but to "have them" and not "buy them." The goal is to neutralize thoughts, untangling them from strong negative emotions. When thoughts are intrusive, mindfulness skills allow them to pass. In one way or another, all therapies address the problem of thoughts.

While there are many ways to cultivate mindfulness, certain meditation techniques directly aim to increase specific mindfulness skills. *Mindful breathing* (Kornfield, *The Inner Art of Meditation*, available from Sounds True Audio, Boulder, Colorado) is a meditative technique that promotes a mindfulness resource by focusing awareness on breath. In a comfortable posture, the individual practices focusing and redirecting attention to the quality and sensations associated with breathing. As attention is directed towards breath, one notices body movements, changes in rhythm and depth, and internal sensations associated with moving of air through the nose, mouth, throat, and lungs. The practice encourages release from overwhelming states. One notices thoughts, emotions, impulses, and fantasies which pass into awareness, redirecting attention back to the sensations and experiences of breath. The focus is on noticing the natural rhythm, allowing breath to change, without interference. This type of meditation has been shown to increase activity in the medial frontal cortex stabilizing the nervous system, quieting negative affect states and anxiety (Davidson, 2003). The psychotherapist may teach the practice in session and encourage practice outside of treatment or incorporate periods of mindful breathing practice into the frame of the therapy session. Mindful breathing can be taught and practiced in a group, although patients with severe ED may need individual support initially to learn how to manage intrusive thoughts.

Mini mindfulness refers to brief more frequent practice of reflection, focusing attention on breath, body sensations, states of consciousness or areas that increase states of wellbeing. This strategy may be better suited for those with severe malnutrition, or comorbid attention deficit disorder, bipolar disorder, or other conditions which disrupt sustained attention, strengthening the brain tracts which modulate and calm overwhelming states or negative emotion in these patients. The therapist may encourage the patient to try mindful breathing or eating exercises in session and suggest a practice of spending a few minutes each day focusing in a particular way, revisiting these exercises.

Body scanning is a simple technique to facilitate mindful states. Body scans start with patients in a relaxed state, often with eyes closed to reduce distractions. The individual is to "check in" with the body, noticing sensations such as tingling, flow, tightening, warmth, moving from extremities to abdomen to heartbeat and breath to head, often experimenting with movement, noticing the impact of initiated movement on the quality and intensity of sensations. The therapist and client can build on this emerging connection to the body in awareness, exploring the links to memories and emotions that might occur and noticing states of calm, changes in posture, muscle tone, or breath (Ogden et al., 2006).

Cubby holing is a technique aimed at increasing the mind's capacity to remain in a reflective mode. The individual practices identifying and naming categories of perceptions, thoughts, and mind states such as: judgments (I am fat); sadness; guilt; fears; doubts (I can't do this); impulses (wanting to exercise); planning (what should I eat?); second guessing (I shouldn't have done that); and body sensations (states of fullness, discomfort). The therapist can use this technique in session to help the individual understand that these states are transient, identifying, naming, and building on states of calm or hope as they emerge.

Oft overlooked is the skill of reflection on the mind's process itself. The therapist cultivates this by recognizing and naming states of consciousness, emotion, motivation, and engagement. Is the client in a joyful or fearful state, is consciousness expanded or contracted? Are they regretful? Loving? Mourning? Naming these states will free up exploration of associated stories, feelings, perceptions, beliefs, and intentions. Confidence and a sense of competence builds as the therapist and client together become more aware of the mind's process. Compassion flows from the awareness of the judging mind. In turn, with practice it becomes easier for the client to let go of unrelenting standards, to let go of identifications and free up the courage to change.

MINDFUL EATING

Eating is a complex behavior, integrating emotional and social needs and desires with physical hunger and self-modulation. While both a physically necessary and deeply meaningful part of life, it is a great challenge for those struggling with ED.

For all of us, eating is a strong neuromodulator. Neurotransmitters and hormones affect hunger, and blood glucose and nutrients are linked to a satiety and pleasure response. Habits, emotion, hunger, cueing, and available energy also guide eating behavior. We are vulnerable to overeating when food is supplied in excess and hoarding when we have been deprived. When we are dis-regulated, our hunger cues sometimes lead us astray. If we have developed a habit of binge eating, our impulses may not accurately guide us to stop eating.

For those who have developed an ED, eating is not guided by coherent brain function. Patients with ED may have blunted or unusual brain reward responses to food (Kaye et al., 2009). Body sensations such as fullness may become frightening as they become linked to beliefs that one is fat, leading to excessive food restriction. Meal-planning becomes difficult as sufferers lose their ability to think about the big picture and focus instead on minute details and measurements. Eating loses its flexibility, becoming characterized by rituals, avoidance, over-focus on nutrients and calorie counting. Timing of meals can become chaotic, with loss of regular feedings. Those prone to binges may eat in response to periods of overwhelming negative emotion or hunger.

As the ED progresses, social aspects of eating become more difficult. Many try to hide their rigidity and binge eating and become increasingly avoidant and dissociated during meals. Meal-related stress interferes with interpersonal communication and the sufferer may avoid eating with others.

Several have described mindfulness-based meditations and eating (Brach, 2003; Kabat-Zinn et al., 1992; Kristeller, 2003; Kristeller & Hallett, 1999). Key elements include facilitating a state of gratitude for the food and guiding a mindful eating exercise. A simple reflection about the work and energy that went into growing, gathering, hunting, or preparing the food, and the fact that we are all interconnected in a web of life, may cultivate a state of appreciation. The exercise involves focusing attention on descriptive qualities associated with the sensory experiences of food, particularly chewing, noticing changes in intensity and quality, texture, touch, and scents, and noticing body sensations associated with the exercise. For most, evaluative thoughts, fears, and impulses typically intrude. The therapist redirects attention on description of sensation, away from judgment, noticing the mind's habit of leaving sensation to get stuck in sabotaging thoughts. Although research into dosing and types of mindfulness intervention are still preliminary, there is increasing evidence that mindful eating exercises help reduce binge eating (Kristeller, 2003; Kristeller & Hallett, 1999). Most patients with ED describe either chaotic or rigid mindsets surrounding eating. Binge eating may occur in either a stereotypic or disorganized way. In AN, food and even fluids are avoided as are the body sensations and emotions that eating can evoke. Rigid rules guide eating, stifling spontaneity and snubbing desire. For patients with BN, mindful eating exercises help regulate the pace of eating, allowing satiety to emerge. Similarly, for patients with AN, therapist-guided mindful eating helps increase awareness and release of the rigid rules that guide the process of eating.

IDENTITY, PERFECTIONISM, AND MINDFULNESS

Patients with ED almost universally struggle with identity concerns. For most with serious, enduring ED, symptoms developed during a critical developmental period when an abstract and independent notion of self-identity was forming. ED patients are particularly vulnerable to the dominant cultural messages overvaluing role achievement and role status and over-focusing on performance and appearance, and are easily seduced into these unhealthy identifications. Longing to control or possess an identity, they will cling to and identify with their illness. They fear a loss of identity if they do not have an ED, or if their body ages, changes, or appears imperfect. Therapists must address the problem of identification actively with the shared goal of non-identification, broadening their view of identity, changing their relationship with their bodies, and confronting unrealistic expectations and fantasies.

Brach (2003) utilizes mindfulness principles to address the *problem of identification*. Buddhism suggests that the healthiest sense of self is one which is created moment to moment. Healthy identity is fluid, open to role transitions. One's identity is as much defined by being and witnessing, as by what one has done or experienced. It is as much defined by values and intentions as it is by roles, experiences, limitations, and the natural talents that shape life. Identity, therefore, is fluid and shifting. The therapist may notice attempts to identify with a particular role or image (i.e., the sick role, the thin one). Together, the therapist and patient can explore ways in which unhealthy identity comes from clinging, and from attempts to control or possess (see Box 23.2). As the therapist and patient investigate questions of identity, they will become aware of how easy it is to cling to and identify with roles, experiences, thoughts, and appearance. Consciousness brings the understanding that identity is fluid. For example, roles change over time. One moves from child to adult to parent. Jobs change, experiences shift. Ambitions and desires change. Without fluid identity, we become stuck, lose resilience, lose our ability to self-regulate in times of change. The ability to release identifications enhances wellbeing.

BOX 23.2

EXPLORING IDENTITY EXERCISE

Ask yourself: Who are you (we) really?

Are you the same as the stories you tell about yourself?

Are you defined by your appearance? The size, shape or weight of your body?

Are you the same as your self-image?

Are you the same as your occupation? Your highest degree of education?

Are you defined by your role in your family?

Are you the same as your thoughts? Your fears?

Are you defined by the traumas you have experienced?

Adapted from Brach (2003).

Perfectionistic personality traits are also associated with an increased risk of developing depression and ED. Most sufferers struggle with unrelenting standards driving the pursuit of unattainable ideals. In short, perfectionists do not accept the fact that we are all flawed. When we experience ourselves as flawed, without acceptance and compassion, we can become driven to avoid and correct our perceived flaws, stuck in anxiety about imperfection. Compulsive comparing occurs when we cling to these perfectionistic ideals and beliefs of deficiency. Compulsive body checking is a sign of clinging to ideals about a perfect body size, shape, and weight and to the stories that our worth and beauty are linked to achieving a particular body weight. Brach (2003) describes these states as the *trance of unworthiness*. She proposes that the comparing mind becomes stuck in a fruitless attempt to avoid pain of feeling unworthy. Focused mindfulness practice which cultivates acceptance and non-judgment, will allow the comparing mind and associated states of shame to pass.

RADICAL ACCEPTANCE

Brach's (2003) mindfulness-based approach to psychodynamic psychotherapy and Linehan's (1993) DBT treatment both incorporate the principle of radical acceptance. This approach encourages acceptance of the inevitability of human flaws, de-emphasizing attempts to control that are driven by judgment and negative self-evaluation. Focusing on controlling objects of our attachment can only lead to dissatisfaction, because our true nature does not allow for permanence: our bodies change constantly as do our relationships. The anxiety associated with life changes can disrupt our wellbeing, shifting our intentions towards control. Radical acceptance (Brach, 2003) is not passive; it means actively accepting what the mind seeks to reject, and it means accepting our vulnerability to become stuck in judging, comparing mind states.

To disrupt their clinging to behaviors that maintain an ED, patients must choose activities that enhance balance, acceptance, and connection with wise/calm mind states over the pursuit of thinness, control, and other states which increase suffering. The therapist should guide the patient to evaluate the degree of influence of the comparing/judging mind, especially in relationship to their judgments of their own body and appearance. The patient should be encouraged to explore rather than avoid self states and beliefs associated with pain and discomfort. These processes allow distress tolerance and compassion to grow. As skills develop, the patient becomes more able to allow the comparing mind to pass, accept discomfort and pain, and wellbeing is enhanced. Restriction, purging, and other efforts to control and punish the body become unnecessary as attention shifts to other, more effective strategies of self-regulation.

In psychological health, we appreciate our bodies and the experiences our bodies allow us. This body is ours for life. We inhabit it, are influenced by it, but not identical to it. We live as its steward. We experience birth, explore the world, develop skills, and touch and explore others through our body. When we overvalue our body image, the body is viewed as an object we possess. Our intention shifts away from acceptance towards control. In those suffering from ED, the body is manipulated, measured, and hated for its flaws. Sensations are a reminder of the body's independence; they are dreaded and avoided as the mind and body become increasingly dissociated.

Treatment must help the patient to recognize that clinging to a body ideal only increases suffering and to shift to a mindset of stewardship. Mindfulness practice encourages gratitude, acceptance of flaws and aging, and the cultivation of compassion. Instead of control, energy can be refocused on caring for those you love, caring for the physical body you inhabit, and caring for the environment. States of self-criticism, shame, and self-hatred are signs of the judging mind. Compassion follows naturally when the therapist is able to shift awareness to the suffering associated with these states. This self-compassion is necessary for the mind to release the stories that sustain judgmental mindset.

THE ROLE OF THE MINDFULNESS PRACTICE OF THE THERAPIST

To integrate mindfulness into the therapy, the therapist must have a mindfulness practice, values, and skills. Especially when patients are very ill, and in a state of disconnection, the therapist's own mindfulness is a powerful tool for maintaining efficacy and alliance. Clinicians must be aware of their own potential for disconnection, be willing to acknowledge mistakes, and initiate repair.

Therapy builds on a capacity for non-judgment, focused attention, and active listening. Signs of loss of therapeutic coherence occur when the therapist is in un-mindful states. Therapists may feel states of judgment, have trouble listening or focusing, or may notice a loss of compassion, boundaries, or frame. They may experience confusion, a loss of sensitivity or tact, or become rigid or scripted in their responses. The therapist's mindfulness will only enhance the therapeutic process.

Working with Transference

Cultivating mindfulness in psychotherapy activates transference. Mindful attention to patients' beliefs, expectations, fears, and desires about the clinician's responses is attention to the transference. Transference is, in turn, a window into the patients' attachment patterns.

When patients are securely attached, the emotionally sensitive responses of their early attachment figures serve to facilitate self-regulation. Secure attachment is associated with healthy reflective function or the caregiver's awareness and acceptance in reaction to the child's communications about mind, body, and emotional states. Securely attached individuals show similar patterns of brain activation, as seen with mindful meditation.

Attachment patterns can be most obvious during inevitable periods of disconnection in therapy. Shame may be associated with early experiences and desires that were responded to without attunement. Insecure attachments occur when early caregiver responses were mis-attuned, being alarming, over-stimulating, rigid, shaming, dismissive, or chaotic. Insecure attachments lead to chronic states of dis-regulation, and a lack of coherence of brain activation. The therapist can use mindfulness strategies to orient the patient to examine their mind—body states. Body scanning and mindful breathing help to repair the therapeutic alliance when states of hypo- or hyper-arousal are overwhelming. Often appearing as avoidance or a lack of communication, negative transference states in patients with ED are common. The relationship with the therapist may be controlled and dreaded like the body and its sensations: restricted, purged, or negated like food. Sometimes, monitoring the affect of the therapist becomes a central interfering concern, for patients' fear that the therapist will judge, reject, or dismiss them. In this state, the patient is shifting away from wise mind into clinging and controlling states. The therapist may point out this shift, offering that, although the patient's attempts to control the therapist's response may be an attempt to avoid painful experience, these efforts also guard against spontaneous states of safety and wellbeing that can come from working closely together. In this way, the therapist urges the patient to use mindful exploration to allow awareness of emotional and consciousness states, decreasing dissociation and increasing interpersonal safety and comfort.

Mindfulness and Countertransference

Countertransference may occur in response to the patient's transference or in reaction to the patient's dis-regulated mind and body/self states. In a state of countertransference, the therapist becomes temporarily deskilled. These states can signal that something important has happened. Countertransference is experienced as a loss of mindfulness, a disconnect or break in role with the patient. Although countertransference is often interfering, awareness and subsequent repair can facilitate the therapeutic work. It may be precisely because patients with ED are prone to be in un-mindful internal and interpersonal states that strong countertransference reactions are so common. Disconnection and discouragement are frequent companions of the work. In an effort to quiet self-critical mind-states, dis-regulated body states and intense negative emotions, patients typically restrict communication, identify with their illness, resist change, cling to unrealistic ideals, and bargain to change expectations. These therapeutic obstacles, along with the chronicity of symptoms, lead many therapists to lose mindfulness, as they are swept away in states of frustration, hopelessness, shame, and even loneliness.

Thus, countertransference can be seen as a temporary loss of wise mind. With strong currents, therapists and other team members may get stuck on the banks of rigidity and chaos. To navigate back to a state of wellbeing, therapists must recognize their own state of dis-regulation. Mindful awareness of attempts to control the patient, will help the therapist navigate away from power struggles, imposing rigid rules, and punitive contracts. When stuck in chaotic states, therapists experience a loss of structure or direction. They may feel bored or confused, respond inconsistently, or experience a loss of boundaries. Mindfulness can guide the therapist back to compassion and curiosity, helping them to re-establish shared expectations and boundaries.

Recognizing the countertransference helps therapists to return to a state of mindfulness. Because strong negative affective experiences will typically trigger countertransference, the therapist should pause, reflect, identify, and explore their own negative affective responses. Shame may be experienced as helplessness, hopelessness, incompetence, or ineffectiveness while anger may be associated with beliefs of being compromised, devalued, or rejected. Sadness or loneliness may be linked to a belief that one is unable to have an impact, and confusion may be related to flooding of emotion or distress due to the belief that the patient is in danger. After recognizing, accepting, and investigating one's own response, the state can be understood in context. Returned to mindfulness, therapists are better able to identify the patient's triggering behaviors and transference. The source of the therapist's own discomfort, clinging, avoiding, identification, or contracted consciousness can be clarified. After these steps, therapists should be better able to reconnect with the patient.

Patients with severe ED symptoms or complex comorbidities often work with an interdisciplinary treatment team that might include a psychiatrist, primary care physician, dietitian, and other specialists. Without a mindful-based frame for intervening, team members may be prone to cling to painful beliefs (such as: "she is getting away with something, she doesn't care, she is disrespecting me, she is not appropriate, shouldn't be here, why should I bother ... she doesn't listen anyway"). Staff may struggle with letting go of their own shame and blame concerns (such as: "I will be held accountable, I am letting the community down, and I am ineffective"). Psychotherapists can be an important resource to other treatment team members, helping make sense of countertransference reactions and re-focusing the team on the patient's experience and wellness goals. Through this process, therapists may model the practice of orienting the work by using mindfulness. Box 23.3 describes some questions to guide this procedure.

Reconnection and repair or development of new interpersonal relationships is an important goal for many patients. Even so, sometimes therapists miss the opportunity for change by accepting or "buying into" negative stories about family relationships without mindful exploration. This can lead to inappropriate exclusion, not exploring the implications of leaving family members out, a failure to negotiate boundaries, or acceptance of overly rigid boundaries with family members. Thus, failure to make available or to provide education or resources to families when appropriate may be a sign of unrecognized countertransference. Conversely, mindful exploration of rigid family stories may allow the patient to change them, allowing them to release abusive ties, to revisit, reshape and repair relationships that had been abandoned, and to increase support resources.

The working through of transference and countertransference can be greatly enhanced by the therapist's mindful awareness. Many ED patients struggle with both un-mindful states related to insecure attachment relationships and to un-mindful states related to effects of starvation of the body and brain. Negative emotion, boredom, loneliness, rigidity and boundary shifts may signal an unmindful state in the therapist. The therapist can follow the steps of recognition, acceptance, investigation, and non-identification to navigate back to wise

---

BOX 23.3

QUESTIONS TO RE-ORIENT TEAM INTERVENTIONS WITH MINDFULNESS

Ask:

Is this action/intervention compassionate and non-judging?

Does my intervention direct the patient towards recovery, move towards understanding and tolerance rather than avoidance of conflict or pain?

Have I helped the patient see the intervention from a big picture view?

If I am setting limits, do they ensure safety and move the patient towards recovery?

mind. Together, the therapist and patient can then explore negative emotion, mind states dominated by judging and comparing thoughts, clinging to the pursuit of the thin ideal, fantasies of control, and identifications with the illness.

SUMMARY AND FUTURE PATHS

Mindfulness skills enhance wellbeing. Mental health clinicians are demonstrating increased interest and experimentation with using these principles to enhance and focus psychotherapy techniques. Although individuals suffering from ED usually lack the capacity to cultivate mindful states, this capacity or skill can be learned with practice. Therapists may incorporate mindful reflection, mindful breathing, mindful eating, or compassion meditation into the frame of therapy. The mindfulness practice of the therapist will enhance therapeutic alliance by mitigating the disregulation in therapy caused by countertransference. Integrating mindfulness into psychotherapy encourages the patient to accept, rather than resist, their own suffering with compassion and non-judgment. With increasing practice, patients will focus attention, be better able to benefit from other aspects of treatment, and restore their health.

Preliminary research shows promise for mindfulness-based psychotherapies in the treatment of ED as well as common comorbid conditions such as anxiety disorders and OCD. Nonetheless, many questions remain about the best methods for integrating mindfulness. Are children and teenagers with ED responsive to mindfulness training? Underweight patients are known to have significant cognitive changes and distortions. Can mindfulness-based treatment strategies help them to override automatic thoughts and misperceptions? How should mindfulness interventions be dosed? At what point in the course of treatment or recovery can they be best utilized? What are the best forms of skill introduction? How enduring are the benefits? Further study of the benefits of mindfulness practice as a therapy enhancing tool will guide clinicians to focus their work and help patients to develop better strategies for self-regulation, let go of unnecessary clinging, actively change their behavior, and move toward recovery.

References

Baer, R., Fischer, S., & Huss, D. (2005). Mindfulness and acceptance in the treatment of disordered eating. *Journal of Rational—Emotive and Cognitive Behavioral Therapy, 23*(4), 315–336.
Brach, T. (2003). *Radical acceptance.* New York: Bantam.
Cloninger, C. R. (2004). *Feeling good: The science of well-being.* New York, NY: Oxford University Press.
Davidson, R. J. (2003). Alterations in brain and immune function produced by mindfulness meditation. *Psychosomatic Medicine, 65,* 564–570.
Eifert, G. H. & Forsyth, J. (2005). *Acceptance and commitment therapy for anxiety disorders.* New Harbinger Publications.
Epstein, M. (2006). *Open to desire.* New York, NY: Gotham.
Hayes, S. C. (1999). *Acceptance and commitment therapy.* New York, NY: Guildford Press.
Kabat-Zinn, J., Massion, A. O., Kristeller, J., Peterson, L. G., Fletcher, K. E., & Pbert, L. (1992). Effectiveness of a meditation-based stress reduction program in the treatment of anxiety disorders. *American Journal of Psychiatry, 149,* 936–943.
Kabat-Zinn, J. (2003). Mindfulness-based interventions in context: Past, present, and future. *Clinical Psychology: Science and Practice, 10*(2), 144–156.

Kaye, W. H., Fudge, J. L., & Paulus, M. (2009). New insights into symptoms and neurocircuit function of anorexia. *Nature Reviews/Neuroscience, 10*, 573–584.

Kornfield, J. (2008). *The wise heart: A guide to the universal teachings of Buddhist philosophy.* New York, NY: Bantam.

Kristeller, J. (2003). Mindfulness, wisdom and eating: Applying a multi-domain model of meditation effects. *Journal of Constructivism in the Human Sciences, 8*(2), 107–118.

Kristeller, J., & Hallett, B. (1999). An exploratory study of a meditation based intervention for binge eating disorder. *Journal of Health Psychology, 4*(3), 357–363.

Linehan, M. (1993). *Cognitive behavioral treatment of borderline personality disorder.* New York, NY: Guilford Press.

Marsh, R., Steinglass, J. E., Gerber, A. J., Graziano-O'Leary, K., Wang, Z., Murphy, D., ... Peterson, B. S. (2009). Deficient activity in the neural systems that mediate self regulatory control in bulimia nervosa. *Archives of General Psychiatry, 66*(1), 51–53.

Phillips, M., Drevets, W., & Rauch, S. L. (2003). Neurobiology of emotion perception II: Implications for major psychiatric disorders. *Biologic Psychiatry, 54*, 515–528.

Ogden, P., Minton, K., & Pain, C. (2006). *Trauma and the body: A sensori-motor approach to psychotherapy.* New York, NY: Norton.

Safer, D. L., Telch, F., & Agras, W. S. (2001). Dialectical behavioral therapy for bulimia nervosa. *American Journal of Psychiatry, 158*, 632–634.

Schwartz, J. (2002). *Mind brain, neuro-plasticity and the power of mental force.* Pennsylvania, PA: Harper Collins.

Segal, Z., Williams, J., & Teasdale, J. (2002). *Mindfulness based cognitive therapy for depression.* New York, NY: Guilford Press.

Siegel, D. J. (2007). *The mindful brain: Reflection and attunement in the cultivation of well-being.* New York, NY: Norton.

Teasdale, J. D. (1999). Metacognition, mindfulness and the modification of mood disorders. *Clinical Psychology and Psychotherapy, 6*, 146–155.

CHAPTER 24

The Use of Holistic Methods to Integrate the Shattered Self

Adrienne Ressler, Susan Kleinman, and Elisa Mott

> *Humpty Dumpty sat on a wall*
> *Humpty Dumpty had a great fall*
> *All the kings' horses*
> *And all the kings' men*
> *Couldn't put Humpty together again* **(Opie & Opie, 1997)**.

The therapeutic efficacy of body/mind and experiential therapies has only recently been recognized despite their long-term applications. Indeed, holistic interventions actually appear as early as recorded time. To wit, Plato's philosophical stance that "the cure of the part should not be attempted without treatment of the whole" (Brennan, 1991, p. 17). Evidence-based research on holistic approaches is limited, compared to substantial anecdotal evidence cited by practitioners in the field (Serlin, 2007). The impetus to blend the "art and science" of these healing methods is nonetheless gaining professional regard. In his preface to *The Art and Science of Dance/Movement Therapy*, Barenblit (2009) recognizes that mental health today is "developing as a complex combination of theories and techniques and a host of multidisciplinary experiences…[promoting] the implementation of different types of expertise that complement each other strategically to enhance prevention, treatment, and rehabilitation in the mental health field" (p. xi).

This chapter will address the art/science gaps of holistic practices as they relate to the treatment of eating disorders (ED). Readers will be invited to take intellectual and emotional risks—risks involving a paradigm shift from conventional cognitive theory about ED treatment to one that is based on holistic approaches incorporating the inherent creative power of the right hemisphere of the brain (see also Lapides, Chapter 3, and McCallum, Chapter 23, regarding updates on neuroscience and mindfulness applications in ED treatment). As Lewis, Amini & Lannon (2000) so eloquently state, "When a person starts therapy, he isn't beginning a pale conversation; he is stepping into a somatic state of relatedness" (p. 168).

Furthermore, the following reciprocal relationships will be examined: the client's body/mind; the client—clinician bond; and the client's manner of dealing with food and the world. Clinicians' own body/mind relationships and their emotional and physiological responses in relationship to others also figure into the therapeutic frame, as they are change agents and catalysts for the client's transformation. Sourcing their body/mind faculties requires a willingness to view clients and the process of treatment through a different lens. A somatically oriented perspective is based on awareness of, and attention to, the client's breathing, body, posture, energy level, use of space, non-verbal communication, voice, speech patterns, language, and capacity for connection. Integrating this wealth of therapeutic data and effectively translating it into healing interventions is the objective of holistic practices.

EATING DISORDERS AND THE SHATTERED SELF

The shattered "self" of ED clients is evident in every aspect of their lives, from their need to maintain a false self to their disconnection from the experience of living in their body. Clients with ED detach from their feelings, keeping their physical movements under control so as to not reveal themselves in any way. They literally give form to their feelings by binging and purging, restricting, or over-exercising, rather than by turning to their inner sensations and experiences for guidance. As a result, their external behaviors often do not match-up with their internal states. The words of T.S. Eliot (1922) resonate with this conundrum:

> I can connect
> Nothing with nothing

Reclaiming the intact self requires that clients transform their shattered self into a state of wholeness and connectedness. Integration is a slow process for the client, as shifting away from what is known or familiar triggers an inordinately high level of anxiety. As they re-establish their relationship with their own self, clients often fear being flooded with the feelings of shame and pain that initially led them to abandon their bodies. But "to be without feeling is to exist in a vacuum, cold and lifeless" (Lowen, 1967, p. 231). When the body's pain is replaced by pleasure, and its despair by positive feelings, clients become more accepting of their body and themselves. Re-introducing a sense of wholeness is necessary in order to access the specific elements required for growth and recovery.

As their unique story unfolds in the treatment process, the client's emerging awareness of all aspects of themselves is of vital importance. Reclaiming a forsaken body requires therapeutic support. Despite the fact that clients are obsessed with their body, they have split off from it and all stored experiences. However, outer movements reflect disavowed internal experiences; how the client walks, breathes, and gestures all say something about who they are and how they live their life. According to the great dancer Martha Graham (1991), "Movement never lies. It is a barometer telling the state of the soul's weather to all who can read it" (p. 4). Therefore, it is imperative that clinicians have somatic attunement skills to detect if sunny skies or thunderstorms are on the client's horizon.

EXPANDING CLINICIAN RESOURCES

Holistic clinicians are challenged to expand their resources by tapping into their somatic as well as their intellectual senses. This involves listening beyond patients' spoken words to hear and understand unexpressed and bodily stored experiences. Integrating and reflecting these disconnected experiences back to clients empowers them to communicate the unspeakable, express the unknown, and recognize the truths that lie beneath the surface.

In the novel, *All the Flowers Are Dying* (Block, 2005), Elaine Scudder, an antique dealer, is called by the NYPD to identify the body of her friend who has been killed. Elaine identifies the murder weapon, an ornately carved dagger, as a piece from a collection which she had sold the previous day to a customer. Dissatisfied with a generic computer-generated image of the suspect, Elaine seeks out the help of a talented artist who possesses the ability to bond with his subjects. He is able to channel their perceptions and emotions into black-and-white reality, allowing his drawing hand to create a vision of what he experiences in the somatic transference process. Upon seeing the new drawing, Elaine says, *"The affect is different. This sketch looks just like him....You know how I can tell? Because I can't stand to look at it; I get sick to my stomach"* (p. 216). The artist conveyed not just what Elaine had seen in the suspect's face, but how she felt about it once she knew what he had done. Sensing Elaine's feelings in himself, the artist was able to express them in his drawing to make the picture "come to life."

This potent exchange between Elaine and the artist precisely parallels what can happen in the therapeutic alliance when both clinician and client pay attention to their "felt-sense," drawing on the body's deeply stored information. This felt-sense is an internal aura encompassing and communicating everything felt and known about a given subject at a given time. With a sudden, new understanding of a previously unclear feeling, comes a bodily change and sense of release—a definite, physical feeling of something shifting or moving within, a tight place loosening (Gendlin, 1981). Like the artist in the vignette above, clinicians have the capacity to capture and make known the essence of the client's "unknowing." When clients are truly available and open, clinicians can help them discover and embrace their inner truths.

UNDERSTANDING THE GESTALT PERSPECTIVE

The term *gestalt* means a configuration, pattern, or organized field having specific properties that cannot be derived from the summation of its component parts (Archer & McCarthy, 2007). As noted, ED clients characteristically present with a lack of wholeness and integration due to their detachment from their internal bodily awareness and their external perceptual experience. They have great difficulty identifying and meaningfully describing their feelings because they are so far removed from their body's sensations and signals. To guide themselves through life, they develop and rely on a "manual" of warped rules and regulations based on their ED distortions.

The *Gestalt Cycle of Awareness and Contact* can serve as a healthier guide for clients as it is based on reality rather than distortion (Gestalt Center of Gainesville, 2009). This cycle helps clients identify what is most important (*figure*) and what is least important (*ground*), thus allowing them to recognize a true want or need, rather than a distorted one. For example, in the distorted client frame, thinness is often the most important (figure) and all efforts are aimed in that direction. Clients struggling with ED often remain stuck in this part of the cycle as the direction they are traveling will not lead them to genuine satisfaction. They have been attempting to fill up inner emotional voids with food, substances, or unhealthy relationships. In order to experience the full gestalt and move through the cycle, they must identify their true wants/needs and subsequently experience receptiveness, satisfaction, and integration. Debbie, a client with BED, was able to identify that her feelings of shame which she had previously attributed to being "fat," actually represented her inability to gain her mother's approval. Recognizing it was her mother's love, not thinness, that she desired, she was able to experience the sadness that was buried beneath the shame. Having made contact with her more genuine feelings and needs, Debbie was willing to engage in treatment and re-direct her goals based on an accurate, reality-based, identification of figure, which was, in essence, a need for unconditional love.

Three important concepts are integral to the process of gestalt therapy: *contact*, *process*, and *experimentation*. Contact refers to the therapeutic relationship and the interaction between clinician and client. Process is the client's awareness of *how* he or she does what he or she does. Experiments are opportunities to take therapeutic risks by behaving, thinking, and feeling differently within the safety of the clinical relationship. Traditional gestalt experiments include an empty chair, giving directives, staying with the feeling, playing projection, and having clients talk to the various parts of themselves (Gestalt Center of Gainesville, 2009). Specific creative experiments designed to assist ED clients in embracing wholeness include principles from yoga, dance/movement therapy (DMT), bioenergetic analysis, focusing, and other body/mind methods.

The Holistic Lens: Paradigm Shift to Connection

The client's most important relationship (figure) is with their ED; everything and everyone else is relegated to ground. No matter how much they attempt to fill up and surround themselves with food, activity, accomplishments, perfection, or rituals, they remain empty, alone, and insatiable—stuck in their wants and needs. The high is never high enough, the scale is never low enough and the mirror image is never good enough. Contrary to the client's perceptions, food is only one part of the ED whole. The whole is the gestalt of their environment and how they navigate it. How the client deals with food is mirrored in how they deal with people and the world.

Anorexia Nervosa (AN)

The thinking of the anorexic is rigid and controlled, physically reflected in the "holding together" of his or her body. Lacking aliveness and responsiveness, their "body becomes an instrument of the will" whose musculature is in a continual state of contraction, a literal defense against falling apart (Lowen, 1967, p. 41). Mind and body are frozen in place. Insistent in a refusal of food, experiences, pleasure, and relationships, the anorexic creates a barricade that prohibits entry or any genuine interpersonal contact. "I need nothing" becomes a stated position in the world. The lines from Paul Simon's song "*I am a Rock*" hauntingly reflect the world of the anorexic:

> I've built walls, A fortress deep and mighty, That none may penetrate.
> I have no need of friendship; friendship causes pain. It's laughter and it's loving I disdain.
> I am a rock, I am an island.
> Don't talk of love, Well I've heard the words before; It's sleeping in my memory.
> I won't disturb the slumber of feelings that have died.
> If I never loved I never would have cried…
> I am shielded in my armor, Hiding in my room, safe within my womb.
> I touch no one and no one touches me.
> I am a rock, I am an island.
> And a rock feels no pain; And an island never cries. (*Simon and Garfunkle's Greatest Hits* (CD) 1965, New York, CBSN).

Clients with AN attempt to control feelings and sensations by restricting movement patterns, allowing little or no flow between body parts. The flow of emotional and sensory experiences is further decreased through their shallow breathing (Rice, Hardenbergh & Hornyak, 1989). Addressing the anorexic's rigid stance literally helps her "unbend," becoming more open and flexible in body and mind. As will be illustrated, yoga and other body-oriented methods can facilitate the process of unlocking the anorexic's affective and somatic states.

Bulimia Nervosa

The client suffering from bulimia nervosa (BN) presents with impulsivity, behavioral undercontrol, and containment issues (Ressler, 2009). Like the AN client, the BN client denies her neediness but senses a "lack" that she is unable to pinpoint (Dana, 1994). The ambivalent approach to her needs, food, and relationships manifests in a "yes, but…" position, indiscriminately taking things in and throwing them back out when they become poisonous. She attempts to keep her boundaries tightly closed in order to appear acceptable and to conceal her perceived undesirable self. However, her boundaries are porous and what she keeps hidden eventually resurfaces, revealing her shaming secrets.

Unlike anorexics, whose bodies and minds are so rigid, those of the bulimic are dynamic and malleable. The bulimic merges or blends—with both people and objects. In a chameleon-like manner, she takes on others' characteristics and opinions as a way to be included or to take on a "like" identity since hers is so malleable. Her presentation may be accompanied with a false smile, animated manner, or over-exaggerated use of gestures as a way of blending in and imparting an image of "normalcy."

The body can be referred to as a "shape shifter" as it can conform to the contours of furniture or objects as if unable to support itself or "stand on its own two feet" (Stark, Aronow & McGeehan, 1989). For example, years ago when pillow furniture was popular, diagnostic distinctions in shape shifting were noteworthy. In therapy groups, bulimic clients loved to be contained by the big, soft pillows that molded to their bodies, whereas anorexic clients sought out straight back chairs.

Binge Eating Disorder (BED)

Longing and yearning make up the emotional environment of binge eaters reflecting desires whose fulfillments are always just beyond their grasp. Internally the binge eater is hollow and desperate, searching to fill the void through something outside of themselves. Binge eating is born out of the frustration of trying to satisfy and protect the diminished self. Although clients may appear to be connected and involved, it is often only on a superficial level. Paradoxically, while the binge eater does not feel worthy enough to choose what they really want, they maintain unrealistic expectations that all of their wishes should be met (Selby, 1994).

The boundaries of the binge eater tend to be wide open in an attempt to take everything in. "It is in the way she says 'yes' to all food even when she is not hungry and even to food she does not particularly want or like. This also holds true for her with people and experiences" (Dana, 1994, p. 59). Teaching BED clients to connect to their body, and to experience "real" sensations and feelings, is a first step on the road to "full" engagement with food and life. Repeated, authentic involvement with their body may reduce desperation and entitlement, helping the client to learn how to satisfy their true wants and needs.

Finally, in the physical dimension, binge eaters tend to restrict the outer space they use in an attempt to feel internally safe. They appear to prefer gestural movement to more full-bodied postural expression, signaling the possibility that shame and fear underlie their self-presentation (Mennuti, Billock-Tropea & Feibish, 2003).

Eating Disorder Not Otherwise Specified (EDNOS)

The most common ED diagnosis, EDNOS, encompasses a broad spectrum of symptoms, which fail to meet the full criteria of the other EDs. Despite their subsyndromal status, EDNOS clients evidence similar levels of psychopathology. Research suggests that "EDNOS is mainly composed of individuals with an ED diagnosis transitioning to no diagnosis [i.e., in remission] or from no diagnosis to an eating disorder diagnosis" (Agras, Crow, Mitchell, Halmi & Bryson, 2010, p. 569). Therefore, clinicians must not be rigid in utilizing interventions based only on a narrow diagnostic perspective. Instead, a broader, transdiagnostic approach will be most effective for holistic healing practices.

YOGA AS A PATHWAY TO WHOLENESS

Incorporating the physical practice and philosophy of yoga into ED treatment provides a gestalt experiment in which clients can improve and integrate their awareness of what they are doing, feeling, needing, and experiencing. This awareness often can lead to ameliorating ED symptoms. In a study of 284 AN patients, those who participated in a daily program of yoga and structured exercise gained 40% more weight compared to controls who were on exercise restriction (Calogero & Pedrotty, 2004). The word yoga literally means "yoke," or the union of body, mind, and spirit. Because a great majority of ED clients are cut off from their bodies, they are living in a world of images removed from their visceral signals and sensations. The inner stillness provided by yoga allows clients to become more aware of emotions and feelings, brings clarity to deeper wants and needs, and makes the completion of the cycle of awareness and contact more attainable. For further information on yogic traditions and postures, see *Yoga, Mind, Body & Spirit: A Return to Wholeness* (Farhi, 2000).

Yoga Postures and the Seven Chakras

Awareness and comfort in the body are encouraged while engaging in a yoga practice. *Asana* (posture) translates to "comfortable seat." One of the first yoga *sutras* (written laws about yoga) states that posture needs to be "steady and comfortable." These definitions are important because they remind us that yoga is not about perfection or outcome but rather acceptance and process. Emphasizing the loving and embracing nature of yoga is essential to ED clients who often strive to achieve the perfect posture.

Each yoga posture works to balance and align the body's seven *chakras* (energy centers). When the chakras are aligned and balanced, so is our emotional and physical state. For example, just as the pineal and pituitary glands control the hormones of the body, the 7th chakra or crown chakra regulates the chakra system (Lilly & Lilly, 2004). The endocrine glands, like the chakra system, maintain balance in the body's function, energy levels, emotional states, and reactions to external and internal stimuli.

Each chakra corresponds with an area of the body, an affirmation, and a color. The first chakra begins at the base of the spine, the body's foundation, representing the seat of safety and survival. The affirmation of the first chakra, "I exist," challenges the underlying struggle of the ED client who feels fearful living in their world which they experience as dangerous and unsafe. For example, neglected by her family, Betty abandoned her "true home," her body, in order to seek shelter, security, and predictability in the rituals and routines of an ED. Utilizing yoga practices which explore the first chakra, allowed her to meet her needs through healthy rather than disordered behaviors. Yogic breathing, which supports the inherent life force of "taking in and letting go," is also associated with the first chakra and the process of elimination.

TABLE 24.1 Chakra Grid

Chakra Number	Chakra Name	Body Location	Yoga Posture	Life Affirmation	Balancing Color
1st	Muladhara	Base of Spine	Mountain Pose	"I exist"	Red
2nd	Svadhisthana	Hips, Lower Back	Goddess Squat	"I want"	Orange
3rd	Manipura	Solar Plexus	Warrior Pose	"I will"	Yellow
4th	Anahata	Heart	Standing Backbend	"I care"	Green
5th	Vishuddha	Throat	Cobra	"I express"	Blue
6th	Ajna	Third Eye	Child's Pose	"I see"	Purple
7th	Sahasrara	Crown of Head	Meditation	"I am"	White

Finally, color has been known to have a significant positive effect on the nervous system (Lilly & Lilly, 2004). Symbolizing vitality and energy, red is the color of the first chakra. Clients wishing to balance the first chakra may find it helpful to incorporate red into their wardrobe or artwork, or tap into the power of their minds by evoking colorful images such as fire or a beautiful sunset.

The remaining six chakras are located along the spine, ending at the crown of the head. These energy centers are activated by the mind (Lilly & Lilly, 2004). Thus, imagery and mental intention are vital to the process of exploring and balancing the energy of the chakras. Table 24.1 shows a guide adapted by Lusk (2005) which incorporates the part of the body and its accompanying personal affirmation (life meaning), balancing color, and chakra area that are representative of each pose.

Clinicians can readily apply the concepts suggested by this framework when working with clients. For example, if a client speaks of her wants and needs, her corresponding felt sense is found in the second chakra. The bridge pose provides an associated balancing posture. Likewise, as she discusses her desire to care for herself and others, she may be aligning her fourth chakra; the standing backbend is the corresponding balancing pose. Poses can be suggested to the client and to the yoga instructor, if a member of the treatment team. Ideally, aligning the chakras helps to maintain emotional and mental balance. The relaxed state provided by balance allows the client to be more mindful of her feelings and sensations and, therefore, better able to identify her true wants and needs. Yoga practices provide opportunities for the client to slow down and enter a state of relaxation, creating a counter-experience to the anxiety emanating from her underlying emotions.

Certain categories of poses are particularly valuable in challenging and reframing the various ED mindsets. Table 24.2 describes these and their countering-balancing yoga postures.

TABLE 24.2 Yoga Healing Chart

ANOREXIC

Mindset	Internal Focus	Closed-off	Never taking in
Life Stance	I need nothing and no one	I expect perfection	I keep rigid boundaries
Yoga Posture	Backbends	Mountain Pose	Warrior Pose

BULIMIC

Mindset	External Focus	Difficulty Setting Boundaries	Regulating and Containing
Life Stance	I do not know what I need	I present an acceptable self	I am unacceptable to myself
Yoga Posture	Forward Bend	Child's Pose	Tree Pose

BINGE EATER

Mindset	External Focus	Wide Open Boundaries	Taking in and not letting go
Life Stance	I need and deserve everything	I am never satisfied	My body represents my shame
Yoga Posture	Shavasana	Tree	Forward Bend

The anorexic client who takes in little, whether it is food or love, often benefits from poses that are heart-opening such as the cobra pose (a backbend). Backbends provide an opportunity for the client to open her heart, her body, and her life to the possibility of being receptive. The anorexic typically has rigid body movements and a frail stature. Warrior pose and chaturanga-dandasana (low push-up) provide her with flexibility and strength. When holding a posture, the anorexic client should be reminded to follow and link her breath with her movements, thus creating a flow (*vinyasa*), the very opposite of rigidity.

The bulimic client is typically externally focused. With both food and relationships, she looks outside herself for love, acceptance, and validation. She has difficulty forming appropriate boundaries and containing her food, energy, and impulses. Thus, poses that allow her to turn inward are most helpful. Child's pose and seat-forward folds are ideal. Given her repetitive and pervasive bulimic cycle of taking in and throwing out, balancing poses (e.g., tree pose) would provide opportunities to experience stability, and peace in body and mind. To offset her tendency to let go or release quickly, she should be encouraged to hold a pose for several minutes.

Similar to the client with BN, the client with BED seeks external support and has difficulty forming boundaries. Unlike the BN client, the binge-eater takes in and contains, but does not throw out or let go. Specifically, she holds her emotions, food, and experiences inside her body. However, the body is not seen as a sacred holding space, but rather as an unworthy vessel whose contents are also not valued. Often quick to trust and form relationships with others, she has difficulty trusting her own body and creating an intimate relationship with herself. Poses in which she can physically "let go" in her body, such as shavasana (final relaxation) and standing-forward folds, may also allow her to "let go" emotionally. The process of letting go can also be facilitated with yogic twists, which symbolically and literally help the client wring out unnecessary tension and stored experiences on both an emotional and physical level.

Dialoguing with clients about their experiences in the various poses is a useful therapeutic tool. Regardless of how they practice (e.g., alone or in a class), reflecting upon their ensuing physical, emotional, and spiritual sensations through journaling, poetry, song, painting, or other forms of self-expression can deepen the impact of their experience. Rachel, a client who struggled with restricted eating patterns, wrote this yogic poem: "I feel strong when the flow of energy causes me to awaken—to become spiritually open—but I become unbalanced when fear of judgment and self-inflicted pressure cause frustration." Rachel felt energy flowing in warrior pose, felt open in a backbend, and was unbalanced in tree pose. Sheila, who struggled with BED found that child's pose brought her a sense of innocence while warrior pose allowed her to feel focused and strong. She wrote this touching poem to read in times of stress and struggle: "In moments of innocence, there is the absence of nervousness. No need pleading for help when I am focused and strong."

DANCE/MOVEMENT THERAPY: EXPLORING THE DANCE OF CONNECTION

Most ED clients tend to focus on their thoughts, while their dance of life is empty. "Ignoring internal states amounts to burying feelings, and the burial site exists in the body itself" (Kleinman & Hall, 2006, p. 3). DMT is defined as "the psychotherapeutic use of movement as a process which furthers the emotional, cognitive, social and physical integration of the individual" (American Dance Therapy Association, 2009). By addressing the body directly, DMT methods help clients recognize that communication is always present even when they are not speaking. Building on their clients' idiosyncratic movements, dance/movement therapists help clients experience feelings, express them through their body language, and identify the connection between what they discover and how it parallels their lives (Kleinman, 2009; Kleinman & Hall, 2005, 2006). Dance/movement therapists, using the cues and signals from their own bodies, embrace and respond to whatever the client presents kinesthetically and emotionally. This process of contact involving the resonance between the client and the dance/movement therapist deepens the expression and communication occurring in the therapeutic relationship.

The Building Blocks

Three concepts that underlie the process of DMT are *rhythmic synchrony* (experimentation), *kinesthetic awareness* (contact/process/experimentation), and *kinesthetic empathy* (contact, experimentation). Essentially, dance/movement therapists get into rhythm with their clients' needs and feelings (rhythmic synchrony). They heighten their physical awareness to their own feeling states in order to understand their responses to the client (kinesthetic awareness). Finally, DMT facilitates simple actions that allow greater understanding of the meaning of clients' struggles on a body level (kinesthetic empathy). These methods can be adapted for use by any clinician to become better attuned to nonverbal nuances. Over time, practice of these techniques will help clinicians discover and learn to "trust their innate ability to 'attend' empathically, respond authentically, and translate non-verbal experiences into cognitive insights" (Kleinman, 2008, p. 2).

Rhythmic Synchrony. Rhythmic synchrony manifests in the clinician's ability to attune to and cultivate connection with clients. When clinicians are in synchrony with clients' rhythms, they are more attuned to their clients' emotions. Moving too fast or slow, or giving directions that are too complex or rapid, can cause clients to "detach [from their feelings] if they become overwhelmed " (Kleinman, 2009, p. 132). Many clients tend to be out of rhythm with themselves in an attempt to avoid emotional connection. It becomes their familiar way of "being," and anything else feels foreign.

> Jane, a bulimic over-exerciser, turned to compulsive running as a way to organize her rhythms. While running straight ahead to a particular point gave her a clear direction to follow, quickly reaching that destination provided her with a false sense of achievement. This mirrored her destructive pattern of psychological coping (e.g., applying rigid, linear solutions to complex, emotionally overwhelming situations to try to feel in control).

Clients like Jane can shift into healthier, more natural rhythms by learning to physically slow down in order to "experience the journey." Following this first step, their emotions may become more acceptable. Collaboration with the treatment team is essential so that they can anticipate changes and help the client move through the surfacing of suppressed feelings and experiences.

The capacity for, and the benefits of, rhythmic synchrony can be understood from a neuroscientific perspective. "This attunement of right-to-right hemisphere may be crucial in establishing the secure attachment environment which may be essential for effective therapy to occur" (Siegel, 1999, p. 290). For example, in Jane's case, understanding the speed or urgency propelling this client was the clinician's first task. Clients like Jane sustain their anxiety and remain powerless to address their feelings by speeding through their lives as if they are aiming for a finish line. When they are encouraged to explore and "stir up" their feelings by literally moving in their body, their level of anxiety can be diminished. Through use of rhythmic synchrony, clinicians attempt to "transfer over" a sense of the client's agitated feelings to their own bodies. Once in synchrony, clinicians can progressively slow client's movements down by half time until the baseline becomes slow enough for feelings and sensations to be safely experienced and expressed. Through this process, the clinician is better able to understand and address the distress that lies beneath the client's anxiety.

> Erin, a trauma survivor with AN, sat in her chair frantically shaking her leg and pinching her arm to keep from "giving in" to experiencing feelings. In order to get Erin unstuck from this repetitive pattern, the clinician needed to help her take some type of action. They began by walking, and Erin took off so quickly that the clinician could barely keep up. The clinician stopped her and provided this feedback. Erin responded, *"That's how I always walk."* She was desperate to outrun a connection with her self. The shaking and pinching symptoms had disappeared with the onset of the activity, but resurfaced in the form of a frantic speeding pace. They resumed walking, with Erin speeding back and forth across the room, distracted, disconnected, and breathless. The clinician then asked Erin to slow down to half the pace, repeating and modifying their walk at increasingly slower speeds. Erin's breathing began to slow, her eyes focused, and she started to laugh and engage in verbal bantering. The clinician sensed that these observable signals were indicators that the reduced pace had allowed for synchronized interaction to occur and for Erin to begin to tolerate "being in her body." Erin was finally able to identify the correlation between the speed with which she moves through life and her level of anxiety.

Kinesthetic Awareness. Kinesthetic awareness represents the ability to experience feelings and sensations inwardly—turning to the body first to discover awareness of feelings before *thinking* "What am I feeling?" (Kleinman, 2009; Kleinman & Hall, 2006). In the following vignette, Erin described what she experienced when the clinician asked her to take a step into the empty space in front of her.

ERIN

> *"I didn't know where to go. I felt afraid to move into the space since it was so large. I didn't trust myself enough to just move in it. When you placed a piece of fabric on the floor in front of me, it gave me a sense of direction. I stepped into the space, stayed there for a few seconds, and then backed out, moving behind the fabric. That space represented the huge amount of shame I feel all the time, and in moving backwards I was attempting to leave the shame behind. However, what was really happening was I was just backing up, still moving the same way, taking the shame with me."*

After this pivotal experience, she explained:

> *"The movement part of the session is always what affects me the most. There was a time when I was not open or comfortable with doing any moving as I would have to connect not only with my body, but with the feelings in my body as well. Sometimes, without even understanding my feelings, a simple movement can help to unbury them."*

Kinesthetic Empathy. Kinesthetic empathy is the ability, on a bodily level, to understand and sometimes experience what others are physically feeling; for example, transfer of heightened somatic sensations including physical tension, body slackness, hypervigilance, or breathing changes. It has significant relevance to the therapeutic relationship and the journey clients and clinicians undertake together (Kleinman, 2008; 2009; Kleinman & Hall, 2005, 2006). The capacity for kinesthetic empathy also appears to have a neurological basis. Attuning to, and sharing the emotions of, others appear to be linked to the function of mirror neurons in the prefrontal cortex (Blakeslee, 2006; see also Lapides, Chapter 3). The following journal entry is an excellent example of kinesthetic empathy and came from the first session in which Erin allowed herself to express anger:

> I WAS able to challenge my thinking yesterday about how I feel about expressing anger. I know this is because you expressed anger WITH me. I KNOW that is why I don't feel any shame or fear in expressing the anger WE expressed yesterday. Since it was something YOU were okay with doing, instead of just watching me do it, that changed my thinking. I don't think I would have felt the same way about this experience if I was the only one expressing anger. I would have felt ashamed and wrong for expressing anger if I was doing it by myself.

The empathic experience between Erin and her therapist served as a catalyst to help Erin normalize her feelings and expand her perspective to embrace a full range of emotional expression.

Integrating the Concepts

Rhythmic synchrony, kinesthetic awareness, and kinesthetic empathy are effective skills only when built upon the foundation of a therapeutic alliance. They are not tools to be used in isolation. The example that follows demonstrates how Cindy, diagnosed with EDNOS, built upon her therapeutic attachment in order to risk being more in charge of her emotional hungers.

CINDY

Cindy requested help connecting with her feelings. In the first session she appeared so uncomfortable in her body that the clinician suggested that she choose the safest space in the room. She selected a space beside a podium, the only place providing cover. Cindy placed two empty chairs beside her to "create boundaries," and indicated for the clinician to move to the space beside the second one. Because Cindy felt so unsafe, the clinician gave her a ball so that she would have something to hold. Using the ball as a neutral tool, the clinician established rapport by asking Cindy simple questions. Cindy soon offered to share the ball with the clinician and they began tossing it back and forth in an easy rhythm. The clinician subtly introduced conversation about Cindy's relationship with herself and how that paralleled her relationship with the ball. Cindy was able to identify themes of comfort and companionship. Acknowledging that she liked working in this way, she said she would like to schedule a second session. In that session, Cindy wanted to move deeper into her feelings to "see further into the picture." Both sat on the floor as per Cindy's instructions. After re-establishing her safe parameters, she was given the ball and asked to practice letting it go so that it would become separate from her. She immediately became aware of a pain in the center of her chest which she identified as sadness. She began to feel tearful through this very simple action. To intensify her emotional connection, the clinician asked her to let go of the ball with her eyes closed. Although her sadness increased, Cindy was willing to take another risk and moved from her safe space into one that was unfamiliar. Closing her eyes, she let go of the ball and scooted herself into the space where the ball had been. Assured that the clinician was taking cues from her and would not push her too far, Cindy tried the experiment despite fearing she would not be able to find the ball after letting it go. Exploring her metaphorical fear of losing the ball in relation to the bigger picture of her life, Cindy connected it to the recent loss of her father. Her tendency to get stuck in her old "safe space" represented her fear of loss, further evidenced by her reluctance to let go of her ED thoughts and behaviors.

LANGUAGE AS A GATEWAY TO THE BODY

As they say: the journey is as important as the destination. Similarly, the manner in which the client structures and delivers sentences is as important as the actual spoken words or phrases themselves. The client who stumbles over words may also be stumbling through life. The young woman who stops and starts in her speech may be replicating that same pattern with all of her relationships, including food. The soft, barely audible voice that the clinician strains to hear may be indicative of someone who feels unworthy or, through her passivity, forces others to take action. Highlighting these observations with the client becomes a touchstone to body awareness and connection. Dimensions of language that have therapeutic relevance include the rhythm and pacing of the delivery, idiomatic phrases, and emotion, action, and trigger words. While it is important that the clinician ask the client to explicitly define the meaning of the word the client is using in order to avoid misunderstanding or misinterpretation, it is equally important for the clinician to read between the lines to discern what is not being said.

The Rhythm and Pacing of the Delivery

> Andrea rarely completed a sentence. She was constantly tripping over her words and stopping mid-thought to rush on to her next "stream of consciousness." The rapid pace of Andrea's speech, just like Erin's speed-walking through life, reflected an attempt to outrun her feelings. Her verbal interchanges had a chaotic, inattentive, jumbled style which paralleled the way she moved through life. During her residential admission for the treatment of BN, she fell and broke her foot. A few weeks later she bumped into a wall, spraining her shoulder. "*I was just thinking so fast I didn't notice where I was going*," she explained. Becoming aware of how her hurried speech, movement, and thinking all kept her "ahead" of herself, Andrea was willing to practice slowing down the rhythm of her breathing and movements, and explored her need to literally "outpace" her fear of staying in the moment.
>
> After several sessions, Andrea was asked if she noticed any difference in the way she was feeling. "*I don't feel like I have to rush to say something*," she responded, indicating her rhythm was now more fluid than frantic. "*It's a way of being with myself that I've almost never experienced so it feels uncomfortable. But I'm going to keep practicing it until it starts to feel a little more familiar—like this is* me *now*."

Gauging whether a client's verbal pacing is rapid, slow, or uneven, and/or their vocal tone high, deep, breathy, muted, or shrill, provides clinicians with additional information. The relationship between sounds and silences is useful to explore as well.

Idiomatic Phrases

Idiomatic phrases are excellent doorways to body awareness. Examples of idioms include "bouncing off the walls," "coming apart at the seams," "smothered with love," "jumping out of my skin," "under her spell," "climbing the walls," "breaking my heart," and "running in place." By exploring the idiom with the client, the clinician can help the language come to life and uncover the true meaning behind the words.

> Grace, while explaining her need to purge, frequently referred to feeling unable to breathe after she binged. The clinician suggested that Grace close her eyes and try to replicate and describe the feeling. "*My chest is getting really heavy and then my throat starts to constrict. I feel like I'm being smothered.*" Asked to imagine an image of herself being smothered, Grace abruptly opened her eyes. "*Love. I'm being smothered with love. I saw myself lying down covered by piles of fluffy pillows with hearts on them. There were so many pillows that they were crushing my chest. I feel like I'm being smothered with love!*"

Grace's purging behavior had been her way of coping with the unwanted affection of a stifling romance. By exploring her idiomatic language, she was able to pinpoint the underlying cause of her dissatisfaction

"Emotion" Words

Words that contain descriptive emotional content are relatively simple to identify. Some examples are: mad, sad, happy, scared, terrified, joyful, agonized, surprised, anxious, disgusted, lost, grieving, yearning, depressed, grateful, loving, hopeless, confused. Also, many words, despite their benign origin, are highly charged with emotion for ED clients. When clinicians explore clients' use of a word for its specific, unique or "loaded" meaning, an effortless pathway to the body may subsequently open.

> As Jillie began sharing her history of BN, she used the word "disgust" numerous times and referred to herself as "disgusting." Asked how she would define "disgust," she replied, "*It means completely unacceptable and dirty—like slime or garbage.*" Agreeing to close her eyes and do an experiment, Jillie followed directions to relax, slow her breathing, raise her hand, and allow it to move to the part of her self where her disgust was housed. Slowly, her hand traversed her body and came to rest on her stomach and diaphragm. Responding to softly-spoken questions, she was able to pinpoint that the disgust took up a "*huge*" amount of space in her body, was close to the surface (as opposed to being secreted deep within her), had the consistency of mucus, and was "*dirty brown*" in color. "*I'm blown away. I never thought of my disgust as being so much a part of me,*" Jillie said later in the session. "*It's such a big part of me and, yet, it doesn't belong. It's taking up so much of me that there's no room left for anything else!*"

Jillie had begun to make an initial connection to her body and the emotions stored there. She recognized that she was engaged in constant attempts to purge the disgust that had overtaken her. In subsequent sessions she engaged in gestalt empty chair work to communicate with the "disgusting" part of herself and the role it continued to play in her life.

In the example below, Carla's powerful expression of her feelings illustrates a counterpoint to the alexithymic position of the anorexic. Despite their inability to identify and express emotions, individuals with AN are often very intense and passionate about things of importance to them.

CARLA

Carla had been in treatment for two decades. In a letter to her therapist, she wrote of the strong connection she had to the word *"bones"* and her relentless anorexic pursuit of perfection.

> It's scary to follow the logic for what makes me feel safe. CONTROL—The thinner I am (or feel inside) the safer I feel. The end conclusion then is that the safest place/size to be…is bones. How did I get to this bizarre place? How in the world do I get out of such a place? All I know is that in some ways it seems that nothing feels as safe as bones; hard, invulnerable, immutable, untouchable, unbruisible [sic], beautiful, perfect bones.

"Action" Words

Certain words can quickly bring to mind a physicalized image or picture. Words such as "stuck", "struggling", "crushed", "falling", "crazy", "frantic", "entombed", "broken", "weak", "in a bind", "cornered", and "torn apart" all carry with them a sense of movement and have very individualized associations. For one client "stuck" was described as *"lots and lots of activity but never getting anywhere, like I'm on a treadmill,"* while for another, it represented a feeling of *"floundering, sinking in quicksand unable to get a footing."* Like the emotion words, action words lead the client to connect with her body and literally allow her to experience the meaning of her language on a physical level. The more the clinician directs the client back to her body, the more real and alive her body becomes. Representations of images can be expressed through many modalities—art, dance, journaling, role play. This particularly poignant, vivid, description accompanied a client's drawing of her representation of the word "stuck." *"I feel like my tears and my pain are stuck in a knot in my chest that is constantly being pulled tighter and tighter but never seems to break."*

The reciprocal relationship between language and movement warrants illustration.

BARRI

> Barri was an aspiring opera singer on scholarship at a prestigious university. She came to group therapy in an attempt to deal with her BN and self-defeating cycle. Each success was followed by a pattern of skipping classes, failing tests, or practicing her librettos in a manner that stressed her vocal cords. After discussing her goals and what was holding her back from realizing them, Barri revealed her background. She was brought up in a tough neighborhood in which graduating from high school was a rarity, and most of her family and friends had been in frequent legal skirmishes. Barri was struggling with the *"yes but…"* position of BN. Her BN symptoms mirrored her ambivalence about her accomplishments: she was desperate to acquire a taste for success but too guilt-ridden to savor the applause. Her struggle involved a great deal of aimless energy and activity. Intellectually, Barri understood this. However, the "felt sense" of this struggle on a body level eluded her.
>
> Barri agreed to do a role-play experiment to "physicalize" her struggle and experience rather than just think about the conflict between the two parts of herself. She drew a line on the floor at the far end of the room

representing her goal of being an opera singer, and chose a group member to stand there and be her "cheerleader." Another member volunteered to be the part of Barri (her "negative self") that kept her from moving forward. They stood together at the other end of the room. Barri provided the script from her own internal dialog: *"You'll never make it." "You think you're so hot." "Don't leave us."* In order to fully experience her struggle and heighten her "stuck" feeling, Barri directed the woman playing her "negative" self to physically restrain her from behind to keep her from reaching the cheerleader.

Soon the room was in chaos. The cheerleader was shouting to Barri that she was a terrific opera singer, and would live *"happily ever after"* if only she reached the other side of the room. Simultaneously, the "negative" self was delivering her lines with emphatic discouragement, while keeping Barri harnessed in the grip of strong arms. Barri struggled, rocking back and forth, literally and figuratively caught in the grip of her conflict. Suddenly, she sprang toward the cheerleader dragging the "negative" self with her, then throwing her off as she reached the goal. *"I did it,"* she exclaimed. Slowly she walked back to her starting place, chattering all the while. *"Do you believe how strong I was? How I wouldn't give up? How committed I am to making it?"* The group was abuzz with Barri's feat.

However, examining her pattern revealed that *yes*, she had reached her goal, *but* while exalting in her victory she had unconsciously turned around and moved herself back to the place where she had begun. Metaphorically, she had taken herself right back to the *"old neighborhood."* Barri repeated the exercise and each time followed the same pattern, not allowing herself to stay in the goal. On the third trial, she became consciously aware of her body automatically starting to turn back. *"It's as if my body feels comfortable with failure. It feels so familiar that it feels like who I am. It's where I belong."* Barri continued to practice building new "body memories" by realizing that instead of literally "turning back" to the familiar, she needed to turn her back on the pull of her past that was keeping her from her dreams.

Trigger Words

"Control," "fat," and "numb" are among the red flags that signal the clinician that powerful associations are lurking beneath the surface. Because we all have our own personal associations to words, language must be examined for its specific meaning to the client. Punishing words make up much of the ED self-talk. "Gross," "fat," "disgusting," "stupid," "squat" and others are reinforced over and over, eventually losing their shock value. By repeating the words until they have been imbedded and imprinted in the brain and body, the client has created a body image that becomes their identity (see Johnston, Chapter 26). It defines how the client perceives herself, how she believes others see her, and how she feels living in her body (Ressler & Kleinman, 2006). One midlife client constantly referred to herself as "short, fat, and ugly." Those three words had been her mantra for so long, and her belief in them so strong, that she unconsciously presented that image of herself to the outside world. Eventually others came to perceive her that way also.

Mirror work can be helpful in changing negative perceptions. One experiment involves having the client first look at herself in her typical scrutinizing manner with narrowed, "hard" eyes, scanning for flaws. The client is then instructed to close and "relax" her eyes, then open them wide and look at herself with eyes that are "soft" and welcoming. The contrasting experiences are then discussed.

The following example illustrates the power of self-talk and its ability to have an effect on the body.

MELISSA

Melissa, a binge eater whose weight had escalated to 225 pounds stated in her first session that she was an "all or nothing" kind of person. *"I am meticulous about everything in my life. As long as I am rigid in my food choices and calorie intake I have no problem. The minute I deviate, I go out of control."* Several times she described herself as trying to "hold it together." *"Hold it together against what?"* the clinician inquired. *"Against my impulsivity, of course. It gets me into all kinds of trouble,"* she explained in exasperation. It became clear that the polarities of rigidity and impulsivity were the two predominant facets of her being, neither of which alone served her well. Asked how she wanted to feel, she settled on the word "balanced," which she defined as being "light, peaceful, relaxed, and satisfied." She was instructed to stand, look at herself in front of the mirror and say out loud: "I am____," filling in each word in turn. She started to smile as she finished the experiment. *"This probably sounds crazy, but I feel lighter inside now. A burden has been lifted. This is the 'real' me!"*

For Melissa, a shift had taken place very quickly. In subsequent sessions she learned simple grounding and centering exercises from yoga and the martial arts to accompany her affirmations. After three sessions, Melissa reported that she had settled on the word "balanced" to be the linchpin of her mantra. In those instances when she repeated it to herself, she experienced an almost-immediate physiological response. *"It seems involuntary. My mouth curves upward in a smile, my feet plant themselves, my diaphragm lifts and I slow my breathing. Somehow my body and my brain finally believe the same message!"*

CONCLUSION

Unlike all the king's horses and all the king's men who were unable to put Humpty Dumpty together again, clinicians can learn to heal and repair the shattered remnants of those who fall far from their place of safety. When holistic methods are coupled with traditional practices to bring underlying ED issues into conscious awareness, the distorted lens through which the client sees and experiences the world shifts into sharper focus. This provides the clarity necessary for a return to wholeness. The integration of these approaches produces a "whole" greater than the sum of its parts and, thus, a more potent and creative treatment process.

Methods linking the pathways between the brain and the body produce a synergy that is at once powerful and mysterious and, to date, not readily measurable. This may in part be due to a preponderance of tools calibrated only to measure the workings of the linear patterns of the left hemisphere of the brain. The power of healing mediated through right brain, experiential methods has yet to be fully realized or documented. The challenge for both clinicians and researchers alike is to not underestimate or dismiss the power of these right brain methods.

This chapter has illustrated how body-oriented therapies can help ED clients heal through integration of body and mind. Although experiential therapies are frequently incorporated into ED treatment programs and often considered transformative by clients in the process of recovery, traditionally trained clinicians lack exposure to these methods in their formal education. Furthermore, clinical and empirical research needs to include and evaluate the effectiveness of these methods in treatment.

While it may be difficult to quantify and measure, clinical experience tells us that the interplay of body-oriented, experiential treatments with traditional, multidisciplinary approaches (psychotherapies, nutritional and medical rehabilitation), creates a powerful and life-altering potential for change. The field needs to better understand how, what, and when to integrate holistic methods in order to close these critical gaps in the science and practice of ED treatment.

References

Agras, W., Crow, S., Mitchell, J., Halmi, K., & Bryson, S. (2010). A 4-year prospective study of eating disorder NOS compared with full eating disorder syndromes. *International Journal of Eating Disorders, 42*, 565–570.

American Dance Therapy Association (2009). *What is dance/movement therapy?* Retrieved from http://www.adta.org/Default.aspx?pageId=378213

Archer, J., & McCarthy, C. (2006). *Theories of counseling and psychotherapy: Contemporary applications*. Upper Saddle River, NJ: Prentice Hall.

Barenblit, V. (2009). Preface. In S. Chaiklin, & H. Wengrower (Eds.), *The art and science of dance/movement therapy: Life is dance* (p. xi). New York, NY: Routledge.

Blakeslee, S. (2006). Cells that read minds. *The New York Times*. January 6. Retrieved from. http://www.nytimes.com/2006/01/10/science/10mirr.html

Block, L. (2005). *All the flowers are dying*. New York, NY: William Morrow.

Brennan, R. (1991). *Alexander Technique: Natural poise for health*. Shaftesbury, Dorset: Element.

Calogero, R., & Pedrotty, K. (2004). The practice and process of healthy exercise: An investigation of the treatment of exercise abuse in women with eating disorders. *Eating Disorders: The Journal of Treatment and Prevention, 12*(4), 273–291.

Dana, M. (1994). Boundaries: One way mirror to the self. In M. Lawrence (Ed.), *Fed up and hungry: Women, oppression and food* (pp. 46–60). London, UK: Women's Press.

Eliot, T. S. (1922). *The waste land*. New York, NY: Boni and Liveright.

Farhi, D. (2000). *Yoga, mind, body & spirit: A return to wholeness*. New York, NY: Owl Books.

Gendlin, E. (1981). *Focusing* (Rev. ed.). New York, NY: Bantam Dell.

Gestalt Center of Gainesville (2009). *Awareness and attention*. Retrieved from http://www.afn.org/~gestalt/about.htm

Graham, M. (1991). *Blood memory*. New York, NY: Doubleday.

Kleinman, S. (2008). *Making the most of your whole self: Being an embodied therapist*. Retrieved from Eating Disorder Hope website. <http://www.eatingdisorderhope.com/whole-self.html>

Kleinman, S. (2009). Becoming whole again: Dance/Movement therapy for those who suffer from eating disorders. In S. Chaiklin & H. Wengrower (Eds.), *The art and science of dance/movement therapy: Life is dance* (pp. 126–142). New York, NY: Routledge.

Kleinman, S., & Hall, T. (2005). Dance movement therapy with women with eating disorders. In F. Levy (Ed.), *Dance/Movement therapy: A healing art* (Rev. ed., pp. 221–227). Reston, VA: The American Alliance for Health, Physical Education, Recreation, and Dance.

Kleinman, S., & Hall, T. (2006). Dance/Movement therapy: A method for embodying emotions. In *The Renfrew Center Foundation Healing Through Relationship Series: Contributions to eating disorder theory and treatment: Vol. 1. Fostering body–mind integration* (pp. 2–19). Philadelphia, PA.

Lewis, T., Amini, F., & Lannon, R. (2000). *A general theory of love*. New York, NY: Vintage Books.

Lilly, S., & Lilly, S. (2004). *Healing with crystals & chakra energies*. New York, NY: Hermes House.

Lowen, A. (1967). *The betrayal of the body*. New York, NY: MacMillan.

Lusk, J. T. (2005). *Yoga meditations*. Milford, OH: Whole Person Associates.

Mennuti, R., Billock-Tropea, E., & Feibish, H. (2003, Summer). Body balance: An innovative treatment model for large women (pp. 4–20). *The Renfrew Center Working Papers, Vol. 1*.

Opie, I., & Opie, P. (1997). *The Oxford dictionary of nursery rhymes* (2nd ed.). Oxford, UK: Oxford University Press. (2nd ed., p. 213).

Ressler, A., & Kleinman, S. (2006). Reframing body-image identity in the treatment of eating disorders. In *The Renfrew Center Foundation Healing Through Relationship Series: Contributions to eating disorder theory and treatment: Vol. 1. Fostering body—mind integration*. Philadelphia, PA.

Ressler, A. (2009, October). *Insatiable hungers*. Lecture presented at The Renfrew Center Foundation 2009. Fall Series, Dallas, TX.

Rice, J., Hardenbergh, M., & Hornyak, L. (1989). Disturbed body image in anorexia nervosa: Dance/movement therapy interventions. In L. Hornyak, & E. Baker (Eds.), *Experiential therapies for eating disorders through body metaphor* (pp. 259—261). New York, NY: Guilford Press.

Selby, T. (1994). Food, need and desire: A postscript. In M. Lawrence (Ed.), *Fed up and hungry: Women, oppression and food* (pp. 226—232). London, UK: Women's Press.

Serlin, I. A. (Ed.). (2007). *Whole person healthcare*. Westport, CT: Praeger.

Stark, A., Aronow, S., & McGeehan, T. (1989). Dance/Movement therapy with bulimic patients. In L. Hornyak, & E. Baker (Eds.), *Experiential therapies for eating disorders through body metaphor* (pp. 121—143). New York, NY: Guilford Press.

Siegel, D. (1999). *The developing mind*. New York, NY: Guilford Press.

CHAPTER 25

Incorporating Exercise into Eating Disorder Treatment and Recovery
Cultivating a Mindful Approach

Rachel M. Calogero and Kelly N. Pedrotty-Stump

The eating disorder (ED) literature is replete with labels and definitions to describe the exercise behaviors often observed in this patient population: "activity anorexia" (Epling, Pierce & Stefan, 1983), "exercise anorexia" (Touyz, Beumont & Hoek 1987), "obligatory exercise" (Davis, Brewer & Ratusny, 1993), "exercise addiction" (Adams & Kirkby, 2002), "exercise dependence" (Veale, 1987), "exercise abuse" (Davis, 2000), "excessive exercise" (Le Grange & Eisler, 1993). The common theme underlying all of these labels is that exercise is dysfunctional and detrimental in the context of ED. In light of this prevailing negative view, there is considerable debate about whether exercise should be promoted or prohibited during treatment (Rosenblum & Forman, 2002). This chapter reviews the evidence for the role of exercise in eating-related pathology and for the use of exercise in ED treatment. "Dysfunctional exercise" (DEX) will be used throughout the chapter to cover the range of terms, labels, and definitions otherwise used to describe this phenomenon in the literature.

This chapter bridges the research–practice gap in several ways. To begin, the harmful role of exercise in eating-related pathology is described, underscoring the imperative of addressing DEX in treatment. Common myths and misconceptions about the function and meaning of exercise in ED treatment, and how these faulty beliefs serve as barriers to implementing exercise in treatment, are also considered. Then, the evidence for a quantity versus quality approach to the conceptualization and operationalization of DEX is evaluated, with an eye toward reconsidering historically narrow definitions of DEX. Published exercise protocols that have been empirically tested in ED programs are reviewed and compared to elucidate the common and unique themes. Finally, an integrative approach to the treatment of DEX within the context of ED treatment is offered, providing a new direction for ED research, treatment, and recovery that relies on a conceptualization of mindful exercise informed by both science and practice. The chapter concludes with future directions and perspectives on how to bridge the research–practice gap in the area of exercise and ED.

THE HARMFUL ROLE OF EXERCISE IN THE CONTEXT OF EATING DISORDERS

Dysfunctional patterns of exercise feature in the etiology, development, and maintenance of EDs (Beumont, Arthur, Russell & Touyz, 1994; Brewerton, Stellefson, Hibbs, Hodges & Cochrane, 1995; Bruch, 1973; Calogero & Pedrotty, 2004; Davis et al., 1997; Touyz et al., 1987). Clinical studies have estimated that the prevalence of disordered eating (DE) among residential ED patients is 33 to 78% (Dalle Grave, Calugi & Marchesini, 2008; Davis et al., 1997, Davis, Kennedy, Ravelski & Dionne, 1994; Katz, 1996; Shroff et al., 2006). It occurs across the spectrum of EDs and is not, as commonly presumed, exclusively an aspect of anorexic pathology (Boyd, Abraham & Luscombe, 2007; Peñas-Lledó, Vaz Leal & Waller, 2002; Solenberger, 2001).

Dysfunctional exercise can precede the onset of the ED, is one of the last symptoms to subside, or develops later as the major behavioral problem, suggesting that it is not merely secondary to weight loss or weight loss attempts (Calogero & Pedrotty, 2004; Crisp, Hsu, Harding & Hartshorn, 1980; Davis, Blackmore, Katzman & Fox, 2005, Davis et al., 1994; Kron, Katz, Gorzynski & Weiner, 1978; Long & Hollin, 1995; Wichstrom, 1995; Windauer, Lennerts, Talbot, Touyz & Beumont, 1993). In addition, not only do high levels of exercise predict longer periods of ED hospitalization (Solenberger, 2001), but a compulsion to exercise at discharge from treatment predicts a quicker relapse and a chronic outcome, at least among AN patients (Strober, Freeman & Morrell, 1997). Other research has shown that AN patients who resumed high levels of exercise (i.e., more than 6 hours of intense exercise per week) within the first 3 months following discharge were more likely to relapse (Carter, Blackmore, Sutandar-Pinnock & Woodside, 2004). In sum, the evidence clearly indicates that DEX is related to ED pathogenesis, can disrupt treatment, can bring about relapse, and occurs across the spectrum of ED diagnoses.

EXERCISE: A NEGLECTED COMPONENT OF EATING DISORDERS TREATMENT

Considering the severity of DEX with regard to ED development, course, and recovery, its neglect in treatment protocols is of special importance to ED professionals. Several misconceptions and problems on the topic of exercise have contributed to this oversight. One misconception is that exercise serves as an obstacle to weight recovery. The common wisdom among ED professionals is that additional physical activity during treatment interferes with patients' weight recovery. The fear of compromised weight gain is reasonable, due to longer and more costly treatment stays. This perspective is guided, however, by the belief that the primary purpose for patients' exercising would be weight loss. It fails to distinguish between supervised, structured physical activity and unsupervised, high-level exercise, which undoubtedly compromises weight gain. In following a weight-based approach, prohibiting exercise is perceived as imperative in ED treatment, but it only appears to be relevant for the low weight patients. For example, whereas underweight patients may be restricted from exercise indefinitely, normal weight patients may be allowed to exercise 5 days per week, yet they may similarly struggle with DE and be equally at risk for quicker relapse.

A second misconception is that DEX will resolve with general treatment. ED professionals rightly agree that DEX is integral in ED pathology (Hechler, Beumont, Marks & Touyz, 2005); yet many also believe that DEX will resolve itself with ED treatment and weight restoration. This passive resolution of DEX within the context of standard treatment is unlikely to occur. The evidence reviewed earlier in this chapter suggests that, as a central feature of ED pathology, *not* specifically addressing DEX during treatment is more likely to compromise treatment and recovery.

Third, the boundaries between healthy versus unhealthy exercise are blurred. Dysfunctional exercise is unique among the constellation of ED symptoms (e.g., vomiting, starvation, or laxative abuse) because of the widely proclaimed health properties associated with the core behavior (regular exercise). Compounding this issue, according to a survey of ED units in the United Kingdom, very few had a written definition of healthy exercise (Davies, Parekh, Etelapaa, Wood & Jaffa, 2008). Given this acclaim, and the wider cultural fervor around 'fitness' (Robison, 2000), discerning healthy versus unhealthy exercise among patients presents unique challenges for ED professionals.

Finally, despite some recommendations (Andersen, Bowers & Evans, 1997; Beumont, Beumont, Touyz & Williams, 1997; Calogero & Pedrotty, 2004), there is a general lack of widely established exercise protocols in ED treatment. Protocols that do exist have not been adequately tested, raising valid concerns among ED professionals about the cost-benefit ratio of systematically implementing these protocols (Hausenblas, Cook & Chittester, 2008; Hechler et al., 2005). Standardized guidelines, such as those published by the American College of Sports Medicine, would not be appropriate because they do not consider individuals with eating and exercise pathology (Corbin, LeMasurier & Franks, 2002). In addition, there is little specific expertise about exercise among ED professionals (Yates, 1991), making it difficult for some professionals to confidently or competently interpret and evaluate existing recommendations.

IT'S NOT THE EXERCISE: QUALITY TRUMPS QUANTITY

Actual frequency, volume, and intensity of exercise appear to be largely unrelated to ED pathology. It is the underlying psychological motivations and beliefs, and not the exercise behavior per se, that warrants more attention in research and practice. In a large study of American college women, Ackard, Brehm & Steffen (2002) found that those who did *not* exercise, but indicated more negative thoughts and feelings about exercise, reported lower self-esteem, greater depressed mood, and more disordered eating compared to women who frequently engaged in exercise without the accompanying emotional commitment. In a more recent study, Cook and Hausenblas (2008) found that exercise dependence, defined by an individual's pathological motivation to exercise, fully explained the link between exercise behavior and eating-related pathology. A qualitative study of British women found that exercise dependence occurred exclusively in the context of eating-related pathology or a full-blown ED, and not in relation to recreational or high-level exercise (Bamber, Cockerill, Rodgers & Carroll, 2000). Other research has shown that an emotional commitment to exercise predicts disordered eating (Davis et al., 1993) and mediates the relationship between perfectionism and dietary restraint (McLaren, Gauvin & White, 2001).

Exercise that is undertaken to escape or regulate negative affect also signals DEX. A study of 21 consecutively admitted AN inpatients found that regulation of negative affect was a major reason for their exercise (Long, Smith, Midgley & Cassidy, 1993). In a sample of female ED inpatients in Belgium, chronic negative affect was identified as a central feature precipitating both the compulsion to exercise and actual exercise behavior (Vansteelandt, Rijmen, Pieters, Probst & Vanderlinden, 2007). These findings are particularly compelling because the women reported on their emotions, cognition, and behaviors at random intervals throughout the day over the course of a week, thus providing a more naturalistic and ecologically valid representation of the link between negative affect and exercise in ED patients. Using exercise to escape negative affect altogether is also not uncommon in non-clinical samples, as demonstrated by De Young and Anderson (2010). These researchers found that exercise motivated primarily by negative affect was associated with more eating-related pathology among both college women and men, independent of exercise frequency, although this pattern was stronger for women (see Figure 25.1).

Other research has shown that exercise motivated by external appearance goals predicts greater eating-related pathology and poor psychological functioning (Adkins & Keel, 2005; Calogero et al., 2009; Maltby & Day, 2001; Strelan, Mehaffey & Tiggemann, 2003), especially exercise undertaken primarily for weight loss and control of body shape (Mond & Calogero, 2009). In several large-scale studies of community populations of American and Australian

FIGURE 25.1 Personal exercise world of a residential eating disorders patient who struggles with DEX. Her drawing reflects key qualities of DE, such as pain and punishment, escaping negative affect, permission to eat, weight/appearance focus, isolation, and interferes with daily functioning. These themes are accentuated by her color choices (red, orange, and yellow) which communicate the intense power and anger she experiences in her exercise world. Reprinted with permission from author. Please refer to color plate section.

women, Jon Mond and colleagues have found no relationship between self-reported frequency of exercise and disordered eating or quality of life (Mond, Hay, Rodgers, Owen & Beumont, 2004; Mond, Hay, Rodgers & Owen, 2006). Instead, they have demonstrated that exercise for the primary purpose of changing appearance (i.e., weight, shape, or body tone), and feeling intense guilt when missing exercise, were the two strongest predictors of disordered eating in community samples, and also significantly distinguished ED patients from healthy participants (Mond & Calogero, 2009).

In addition to the qualities highlighted above, other dimensions of DEX have been reported among ED patients. Based on clinical observations and anecdotal reports from female ED inpatients, we have extended the scope of DEX to include the following exercise functions or attitudes: self-harm, identity maintenance, permission to eat, bodily disconnection, rigid isolation, dread, all or nothing approach, and when physical well-being and/or safety are compromised (Calogero & Pedrotty, 2004, 2007). Empirical evidence for some of these patterns demonstrated that exercise characterized by self-harm, identity maintenance, weight loss focus, and an outcome orientation (referred to as *mindless* exercise) predicted more eating-related pathology (measured by EDE-Q), higher body shame and chronic body monitoring, more distorted exercise beliefs, and less body responsiveness and body appreciation (Calogero et al., 2009). Other researchers have found that exercise characterized by identity maintenance, affect regulation, rigid routines, and exercising when physically compromised clearly distinguished ED patients from a non-clinical comparison group (Boyd et al., 2007). Similarly, systematic interviews with a variety of undiagnosed women found that exercise in the context of ED pathology (high levels of eating-related pathology or full-blown ED vs. healthy women) was characterized by patterns of self-harm, focus on weight loss and caloric compensation, surreptitious activity, and largely controlled day-to-day living (Bamber et al., 2000). In sum, the evidence consistently shows that irrespective of the volume or frequency of exercise activity, the *quality* of the exercise—or exercise mindset—is what links DE to eating-related pathology (see Figure 25.2).

EVIDENCE FOR EXERCISE PROTOCOLS IN EATING DISORDERS TREATMENT

Beumont and colleagues (1994) pioneered the use of exercise in ED treatment by incorporating a structured anaerobic exercise program into patients' total treatment plans based on the following rationale: (a) total restriction of physical activity is an ineffectual policy that cannot be fully instituted; (b) distorted beliefs about exercise need to be challenged and exercise education is required; and (c) real-world pressures to exercise can be confronted more successfully when patients have been exposed to a healthy model for exercise. They argued cogently for the inclusion of physical activity and exercise counseling in ED treatment, and they emphasized the importance of exercise across ED diagnoses (Beumont et al., 1997). Despite this call for exercise in ED treatment, there is a dearth of research that evaluates exercise in ED treatment, and virtually no published research on the effectiveness of exercise programs that incorporate Beumont's recommendations. The scientific evidence that does exist has been criticized on various methodological grounds, including the use of relatively small sample sizes, the lack of appropriate comparison groups, and/or the reliance on

FIGURE 25.2 Personal exercise world of a residential eating disorders patient who struggles with DEX. Her drawing reflects the core experience of imprisonment, as illustrated by the prison bars and handcuffs that confine her. She identifies many qualities of DEX in her exercise world: source of identity, fear and isolation, dangerous, escaping feelings, and control over her life. We can see that what imprisons her is not the behavior, but her mindset. Yet, she also communicates her awareness of an alternative exercise world, where she is free of these restraints, as depicted by the hopeful ray of sunshine in the corner. Reprinted with permission from author. Please refer to color plate section.

non-randomized assignment. Although these methodological weaknesses are difficult to overcome in real treatment settings, and to some degree are offset by the ecological validity of the findings, these issues represent a critical research-practice gap that requires more attention.

Exercise Protocols in Anorexia Nervosa Patients

An evaluation of the effectiveness of Beumont and colleagues' program in a sample of 39 female AN inpatients in Australia found that participation in structured exercise did not interfere with weight gain compared to a non-exercising control group (Touyz, Lennerts, Arthur & Beumont, 1993). Exercising patients engaged in a structured anaerobic exercise program 3 hours per week throughout their treatment stay which included: stretching, posture enhancement, weight training, social sport, and occasional low-impact aerobics. Eligibility for participation required a body mass index (BMI) of at least 14, weekly weight gain (1+ kg), and no significant medical complications that would preclude exercise. The exercise program emphasized exercise education, challenging distorted exercise beliefs, and practicing healthy forms of exercise. Exercising patients showed an average increase of 1 kg per week over 4- and 6-week refeeding periods and improved program compliance. No other studies were located that reported on whether patients' exercise attitudes and behaviors improved following participation in this program.

A different approach was implemented and tested by Long and Hollin (1995). In this case, an adjunctive exercise intervention was implemented immediately following inpatient treatment for the purpose of relapse prevention. Participants included six fully weight-restored AN outpatients (one male) in the UK who were excessive exercisers. The exercise program did not include structured or supervised physical activity, but instead followed a two-phase treatment approach to moderate exercise. In the first phase, individual or group sessions included exercise education, making self-motivational statements, identifying the perceived costs and benefits of changing exercise, and setting appropriate treatment goals. In the second phase, individual sessions incorporated a variety of cognitive-behavioral techniques to help self-monitor exercise, restructure distorted cognitions around exercise, and manage exposure and response prevention. On average, 12 days (range 8–12) of outpatient treatment were focused on DEX, and 8.3 sessions (range 4–12) were devoted to exposure and response prevention. At the 4-year follow-up, all patients had maintained their weight and only two patients had a poor outcome with respect to exercise. That is, although the patients reported similar levels of exercise behavior at follow-up, the two poor outcome patients continued to exercise to control weight and shape, to regulate negative affect, and dreaded exercise. In contrast, the good outcome patients reported using exercise for social reasons and general fitness and they did not change their schedule to accommodate exercise. It is noteworthy that although the poor outcome patients continued to exercise at their pre-treatment levels at follow-up, they had not returned to the particular sports that were targeted in the exposure and response prevention sessions.

The use of an exercise program over a 3-month period was evaluated by Thien, Thomas, Markin & Birmingham (2000) with 16 AN outpatients (one male) in Canada to determine whether randomly assigned participation in exercise could improve patients' quality of life without interfering with weight gain. The exercise protocol involved a graduated reintroduction of activity with seven levels. Exercising patients began with stretching three times per week and exercise levels were increased based on percent ideal body weight and percent body fat. Higher levels incorporated higher intensity exercise, including isometric work, low-impact aerobics, and resistance training. The exercise activities were not supervised, but they were highly structured, periodically monitored, and revised (levels increased or decreased) based on progress toward treatment goals. Patients were clearly instructed about the type, intensity, and frequency of exercise at each level, and wore heart-rate monitors to check intensity. The findings showed that exercise participation did not interfere with weight gain as indicated by BMI and percent body fat values. In addition, exercising patients reported better quality of life than did controls (i.e., functioning at work, daily and social activities, energy level), although these differences were not statistically significant.

Szabo and Green (2002) evaluated resistance training as an adjunctive treatment for AN inpatients in South Africa. Patients were randomly assigned to participate in resistance training or a control group for 8 weeks, and a healthy exercising control group was included for comparison. All patients were within 15–20% of their target discharge weight. Resistance training consisted of two alternating exercise schedules that included a combination of 2.5 kg dumbbells, elastic bands, and body weight to target a wide range of muscle groups throughout the body. Compared to healthy exercising controls, body fat and body weight significantly increased and ED pathology significantly decreased from baseline for the exercising and non-exercising AN groups, and depression decreased among non-exercising AN.

Thus, while exercise participation did not interfere with weight gain, the exercising and non-exercising AN groups did not significantly differ from each other on these outcomes.

Tokumura, Yoshiba, Tanaka, Nanri & Watanabe (2003) evaluated stationary bike riding in adolescent AN patients in Japan during the convalescent phase of inpatient treatment to improve their exercise capacity and to alleviate emotional stress. Eligibility required patients to have achieved a medically stable weight and increase their body fat by 25% from admission. Patients engaged in 30 minutes of supervised cycling at their individualized anaerobic threshold five times per week, for about 10 months. At the 1-year follow-up, patients who exercised had a significantly higher BMI and improved exercise capacity relative to their baseline, whereas a non-exercising control group did not. Participation in this exercise protocol did not interfere with weight gain or the onset or resumption of menstruation, and it did not produce relapse. Although measures of emotional stress were not indicated, the researchers reported that emotional stress was reduced as well.

Exercise Protocols in Bulimia Nervosa Patients

In one of the only studies to examine the use of exercise with bulimia nervosa (BN) patients, Sundgot-Borgen, Rosenvinge, Bahr & Schneider (2002) compared physical exercise (mixed aerobic and non-aerobic) to cognitive-behavioral therapy (CBT) on a range of outcomes in female BN outpatients in Norway who were randomly assigned to exercise, CBT, nutritional counseling, or wait-list control. Exercise was described as moderate levels of aerobic and anaerobic activities over a period of 16 weeks designed to promote physical fitness, to reduce feelings of fatness, bloating, and distention associated with eating, and to reduce ED pathology. Exercising patients participated in weekly 1-hour sessions with a qualified fitness instructor who did not engage in eating-related discussions. They were also advised to exercise unsupervised twice weekly for at least 35 minutes. Compared to baseline, exercising patients reported significantly lower drive for thinness, fewer bulimic symptoms, and less body dissatisfaction at the 6- and 18-month follow-up. In addition, exercise patients showed greater improvement than CBT patients on drive for thinness and bulimic symptoms at both follow-ups. Not surprisingly, exercise capacity (as measured by peak oxygen consumption) was significantly improved in the exercise patients relative to all other conditions at the 18-month follow-up. The researchers also reported significant reductions in percent body fat for the exercise patients compared to the other groups at the 18-month follow-up. However, at 18 months, patients were regularly exercising with concomitant vomiting, which is not only dysfunctional, but life-threatening. The markers of success associated with exercise protocols for BN patients require further consideration.

Exercise Protocols with Anorexia Nervosa, Bulimia Nervosa, and Eating Disorder Not Otherwise Specified (EDNOS) Patients

Similar to the philosophy underlying Beumont et al.'s (1994) program, we implemented an exercise program specifically targeting DEX within the context of the ED by providing residential ED patients with opportunities to practice and process alternative experiences with exercise before discharge (Calogero & Pedrotty, 2004). Core features of the program include: disrupting the link between exercise and weight loss, providing exercise education, challenging distorted exercise beliefs, supervising exercise sessions, and guided processing of exercise experiences during and after the activity. Exercise is based on a graded 3-level system that varies in frequency and intensity, and includes a combination of guided stretching, posture and alignment work, yoga, Pilates, partner exercises, resistance training, stability ball, low impact aerobic activity, strength training (2.5–5 kg), and recreational activities (e.g., skipping, bat and ball, nature walks, musical pillows). Participation is voluntary and eligibility requires that patients are: not on bed rest (or modified bed rest); medically cleared; and agree to work on their exercise issues within the program. Patients advance to the next level based on clearance from both the treatment team (including medical clearance) and the exercise program coordinators.

An evaluation of the effectiveness of this exercise program in a non-randomized mixed sample of 254 residential ED patients (AN, BN, EDNOS) in the USA found that, after an average of 4 weeks' participation in supervised, structured exercise, the exercising AN patients gained significantly more weight than the non-exercising AN control patients (Calogero & Pedrotty, 2004). All exercising patients also reported significantly less ED pathology and DEX at discharge compared to those who did not participate. Specifically, emotional commitment to exercise, excessive involvement with exercise, and rigid exercise were significantly lower at discharge.

CRITICAL THEMES OF EXERCISE PROTOCOLS IN EATING DISORDERS TREATMENT

Based on the review above, it is clear that the use of exercise in ED treatment is happening on a global scale, which underscores the need for established guidelines and formal evaluations of their effectiveness in ED treatment. There were many notable differences in the methodology followed across the exercise protocols, making it difficult to directly compare effects. The nature of the exercise, the degree of supervision, the indicators of success, the rationale for using exercise in ED treatment, and whether DEX was directly targeted varied widely across the studies. In addition, researchers did not randomly assign patients to exercise or control groups. Personality and motivational factors could thus be confounding the interpretation of any significant effects of exercise participation. Several critical themes emerged from the review of these protocols.

First, supervised exercise did not interfere with weight gain and it reduced ED pathology. Regardless of the type or intensity of exercise it did not interfere with weight gain/maintenance during weight recovery/maintenance periods. When ED pathology was assessed, symptomatic behaviors were markedly reduced after participation in exercise, even up to four years following the intervention. In each of the exercise protocols, actual BMI and caloric intake were closely monitored by the treatment team and exercise activities were monitored to some degree. Thus, structured exercise can be used with significant profit in ED inpatient and outpatient treatment without interfering with weight gain/maintenance.

Second, female AN patients were predominantly targeted. None of the exercise protocols included an evaluation of male ED patients. Moreover, only two studies examined BN patients (Calogero & Pedrotty, 2004; Sundgot-Borgen et al., 2002) and only one study examined EDNOS patients (Calogero & Pedrotty, 2004). None of the studies examined binge eating disorder patients, children, older adults, or elite athletes. Thus, despite the cross-cultural scope of these exercise protocols, the documented effects of exercise in ED treatment cannot be widely generalized.

Third, few protocols targeted DEX directly. Findings from the three programs that did specifically target DEX showed significant decreases in: emotional commitment to exercise; obligatory exercise attitudes toward exercise; appearance-based motives for exercise; and distress over missing exercise sessions (Calogero & Pedrotty, 2004; Long & Hollin, 1995; Touyz et al., 1993). Those programs that did not target DEX used exercise to facilitate treatment and/or to improve overall recovery. Most programs did not implement any of Beumont et al.'s (1994) recommendations that focused on changing patients' relationship with exercise.

Fourth, exercise participation was based predominantly on weight. Despite the evidence that exercise does not interfere with weight gain, exercise participation was typically based on the achievement of a particular weight, without consideration of the underlying exercise pathology. Many treatment programs that incorporate exercise rely on BMI values to determine appropriate activity levels, although the values used to set these levels vary markedly (Davies et al., 2008). This relatively common approach reinforces the idea that exercise is only about weight—weight loss, weight maintenance, and weight control. From the patients' perspective, this policy could be interpreted to mean that they are now in a position where they can afford to lose weight or that they need to exercise at this weight. It is critical to question why exercise is often considered dangerous or dysfunctional only in the context of low weight (vs. normal or higher weight). The answer to this question has real implications for how we utilize exercise in the treatment of all ED.

TAKING A MINDFUL APPROACH TO EXERCISE

Most of the abovementioned exercise protocols neglected the mind–body connection. They did not target the problematic quality of the exercise—the negative exercise mindset. Yet, this may be where exercise can make its greatest impact in ED treatment. Non-verbal movement-based groups have been highly recommended in ED treatment (Beumont et al., 1994), but rarely included in the management of DEX (Hechler et al., 2005). We consider mindful awareness during exercise to represent a key phenomenological shift in how patients move and feel in their bodies. Thus, we recommend that ED professionals broaden their conceptualization of DEX as described earlier to include a wider spectrum of mindless and dysfunctional patterns of exercise that disconnect the mind and body (e.g., self-harm, identity maintenance, appearance motives).

Cultivating Mindfulness in Exercise

Drawing from our clinical experiences and seminal work on mindfulness (Langer, 1989; Siegel, 2007), we have come to conceptualize DEX on a continuum that ranges from more *mindless* exercise to more *mindful* exercise (see Table 25.1). These opposing dimensions of exercise capture the variety of labels, definitions, and qualities of exercise described in the extant literature, while offering healthier, positive practices to replace DEX. Mindful exercise encompasses any movement that is done with attention, purpose, self-compassion, acceptance, awareness, and joy. It is focused on the *process* of becoming more connected, healthier, and stronger, whereas mindless exercise is often appearance-based and focused on outcomes.

TABLE 25.1 Qualities of Mindful versus Mindless Exercise

Mindless Exercise	Mindful Exercise
Orientation to past and/or future	Orientation to the present moment
Focus on external outcomes (calories burned)	Attention to internal process (breathing)
Injures and depletes the body	Rejuvenates the body
Disrupts mind–body connection	Enhances mind–body connection
Exacerbates mental and physical strain	Alleviates mental and physical strain
Brings pain, dreaded	Provides pleasure, joy, fun

Mindfulness-based practices promote well-being and self-care without increasing reactivity to stress and other triggers (Siegel, 2007; Wall, 2005; see also Ressler et al., Chapter 24). In addition, mindfulness practices decrease cortisol levels, thereby changing the neurophysiology of the body to facilitate healing (e.g., Monnazzi, Leri, Guizzardi, Mattioli & Patacchioli, 2002; Schell, Allolio & Schonecke, 1994; West, Otte, Geher, Johnson & Mohr, 2004). Utilizing a new scale that measures mindful exercise (Calogero et al., 2009), recent research has demonstrated a link between a mindful exercise mindset and improved ED pathology and psychological functioning.

Yoga is a particularly powerful tool for cultivating mindful exercise. The aim of yoga is to fully experience the present moment, encouraging attunement to internal sensations versus external stimuli (see Ressler et al., Chapter 24). Yoga serves as a metaphor for life in many different ways. A yoga practice can be created to include poses (or asanas) that address particular psychological or emotional challenges. Some asanas emphasize balance or flexibility, whereas others may focus on openness and trust, or acceptance. Thus, yoga may help in ED treatment and recovery because it teaches specific tools to facilitate healing.

Evidence for the benefits of yoga in ED treatment is largely anecdotal (Boudette, 2006; Douglass, 2009; Wyer, 2001); however, some empirical research with non-clinical samples has been conducted. Daubenmier (2005) found that regular yoga participants had significantly greater body awareness, responsiveness, and satisfaction and less self-objectification compared to non-yoga participants. Dale et al. (2009) found significant improvements in psychological well-being and mood and less negative emotionality after an intensive 6-day yoga workshop among a sample of women with histories of EDs, and these effects were maintained up to 1 month following the workshop. More recently, a randomized controlled clinical trial found that incorporating individualized yoga as an adjunctive therapy in ED outpatient treatment significantly reduced ED pathology compared to a no-yoga control group, and did not interfere with weight gain/maintenance (Carei, Fyfe-Johnson, Breuner & Brown, 2010). In contrast, Mitchell Mazzeo, Rausch & Cooke (2007) did not show significant benefits of yoga compared to a dissonance-based intervention with respect to ED pathology, although no harm was shown either. A key difference between Mitchell et al. (2007) and the other studies is that Mitchell and colleagues studied a brief yoga intervention among women largely new to yoga; whereas the studies that had demonstrated positive effects included participants familiar with yoga. However, none of these studies examined the effect of yoga on DEX *per se*. Therefore, the effectiveness of yoga for reducing DEX cannot be inferred from these findings alone. More systematic research is needed to clarify these findings, especially which type of yoga is most appropriate for reducing DEX in ED treatment, and which mechanisms underlie its potential positive impact. Yoga is a popular and promising intervention, but the idea of a mindfulness-based approach to DEX applies to all types of physical activity. The principles and techniques that characterize yoga (and mindful awareness more generally) could be—and we would argue should be—applied to any exercise protocol.

Cultivating Mindful Exercise in Patients

Although most treatment programs do not systematically address DEX (Davies et al., 2008; Hechler et al., 2005), those programs that do tend to rely heavily on psycho-education, challenging distorted beliefs, and self-monitoring—standard tools from the CBT toolbox. Programs that have used these techniques have reported positive changes with respect to exercise quantity and quality. As reviewed above, current data support incorporating supervised mindful exercise into ED treatment in conjunction with CBT. Calogero and Pedrotty (2007) suggest additional ways that ED professionals can build their practice to cultivate mindful exercise.

Hechler et al. (2005) found that 95% of ED professionals (primarily psychiatrists) surveyed assessed patients' exercise history during initial evaluation. Learning more about the contexts in which exercise occurs maximizes this aspect of assessment. Asking patients about various exercise-relevant contexts that characterize their exercise (e.g., physical, environmental, historical, emotional, social, cognitive, cultural) provides a more comprehensive picture of the nature and meaning of their DEX—the quality as well as quantity (e.g., Calogero & Pedrotty, 2007; Otis & Goldingay, 2000; Prichard & Tiggemann, 2005; Rejeski & Thompson, 1993; Taylor, Baranowski & Sallis, 1994; Trost, Owen, Bauman, Sallis & Brown, 2002). For example, exercise without proper nourishment, with concomitant laxative abuse, or in unsafe locations, characterizes physical contexts of DEX. Raising patients' consciousness around the role and patterns of DEX in their lives challenges their mindless exercise mindset.

"Drawing an exercise world" is a powerful non-verbal technique for cultivating awareness and facilitating communication around exercise issues (see Figures 25.1 and 25.2). Patients are asked to draw their exercise world—the people, places, things, thoughts, feelings, and contexts that represent their exercise as they experience it. Examining their drawings can help both patients and therapists better understand the present mindsets guiding the exercise. It is not uncommon for patients to be unaware of their exercise issues until they begin to draw their exercise world. Ideally, specific components of the patient's exercise world are identified as targets for change during treatment. It is also beneficial to ask patients to draw another exercise world at a later stage of treatment to evaluate the extent to which their DEX has changed as a result of the exercise interventions. The therapeutic impact of drawing an exercise world, as well as using it as a marker of change, is a subject for future empirical investigation.

Cultivating Mindfulness in Treatment Teams

Resolving DEX requires a treatment team (e.g., physicians, psychiatrists, therapists, nutritionists, nurses, coaches, counselors, exercise specialists). Becoming more mindful of the unique role each team member plays in the management of DEX improves the quality of care around exercise. The treatment team should send clear and consistent messages about the purpose of exercise, how to create a balanced program of activity, and how participation in exercise depends on safety. In order to facilitate treatment of DEX in the context of ED treatment, the treatment team must implement exercise protocols with a shared understanding of the purpose of the exercise and how they will support it. Supporting exercise in ED patients does not mean aligning or colluding with the ED, although it can feel this way if patients are not changing how they think about and practice exercise. However, focusing on weight to determine the prescription or prohibition of exercise *does* collude with the ED, and may actually demand more policing of surreptitious exercise by staff. Moreover, a weight-based exercise policy represents an all or nothing approach to exercise that mimics (not challenges) ED patients' relationship to exercise. Being mindful of the messages communicated to patients regarding policies and programs around exercise is critical to the treatment of DEX.

Addressing DEX provides a unique opportunity to more fully develop the therapeutic alliance. Conveying to patients that the intention is not to take exercise away, but to help them redefine and discover a new relationship with it, creates a new link in the therapeutic connection. Indeed, giving them opportunities to change, and not give up, their exercise practices lets them feel more supported and empowered in the treatment process. Patients are correct when they say that exercise is good for them (unlike other symptomatic behaviors). The difference is that their approach toward exercise, and the quality of their exercise, does not facilitate their health in the way that regular exercise is meant to do. As mentioned above, there are a variety of tools and interventions that can be used during therapeutic sessions to stimulate this process. In addition, patients often appreciate the sharing of personal exercise experiences. Self-disclosure about exercise is another way to build a bridge with clients and develop a dialog around shared interests and/or mutual confusion about exercise. When executed appropriately and mindfully, self-disclosure deepens the therapeutic alliance and assures clients that clinicians do "get it" (Bloomgarden & Mennuti, 2009; see also McGilley and Szablewski, Chapter 12).

FUTURE DIRECTIONS AND PERSPECTIVES

The topic of exercise in the international ED community is receiving necessary and overdue recognition. This is a promising and critical shift in ED research and treatment. However, current approaches to exercise in the context of ED remain largely unbalanced. Much of the focus is on reducing DEX, with less attention given to replacing dysfunctional practices with healthier positive ones. Clearly, more work is needed.

Systematic evaluation of the effectiveness of current interventions targeting DEX is urgently needed. One important empirical question is to what extent DEX can be resolved in ED patients if not addressed in combination with other ED therapy. More qualitative research is needed to better inform our quantitative designs (see McGilley and Szablewski, Chapter 12). For example, more information is needed about patients' perspectives on exercise, including how they perceive it to be viewed and managed within ED programs. To date, research offers virtually no information about DEX in the treatment of men, children, or elite athletes, highlighting a critical research–practice gap. In addition, much of the attention around DEX has centered on the AN experience. To better inform practice, DEX across the ED spectrum requires greater recognition and investigation. Importantly, research is needed on the long-term impact of an exercise program during ED treatment on ED relapse and recovery.

In closing, caution is recommended. It is imperative to standardize exercise protocols implemented in ED treatment, but ED professionals do not yet have a shared understanding of the problem. ED professionals do not agree on the nature or definition of DEX, the goals for exercise programs in ED treatment, the quality and quantity of exercise that is helpful (and why), who should facilitate these programs, who should participate in these programs, and how positive outcomes should be defined. There is a wealth of clinical wisdom around exercise in ED treatment that could be shared and documented to help build a sturdier bridge between science and practice on this topic. As researchers and practitioners, we must proceed mindfully, paying critical attention to both qualitative and quantitative perspectives that inform our use of exercise in ED treatment.

References

Ackard, D. M., Brehm, B. J., & Steffen, J. J. (2002). Exercise and eating disorders in college-aged women: Profiling excessive exercisers. *Eating Disorders: The Journal of Treatment and Prevention, 10*, 31–47.

Adams, J., & Kirkby, R. J. (2002). Excessive exercise as an addiction: A review. *Addiction Research & Theory, 10*, 415–438.

Adkins, E. C., & Keel, P. K. (2005). Does "excessive" or "compulsive" best describe exercise as a symptom of bulimia nervosa? *International Journal of Eating Disorders, 38*, 24–29.

Andersen, A. E., Bowers, W., & Evans, K. (1997). Inpatient treatment of anorexia nervosa. In D. M. Garner, & P. E. Garfinkel (Eds.), *Handbook of psychotherapy for anorexia nervosa and bulimia* (pp. 327–353). New York, NY: Guilford Press.

Bamber, D., Cockerill, I. M., Rodgers, S., & Carroll, D. (2000). "It's exercise or nothing": A qualitative analysis of exercise dependence. *British Journal of Sports Medicine, 34*, 423–430.

Beumont, P. J. V., Arthur, B., Russell, J. D., & Touyz, S. (1994). Excessive physical activity in dieting disorder patients: Proposals for a supervised exercise program. *International Journal of Eating Disorders, 15*, 21–36.

Beumont, P. J. V., Beumont, C. C., Touyz, S. W., & Williams, H. (1997). Nutritional counseling and supervised exercise. In D. M. Garner, & P. E. Garfinkel (Eds.), *Handbook of treatment for eating disorders* (2nd ed., pp. 178–187). New York, NY: Guilford Press.

Bloomgarden, A., & Mennuti, R. B. (2009). *Psychotherapist revealed: Therapists speak about self-disclosure in psychotherapy*. New York, NY: Routledge.

Boudette, R. (2006). How can the practice of yoga be helpful in recovery from an eating disorder? *Eating Disorders: The Journal of Treatment and Prevention, 14*, 167–170.

Boyd, C., Abraham, S., & Luscombe, G. (2007). Exercise behaviours and feelings in eating disorder and non-eating disorder groups. *European Eating Disorders Review, 15*, 112–118.

Brewerton, T. D., Stellefson, E. J., Hibbs, N., Hodges, E. L., & Cochrane, C. E. (1995). Comparison of eating disorder patients with and without compulsive exercising. *International Journal of Eating Disorders, 17*, 413–416.

Bruch, H. (1973). *Eating disorders: Obesity, anorexia nervosa, and the person within*. New York, NY: Basic Books.

Calogero, R. M., & Pedrotty, K. N. (2004). The practice and process of healthy exercise: An investigation of the treatment of exercise abuse in women with eating disorders. *Eating Disorders: The Journal of Treatment and Prevention, 12*, 273–291.

Calogero, R. M., & Pedrotty, K. N. (2007). Daily practices for mindful exercise. In L. L'Abate, D. Embry, & M. Baggett (Eds.), *Handbook of low-cost preventive interventions for physical and mental health: Theory, research, and practice* (pp. 141–160). Amsterdam: Springer-Verlag.

Calogero, R. M., Pedrotty, K. N., Menzel, J., Thompson, J. K., Wood, K., & Levine, M. P. (2009). Mindful versus mindless exercise in eating disorders pathology: Development and validation of the Mindful Exercise Mindset Scale. Manuscript in preparation.

Carei, T. R., Fyfe-Johnson, A. L., Breuner, C. C., & Brown, M. A. (2010). Randomized controlled clinical trial of yoga in the treatment of eating disorders. *Journal of Adolescent Health, 46*(4), 346–351.

Carter, J. C., Blackmore, E., Sutandar-Pinnock, K., & Woodside, D. B. (2004). Relapse in anorexia nervosa: A survival analysis. *Psychological Medicine, 34,* 671–679.

Cook, B. J., & Hausenblas, H. A. (2008). The role of exercise dependence for the relationship between exercise behavior and eating pathology: Mediator or moderator? *Journal of Health Psychology, 13,* 495–502.

Corbin, C. B., LeMasurier, G., & Franks, B. D. (2002). Making sense of multiple physical activity recommendations. *President's Council on Physical Fitness and Sports Research Digest, 3,* 1–8.

Crisp, A. H., Hsu, L. K. G., Harding, B., & Hartshorn, J. (1980). Clinical features of anorexia nervosa. *Journal of Psychosomatic Research, 24,* 179–191.

Dale, L. P., Mattison, A. M., Greening, K., Galen, G., Neace, W. P., & Matacin, M. L. (2009). Yoga workshop impacts psychological functioning and mood of women with self-reported history of eating disorders. *Eating Disorders: The Journal of Treatment and Prevention, 17,* 422–434.

Dalle Grave, R., Calugi, S., & Marchesini, G. (2008). Compulsive exercise to control shape or weight in eating disorders: Prevalence, associated features, and treatment outcome. *Comprehensive Psychiatry, 49,* 346–352.

Daubenmier, J. J. (2005). The relationship of yoga, body awareness and body responsiveness to self-objectification and disordered eating. *Psychology of Women Quarterly, 29,* 207–219.

Davies, S., Parekh, K., Etelapaa, K., Wood, D., & Jaffa, T. (2008). The inpatient management of physical activity in young people with anorexia nervosa. *European Eating Disorders Review, 16,* 334–340.

Davis, C. (2000). Exercise abuse. *International Journal of Sport Psychology, 31,* 278–289.

Davis, C., Blackmore, E., Katzman, D. K., & Fox, J. (2005). Female adolescents with anorexia nervosa and their parents: A case-control study of exercise attitudes and behaviours. *Psychological Medicine, 35,* 377–386.

Davis, C., Brewer, H., & Ratusny, G. (1993). Behavioural frequency and psychological commitment: Necessary concepts in the study of excessive exercising. *Journal of Behavioural Medicine, 16,* 611–628.

Davis, C., Katzman, D. K., Kapstein, S., Kirsch, C., Brewer, H., Kalmbach, K., … Kaplan, A. S. (1997). The prevalence of high-level exercise in the eating disorders: Etiological implications. *Comprehensive Psychiatry, 38,* 321–326.

Davis, C., Kennedy, S. H., Ravelski, E., & Dionne, M. (1994). The role of physical activity in the development and maintenance of eating disorders. *Psychological Medicine, 24,* 957–967.

De Young, K. P., & Anderson, D. A. (2010). Prevalence and correlates of exercise motivated by negative affect. *International Journal of Eating Disorders, 43,* 50–58.

Douglass, L. (2009). Yoga as an intervention in the treatment of eating disorders: Does it help? *Eating Disorders: The Journal of Treatment and Prevention, 17,* 126–139.

Epling, W. F., Pierce, W. D., & Stefan, L. (1983). A theory of activity-based anorexia. *International Journal of Eating Disorders, 3,* 27–46.

Hausenblas, H. A., Cook, B. J., & Chittester, N. I. (2008). Can exercise treat eating disorders? *Exercise and Sports Medicine Review, 36,* 43–47.

Hechler, T., Beumont, P., Marks, P., & Touyz, S. (2005). How do clinical specialists understand the role of physical activity in eating disorders? *European Eating Disorders Review, 13,* 125–132.

Katz, J. L. (1996). Clinical observations on the physical activity of anorexia nervosa. In W. F. Epling, & W. D. Pierce (Eds.), *Activity anorexia: Theory, research, and treatment* (pp. 199–207). Mahwah, NJ: Lawrence Erlbaum Associates.

Kron, L., Katz, J. L., Gorzynski, G., & Weiner, H. (1978). Hyperactivity in anorexia nervosa: A fundamental clinical feature. *Comprehensive Psychiatry, 19,* 433–439.

Langer, E. J. (1989). *Mindfulness.* Reading, MA: Perseus Books.

Le Grange, D., & Eisler, I. (1993). The link between anorexia nervosa and excessive exercise: A review. *European Eating Disorders Review, 1,* 100–119.

Long, C. G., & Hollin, C. R. (1995). Assessment and management of eating disordered patients who overexercise: A four-year follow-up of six single case studies. *Journal of Mental Health, 4,* 309–317.

Long, C. G., Smith, J., Midgley, M., & Cassidy, T. (1993). Over-exercising in anorexic and normal samples: Behaviour and attitudes. *Journal of Mental Health, 2,* 321–327.

Maltby, J., & Day, L. (2001). The relationship between exercise motives and psychological well-being. *The Journal of Psychology, 135,* 651–660.

McLaren, L., Gauvin, L., & White, D. (2001). The role of perfectionism and excessive commitment to exercise in explaining dietary restraint: Replication and extension. *International Journal of Eating Disorders, 29,* 307–313.

Mitchell, K. S., Mazzeo, S. E., Rausch, S. M., & Cooke, K. L. (2007). Innovative interventions for disordered eating: Evaluating dissonance-based and yoga interventions. *The International Journal of Eating Disorders, 40,* 120−128.

Mond, J. M., & Calogero, R. M. (2009). Excessive exercise in eating disorder patients and in healthy women. *Australian and New Zealand Journal of Psychiatry, 43,* 227−234.

Mond, J. M., Hay, P. J., Rodgers, B., Owen, C., & Beumont, P. J. V. (2004). Relationships between exercise behavior, eating-disordered behavior and quality of life in a community sample: When is exercise "excessive"? *European Eating Disorders Review, 12,* 265−272.

Mond, J. M., Hay, P. J., Rodgers, B., & Owen, C. (2006). An update on the definition of "excessive exercise" in eating disorders research. *International Journal of Eating Disorders, 39,* 147−153.

Monnazzi, P., Leri, O., Guizzardi, L., Mattioli, D., & Patacchioli, F. R. (2002). Antistress effect of yoga-type breathing: Modification of salivary cortisol, heart rate and blood pressure following a step-climbing exercise. *Stress & Health: Journal of the International Society for the Investigation of Stress, 18,* 195−200.

Otis, C., & Goldingay, R. (2000). *The athletic woman's survival guide.* Champaign, IL: Human Kinetics.

Peñas-Lledó, E., Vaz Leal, F. J., & Waller, G. (2002). Excessive exercise in anorexia nervosa and bulimia nervosa: Relation to eating characteristics and general psychopathology. *International Journal of Eating Disorders, 31,* 370−375.

Prichard, I., & Tiggemann, M. (2005). Objectification in fitness centers: Self-objectification, body dissatisfaction, and disordered eating in aerobic instructors and aerobic participants. *Sex Roles, 53,* 19−28.

Rejeski, W. J., & Thompson, A. (1993). Historical and conceptual roots of exercise psychology. In P. Seraganian (Ed.), *Exercise psychology: The influence of physical exercise on psychological processes* (pp. 3−35). New York, NY: Wiley.

Robison, J. I. (2000, Sept/Oct). Do we really need to exercise and eat low fat to get into heaven? *Healthy Weight Journal, 74−75.*

Rosenblum, J., & Forman, S. (2002). Evidence-based treatment of eating disorders. *Current Opinion in Pediatrics, 14,* 379−383.

Schell, F. J., Allolio, B., & Schonecke, O. W. (1994). Physiological and psychological effects of Hatha-Yoga exercise in healthy women. *International Journal of Psychosomatics, 41*(1−4), 46−52.

Shroff, H., Reba, L., Thornton, L. M., Tozzi, F., Klump, K. L., Berrettini, W., ... Bulik, C. M. (2006). Features associated with excessive exercise in women with eating disorders. *International Journal of Eating Disorders, 39,* 454−461.

Siegel, D. (2007). *The mindful brain: Reflection and attunement in the cultivation of well-being.* New York, NY: W.W. Norton and Company.

Solenberger, S. E. (2001). Exercise and eating disorders: A 3-year inpatient hospital record analysis. *Eating Behaviors, 2,* 151−168.

Strelan, P., Mehaffey, S. J., & Tiggemann, M. (2003). Self-objectification and esteem in young women: The mediating role of reasons for exercise. *Sex Roles, 48,* 89−95.

Strober, M., Freeman, R., & Morrell, W. (1997). The long-term course of sever anorexia nervosa in adolescents: Survival analysis of recovery, relapse, and outcome predictors over 10−15 years in a prospective study. *International Journal of Eating Disorders, 22,* 339−360.

Sundgot-Borgen, J., Rosenvinge, J. H., Bahr, R., & Schneider, L. S. (2002). The effect of exercise, cognitive therapy, and nutritional counseling in treating bulimia nervosa. *Medicine & Science in Sports & Exercise, 34,* 190−195.

Szabo, C. P., & Green, K. (2002). Hospitalized anorexics and resistance training: Impact on body composition and psychological well-being. A preliminary study. *Eating and Weight Disorders, 7,* 293−297.

Taylor, W. C., Baranowski, T., & Sallis, J. F. (1994). Family determinants of childhood physical activity: A social-cognitive model. In R. K. Dishman (Ed.), *Advances in exercise adherence* (pp. 249−290). Champaign, IL: Human Kinetics Publishers.

Thien, V., Thomas, A., Markin, D., & Birmingham, C. L. (2000). Pilot study of a graded exercise program for the treatment of anorexia nervosa. *International Journal of Eating Disorders, 28,* 101−106.

Tokumura, M., Yoshiba, S., Tanaka, T., Nanri, S., & Watanabe, H. (2003). Prescribed exercise training improves exercise capacity of convalescent children and adolescents with anorexia nervosa. *European Journal of Pediatrics, 162,* 430−431.

Touyz, S. W., Beumont, P. J. V., & Hoek, S. (1987). Exercise anorexia: A new dimension in anorexia nervosa? In P. J. V. Beumont, G. D. Burrows, & R. C. Casper (Eds.), *Handbook of eating disorders: Part 1. Anorexia and bulimia nervosa* (pp. 143−157). Amsterdam: Elsevier.

Touyz, S. W., Lennerts, W., Arthur, B., & Beumont, P. J. V. (1993). Anaerobic exercise as an adjunct to refeeding patients with anorexia nervosa: Does it compromise weight gain? *European Eating Disorders Review, 1*, 177–181.

Trost, S. G., Owen, N., Bauman, A. E., Sallis, J. F., & Brown, W. (2002). Correlates of adults' participation in physical activity: Review and update. *Medicine and Science in Sports and Exercise, 34*, 1996–2001.

Vansteelandt, K., Rijmen, F., Pieters, G., Probst, M., & Vanderlinden, J. (2007). Drive for thinness, affect regulation and physical activity in eating disorders: A daily life study. *Behaviour Research and Therapy, 45*, 1717–1734.

Veale, D. M. W. (1987). Exercise dependence. *British Journal of Addiction, 82*, 735–740.

Wall, R. B. (2005). Tai Chi and mindfulness-based stress reduction in a Boston public middle school. *Journal of Pediatric Health Care, 19*, 230–237.

West, J., Otte, C., Geher, K., Johnson, J., & Mohr, D. C. (2004). Effects of Hatha yoga and African dance on perceived stress, affect, and salivary cortisol. *Annals of Behavioral Medicine, 28*, 114–118.

Wichstrom, L. (1995). Social, psychological, and physical correlates of eating problems: A study of the general adolescent population in Norway. *Psychological Medicine, 25*, 567–579.

Windauer, U., Lennerts, W., Talbot, P., Touyz, S., & Beumont, P. J. V. (1993). How well are "cured" anorexia nervosa patients? An investigation of 16 weight-recovered anorexic patients. *British Journal of Psychiatry, 163*, 195–200.

Wyer, K. (2001). Mirror image: Yoga classes at the Monte Nido clinic are changing how women with eating disorders see themselves. *Yoga Journal*, 70–73.

Yates, A. (1991). *Compulsive exercise and the eating disorders: Toward an integrated theory of activity*. New York, NY: Bruner-Mazel.

CHAPTER 26

Body Talk
The Use of Metaphor and Storytelling in Body Image Treatment

Anita Johnston

Once upon a time, there was a king who had been on a hunting expedition in a far away land. Upon his return to the kingdom, he was greeted by his most loyal and faithful companion, a dog who he had raised since she was a puppy. The king's joy at reuniting with his "best friend" quickly turned to concern when the dog began to behave in the most unusual manner. Rather than wagging her tail playfully as she usually did, the dog kept turning 'round and 'round in circles, baring her teeth, snarling, and barking at the king. She would run away from the king as he approached her, stop a short distance away, and then return, barking even more loudly than before and running in large circles around him. Then she would repeat this action, over and over. The king was troubled with this odd behavior and he figured that, in his absence, his dog had gone mad. So he began to chase after the dog as it ran towards the castle, following it down a narrow, twisting corridor, which eventually led to the nursery of the king's first born. When the dog stopped abruptly at the entrance to the nursery and faced the king, he noticed with alarm that her muzzle was covered with blood. And as he peered inside the nursery he was horrified to see it in complete disarray: the walls were splattered in blood, and the baby's cradle was overturned! The king immediately became enraged at having been betrayed by his "best friend," swiftly drew his sword, and with the full force of his fury and anger, he plunged it into the dog's heart. Less than a moment later, he heard his baby cry and he rushed over towards the overturned cradle. There, beneath the bloody carcass of a wolf, he found his infant daughter—completely unharmed.

Like the king, a woman struggling with body image issues tells herself a story ("I am too fat" or "I am ugly"), but fails to question what is real and true. She is not conscious of the inner stories that are causing her great suffering. Unaware that she is blaming the wrong aspect of herself for her misery, she turns against her body the way the king turned against his dog. Because her physical appearance is the first thing she scrutinizes when she senses something is wrong, she turns the full force of her fear, disgust, and rage against it. Not taking the time to fully examine the whole picture, she responds, instead, to the initial thought that comes into her mind when distressed ("I am fat"). She immediately assumes it is her body, the home of her appetites and instincts, which is the enemy that has caused her the pain with which she struggles, and develops a complicated narrative to support that, just as the king did.

This chapter proposes the use of ancient teaching stories, myths, and folk tales as a means of investigating and exploring the issues necessary for recovery and for developing a healthy relationship with the body. According to Campbell (1988), myths guide us in the "experience of being alive" and are sustained in the human imagination over long periods of time because they ring true to our shared experience. As such, they are told to enlighten us about our inner realities and can take us to deeper truths that can inform, instruct, and heal. These "wisdom tales" can be used as vehicles to transport clients to the heart of the matter, to a greater understanding of their struggles by allowing them to examine their own inner stories from another perspective. Through these stories, the dilemmas with which clients struggle can be portrayed by different characters, in a different place and time, so that clients can recognize their struggle for what it is, before their "resistance" has had a chance to come into play.

What if the king had taken a moment to question the situation rather than believe the first story that came into his mind? What if he had assumed that the dog's "crazy" behavior had meant something other than, "There must be something wrong with my dog?" What if the king were as loyal to his dog as she was to him? What if he had stopped to decipher what the dog was trying to say and trusted her instinctual nature? Could the tragedy have been averted? Could the mistake have been prevented? Could the king have spared himself untold grief? And, similarly, can the use of metaphor and myth disrupt the power of eating disordered thoughts and body image dissatisfaction?

ABOUT BODY IMAGE AND EATING DISORDERS

Once an esoteric psychological concept, body image is now a household word, seen as a normative problem for most Western women, in part due to the ever-expanding role of the mass media promulgating unreasonably thin bodies as the standard of beauty, and to the many changes in women's sociocultural roles (Maine & Kelly, 2005; Wolf, 2002). Since 2004, a professional journal (*Body Image: An International Journal of Research*) has been devoted to the research and understanding of its function in psychological and physical well-being, and articles abound in numerous other, more generic mental health publications. It is a "hot topic" among graduate students in psychology, resulting in countless theses and dissertations each year. Still, eating disorder (ED) experts and programs struggle to help clients address the ruinous power negative body image holds in their lives.

According to Freedman (1988), one of the early contributors to the treatment of body image disturbances, body image is "an inner view of our outer self…body image can feel as real as the body itself…a constant source of strength or a chronic cause of pain" (p. 8). Hutchinson (1985), another pioneer in this area, describes body image as the "piece of psychological space where your body and mind come together" (p. 48). She explains that it is deeply personal, reflecting how the owner sees her body, and not necessarily how others do. It can be both dynamic and flexible, but also stable and fixed; an individual can be either painfully aware of her body image or totally unconscious. Kearney-Cooke (1989), in her seminal work on body image development and treatment, describes how body image may actually begin before birth with parental projections about the gender and physical characteristics and appearance of their anticipated baby. In her view, body image is an organic construct, as experiences throughout life affect it. These experiences can either be normal developmental changes (such as puberty) or traumatic and unexpected (like childhood sexual abuse). She explains that having a sense of mastery of one's body may generalize to a healthy sense of personal mastery.

A distorted body image, and overvaluation of the importance of weight and shape to self-esteem and identity are central aspects to clinical ED. Prompted by their diagnostic relevance, numerous assessment tools have been developed for both empirical and clinical practice, including the Body Image Disturbance Questionnaire (Cash, Phillips, Santos & Hrabrosky, 2004b), the Appearance Schemas Inventory (Cash, Melnyk & Hrabrosky, 2004a), the Body Image Guilt and Shame Scale (Thompson, Dinnel & Dill, 2003), and the Body Checking and Avoidance Questionnaire (Shafran, Fairburn, Robinson & Lask, 2004). Body image is often the last remaining symptom for those in recovery from an eating disorder and may trigger relapse for some (Costin, 2009). In their review of literature related to body image, Heinberg and Thompson (2006), note that, despite much research, little of this has translated into specific treatment of body image issues related to ED and that empirically based treatments and manualized care for ED "often fail to fully target the body image component" (p. 89). According to Wilson, Grilo & Vitousek (2007), the most researched or evidence-based treatment modality for bulimia nervosa, cognitive-behavioral therapy, is only effective in 30–50% of cases treated. Thus the gap between research and practice grows, a dangerous gap in light of the costs of ED to health and well-being.

Developments in neuroscience (see Lapides, Chapter 3) have shown that the left hemisphere of the brain is the rational, conscious, language-based arena, while the right hemisphere is where early, preverbal experience rests and is the center for both empathy and regulation of affect. It is in the subcortical areas of the right hemisphere where ED are triggered and where "bad body image" stories and disordered eating arise. It makes sense that metaphor and myths, reflecting right brain activity, can help the conscious brain to address the powerful scripts leading to problems like ED. Neuroscience also demonstrates that adult brains, which contain the physical signature of our thoughts, are not fixed and incapable of change. Rather, they are adaptable, plastic, ever changing, and capable of enormous transformation. Changing thoughts can rewire brains, forming new neural connections. Through conscious exploration of links between certain thoughts and feeling states, old ways of responding can be unlearned and their actual neuronal links weakened, and new ways of thinking and responding can replace them (Doidge, 2007).

METAPHOR AND HEALING

Traditional tales, like that of the king's story, are best understood as metaphor, which Jung called "the healing symbol" because of its ability to provide us with images that can transform unconscious patterns into forms easily assimilated into conscious awareness (Woodman, 1993). Jung believed that metaphors affect the person on three levels: the mental, where we interpret meaning; the imaginative, where the actual transforming power resides; and the emotional, where we connect to the feelings embodied in the metaphor. He believed that metaphor functions *simultaneously* on these three levels, and because of this, a deep, more immediate connection to the psyche can be made (Woodman, 1993). For example, through the use of metaphor in the teaching story about the king and the dog, the client can mentally understand how it is possible to make errors in judgment. At the same time, the client's imaginative abilities are activated as she creates a picture in her "mind's eye" of the king, the dog, the nursery, the overturned cradle, etc. She can then begin to imagine ways in which she, like the king, might have made erroneous assumptions that have caused her great pain.

It is this imaginative function that can bring about the "in-sight" necessary for change (Woodman, 1993). Finally, since the metaphoric nature of the story is designed to activate emotion, the client can literally feel the emotions of surprise, anger, horror, or remorse embedded in the story while she makes a connection between the king's situation and her own emotions. While all of this is happening simultaneously, the therapist can then help her identify the true source of her pain and recognize the ways in which she has been attacking her body rather than challenging her inner critic. Once these unconscious patterns are brought into conscious awareness, the client can begin to develop the skills necessary for dealing with a fierce inner critic.

Metaphor can transform the perception of seemingly meaningless patterns of thought and behavior into an awareness that these patterns hold and reveal deeper truths (Woodman, 1993). This transforming function is a universal feature of metaphor that works similarly in myths, folk tales, dreams, and our own inner stories. It can give meaning to the struggle with negative body image and provide the awareness necessary for change and healing. Through the imagery of metaphor, the existence of deeper meanings that lie hidden beneath clients' attitudes towards their bodies can be revealed, helping them to understand how their negative thoughts about their bodies can actually be an attempt to direct their attention to the true source of their problem.

The woman struggling with body image or eating issues believes the problem lies in her physical appearance and does not realize that the real culprit, whom she cannot yet see, is her inner critic, the voice from within that tells her she is not thin, toned, or beautiful enough. It is her inner critic that has been stalking and preying upon her inner child and her creative spirit, the source of her greatest joy and the aspect of herself which holds the promise of genuine happiness. This insidious inner critic lies in wait, attempting to ambush her newborn ideas, feelings, and desires by insisting on perfection, decrying her worthiness, or accusing her of being weak for having needs and emotions. It is that voice that says: "You can't have it because you are too fat. When you lose weight, then you will deserve to be happy. You are not lovable unless you are thin and toned." She does not understand that her body is not the source of the distress she is experiencing; instead, it is the messenger, trying desperately to communicate to her through sensations, instincts, intuitions, or emotions, telling her when something is wrong. She cannot remember that her body is her most loyal friend and protector. Nor can she see or appreciate how her body remains truly dedicated in its service to her soul, her essential self, to who she really is. In sum, devoid of embodied understanding and insight, she falsely believes that it is her body that has betrayed her.

The inner critic is supported by a background chorus of voices in today's collective culture telling women that all their problems will be solved and their pain will be eliminated, if only their bodies were different. Billions of dollars are being thrown to the pack of media wolves that feed on this story. Women everywhere hear this false tale, told with a subversive intent to undermine their positive self-image in order to sell some product wrapped in the sheep's clothing of fashion, health, or fitness. It is almost impossible not to be impacted by this daily barrage of beauty bargains broadcast insidiously on television, the Internet, and billboards, and in movie theaters and magazines in the supermarket check-out line. Many women unwittingly join in the chorus, bonding with one another as they bemoan their physical imperfections and exchange disparaging comments about their appearance.

Those struggling with negative body image suffer terribly, not simply because they have bought products promoted in the media which have not delivered on their promise, but because they have bought into the story that they need to look thinner, younger, more this way and less that way in order to be happy, successful, and loved. Consequently, they participate in destructive and sometimes life-threatening behaviors in a futile attempt to get their bodies to conform to an "ideal" image that, however relentlessly promoted, is essentially unobtainable. This is as tragic a tale as can be told—no less brutal or violent than a Grimm's fairy tale.

It is a dreadful dilemma when women lose sight of their true identity, who they really are as human beings and as souls in a physical body, and instead over-identify with only one aspect of their psyche—the voice of the inner critic. It is a terrible tragedy when that voice becomes amplified, drowning out the voices of logic and reason and turning against the physical foundation of their lives, the very bodies they will inhabit throughout their entire lifetime. For those women who struggle with ED and negative body image, the attack is relentless, and is especially destructive to their feminine souls.

According to Jungian analyst Marion Woodman (1985), "The soul will not naturally reject its body image any more than the mother's breast will naturally reject her baby. Where rejection occurs, something has gone seriously wrong" (p. 56). What we feel in the pit of our stomach at the conclusion of the story about the king and the dog is that something has gone seriously wrong. According to White (2007), when someone is struggling with negative body image, "The soul has been sucked out, locked outside the body, and is desperately trying to get back in—by objectifying the body."

Part of our job as clinicians is to help to restore the soul to its rightful place, its birthright. We face many daunting questions and challenges. How can we assist in the embodiment of the soul and help inoculate our clients against the culture's pervasive objectification of the body? What will it take to awaken girls and women with negative body image to the truth of who they really are? How can we prevent boys and men from joining their ranks? How can we help them re-member the relationship they once enjoyed with their bodies as young children before they internalized the destructive lies of the collective culture and engaged in mental dis-memberment of certain body parts? And how can we help them reconnect in a joyful way with their bodies—teach them to love, not kill, the messenger—to honor and listen to the body's wisdom? Adding the use of metaphor, with its access to the right hemisphere of the brain, the home of preverbal experience, empathy, and affect regulation, to our clinical approaches to those suffering from ED, may be a powerful and transformative tool.

LEARNING THE LANGUAGE OF SYMBOLISM AND METAPHOR

Unless those with negative body image issues understand the language of symbolism and metaphor, they can make the mistake of taking the images in their mind's eye literally, assuming they are having negative thoughts about their bodies because something is wrong with their physical appearance. They must learn to examine thoughts about their bodies carefully and consciously, translating the symbolic language of the thoughts and images that appear, bringing what has heretofore been unconscious into conscious awareness. In this way, they can connect to the deeper truth of their experience and recognize what it is that is really begging for their attention. Otherwise, they risk being like the king, making incorrect assumptions, leading to destructive behaviors and untold misery.

An ancient Zen saying instructs us that, in order to see the moon, we must allow our gaze to go beyond the finger pointing to it. Negative thoughts about one's body are like the finger pointing towards the moon. To see the light, we must look beyond what is merely pointing to it.

Because the causes of body image disturbance are unconscious and deeply rooted in nonverbal experience, those struggling with negative body image find it difficult to believe that their physical appearance is not the real problem. Using the metaphor of the "red herring" can help to introduce the idea that what initially appears to be the source of their misery may not, in fact, be the real cause.

Those who are caught up in obsessing about their bodies, scrutinizing every flaw, and agonizing over every half pound, are being distracted from the real issues that are causing them pain. Unable to distinguish between their inner symbolic imagery and their outer material world, they become fixed on the physical, concrete representations of the "flaws" and "heaviness" of their lives as it appears in matter and in their bodies. As long as they keep their focus on their physical appearance, the real problems creating feelings of heaviness and deficiency can never be addressed and resolved. Their pain continues and their obsessions intensify.

In addition to being the language of our folktales, poetry, and dreams, metaphor is the language of our bodies. We have "gut" reactions and "heartaches," experience nuisance as a "pain in the neck" and get "sick to our stomachs" when disgusted. Our bodies talk to us and for us in this symbolic language. In our contemporary culture where physical appearance is overvalued and emotional literacy devalued, metaphoric language is common in the communication between teenage girls, as they use disparaging comments about their bodies when they are actually trying to communicate about some inner angst. When listening to their conversations with an inner ear, one that is attuned to the language of metaphor, "My thighs are huge and disgusting," can be heard as an attempt to talk about overwhelming feelings of shame. "Does my butt look big?" may be heard as a plea for encouragement when feeling insecure. "I hate my flabby stomach" might actually be an unconscious expression of fear and loathing about becoming a woman. When working with negative body image and ED, the language of metaphor can be used to communicate the complex issues that, more often than not, underlie obsessions with physical appearance, eating, and food.

Rudolf Steiner, philosopher and educator whose work is the foundation of the Waldorf Schools, stated that the intellectual powers of rational thought, judgment, and critical thinking need to be rooted in the ground of feeling and imagination in order for them to develop properly (Richards, 1980). Without imagery, which he referred to as "living pictures" seen in the "eye of the mind," complex intellectual concepts cannot be fully grasped. He believed that young children need to develop their imaginative function and enhance their connection to feeling states in their bodies and in their emotional experiences. According to him, this is necessary in order to prepare children for proper development of their intellectual faculties. For this reason, in Waldorf pre-schools and kindergartens, physical and artistic exercises are used to teach simple concepts like the alphabet or counting. "Picture shirts" and

"Disney talk" are forbidden because of the belief that constant exposure to prefabricated images created by others can interfere with young children's developing imagination and impair their ability to fully grasp the complex intellectual ideas they will be taught later.

When working with those who struggle with negative body image, we often see how their critical faculties are overdeveloped and are not grounded in a connection to their physical or emotional experience. They are capable of discerning every flaw in their bodies and they become obsessed with these imperfections, but remain unaware of their overly inflated self-criticism. They are irrational, harshly judgmental, and excessively critical toward their bodies. Consequently, they are extremely vulnerable to the images presented by the media about what their bodies should look like, and their capacity to create an accurate picture of their bodies in their mind's eye is impaired. They are disconnected from their bodily senses and oblivious to emotions being the cause, rather than the result, of their poor body image. Their intellectual understanding of the discomfort they are experiencing is not fully grasped and their sense of their bodies is distorted. In sum, their suffering is caused by distortions in their imagination, not by the size and shape of their bodies.

Metaphor can be an exquisite means of cultivating a client's ability to imagine and *feel* the inner truth of their body obsessions. According to Woodman (1993), storytelling and the use of metaphor have more of an immediate impact than abstract analysis when working with ED and distorted body image because of its physical impact:

> So long as it's theory, it's removed from the actual feeling…if I put it in a story form or use images, the mind may not hear it, but the body responds. And if it's reverberating in the body, sooner or later, it's going to get through to consciousness (p. 53).

By teaching the language of metaphor, with its inherent imaginative function, therapists can help clients to cultivate their imagination and use it constructively, and to find and decipher the metaphors in their eating and body image issues, and in their lives, rather than using it to imagine what is wrong with them. By becoming more proficient with this language, clients can begin to see how their negative thoughts about their bodies can be used like a metal detector at the beach, to uncover deeper, more meaningful truths about themselves and their lives. As they examine the self-critical and judgmental thoughts that have taken them over, they can recognize and feel their deeper emotions in the process. They can learn to look beyond the finger pointing to the moon and find the illumination that will shed light on their true struggles.

Dream analysis explores the imagery found in stories our minds tell as we sleep. If dream images are interpreted literally rather than symbolically, however, the true meaning of the dream can be lost. A simple, too literal interpretation of dreams can be considered ludicrous, even dangerous. For example, a dream about having an affair with your accountant does not mean you literally should have sexual intercourse with that person. If, however, the dream is looked at and understood symbolically, it can be interpreted as directing you towards a "loving union" with the part of your self that needs to be "accountable." Just as we understand the distortion that can occur if dream images are interpreted literally rather than symbolically, we can understand how those who have taken too literal an interpretation of the images about their bodies in their waking mind's eye have missed the deeper truth to be found in their symbolic message. Understanding the metaphoric language of imagination—images found in the eye of the mind—can free clients from false beliefs.

Utilizing metaphor in the exploration of a "fat attack" can be instructive. A "fat attack" is when someone struggling with eating or body image issues wakes up in the morning and "feels" as though she has gained 20 pounds overnight. Now, with her logical mind, she may understand that this is not possible, but because her critical thought processes dominate and are not adequately rooted properly in the "ground of imagination," she believes that because she "feels" fat, she must "be" fat. Irrationally, she thinks that because she feels badly and has the thought, "I am so fat," she must be feeling badly *because* she is fat. The "fat attack" feels real because the emotional pain it causes is real. She doesn't realize that the actual source of her pain is embedded in her ideas about what it *means* to be fat and the associations she has made to the word fat. If, however, she can recognize the metaphoric language of "feeling fat," she can begin to decipher its real meaning and discover the true function the "fat attack" has served in her life.

TRANSLATING A "FAT ATTACK" INTO METAPHORICAL LANGUAGE

In the language of metaphor, a "fat attack" can be understood as a "fear attack." The feeling state that rises up in the body is actually one of fear. The question then becomes: "Fear about what?" The answer lies in the symbolic meaning of "fat" to that individual. Fat may be seen as "insulation" from the outside world, and the associated fear comes from feeling too vulnerable and needing more protection. Questioning from what the client needs protection can lead to exploring the areas in her life where she does not feel safe, or where she feels her personal space has been invaded. She might want to look at who has trespassed in her physical, emotional, or spiritual space.

Maybe fat represents fear of being too visible and thus vulnerable to criticism or verbal attack. In what situations and by whom, has revealing her unique perspective or her true feelings been discounted or rejected? Or maybe the curves that fat brings to her body can be symbolic of her feminine sexuality. Is she concerned about her sexual instincts becoming dangerously out of control, or that others may objectify her, desire her in ways she doesn't want to be desired, or take advantage of her inability to say "no" clearly and directly? Has she used her fat to speak for her, to say the "no" she could not verbalize directly? This line of inquiry can lead a client towards the real source of her pain and fear and can then point her towards the development of skills necessary for providing the protection she needs: assertive communication, limit setting, boundaries, etc.

Perhaps she has an inner story that the curves in her body created by fat cells are what led to the sexual abuse she experienced, not realizing (like the king) that she is erroneously blaming the shape of her body rather than directing her anger towards the real culprit. For those who have experienced sexual abuse as children, this is not unusual. Because fear and shame of abuse are encoded in a physical experience, developing minds can easily see their physical bodies as the betrayer, rather than blaming the actual perpetrator. This can be especially true if the perpetrator was a loved one and someone with whom they could not "afford" to be angry. In a child's mind, "feeling bad" can seem the same as "being bad." This erroneous belief may have been the safest assumption to make while they were still young, vulnerable, and dependent upon the abuser. However, this does not make it accurate.

Clients in this situation need to be assured that the inner stories they created may have been the very best conclusions a young child could have come up with; believing them may have, in fact, saved their lives or kept them connected to those who were essential for their early survival. But, as adults, they are capable of more apt interpretations of traumatic events. Stories created with the mind of a child need to be later examined with the mind of an adult to see if they are, in fact, logical, truthful, sensible, and accurate. Is it true that if bad things happen it means we are bad? Is it true that to feel bad is the same as to be bad, to feel fat is the same as to be fat, to feel bad is to be fat? Are these words logically interchangeable?

Freud (1950/1985) observed that events, rather than leaving permanent memory traces in our minds, are altered in our memories by subsequent events and "retranscribed." Thus, events can take on new and different meanings many years after they initially occurred, as we continue to alter our memories of the events and the way in which we interpret them. Memories are constantly reconstructed, much in the same way that countries write and re-write historical events. Because our modern Western culture perpetuates the idea that fat is bad, a traumatic incident where an abused child felt bad and believed she was bad can be retranscribed with the mind of an adolescent, overlaying the old story line with one that equates "fat" with "bad." The new story can now be heard as, *"if I feel bad, I must be fat."* According to Freud, in order for memories to be changed they must be brought into conscious focus and awareness. Unfortunately, many traumatic memories or events that happened in childhood are not easily accessible, so the inner stories that have been constructed around them are not revisited and corrected in adulthood.

To untangle the story line, it helps to have the client recognize, with her adult logic, that feeling "bad" is a *signal* from her emotional guidance system that something is amiss. Her task is to become a good detective and not automatically assume that feeling bad/fat is the same as being bad/fat; that if she is feeling threatened, the cause of her feeling is not her body—although she may *experience* it in her body. Her instincts, emotions, and intuitions, which reside in her body, are messengers, trying to alert her to a problem that needs her attention. To avoid making the same error as the king, she needs to suspend immediate judgment as she follows the twisting corridor towards the real source of her discomfort, and use her curiosity to discern where the real trouble lies. She needs to discover that the real problem is in a thought process which links feelings of distress with being fat.

BREAKING WITH TRADITION

There is the tale of the young couple who were newlyweds. One evening, as they were fixing dinner, the husband, who was preparing a salad, asked his wife why she was cutting the ends off of the roast as she was putting it into the pan. She replied that she remembered that her mother made the most wonderful roast and always cut the ends off—just as she was doing. One evening, the couple had her mother over for dinner and the curious husband asked her why it was that when she cooked a roast she cut the ends off. She answered that she really didn't know. This was simply how her mother had taught her to do it and she had always done it that way. At Christmas time, when the entire family was gathered around the dinner table, the young husband happened to be seated next to the grandmother. As the course of conversation turned to family recipes, he turned to her and asked why she always cut the ends off of a roast. She replied, *"Well, I never seemed to have a pan quite big enough."*

In grandma's day, making do with what you had made sense. A quick trip to the supermarket wasn't an option. Although times had changed, no one stopped to question the story, assuming that there was still a good reason for the behavior. So it is with some of the stories that live unexamined in our minds, many of which have been passed on from generation to generation. Body image distress has become normative for Western women, a powerful and unrecognized hand-me-down from one female generation to the next (Maine & Kelly, 2005). Breaking with tradition is a big step, and breaking with the tradition of feeling fat and inferior is particularly difficult for women.

Sometimes when exploring the inner story about what fat means to those struggling with body image issues, it is worthwhile to follow the story line to its origins: When did you first get the idea that fat is bad? Who told you? How old were you? It is not unusual to find that these inner stories about fat were descriptions they once heard about themselves from others.

According to body image specialists, negative body images absorbed from childhood teasing and taunts can easily take root in the minds of children between the ages 8 to 16, with long-lasting effects way into adulthood (Cash & Pruzinsky, 2002; Thompson, 1990). Once engraved in the brain, these images can affect countless situations in their adult lives. How can they stop these old images from continuing to affect them on a daily basis? How can the therapist get access to these images encoded in the mind's eye so they can be replaced with a kinder, more accurate perception of the self? One way is through the imagery found in parables and storytelling.

> Once, deep in the forest there was a nightingale singing and singing and singing high in the tree-tops. As she was taking great pleasure in hearing her sweet and throaty melodious song reverberate throughout the entire forest, along came a crow who began to berate her for making too much noise. *"Cut the racket!"* he called out. *"Oh,"* replied the nightingale, *"This is not noise. We nightingales are known the world over for our beautiful song. Listen!"* and she began to sing even more enthusiastically. *"You sound horrible!"* exclaimed the crow, *"That is the worst example of singing I have ever heard. My voice is by far greater than yours!"* The indignant nightingale became infuriated over the crow's rude interruption and the harsh criticism he had heaped upon her. And the two birds began to quarrel heatedly. Shortly thereafter, along came a pig, who proclaimed, *"Look, you two, I am a connoisseur of fine music. Why don't we settle this matter once and for all by having a contest for which I will be the judge?"* The crow and the nightingale agreed and so the nightingale sang her lovely song and the crow cawed loudly. Immediately, the pig declared the crow the winner and the nightingale burst into tears. *"I can't believe you!"* said the crow with disdain, *"Not only are you a terrible singer, but you are a lousy loser to boot!"* *"I'm not crying because I lost,"* said the nightingale, *"I am crying because a pig was my judge."*

The moral of this story is: Consider the source. This is exactly what an individual struggling with body image issues must do. First, she must identify those interior stories with themes derived from adolescent or childhood experiences that she is still carrying around (e.g., "I am too fat, too flat, too ugly") and resurrect painful memories of being teased (about having "thunder thighs" or a "bubble butt" and being called names like "cow" or "fat pig"). Next, these stories and memories must be processed through a discerning filter, examining exactly where they came from, who said these words, and in what context. She needs to scrutinize the source with the same tenacity with which she has attacked her appearance.

Therapists can facilitate this re-examination process by asking directive questions, enabling the client to understand what else was happening and how her unspoken story functioned in response to these experiences. For example, if a client was teased by an older sibling, the following types of questions can expand the client's perspective: "How old was your brother when he called you a fat cow?" "How old were you?" "Did you idolize him and believe every word he said?" "Do you still?" "What do you think might have been going on with him when he said such things?" "What might he have been trying to protect himself against by attacking you?" "What feelings might he have been trying to express and didn't know any other way?" "Could there be another way of understanding why he said such things?"

If peers were the source of disparagement, another line of questioning may be in order: "Who were the girls/boys who put you down for your appearance?" "What can you imagine was going on with them that they felt a need to do so?" "How self-aware were they about their own insecurities and how articulate were they about their feelings at that age?" "What do you know now about their possible motives that you didn't know then?" "How could their feelings of inadequacy have affected their comments?" "What might have been their relationship to their own bodies?" "Is it possible that in an attempt to find relief from their own pain, they were projecting their own issues onto you?"

Finally, questions to pose to those whose parents were the sources of body derision include: "What was going on with your father when he started criticizing your weight?" "How old were you?" "Did he, himself, struggle with a weight issue?" "What weight prejudices/misperceptions had he been raised with?" "In what way might he have felt threatened by the curves of a woman, the power of feminine sexuality?" "Was he trying to protect you or himself from something?" "Did he believe, in some distorted way, that his criticism was helpful?" "What was your mother's relationship to her own body?" "How did she feel about the natural curves in a woman's body or her feminine sexuality?" "Did she celebrate or fear them?" "Did she diet, have body image issues, or make disparaging remarks about her body or those of others?"

While "considering the source" of repeated body disparagement can promote body image healing, many of those struggling with body image want to "forget" painful experiences without realizing how profoundly they continue to affect them. Much like when watching the news on TV, we tend to listen only to the reporters as they deliver the pressing new stories of the day. We ignore those stories we consider less important, "the crawl" scrolling at the bottom of the TV screen. There are stories that may not be in the forefront of our minds but, nevertheless, significantly influence and affect our body image. By ignoring them, we put ourselves at risk.

Unearthing childhood stories which were encoded with the mind of a child, and then examining them with the mind of an adult can be an essential part of the recovery process. The schoolyard taunts, the hurtful comments made by girlfriends, boyfriends, or family members, can take on very different meanings when re-examined with the mind of an adult. Ignoring or trying to let go of a painful story can prove to be a daunting task; however, once a story is fully examined in the light of consciousness and its erroneous conclusions are revealed, the story, itself, can let go of its destructive grip on the psyche.

A NEW LENS FOR FOOD AND FAT

Therapists who are adept with metaphor can assist their clients in looking at their eating and body image issues through a different lens, one that can help them see the deeper issues that lie beneath the more obvious struggles. In her writings about ED and addiction, Woodman (1993) suggests that "food is a metaphor for mother" (p. 115). If we think of mother as a verb, *mothering*, we can understand it metaphorically as providing nurture and comfort. Thus, we refer to certain foods as comfort foods. We use food to soothe ourselves, as though we recall at some deep level in our psyches our very first experience of being soothed and comforted, either at our mother's breast or with the bottle. We can use the metaphor of mothering to help those struggling with ED see that their struggle with food and eating is actually a struggle around obtaining just the right amount of "mothering" they desire, not too much, nor too little. Through metaphor, they can begin to understand that food can be a symbol, no less iconic than the American flag. At the symbolic level, they can understand that neither food nor their actual mothers can provide them with all the mothering they need any more than collecting thousands of American flags will bring them all the freedom they desire.

By working with food metaphorically, clinicians explore the ways in which clients tend to "mother" themselves and can help them create an "inner mother" that is neither overly indulgent ("Chocolate for breakfast? Whatever you want, dear") nor harshly critical ("Chocolate for breakfast? What is wrong with you?"); but rather, curious and supportive ("Chocolate for breakfast? What's that about?"). Those attempting to recover from an ED need to recognize the metaphoric nature of their hunger and learn to discern when it is literal food they need as fuel for their body and when they need food as emotional sustenance to soothe and comfort their soul (e.g., attention, affection, appreciation, acknowledgment). They need to understand the symbolic nature of mothering, so they can feed and nourish themselves accordingly, and cultivate nurturing relationships with others: friends, pets, spouses, or Mother Nature.

When working with metaphor, clinicians need to avoid getting stuck on the literal, physical object and instead keep attention on the metaphoric function of that object to see if there is an analogy that works within the psyche that may bring clients to a deeper truth about their struggle. In that regard, I work with fat as a metaphor for *fathering*, which can be understood metaphorically as providing security and protection. For our ancestors, the function of fat was to provide security and protection against famine. It was tangible evidence of a successful provider. Although in our modern culture, it appears as though money has replaced that function, the fact remains that in our bodies, it is fat that provides insulation to keep us warm, prevents the loss of body heat, cushions us from blows, and keeps us from depleting other valuable energy sources in the body. On a very basic physical level, it provides security and protection, just as the father has traditionally done.

Those struggling with body image issues have an aversion against "too much" fat on their bodies. It is helpful for the therapist to question, "What does that mean metaphorically?" If the function of fat as insulator or protector can be explored, these clients might find some deeper meaning in their body fat rather than it being a source of self-loathing or an indication of failure. For example, those who have struggled with compulsive over-eating and yo-yo dieting often report with dismay that whenever they have lost weight, they have gained it back. Essentially, they do not have a strong enough inner father that can say "no" often and appropriately enough to keep them feeling safe. If they can get clear about what it is they need extra protection from, and understand the ways in which their fat mediates between them and the outside world, they can then develop adequate coping tools to replace this function. They can appreciate their need for stronger boundaries and develop ways to protect themselves from those that would intrude upon their physical, emotional, or spiritual space. They can begin to develop an "inner father" that can anticipate dangerous situations and provide them with the skills necessary to maneuver through life safely and securely.

Those who struggle with anorexia nervosa are fearful of even the smallest amount of fat on their bodies and are locked into a perception of fat as the enemy and their bodies as the battleground. Their inner father says "no" to everything to keep them safe, and their negative body image is overwhelming. Here it can be helpful for therapists to question, "What is so frightening about fat?" "Why is it so abhorrent that they want none of it at all?" "What might they be saying metaphorically?" "If fat is a symbol for fathering, could they be saying they have too much fathering?" "Is it possible that because of early traumatic experiences they have dealt with feelings of extreme vulnerability by developing an overly protective 'inner father' who says no to everything that might possibly be dangerous (e.g., intimate relationships, emotional expression, sexuality)?" By exploring the symbolic nature of fat as father, they might discover an inner father who has gone overboard in creating an excessively overprotective environment for them, giving them a litany of stringent rules designed to control their appetites, make them invisible, and curb their participation in a world that does not seem safe. Or they might discover they are constantly being put "on restriction" by an excessively punitive "inner father" scolding them harshly for daring to indulge in pleasure or for making mistakes. Once they can understand that their fear of fat could be an attempt to control and avoid negative experiences, they can begin to address the real problems. They can understand that fat attacks are a signal that they feel unsafe and insecure and that developing skills to create feelings of safety and protection is the solution.

Recovery from negative body image requires developing an understanding of what fat and weight mean to the individual struggling with them. The use of metaphor can bring breadth and depth to this understanding, enabling the individual to look at the meaning they give to fat and weight from a broader perspective, allowing them to get to the deeper truth of their inner experience. Through understanding metaphoric language, they can begin to recognize that their negative thoughts are directing them towards a greater understanding of who they really are and what they really want and need.

FINAL THOUGHTS ON THE TRANSFORMATIVE POWER OF METAPHOR

As therapists encourage their clients to view their experiences with food, fat, dieting, and body image through the lens of metaphor, using the functions that are now available to them with their mature adult minds, they can help them identify the story lines that lie hidden beneath their struggles. They can help them see that, because they have interpreted their negative experiences and thoughts about their bodies literally, rather than symbolically, they have come to faulty conclusions about what is really wrong in their lives. The stories they created have distracted them from where the real problems are located in their lives and thus, where the solutions are to be found. Their literal focus on their weight and body shape has interfered with finding the freedom they seek—causing, rather than relieving, their pain and suffering.

Through storytelling, clinicians can teach the language of metaphor, the very language that can be found in clients' self-images, bodies, relationships with fat and weight, and in the stories of their lives. By examining these stories in a therapeutic setting and helping clients find the symbolism in their body image struggles, therapists can assist in bringing unconscious issues and memories into consciousness where they can be addressed and resolved. Clients can then learn the life skills they need in order to have lives that are free from struggles with food, fat, and dieting.

The task of the clinician, ultimately, is to help the client struggling with a body image disorder to understand that her recovery lies in changing how she sees herself, and not in changing how she looks. Freedom is to be found in changing her *thoughts* about her body, not in changing her body. Through the use of metaphor, therapists can then assist the client in shifting her focus from how she *looks* to how she *sees*.

References

Campbell, J. (1988). *The power of myth*. New York, NY: Doubleday.

Cash, T. F., Melnyk, S. E., & Hrabrosky, J. I. (2004a). The assessment of body image investment: An extensive revision of the Appearance Schemas Inventory. *International Journal of Eating Disorders, 35*, 305–316.

Cash, T. F., Phillips, K. A., Santos, M. T., & Hrabrosky, J. I. (2004b). Measuring negative body image: Validation of the Body Image Disturbance Questionnaire in a nonclinical population. *Body Image: An International Journal of Research, 1*(4), 363–372.

Cash, T. F., & Pruzinsky, T. (Eds.). (2002). *Body images: A handbook of theory, research, and clinical practice*. New York, NY: Guilford Press.

Costin, C. (2009). The embodied therapist: perspectives on treatment, personal growth, and supervision related to body image. In M. Maine, W. N. Davis, & J. Shure (Eds.), *Effective clinical practice in the treatment of eating disorders: The Heart of the Matter* (pp. 179–192). New York, NY: Brunner-Routledge.

Doidge, N. (2007). *The brain that changes itself*. New York, NY: Penguin Books.

Freedman, R. (1988). *Bodylove: Learning to like our looks and ourselves*. New York, NY: Harper & Row.

Freud, F. (1985). *The complete letters of Sigmund Freud to Wilhelm Fliess, 1887–1904*. (J. M. Masson, Ed. & Trans.). Cambridge, MA: Belknap Press. (Original work published in 1950).

Heinberg, L. J., & Thompson, J. K. (2006). Body image. In S. Wonderlich, J. E. Mitchell, M. de Zwaan, & H. Steiger (Eds.), *Annual review of eating disorders, Part 2, 2006* (pp. 81–96). Oxon, UK: Radcliffe Publishing.

Hutchinson, M. G. (1985). *Transforming body image: Learning to love the body you have*. Freedom, CA: Crossing Press.

Kearney-Cooke, A. (1989). Reclaiming the body: Using guided imagery in the treatment of body image disturbances among bulimic women. In L. M. Hornyak, & E. Baker (Eds.), *Experiential therapies for eating disorders* (pp. 11–33). New York, NY: Guilford.

Maine, M., & Kelly, J. (2005). *The body myth: Adult women and the pressure to be perfect*. Hoboken, NJ: John Wiley & Sons.

Richards, M. C. (1980). *Toward wholeness: Rudolf Steiner education in America*. Middletown, CT: Wesleyan University Press.

Shafran, R., Fairburn, C., Robinson, P., & Lask, B. (2004). Body checking and its avoidance in eating disorders. *International Journal of Eating Disorders, 35*, 93–101.

Thompson, J. T. (1990). *Body image disturbance: Assessment and treatment*. New York, NY: Pergamon Press.

Thompson, T., Dinnel, D. L., & Dill, N. J. (2003). Development and validation of a Body Image Guilt and Shame Scale. *Personality and Individual Differences, 34*, 59–75.

White, F. (2007, September). Body image and exercise resistance. *Tending the feminine psyche*. Symposium conducted by Inner Escapes Workshops, Hilo, Hawaii.

Wilson, G. T., Grilo, C. M., & Vitousek, K. M. (2007). Psychological treatment of eating disorders. *American Psychologist, 62*(3), 199–216.

Wolf, N. (2002). *The beauty myth: How images of beauty are used against women*. New York, NY: William Morrow.

Woodman, M. (1985). *The pregnant virgin: A process of psychological transformation*. Toronto, Canada: Inner City Books.

Woodman, M. (1993). *Conscious femininity: Interviews with Marion Woodman*. Toronto, Canada: Inner City Books.

PART VI

BRIDGING THE GAP: FUTURE DIRECTIONS

27 *The Research—Practice Gap* 459
28 *Call to Action* 479

CHAPTER 27

The Research–Practice Gap
Challenges and Opportunities for the Eating Disorder Treatment Professional

Judith D. Banker and Kelly L. Klump

The field of eating disorders (ED) is experiencing a research–practice gap with far-reaching implications for the quality of patient care, research, prevention, and education. Important research findings are not making their way into clinical practice and vital clinical observations are not impacting the direction of research. While researchers seek ways to disseminate their findings to practitioners, practitioners express concern about the extent to which research findings can be applied to their practice. According to the Institute of Medicine, a 17-year lag exists between the publication of new knowledge from randomized controlled trials (RCTs) and its integration into treatment practices (Institute of Medicine, 2001). This extensive gap is intolerable in the field of ED, where the mortality rate, the burden of illness, and the lack of specialized treatment necessitate an urgent, effective response from the professional community and those responsible for health care policy. Implementation of strategies to bridge the research–practice gap is urgently needed. But to be relevant and impactful, these strategies must develop from a clear understanding of the causes for research–practice tensions.

Emerging in the 1970s, the field of ED was founded on the observations, theoretical models, and research of clinical scholars. In 1978, Hilde Bruch published descriptions of her work with young patients with anorexia nervosa (AN) in her book *The Golden Cage*. At the same time, our field discovered the late nineteenth century writings of the British physician William Gull and the French psychiatrist Ernest Lasègue, who wrote about their encounters with self-starving patients (Gull, 1868; Lasègue, 1873). Finally, Gerald Russell's paper first described and named bulimia nervosa (BN) (Russell, 1979). Since these early observations, the interplay of theory, research, and practice that provided the foundation of our field has given way to an overriding emphasis on the primacy of RCTs over clinical observation and expertise. Related fields, such as medicine and psychology, have embraced Evidence-Based Practice (EBP), shifting toward empirical research as the primary source of knowledge.

The adoption of EBP by the mental health field has generated significant controversy and disagreement over the relative value placed on research evidence versus clinical expertise. For example, although the ED research and practice guidelines promote the use of EBP, clinical practitioners show limited utilization of such practices, especially the use of manualized treatments (Crow, Mussell, Peterson, Knopke & Mitchell, 1999; Mussell et al., 2000; Wilson, 1998).

The divide between research and practice is not unique to ED. In fact, it is commonplace in fields with both an applied and basic science component—fields as seemingly far flung as software engineering, agriculture, and education (McConnell, 2002; McCown, 2001; Donovan, Bransford & Pellegrino, 2007). The field of ED is in a particularly challenging position in regard to managing this divide. The complex nature of ED as serious mental and physical illnesses with sociocultural, psychological, and genetic influences (Klump, Bulik, Kaye, Treasure & Tyson, 2009), requires us to draw upon research findings and clinical/practice expertise from a wide range of fields, such as medicine, neuroscience, psychology, epidemiology, education, and public health. Thus, not only must we contend with the tensions between ED researchers and practitioners, we must contend with the implications of research–practice tensions within this wide array of professions.

This chapter reviews evidence regarding the research–practice gap in the field of ED and explores possible explanations for the gap. It also provides clear, practical steps clinicians can follow to integrate research within their practice, thus enhancing the quality of their clinical work and informing new research directions. To provide a framework for these steps, we describe the process the Academy for Eating Disorders (AED) has followed to address research–practice tensions within its professional membership. We end by discussing actual and potential individual, institutional, and organizational challenges and opportunities that practitioners may encounter in the attempts to bridge the research–practice divide.

EVIDENCE FOR A RESEARCH–PRACTICE DIVIDE IN THE EATING DISORDERS FIELD

The ongoing debate over the utilization of empirically supported treatments (EST) by clinicians treating ED demonstrates the research–practice gap. Despite empirical support for the use of family-based therapy for adolescent AN (Le Grange & Eisler, 2009), cognitive behavior therapy (CBT) in the treatment of BN (Agras, Walsh, Fairburn, Wilson & Kraemer, 2000; Fairburn, Marcus & Wilson, 1993), and research supporting interpersonal psychotherapy (Fairburn, Kirk, O'Connor & Cooper, 1985; McIntosh, Bulik, McKenzie, Luty & Jordan, 2000; Agras et al., 2000), clinicians in community settings do not typically use these methods as their primary treatment approach (Arnow, 1999; Crow et al., 1999; Mussell et al., 2000; Wilson, 1998).

The most common reason cited by practitioners for not using EST as their standard intervention is that they are not adaptable for use with clients with multiple comorbidities or more severe symptoms, and that little guidance is provided about how to adapt EST to the needs of their individual clients (Haas & Clopton, 2003). On the other hand, researchers investigating practitioners' use of EST have speculated that lack of research training hobbles clinicians' ability to translate research into practice. Further, it is suggested that a lack of training and supervision in conducting EST (Crow et al., 1999), and perhaps even reactionary attitudes that lead to rigid clinging to familiar practices (Fairburn, 2005b), are obstacles to the learning and adoption of EST. Attempts to increase the dissemination of EST have focused on the publication of treatment manuals (Agras & Apple, 1997; Lock, Le Grange, Agras & Dare, 2001; Fairburn et al., 1993), conference workshops on treatment methods, and treatment plenary lectures. However, based on studies of treatments provided in clinical settings, these methods may not be effective (Tobin, Banker, Weisberg & Bowers, 2007; Von Ranson & Robinson, 2006).

The research–practice impasse has at least two essential levels: clinicians are not using EST, and researchers are not finding ways to facilitate the dissemination of EST to clinicians. Understanding the causes of this impasse will help identify the efforts and strategies needed to bridge the divide. The AED, a global multidisciplinary professional association comprised of research scientists, clinical practitioners, clinician–researchers, educators, and activists, has taken the lead in elucidating the causes for a research–practice divide, and in developing steps to foster a research–practice partnership. We describe below the advances the AED has made in this area thus far, and their implications for practitioners.

CAUSES OF THE GAP

Setting the Groundwork: The Academy for Eating Disorders Research–Practice Initiative

As a global multidisciplinary association, the AED includes both applied and basic science realms, making it subject to a research–practice divide. Acknowledging the impact of research–practice tensions on its ability to promote excellence in research, education, treatment, and prevention, the AED developed a layered strategy for promoting the ongoing integration of research and practice and a specific action plan (Banker & Klump, 2009). The first step was to determine the causes of this gap.

The authors, as recent past presidents of the AED, long-time friends and colleagues, and mutual representatives of each "side" of this debate (i.e., Judith Banker is a practicing clinician and founding director of an outpatient treatment center; Kelly Klump is a tenured faculty member at a research university), have been intricately involved in the AED's efforts to understand and bridge this gap. They worked with other AED officers to develop the AED Research–Practice Committee (RPC),[1] whose mission is to directly address the research–practice gap within the AED and field at large. In particular, the RPC seeks to advance the research–practice partnership by understanding the causes for the divide and by promoting interactive learning, researcher–practitioner dialog, and a balanced research–practice perspective. Specific RPC activities targeted at these objectives include:

- **Informal Surveys to AED Membership.** Most important was the 2007 survey sent to AED colleagues in North America, Europe, Australia, and the UK, asking their views on the causes and potential resolutions to the research–practice divide. Half of this group was involved in both research and clinical work. The remaining half was split evenly between clinical practice and research. Findings were invaluable for our initial understanding of the research–practice divide

[1] The 2009–2010 Members of the AED Research–Practice Committee include: Drew Anderson, Angela Favaro, Debbie Katzman, Isabel Krug, Michael Levine, Bob Palmer, Susan Paxton, Jill Pollack, Dana Satir, Howard Steiger, and, Co-Chairs Judith Banker and Kelly Klump.

- **Global and Regional Think Tank Sessions.** These interactive sessions are held annually at the AED International Conference on Eating Disorders and at other international (e.g., 2009 London International Eating Disorders Conference) and regional (e.g., European Council on Eating Disorders 20th Anniversary Meeting) conferences. The goal is to facilitate interaction and problem-solving between researchers and practitioners
- **AED Research–Practice Listserv.** This email listserv was created to promote discussion and collaboration among AED clinicians, researchers, and clinician–researchers regarding research–practice issues, projects, and new directions for bridging the gap
- **AED Newsletter Columns.** The RPC hosts two regular columns in the AED's member newsletter, the *AED Forum*. These columns, described below, are aimed at increasing research–practice collaborations and dialog to foster better science, practice, and science-practice integration in our field
  - *New Hypotheses*: An informal format for AED members to share their new ideas, theories, and/or hunches about research, education, treatment, and prevention in our field. Ideally, this initiative will encourage new ideas in research–practice integration and avenues for additional collaborations
  - *Models for Research–Practice Integration*: This forum allows AED members to describe real world examples of methods and processes they use in their own practices, laboratories, and programs to integrate research and practice
  - *Clinical Data Network*: This on-line network links clinicians who have valuable treatment data to researchers who may be interested in using the data for formal or informal research trials. Through this initiative, the RPC hopes to help forge a very tangible and vital form of research–practice partnership.

AED "Data"

Through the AED survey and RPC activities, a clearer picture has emerged of the factors perpetuating the research–practice gap. These more informal data supplement empirical research described above and add more pieces to the puzzle. To date, the RPC information comes primarily from ersatz qualitative research methods: focus group discussions, surveys, and interviews. Interestingly, qualitative or naturalistic research has been proposed by some as an approach that may bridge the gap by bringing research into natural settings (Goodheart, 2006; Kazdin, 2008).

A striking uniformity and overlap was apparent in the tone and content of the responses gathered from our brief survey, online listserv discussions, and in-person interactive discussion sessions with international groups. The feedback suggests three main categories of factors that, together and separately, perpetuate the research–practice divide. These include: (i) relational, attitudinal, and systemic factors; (ii) ineffective knowledge transfer and translational models and systems; and (iii) a lack of a coherent strategy for promoting research–practice partnership.

Relational, Attitudinal, and Systemic Factors

Relational factors. Research–Practice gaps can present challenges to cultivating respect and appreciation between the groups for the value of their mutual contributions. Responses to the AED survey revealed that clinicians clearly felt "under attack," including references to "a faith-based assault against clinicians" and the view that they are perceived as "unscientific charlatans." This felt lack of respect may stem from the increased focus on EBP, contributing to a perceived power differential between clinicians and researchers, and a tendency for researchers to be more commonly at the podium espousing the need for clinicians to change (rather than the reverse). Nonetheless, researchers also reported experiencing a lack of respect, noting that their data fail to be translated into practice and that their research findings are dismissed by clinicians who question their relevance.

Survey respondents and discussion group attendees also reported that the use of specialized jargon further separates the worlds of researchers and clinicians. The precise, scientific language of research data and statistical analyses is reportedly unrecognizable to clinicians whose professional jargon is frequently laced with references to emotional, intrapsychic, and relational dynamics.

Attitudinal Factors. Interestingly, the use of different professional jargon also reinforces the perception of non-intersecting attitudinal and value systems, with clinicians valuing the sometimes nebulous art and nuance of the treatment process, and researchers valuing the concrete, quantifiable, and measurable world achieved in a laboratory. Clinicians report that they tend to value learning from other clinicians more than learning from the research literature or researchers, with some describing a mandated use of manuals "limiting" and "demeaning," and a not-so-subtle abrogation of clinical wisdom and observation. Others note the limitations of manuals, given the ongoing complexity and changeable nature of psychotherapy and the importance of being able to draw on a variety of approaches in order to manage multiple streams of information about the patient. Researchers, on the other hand, object that treatment manuals are not intended to be used in a rigid, step-by-step way, and that manualized approaches have demonstrated excellent results. Further, some researchers have argued that clinicians hold a stereotyped view of clinical trials as using patient participants who are easier to treat than patients in clinical practice, when more recent RCTs have become quite inclusive.

Perhaps the most salient attitudinal factor contributing to a divide between researchers and practitioners is the difference in their views about what constitutes valid evidence. Clinicians tend to place greater value on evidence born from clinical observation and experience. This form of evidence (clinical "know-how," unwritten, practical, contextual knowledge) is referred to as "tacit knowledge" in the field of knowledge transfer. Conversely, researchers tend to value "explicit knowledge" or the evidence derived from structured research trials. This differential emphasis on evidence type perpetuates the gap and the accompanying lack of mutual respect between clinicians and researchers. Clinicians continue to feel as if their knowledge and experience is devalued, and researchers continue to feel as if clinicians dismiss their empirical knowledge base. Attempts to forge a détente within this climate and to increase the sharing of information about preferred sources of evidence would likely be futile.

Systemic Factors. Structural, institutional, political, and economic factors are powerful contributors to a lack of understanding and mutual appreciation between researchers and practitioners. In many cases, the two groups simply do not cross paths other than at annual conferences or other large professional gatherings. Both report that graduate education tends not to support the science–practitioner model, instead reinforcing an "either-or" approach. Clinicians also lament the lack of training opportunities in simple treatment research methods and statistical analyses, and a lack of standardized measures or tools to use to evaluate the effectiveness of their treatments. Researchers further report that funding sources do not provide resources to support clinician–researcher collaborations or to support the training and supervision of clinicians in the practice of EST.

The practical time and economic demands of a professional career discourage many professionals from broadening their scope and appreciation of clinical or research knowledge. Researchers noted that clinical care is not valued or rewarded at their academic institutions, where publishing and grant funding are the standards by which they are evaluated. By the same token, clinicians report having neither the time nor the resources to stay abreast of research as the demands of patient care take precedence. In addition, the nature of the work engaged in by practitioners and researchers is driven by fundamentally different foci. Psychotherapy or patient care focuses on the subjective needs and experience of the individual patient, on the therapy alliance, and effecting changes in the patient's life. Research is focused on objective measures, controlled conditions, statistically significant differences, and replicability of results. The small overlap between these different areas provides little basis for common ground.

Ineffective translation and transfer of knowledge

An AED Global Research–Practice Think Tank Session convened at the 2008 AED's International Conference on Eating Disorders. Over 100 attendees engaged in dialog and about the causes and solutions for the research–practice gap. The dialog centered spontaneously on the theme of "translation," from practice to research and from research to practice. In particular, attendees focused on the AED annual conference keynote and plenary presentations as forums that reflect the current research–practice biases in the field. Attendees contended that the research–practice gap is reinforced by the fact that these presentations are difficult to translate to clinical practice. Participants cited a lack of training for clinicians and researchers in research translation (i.e., identifying the clinical implications and applications of research) and a lack of in-depth description of treatment methods as limiting factors in the extent to which these conference presentations can be translated into clinical practice.

Lack of strategies for promoting research–practice partnerships

In addition to relational/attitudinal/systemic factors and ineffective knowledge transfer, participants in our AED RPC activities also cited the lack of a coherent strategy for addressing the gap as a clear contributor to the research–practice divide. Participants believed that, to date, there have been relatively few attempts to close the gap. Indeed, most thought that the gap was widening rather than narrowing and that it would continue to do so without systematic, targeted efforts to overcome the divide.

STRATEGIES FOR BRIDGING THE GAP

Many factors contribute to the research–practice gap, creating significant barriers to interaction and collaboration between researchers and practitioners. Efforts to address these will require substantial initiatives that target funding sources, changes in institutional philosophies and priorities, and the development of expedient, accessible channels for research–practice interaction and information exchange. Within our RPC activities, the suggestions for how to address the gap were remarkably consistent. Participants felt that a systematic promotion of institutional and organizational support for clinician–researcher interaction and collaboration was critical. A common suggestion was for the AED annual conference to create forums for open and honest discussions between clinicians and researchers. Both clinicians and researchers called for a stronger melding of data where, ideally, presenters of empirical data would discuss clinical applications. Respondents also reported a desire for greater prominence of clinical presentations, to augment and complement the typically research-focused plenaries and keynote addresses. Finally, they suggested smaller, informal group discussions where professionals could discuss the "evidence" they value, the reasons for their opinions, and the ways in which the different forms of evidence inform, rather than contradict, each other. This coming together of the "minds" would serve to mitigate perceived power differentials or implied value placed on one type of evidence over another.

In addition to these conference-focused activities, participants and the RPC discussed broad-based strategies that might address the gap in the field at large, resulting in the RPC decision to develop two sets of guidelines. The first is a general set of guidelines (The AED General Guidelines for Research–Practice Integration) intended to be used by clinicians and researchers in their everyday practices and institutions. The second is an action plan to specifically target areas within the AED structure and culture that could help to promote research–practice partnership within the membership (the AED Action Plan). The RPC felt that this dual approach would maximize research–practice integration at both an organizational and individual level.

Research–Practice Integration Strategies for the Treatment Professional

The principles and strategies described in these documents (see Appendix A) address ways to foster research–practice integration on an organizational or field-wide basis. Readers should review these documents (to view the AED Action Plan go to www.aedweb.org) to assess their application to their work/practice setting. Meanwhile, Box 27.1 summarizes five concrete ways clinicians can apply the AED Guidelines for Research–Practice Integration to develop channels for contributing their clinical observations and experience to the knowledge base in our field and for accessing and utilizing the current knowledge base in their work.

Step 1: Join research–practice networks. Enhancing communication, collaboration, and understanding between researchers and practitioners can be facilitated by seeking out or developing networking opportunities within regional communities such as Pennsylvania's Practice–Research Network (Borkovec, 2004; Borkovec, Echemendia, Ragusea & Ruiz, 2001), professional associations (e.g., the AED), or even online formats (e.g., AED Research–Practice Listserv). Increased interaction and dialog between researchers and practitioners can lessen the use of jargon and help to develop a more unified language for describing the complex issues related to ED treatment, research, prevention, and education. Propinquity can also foster mutual respect and understanding and establish partnerships to promote translation of practice into research and research into practice.

Step 2: Gain access to treatment research. ED treatment involves on-going psychoeducation for the patient and their family members about all aspects of etiology, treatment, and prevention. The complex challenges of ED treatment also require clinicians to expand

> **BOX 27.1.**
>
> **FIVE KEY STEPS FOR INTEGRATING RESEARCH INTO CLINICAL PRACTICE**
>
> 1. Join a practice–research professional network, multidisciplinary professional association and/or consultation group.
> 2. Gain access to current treatment research and/or summaries.
>   Journals
>   International Journal of Eating Disorders (IJED) – www.aedweb.org
>   AED Annual Review of Eating Disorders – www.aedweb.org
>   European Eating Disorders Review (EEDR) – www.aedweb.org
>   Eating Disorders Review – www.gurze.com
>   Websites
>   The Cochrane Collaboration: www.cochrane.org
>   National Institute of Health–MEDLINE/PubMed: www.nlm.nih.gov (follow link for Health Care Professionals)
>   Google Scholar: www.google.com (enter *Google Scholar* as search term)
> 3. *Translate research to practice* by critically evaluating current and new treatment approaches in light of available evidence.
> 4. *Translate practice to research* by honing and communicating clinical observation skills via case study reports.
> 5. Incorporate simple research methods and treatment outcome measures into everyday practice.

and improve their treatment tools and to stay abreast of key findings to ensure they are offering patients and their families the best available treatments and information.

However, for many clinicians, the time it takes to access and read research articles is a challenge. Fortunately, a trend in current research publications, in response to the need for research translation, is the provision of lay language summaries and/or clinical implication sections with treatment or clinically-oriented research articles. The *International Journal of Eating Disorders* (IJED), for example, provides jargon-free summaries of each article published along with periodic review articles on current topics in research. Further, IJED research articles include a paragraph that discusses the clinical implications of the research described. A simple perusal of the article abstract, clinical implications summary, and/or lay language summary can provide a clinician with valuable updates on research trends and breakthroughs.

In general, IJED, and other journal review articles and meta-analyses addressing treatment research, are expeditious sources of information about effective treatments. Review articles summarize the findings of the research literature in a particular topic area, providing the reader with an up-to-date overview of effective treatments, while a meta-analysis seeks to draw conclusions about a particular treatment approach by combining the results of existing studies.

Box 27.1 includes other journals and websites oriented toward the practicing clinician, offering summaries or translations of the latest treatment research findings, or access to professional research databases. For example, the Cochrane Collaboration provides free access to "plain language summaries" of all Cochrane systematic reviews. In addition, individual studies can be accessed through a professional literature database called MEDLINE/PubMed on the National Institute for Mental Health website or via Google Scholar. Finally, attending conferences, workshops, or meetings (or purchasing their program CDs), where the latest ED research is reviewed, can familiarize practitioners with current findings while accruing necessary continuing education credits.

Step 3: Translate research to practice. This can be done by critically evaluating current and new treatment approaches in light of available evidence. Gaining access to information on current treatment research findings is half the battle. Comprehending the information, discerning its reliability, and then applying it to clinical practice is an even more daunting challenge for most clinicians. Although graduate programs in clinical psychology typically include extensive training in research and statistical methods, many clinicians treating people with ED are trained in disciplines that do not. Even those clinicians with a research background may find that their ability to interpret statistical analyses and translate and critically appraise research methods has grown a bit rusty. In fact, lack of training in translating research and statistical information into practice has been cited in our RPC surveys and activities as a key factor that prevents clinicians from staying current on research findings. Professional conferences and training programs are excellent ways to learn or refresh skills in research and statistical methods. The AED website (www.aedweb.org) posts information about clinical and research training programs and workshops offered around the world

The application of research findings to clinical practice includes training in the implementation of EST, but it also includes the domain of clinical innovation and ongoing review of treatment practices. In applying EST to community-based practice, clinicians are forced to adapt, discard, and innovate treatment approaches to address the exigencies of individual patient care. In his comprehensive review, Fairburn (2005a) concludes that evidence-based treatment of AN is "barely" possible. Further, he states, "...until promising new treatments have been developed (with preliminary data to support them), it would be premature to embark upon further costly and time-consuming RCTs" (p. S29). Such significant holes in our knowledge base underscore the critical importance of the ongoing evaluation and adaptation of clinical practice in light of available evidence, including the evidence gleaned from disciplined clinical innovation and observation.

Step 4: Translate practice to research. Hone and communicate clinical observation skills via case studies and/or series. Disciplined clinical observation has led to important advancements, not the least of which is the identification of different constellations of ED symptoms. In his essay on the 25th anniversary of Gerald Russell's first description of BN, Bob Palmer (2004) states:

> In psychiatry, even in the age of multivariate analysis and molecular medicine, there is an undiminished role for clinical observation and the detailed study of phenomena and mental states. We need to celebrate and to cherish the contribution of thoughtful and creative clinicians... (p. 448).

This type of disciplined clinical observation can be conveyed via individual case studies and case series. A case study or case series is the systematic evaluation of an individual treatment case (study) or a small number of treatment cases (series). Information can include clinical observations, interviews, past records, and/or measurement tools such as assessment tools or psychological tests. Clinicians can describe in a narrative report the possible causes of the problem, how they formulated or conceptualized the case, the arc of the patient's response to treatment, and the treatment provided (Kazdin, 2007). Indeed, clinicians can contribute to the knowledge base in our field by recording careful observations culled from clinical practice that can provide ideas and hypotheses for single case research. In addition, case studies can provide detailed descriptions of important exceptions to current accepted theories or principles, the rare, unexpected event or outlier, or "black swan" (Westen, 2007) that can overturn or revise these beliefs.

In cultivating their powers of observation, however, clinicians must guard against natural human cognitive pitfalls such as confirmation biases, defined as "the seeking or interpreting of evidence in ways that are partial to existing beliefs, expectations, or a hypothesis in hand" (Nickerson, 1998, p. 175). To that end, clinicians should exercise care to ensure that they:

1. Remain aware of any observations that might contradict their theories or closely-held views, and that they do not simply attend to those observations that confirm or reinforce their current view or theory.
2. Test other interpretations of the evidence or observation besides their favored view.
3. Practice generating more than one inference from their observations to avoid shaping the evidence to fit a favored theory. For example, try interpreting the evidence from different points of view or theoretical frameworks (e.g., "Did the patient have an angry outburst today because of this reason or that reason?").
4. Inquire as to the patient's subjective experience or point of view regarding a particular interpretation or theory about the observation.

Step 5: Incorporate simple research methods and treatment outcome measures into everyday practice. Our field is missing vital information about the quality and effectiveness of treatment as it is delivered in community settings. Clinicians can make a critical contribution to the general knowledge base and research agenda by providing information about how treatment is conducted in their individual practices and programs. In addition to a need for more information about effective treatments for all ED, our field needs to better understand the mediators and mechanisms of change, those essential aspects of our treatments and treatment relationships that make therapy work (Kazdin, 2008). For example, we know that CBT has been shown to produce positive change in the treatment of people with BN. But we have not peeled away the layers from this approach to understand what aspects of CBT are fundamental to producing change or which aspects are superfluous. Also, many therapists are convinced of the power of a strong therapeutic alliance to effect change and to promote treatment adherence. Yet scant data have been gathered on how the alliance works (Kazdin, 2007).

In addition to contributing to knowledge in the field, the inclusion of assessment instruments in clinical practice provides clinicians with valuable feedback about the effectiveness of their own treatments, programs, and practices. This information can be critical to adapting one's therapeutic approach and matching treatment techniques to the needs of individual patients. Table 27.1 outlines information on simple measurement tools that can be easily delivered in a clinical setting to measure treatment outcome related to ED symptoms, patient quality of life, self-esteem, and mood.

TABLE 27.1 Treatment outcome measurement tools

Measurement tool	Type	Completion time	Availability
Bulimic Investigatory Test, Edinburgh (BITE)	Self-report. 33-item scale. Measures presence and severity of bulimia nervosa symptoms.	10 minutes	Available at no cost at aedweb.org Also in Appendix I, Henderson, M. and Freeman, C. P. L. (1987). A self-rating scale for bulimia: The 'BITE', *British Journal of Psychiatry, 150,* 18–24.
Eating Attitudes Test-26 (EAT-26)	Self-report. 26-item scale. Screens for ED-related symptoms & concerns.	5–10 minutes	Available at no cost at http://psychcentral.com Garner, D. M., Olmsted, M. P., Bohr, Y., and Garfinkel, P. E. (1982). The Eating Attitudes Test: Psychometric features and clinical correlates. *Psychological Medicine, 12,* 871–878.
Eating Disorders Examination (EDE)	Clinical interview. 62-item scale. Assesses ED behaviors & attitudes over past 4 weeks. Can generate ED diagnosis.	30–60 minutes	Available at no cost at aedweb.org Fairburn C. G., and Cooper, Z. (1993). The eating disorder examination (twelfth edition). In: C. G. Fairburn & G. T. Wilson (Eds.), *Binge eating: Nature, assessment and treatment.* (pp. 317–360). New York: Guilford Press. Version for children and adolescents: chEDE also available online. Waugh, R. J., Cooper, P. J., Taylor, C. L., and Lask, B. D. (1998). The use of the eating disorder examination with children: A pilot study. *International Journal of Eating Disorders, 19,* 391–397.
Eating Disorders Examination Questionnaire (EDE-Q)	Self-report. 28-item scale. Self-administered version of the EDE.	10 minutes	Available at no cost at aedweb.org Fairburn, C. G., and Beglin, S. J. (1994). Assessment of eating disorder psychopathology: Interview or self-report questionnaire? *International Journal of Eating Disorders, 16,* 363–370. Appendix II in Fairburn, C. G. (2008). *Cognitive behavior therapy and eating disorders,* New York: Guilford Press.

(Continued)

TABLE 27.1 Treatment outcome measurement tools—cont'd

Measurement tool	Type	Completion time	Availability
Eating Disorders Inventory 3 (EDI-3)	Self-report.	20–30 minutes	EDI-3 Introductory Kit can be ordered from www3.parinc.com. Cost $272.
	91-item scale. Assesses ED symptoms, and ED-related interpersonal and general psychological issues		Garner, D. M. (2004). *Eating disorder inventory (3rd ed.)* (EDI-3). Lutz, FL: Psychological Assessment Resources.
Eating Disorders Quality of Life Scale (EDQLS)	Self-report.	10–15 minutes	Free for publicly funded or non-profit organizations, but there is a fee for the manual. Registration is required at edqls.com to obtain test.
	40-item scale. Measures quality of life of adolescents and adults with EDs.		
Quality of Life Enjoyment and Satisfaction Questionnaire (Q-LES-Q)	Self-report.	10 minutes	Available at no cost at outcometracker.org.
	16-item scale. Assesses overall happiness and functioning, in particular for those suffering from depression.		Ritsner, M., Kurs, R., Gibel, A., Ratner, Y., and Endicott, J. (2005). Validity of an abbreviated Quality of Life Enjoyment and Satisfaction Questionnaire (Q-LES-Q-18) for schizophrenia, schizoaffective and mood disorder patients. *Quality of Life Research, 14*, 1693–1703.
SF-36, SF-12, SF-8	Self-report or interview.	5–20 minutes depending on version used	Available at no cost in over 120 language translations at iqola.org
	36/12/8-item scale. Measures functional health and well-being from patient's point of view.		Ware, J. E., Kosinski, M., and Keller, S. D. (1994). *SF-36 Physical and Mental Health Summary Scales: A user's manual.* Boston, MA: The Health Institute, New England Medical Center.
State Trait Anxiety Inventory (STAI)	Self-report or interview.	10 minutes	Starter kit with manual available for $40 (tests $1 each) at mindgarden.com
	40-item scale. Differentiates between "state" and "trait" anxiety; helps distinguish between anxiety and depression.		Children's version also available.
			Spielberger, C. D. (1983). *Manual for the State-Trait Anxiety Inventory (STAI).* Palo Alto, CA: Consulting Psychologists Press.

TABLE 27.1 Treatment outcome measurement tools—cont'd

Measurement tool	Type	Completion time	Availability
Beck's Depression Inventory (BDI-II)	Self-report.	5–10 minutes	Starter kit (includes manual and 25 record forms) available for $109 at pearsonassessments.com
	21-item scale. Assesses presence and severity of depression symptoms in adolescents and adults.		Beck, A.T., Brown, G., and Steer, R.A. (1996). *Beck Depression Inventory II manual.* San Antonio, Texas: The Psychological Corporation.
Rosenberg's Self-Esteem Scale	Self-report.	5 minutes	Free and widely available online and at aedweb.org
	10-item scale. Measures self-esteem.		Rosenberg, M. (1965). *Society and the adolescent self-image.* New Jersey: Princeton University Press, Princeton.

All scales can be used as intake assessment tools and, when administered at intervals and/or at the end of treatment, as measures of treatment outcome and/or effectiveness.

INTEGRATING RESEARCH AND PRACTICE—OPPORTUNITIES AND CHALLENGES

Opportunities

The recommendations and strategies aimed at bringing about core attitudinal shifts in views about what constitutes valid evidence are among the most important steps in the AED Guidelines. The Guidelines support a broadened definition of evidence outlined in the American Psychological Association's Presidential Task Force on Evidence-Based Practice (Levant, 2006). This includes multiple types of evidence (e.g., efficacy, effectiveness, cost-effectiveness, cost-benefit, epidemiological, treatment utilizations) that contribute to effective clinical practice and multiple sources (e.g., clinical observation, qualitative research, systematic case studies, RCTs, meta-analysis). The expanded definition creates opportunities for clinicians to contribute to the knowledge base and for research to address questions that cannot be answered via RCTs. This lays the groundwork for vital clinician–researcher collaborations to facilitate efforts to develop clinically-relevant, applicable research and to determine the effectiveness of EST.

Furthermore, this shift to a broader definition of reliable evidence opens the door to clinically-driven qualitative research that can contribute valuable information and data about: what treatments clinicians are using in their practices; their effectiveness relative to EST; how treatment decisions are made and why; and how these decisions influence the quality of the treatment. At the time of writing, we have very little information about what aspects or components of ED treatment actually result in positive change in the patient's quality of life or symptomatology.

The more we learn about the essential aspects of a given treatment, the easier it will be to translate and disseminate these treatments. The complex demands of ED treatment require clinicians to work as efficiently, effectively, and flexibly as possible: whittling available, recommended treatments down to their core mechanisms in order to promote therapeutic change in any number of dimensions. Clinicians have invaluable, untapped information about the nature of the treatments they use, the role of the therapeutic alliance and other variables, methods for integrating treatments, the rationale for using one treatment over another in any given situation, and a whole host of other treatment-related issues. This information, if communicated clearly and in disciplined, standardized formats, can serve to inform and guide future ED research and formal treatment recommendations and guidelines.

Challenges

While our field remains stalemated over the use of EST, major holes continue to exist in our knowledge base about what treatments to use when EST fail. We have data to support the use of family based treatment for adolescents with AN (Le Grange, Lock, Loeb & Nicholls, 2010) and CBT for adults with BN (Agras et al., 2000; Fairburn et al., 1993), but we lack information about what therapies constitute first line treatments of choice for adults with AN and for children, adolescents, and adults with Eating Disorder Not Otherwise Specified (EDNOS), the largest diagnostic category of all. Further, our current knowledge does not include information about the effectiveness of EST in community settings or about what the true mechanisms of change are.

For example, research findings in our field are typically based on "efficacy trials" or RCTs, rigorously structured paradigms that are used to compare one treatment against another, or against no treatment. Participants with a specific disorder, selected according to specific criteria, are randomly assigned to one of these conditions. Those with certain co-existing medical or psychological conditions, such as suicidality, substance abuse, or particular medical diseases, may be excluded from the study. The clinicians providing the treatment in these trials are typically trained and closely supervised in the application of manualized treatments.

Given the less controlled conditions of everyday practice settings, it is perhaps not wise to assume that treatments with demonstrated efficacy are automatically generalizable or effective in a natural or community-based setting. Effectiveness studies are designed to examine the generalizability of EST: however, mechanisms have heretofore not been readily available for translating EST to clinical settings where their applicability and benefits can be measured. Thus, we lack information about this important aspect of EST (Haas & Clopton, 2003; Thompson-Brenner & Westen, 2005).

Further, little progress has been made in our field in the area of change process research, which is designed to identify the mechanisms responsible for actually effecting change in psychotherapy treatment. Without understanding the "active ingredients," we can only speculate about what aspects of any particular treatment actually produce change.

CONCLUSIONS

The research–practice divide has kept the ED field from effectively addressing key issues. Valuable research findings are not integrated into clinical practice, and valuable practice experience and observations do not impact research directions. This, of course, ultimately results in serious consequences for the individuals and families who are suffering the effects of these devastating illnesses. ED remain highly stigmatized illnesses (Katterman & Klump, 2010). Misinformation, inadequate treatment and research funding, and discrimination in the treatment of people with these illnesses and their families are common (Klump et al., 2009). Given all that is at stake, we simply cannot afford to continue ignoring clinical or research evidence and burying our heads in dogmatic sand.

A strong research–practice partnership will help produce unified public messages by fostering the mutual exchange of clinical/practice knowledge and research findings, thus integrating our perspectives, enhancing our knowledge base, and improving the quality of our work. Moreover, the strategies used to facilitate research–practice integration can also be applied to efforts to bridge the gap between the public's understanding of ED and the current knowledge in our field.

It seems self-evident that to enhance our knowledge base, it is essential to promote collaboration and interchange between ED researchers and practitioners. Forging a dialectic between research and practice will result in a synthesis of research and clinical evidence that could translate into significant breakthroughs for our patients and for our treatment, research, and prevention efforts. There are simple yet important steps individual practitioners and treatment facilities can take to integrate research and practice within their own settings. By gradually implementing these, clinicians can make strides in improving the quality of their treatment and in contributing vital clinical observations and hypotheses to influence the direction of future research and, ultimately, improving the lives of people with ED and their families.

References

Agras, W. S., & Apple, R. (1997). *Overcoming eating disorders: A cognitive-behavioral treatment for bulimia nervosa and binge-eating disorder: Therapist guide.* U.S.: Graywind Publications, Inc.

Agras, W. S., Walsh, T., Fairburn, C. B., Wilson, B. T., & Kraemer, H. C. (2000). A multicenter comparison of cognitive-behavioral therapy and interpersonal psychotherapy for bulimia nervosa. *Archives of General Psychiatry, 57,* 459–466.

Arnow, B. A. (1999). Why are empirically supported treatments for bulimia nervosa underutilized and what can we do about it? *Journal of Clinical Psychology/In Session: Psychotherapy in Practice, 55,* 769–779.

Banker, J. B., & Klump, K. L. (2009). Research and clinical practice: A dynamic tension in the eating disorder field. In I. F. Dancyger & V. M. Fornari (Eds.), *Evidence based treatments for eating disorders: Children, adolescents and adults* (pp. 71–86). Hauppauge NY: Nova Science Publishers, Inc.

Borkovec, T. D. (2004). Research in training clinics and practice research networks: A route to the integration of science and practice. *Clinical Psychology: Science & Practice, 11,* 211–215.

Borkovec, T. D., Echemendia, R. J., Ragusea, S. A., & Ruiz, M. (2001). The Pennsylvania Practice Research Network and future possibilities for clinically meaningful and scientifically rigorous psychotherapy research. *Clinical Psychology: Science and Practice, 8,* 155–168.

Bruch, H. (1978). *The golden cage: The enigma of anorexia nervosa.* New York, NY: Vintage.

Crow, S. J., Mussell, M. P., Peterson, C. B., Knopke, A. J., & Mitchell, J. E. (1999). Prior treatment received by patients with bulimia nervosa. *International Journal of Eating Disorders, 25,* 39–44.

Donovan, M. S., Bransford, J. D., & Pellegrino, J. W. (Eds.), (2007). *How people learn: Bridging research and practice.* Committee on Learning Research and Educational Practice, Commission on Behavioral and Social Sciences and Education, National Research Council, National Academy Press, Washington, DC.

Fairburn, C. G. (2005a). Evidence-based treatment of anorexia nervosa. *International Journal of Eating Disorders, 37,* S26–S30.

Fairburn, C. G. (2005b). Let data guide treatment of eating disorders, Guest Editorial. *Clinical Psychiatry News, 7.*

Fairburn, C. G., Kirk, J., O'Connor, M., & Cooper, P. J. (1985). A comparison of two psychological treatments for bulimia nervosa. *Behavior Research and Therapy, 24,* 629–643.

Fairburn, C. G., Marcus, M. D., & Wilson, G. T. (1993). Cognitive behavioral therapy for binge eating and bulimia nervosa: A comprehensive treatment manual. In C. G. Fairburn & G. T. Wilson (Eds.), *Binge eating: Nature, assessment, and treatment* (pp. 361–405). New York: The Guilford Press.

Goodheart, C. D. (2006). Evidence, endeavor, and expertise in psychology practice. In C. D. Goodheart, A. E. Kazdin, & R. J. Sternberg (Eds.), *Evidence-based psychotherapy: Where research and practice meet* (pp. 37–61). Washington, DC: American Psychological Association.

Gull, W. W. (1868). The address in medicine. *Lancet, 2,* 171–176.

Haas, H. L., & Clopton, J. R. (2003). Comparing clinical and research treatments for eating disorders. *International Journal of Eating Disorders, 33,* 412–420.

Institute of Medicine (2001). *Crossing the quality chasm: A new health system for the 21st century.* Committee on Quality of Healthcare in America. Institute of Medicine, National Academy Press, Washington, DC.

Katterman, S. N. & Klump, K. L. (2010). Stigmatization of eating disorders: A controlled study of the effects of the television show Starved. *Eating Disorders: Journal of Treatment and Prevention, 18*(2), 153–164.

Kazdin, A. E. (2007). Mediators and mechanisms of change in psychotherapy research. *Annual Review of Clinical Psychology, 3,* 1–27.

Kazdin, A. E. (2008). Evidence-based practice: New opportunities to bridge clinical research and practice, enhance knowledge base, and improve patient care. *American Psychologist, 63,* 146–159.

Klump, K. L., Bulik, C. M., Kaye, W. H., Treasure, J., & Tyson, E. (2009). Academy for Eating Disorders position paper: Eating disorders are serious mental illnesses. *International Journal of Eating Disorders, 42,* 97–103.

Lasègue, E. C. (1873). De l'anorexie hystérique. *Archives Générales Médecine, 21,* 385–403.

Le Grange, D., Lock, J., Loeb, K., & Nicholls, D. (2010). Academy for Eating Disorders position paper: The role of the family in eating disorders. *International Journal of Eating Disorders, 43*(1), 1–5.

Le Grange, D., & Eisler, I. (2009). Family interventions in adolescent anorexia nervosa. *Child and Adolescent Psychiatry Clinics of North America, 18,* 159–173.

Levant, R. F. (2006). APA Presidential Task Force report on evidence-based practice, in psychology. *American Psychologist, 61,* 271–285.

Lock, J., Le Grange, D., Agras, W. S., & Dare, C. (2001). *Treatment manual for anorexia nervosa: A family-based approach.* New York, NY: Guilford Press.

McConnell, S. (2002). Closing the gap. From the Editor. *IEEE Software, 19,* 3–5.

McCown, R. L. (2001). Learning to bridge the gap between science-based decision support and the practice of farming: Evolution in paradigms of model-based research and intervention from design to dialogue. *Australian Journal of Agricultural Research, 52,* 549–572.

McIntosh, V. V., Bulik, C. M., McKenzie, J. M., Luty, S. E., & Jordan, J. (2000). Interpersonal psychotherapy for anorexia nervosa. *International Journal of Eating Disorders, 27,* 125–139.

Mussell, M. P., Crosby, R. D., Crow, S. J., Knopke, A. J., Peterson, C. B., Wonderlich, S. A., & Mitchell, J. E. (2000). Utilization of empirically supported psychotherapy treatments for individuals with eating disorders: A survey of psychologists. *International Journal of Eating Disorders, 27,* 230–237.

Nickerson, R. S. (1998). Confirmation bias: A ubiquitous phenomenon in many guises. *Review of General Psychology, 2,* 175–220.

Palmer, R. (2004). Bulimia nervosa: 25 years on. *British Journal of Psychiatry, 185,* 447–448.

Russell, G. F. M. (1979). Bulimia nervosa: An ominous variant of anorexia nervosa. *Psychological Medicine, 9,* 428–448.

Thompson-Brenner, H., & Westen, D. (2005). A naturalistic study of psychotherapy for bulimia nervosa, Part 2: Comorbidity and therapeutic outcome. *Journal of Nervous and Mental Disorders, 193,* 573–594.

Tobin, D. L., Banker, J. D., Weisberg, L., & Bowers, W. (2007). I know what you did last summer (and it was not CBT): A factor analytic model of international psychotherapeutic practice in the eating. *International Journal of Eating Disorders, 40,* 754–757.

Von Ranson, K. M., & Robinson, K. E. (2006). Who is providing what type of psychotherapy to eating disorder clients? A survey. *International Journal of Eating Disorders, 39,* 27–34.

Westen, D. (2007). Discovering what works in the community: Toward a genuine partnership of clinicians and researchers. In S. G. Hofmann & J. Weinberger (Eds.), *The art and science of psychotherapy* (pp. 3–29). New York, NY: Routledge.

Wilson, G. T. (1998). The clinical utility of randomized controlled trials. *International Journal of Eating Disorders, 24,* 13–29.

APPENDIX A

GUIDING PRINCIPLES FOR RESEARCH–PRACTICE INTEGRATION IN THE FIELD OF EATING DISORDERS

Developed by the AED Research–Practice Committee (2009)

Introduction

A knowledge base built on information from research and practice is critical for providing the highest quality patient care. Therefore, the AED developed these general guidelines to help strengthen research–practice integration in the field of eating disorders.

Background

Principles from the fields of knowledge transfer, innovation diffusion, and Evidence Based Medicine (EBM) were applied in the development of these guidelines. A unifying framework was borrowed from the field of education, which has been addressing its own research–practice gap for over a decade by integrating knowledge transfer and innovation diffusion principles into its strategic approach (Banker & Klump, 2009; Donovan, Bransford & Pellegrino, 2007; Warford, 2005; Love, 1985).

Implementation

The Academy for Eating Disorders is committed to supporting and enacting the guiding principles outlined above. As an organization, we are convinced that long-range, systemic changes in the way in which research and practice are conceptualized and integrated will result in a stronger knowledge base from which to prevent, research, and treat eating disorders. To that end, it is our hope that eating disorder professionals and other eating disorder organizations will join us in endorsing these guidelines, and adopting and implementing them within their practices, groups, organizations, and institutions worldwide. As part of our strategic plan, the AED has developed a specific action plan designed to enact these guidelines within our own organization and the eating disorders field at large.

GUIDING PRINCIPLES FOR RESEARCH–PRACTICE INTEGRATION

PRINCIPLE #1: Research–practice integration will require fundamental attitudinal, relational, and systemic changes

- Eating disorder professionals should recognize that scientific data and clinical observation, judgment and experience (i.e., tacit knowledge) contribute to the knowledge

base in our field. This recognition will support the respectful dialog and communication that is critical to true research–practice integration in our field
- Training programs, conferences, and workshops must emphasize communication and collaboration between researchers and practitioners to allow clinical data and observations to reach researchers and research findings to reach practitioners. These training settings must place a strong emphasis on the value of empirical data and clinical observation and provide hands-on opportunities for research–practice integration
- Conferences and workshops in our field must strive to model research–practice integration by ensuring that all conference activities integrate research and practice through the inclusion of empirical data, clinical observations, and information on clinical implications of the work
- Advocacy efforts should focus on generating research funding mechanisms that support the dissemination of research findings into clinical practice as well as direct testing of clinical observational data in empirical studies. These funding mechanisms must emphasize researcher–clinician partnerships and explicitly acknowledge the value provided by both types of expertise.

PRINCIPLE #2: Research findings and clinical practice information need to be organized and communicated to practitioners and researchers (respectively) in a way that is easy to comprehend and to integrate into their thinking

- Empirical and clinical articles, presentations, and conference abstracts, should provide plain language summaries that limit the use of jargon and enhance interpretability by researchers, clinicians, and clinician–researchers alike
- Across all forms of media, value should be placed on the unique information that can be obtained from empirical research as well as clinical observation. Perhaps more importantly, media pieces on the integration of empirical research with clinical practice should be a top priority for all forms of media in our field
- To facilitate comprehension of research and clinical findings and techniques, professional training and education activities should include interactive, participatory learning methods including mentoring, supervision, simulation, role-play, and the use of small discussion or work groups.

PRINCIPLE #3: Building research–practice integration requires a long-range commitment and a consistent and sustained strategic approach

- Research–Practice integration must remain a top priority in the field in order to enact and sustain the changes outlined in these principles
- Changes in training programs, conferences, and workshops should be continually evaluated and, if necessary, revised to ensure that the goals of research–practice integration are achieved
- The principles outlined in this document must also be evaluated and, if necessary, revised to ensure that the strategic plan for research–practice integration in our field remains current, accurate, and effective.

References

Banker, J. B., & Klump, K. L. (2009). Research and clinical Practice: A dynamic tension in the eating disorder field. In I. F. Dancyger & V. M. Fornari (Eds.), *Evidence based treatments for eating disorders: Children, Adolescents and adults* (pp. 71–86). New York, NY: Nova Science Publishers, Inc.

Donovan, M. S., Bransford, J. D., & Pellegrino, J. W. (Eds.), (2007). *How people learn: Bridging research and practice, Committee on Learning Research and Educational Practice.* Commission on Behavioral and Social Sciences and Education, National Research Council, National Academy Press, Washington, DC.

Love, J. M. (1985). Knowledge transfer and utilization in education. *Review of Research in Education, 12*, 337–386.

Warford, M. K. (2005). Testing a diffusion of innovations in education model (DIEM). *The Innovation Journal: The Public Sector Innovation Journal, 10*(3), Article 32.

CHAPTER 28

Call to Action

As our ideas about the many gaps between research and practice evolved, we decided to give our authors an opportunity to call the field to action on important issues. We hope that readers will answer these calls with a spirit of activism and a commitment to advance the efforts to prevent, understand, and treat the broad spectrum of eating disorders (ED) we see today and to improve the lives of those affected by them.

BRIDGING THE GAP: THE OVERVIEW

What About Gender? Douglas W. Bunnell

The range of topics addressed in this book is testament to the complexity of ED. As we deepen our understanding about the myriad medical, family, nutritional, spiritual, and psychological aspects of these disorders, it is important that we keep an eye on gender. Both men and women develop ED and, for the most part, the clinical picture is similar for both. We need, however, to dig deeper and explore the many ways in which the experience varies between genders.

Some of these differences are clearly rooted in biology. Anxiety and risk factor traits such as perfectionism, obsessionality, harm avoidance, depression, and impulsivity, will look different to a clinician sitting with a female or male patient. What clinicians see, however, is not biology; they see the individual's unique adaptation to his or her biology. This adaptation is shaped by experience and culture, particularly cultural gender influences.

We need to better appreciate these influences in order to lower barriers to treatment for men with ED and to refine our methods for engaging them in treatment. Clinical training programs should ensure that trainees have opportunities to work with both sexes. For our female patients, paying close attention to gender forces will keep us focused on developing a full explanation for why so many women suffer from these disorders. Genetics and biological research may be a valuable link to help us understand the intersection of the personal and the cultural (Bulik, 2009).

All of us need to be addressing these issues outside the office as well. Activism and advocacy about gender forces is still important. Changing the culture can change the risk for ED.

Treatment Implications From the Neurological Perspective Francine Lapides

The treatment of ED patients must encompass interventions that access both the left hemisphere (LH) and the right hemisphere (RH) in an exquisite blend designed to activate and change both hemispheres. The most successful therapists and therapeutic approaches supplement the logical, cognitive, and observational/behavioral approaches (such as patient education, self-monitoring, dietary records, shape- and weight-checking restraints, and verbal assessment questionnaires), all functions based in the LH, with those of the RH that activate the deeper relational, psychophysiological substrates of trauma and affect dysregulation.

The RH stores the templates of attachment and the traces of old relational wounds. Thus, it is in the deep subcortical RH that surges and/or plunges of dysregulated energy occur, prompting the search for an external modulator to bring the energy back to tolerable levels. It is the RH, not the LH, that triggers non-hunger-based consumption, or rigid avoidance of food and the turn to purging, weight-checking, or obsessive exercise.

To effectively address ED, we must do more than restructure our patients' beliefs, habits, and behaviors related to food; we must also activate and heal the unconscious, implicit, procedural, body-based territory of the subcortical RH through interventions that bring the client affectively back into their corporeal selves. This promotes healing by restructuring the patient's deepest sense of self, the very templates of "how to be" in one's own body and in intimate relationships.

To accomplish this, we have to find deeply relational interventions, like short and long-term affective-psychodynamic approaches, that can allow attachment intimacy in the patient/therapist bond to engage and then, with sufficient opportunities for repetition, restructure these deeply relational RH centers in the brain. We must simultaneously intervene with somatic-based interventions that allow our patients to re-own and re-occupy their own bodies in a way that is safe and affectively tolerable.

A Bolder Model Michael Levine and Margo Maine

Eating disorders are multi-determined, reflecting a complex interplay of biopsychosocial factors. Despite the current emphasis on their biogenetic roots, no evidence supports the argument that the majority of cases are attributable to genetics (Levine, 2009; Smolak & Levine, 1994; Smolak, Levine, & Murnen, 2006). While genetic vulnerability significantly increases risk (Bulik, 2004), most cases develop in those at moderate to lower risk because they are simply much more numerous (Austin, 2001). To date, a mountain of evidence supports the impact of sociocultural variables as a significant risk factor for ED. Culture is a risk factor that we *can* and *should* address.

Our treasured colleague, the late Lori Irving (1962–2001), challenged specialists in the ED field to transcend the venerable "scientist-practitioner" model (i.e., the *"Boulder"* model) and replace it with a broader, deeper approach that she called a *"Bolder"* model of the *scientist-activist-and-practitioner of what she or he preaches* (Irving, 1999). *The Bolder Model* stresses the importance of personal and collective agency as intertwined goals for prevention specialists as well as for their various "target" audiences.

Thus, Irving asked ED professionals to take a public and vocal stand about the individual and collective choices we make as participants in and constructors of our culture. In turn, we ask the field to spend less time arguing about the role of genetics and the political necessity of a biopsychiatric approach in favor of much more time spent proactively working to make contemporary culture a safer, healthier place for women and for all at risk for ED. We challenge you, just as we continually challenge ourselves, to work *boldly* to transform the sociocultural context that contributes to unhealthy body image (BI) and disordered eating (DE): confronting the misguided tactics of the "war on obesity"; creating widespread media literacy campaigns so people of all ages can decode and resist these toxic messages; confronting cultural attitudes such as weightism, and rigid, unrealistic ideals of beauty; and working with mass media and the business community to develop healthier, more interesting, and more productive models and images. *The Bolder Model* means that we as professionals can no longer just *"talk the talk,"* we must *"walk the walk."* Join us. *Be Bolder.*

BRIDGING THE GAP: DIAGNOSIS AND TREATMENT

Hippocrates Revisited Edward P. Tyson

In my chapter, I attempted to illuminate and inform the gaps in medical practice that place our patients at increased risk of physical compromise. The primary gap is in the actual management of medical complications associated with ED, and the lack of adequate training for medical professionals in this specialty area. Eating disorders are not part of the required curriculum for most physicians during training, and what training is provided is often insufficient relative to the potential medical complexities involved in these illnesses. Non-physician medical providers (e.g., nurse practitioners, physician assistants) receive even less advanced training in ED, often as little as a one hour lecture devoted to the topic. Clearly, specialized medical training opportunities in ED, such as fellowships, will provide a much needed bridge in the treatment of ED.

A less obvious (except to those who suffer this), but equally concerning, gap involves medical providers' own biases, ignorance, or countertransference in relationship to those who suffer from ED. In addition to the sociocultural prejudices that play into the etiology of ED, our patients face stigmatization from medical providers with potentially devastating implications. In the course of one day, I heard of two such instances. The first was a 12-year-old boy whose pediatrician and therapist quickly withdrew from care when they realized it was an "ED-like problem." They did not refer him elsewhere for care. In the second case, a young woman was being "treated" at a medical hospital where they don't provide emotional support because "they don't treat *those* problems." Tragically, when they restrained her to keep her from pulling out her feeding tube, the patient was retraumatized, as she re-experienced being held down while previously being beaten and raped (after sneaking out of another hospital).

In attachment research, disorganized attachment can occur to a child experiencing a biologic paradox in which the person (parent) who is supposed to be the protector is the one who hurts the child. Whether by lack of adequate training, overt prejudice, or benign neglect, our patients are at risk of being harmed by the very system that is designed to provide their care. The words of Hippocrates bear repeating: *First, do no harm.* Our patients' lives literally depend upon it.

Puzzling Out Where to Go from Here Martha M. Peaslee Levine and Richard L. Levine

With society's emphasis on the obesity epidemic, energy will continue to be directed towards medications that curb excess weight by controlling cravings, impulsive eating, and other addictive behaviors. While we must address the health burdens of obesity, practitioners need to see the total picture. Just as individuals struggling with AN can flip to BN, individuals receiving treatment for obesity can swing from "black" to "white" in their thinking and gravitate toward DE. They have a challenge finding the "gray" zone between restricting (being "good") and indulging (being "bad"). Do medications have a place in developing more fluid thinking?

What is the role of neurotransmitters in perfectionism and rigidity? These traits perpetuate ED symptoms. Can medications shift this inflexibility? Trauma also influences ED. What areas of the brain are particularly affected? Is there a way to more specifically target the resulting symptoms? Efforts have been focused on helping veterans affected by vivid dreams from post-traumatic stress disorder (PTSD). Can we use a crossover of techniques to help our patients?

All of these questions relate to the overall puzzle of brain development and functioning. Have the increased use of electronic devices and the frenetic pace of media with its images, texting, and video games shifted neural development? Clearly, we will not be surrendering these tools. Hopefully, the global interconnection offered by these very devices can increase communication and research efforts in the scientific community. Together we can puzzle out the challenges of ED and increase the available treatment tools.

Extending the Frame of Research and Practice Nancy L. Cloak and Pauline S. Powers

As we noted in our chapter, ED symptoms are frequently intransigent, and evidence-based treatments, while clearly effective, often fall short of producing full recovery, and may be difficult to implement in "real world" settings (Lowe, Bunnell, Neeren, Chernyak & Greberman, 2010). Most treatment research addresses single interventions in fairly homogeneous patient populations, while most ED patients are treated with multiple interventions at various levels of care. Also, there is the problem of relapse, even after apparently effective treatment.

We would challenge the field to put greater emphasis on different paradigms for treatment research, and to look more closely at bridging the gap between formal treatment and self-care. First, we think there is a critical need to more closely examine current treatment practices that involve multi-modal, multi-disciplinary interventions within inpatient, residential, partial hospital, and intensive outpatient settings. One alternative research paradigm with potential applicability is the use of practice-based evidence for clinical performance improvement (Horn & Gassaway, 2007), in which multivariate associations between treatments and outcomes are examined in a naturalistic setting. Another is the use of qualitative research designs that explore patients' perceptions of the most effective elements in their past or current treatment experiences (Bell, 2003).

Second, we would encourage researchers and clinicians to address the question of what patients need in order to transform treatment experiences into self-care. One beginning effort in this direction is the development of community-based support centers, such as USF Hope House in Tampa and Sheena's Place in Toronto. This is another area in which qualitative research could be very useful. We challenge ourselves and our colleagues to look at what does and does not work about what we do now, and to listen to what patients tell us about what works and why.

The Gaps Within the Gaps: Imagination to Imaginaction Beth Hartman McGilley

> *Every step toward action, every response to a call necessitates a leap of faith and is done without knowing the outcome. It is as...Kierkegaard described, the epitome of anxiety meeting courage....A part-time effort, a sorta-kinda commitment...won't suffice....In making the leap from vision to form, you will be tested....Every devil in hell will come out to meet [you]. Only when you try your vision in the world can you test whether it's true.*
> **(Levoy, 1997, p. 11)**

Collaborating in this book was an answer to a call. The science/practice gap had become such an issue in my practice, and in our field, that I could no longer sit idly and ponder or bemoan the rift. Lecturing on it was a warm-up; writing about it was practice. Now it is time to translate this training from imagination, to what DeWitt Jones calls *imaginaction* (DVD, 2002, Extraordinary Visions, available at http://www.dewittjones.com/dvd_ev.htm).

In order for research to be relevant to practice, professionals in both camps must be willing to cross-talk and cross-train. The demands of our work create insular, isolating environments such that all we can seemingly do is publish or perish, or manage our patients' insurance-mangled care. As noted in the introduction, this is no longer enough. Three things would make a difference: joining multidisciplinary ED organizations; reading the scientific literature; and attending clinical conferences. By that I mean: join, read, and attend "the other camp's" organizations, journals, and meetings! As a field, we are less short on "data" than we are on dialog. We must create forums (e.g., the Academy of Eating Disorders listserve) specifically for experts of both domains to become conversant in the other's language, and to create opportunities for clinically relevant collaboration. As I tell my patients, "no one ever gets well being comfortable." Nor will our field improve outcomes if we settle for the comfort of our solitary domains.

BRIDGING THE GAP: SPECIAL POPULATIONS

Bridging the Gap between Eating Disorders and Substance Abuse Disorders Amy Baker Dennis and Bethany Helfman

We would like to make a few recommendations on how the ED and substance use disorder (SUD) professional communities could begin to address the gap between these two fields.

1. *Effective concurrent treatments for patients with comorbid ED/SUD need to be developed and disseminated to both fields.* This would require substantial funding, identifying experienced outcome researchers from both fields who would be willing to collaborate, and a considerable amount of cross training between disciplines and specialties. Clinicians in the ED field are often not well trained in the diagnosis, assessment, or treatment of SUD and are ill prepared to treat this population concurrently. Similarly, SUD clinicians are not typically trained in the treatment of ED or Axis II pathology, and may not be familiar with standard psychopharmacological interventions.
2. *There is a significant gap between identifying evidence-based treatments (EBT) and the dissemination and implementation of these interventions.* This problem is particularly evident in the SUD field where, despite large federally funded research initiatives, published practice guidelines, and manualized EBT, a majority of SUD programs fervently cling to the use of didactic/psycho-educational programming in spite of data suggesting that this intervention may be ineffective (Miller, Wilbourne & Hettema, 2003). In the ED field, studies suggest that despite the availability of EBT, only 6–15% of ED specialists employ them regularly, citing the lack of available training and supervision opportunities (Haas & Clopton, 2003; Mussell et al., 2000; Tobin, Banker, Weisberg & Bowers, 2007).
3. *Practicing clinicians, like our patients, have great difficulty with change, and tend to utilize the techniques that they are comfortable with regardless of their efficacy.* Change requires staying apprised of current research, a willingness to learn and practice new techniques, receiving expert supervision, and evaluating the efficacy of our efforts. Attending a workshop is not enough.
4. *Training in EBT should begin in graduate/medical school and then be continued throughout one's professional career.* Ongoing supervision of newly acquired skills is crucial. Incentive programs for the adoption of EBT are on the horizon. Credentialing organizations and insurance companies in some states are moving toward requiring the use of EBT.
5. *Formal connections between the ED/SUD professional communities are essential.* Philosophical differences and disagreements on treatment approaches have interfered with our ability to effectively serve this comorbid group. We recommend the formation of a special interest group at the Academy for Eating Disorders (AED) and at a similar national SUD professional organization that could collaborate on developing useful protocols and cross training programs.

Trauma and Eating Disorders Diann Ackard and Tim Brewerton

Due to the significant physical and psychological harm associated with ED, as well as the sequelae to traumatic events, it is imperative that we close the gap between research and clinical practice. This goal can be achieved by developing and evaluating the efficacy of manual-based treatments for patients diagnosed with both an ED and PTSD. New treatments for this combination of disorders should draw on interventions with proven effectiveness in treating an ED in combination with PTSD and other trauma-related conditions. These new treatments should then be further tailored to address the needs of the comorbid disorder(s).

We also implore the clinical and research communities to utilize a broader definition of "trauma" than the one described by the diagnostic criteria for PTSD in the current diagnostic classification. Substantial evidence shows that many events (such as bullying and teasing, discrimination, harassment, and witnessing violence) cause distress or dysfunction to an individual, and are associated with PTSD and ED symptoms. However, such events would not currently be included in the DSM-IV definition of PTSD, and this omission may be viewed as minimization of these experiences and ultimately sabotage successful treatment.

We urge clinicians in the ED field to routinely query their patients about traumatic experiences and their effects, and we appeal similarly to clinicians in the field of PTSD to assess the eating habits and weight/shape/body experiences of their patient population.

Eating Disorders and Self-Inflicted Violence: Next Steps in Bridging the Gap
Kim Dennis and Jancey Wickstrom

When such serious conditions as ED and self-inflicted violence (SIV) interact, the gap between research and practice is significant and concerning, but, at the same time, creates a great opportunity for advancement. While the research on ED/SIV is generally lacking, it is especially sparse with regard to the adolescent population, at best providing mixed messages and unclear clinical implications. We can systematically begin to bridge these gaps with: randomized, controlled medication trials; functional brain imaging studies (particularly with respect to reward circuitry); and randomized clinical trials comparing various patient populations (e.g., adults vs. adolescents, or obese teens), diagnostic groups (e.g., EDNOS), and treatment approaches (e.g., unified treatment, DBT). Much of the difficulty in designing such trials lies in the lack of a unified definition of SIV or self injury. Epidemiologic studies and the research suggested above may lead us to better understandings of SIV (possibly as a disorder in and of itself) and its co-occurrence with ED.

In addition, lessening the stigma associated with both ED and SIV is an important step in helping identified patients and their families recover from these devastating illnesses. Encouraging this patient population to engage with 12-step recovery communities available for those with ED and SIV, and even to start meetings in their home communities, will add to the level of recovery support available to all. Finally, compassionate therapeutic relationships will always be at the root of healing, as is caregivers' willingness to be engaged in their own journeys of growth. Recognition of the importance of continued support and education for caregivers in this difficult clinical area will help both patients and caregivers achieve greater success in their treatment and recovery. Attention to these areas will support further advancement in the treatment of patients with ED and SIV.

The Weight of the World Margo Maine

Women carry the weight of the world on their shoulders. Ironically, to keep them "in their place," contemporary culture has given weight and appearance deep significance as signs of female self worth and personal adequacy, in effect disempowering women while breeding and idealizing ED.

After half a century of increasing rates of ED, we now know that these problems occur across the lifespan (Lewis & Cachelin, 2001), but adult women are far less likely to be diagnosed or treated. According to Becker, Eddy & Perloe (2009), "an unacceptably large percentage of eating disorders may go unrecognized in primary care or other specialty settings" due to "suboptimal clinical attunement to and recognition of eating disorder symptoms" (p. 611). Assessing treatment barriers for Mexican-American and European-American women with ED in the USA, Cachelin and Striegel-Moore (2006) concluded that "eating disorders largely go undetected and untreated" (p. 160). Of those who *sought* treatment, less than half were diagnosed or referred, with minority women, in particular, undetected.

It is time to expect all providers in health care and mental health, generalists and specialists alike, to recognize that adult women also suffer ED. During physical and mental health visits, several simple screening questions can be added:

- Has your weight fluctuated during your adult years?
- Are you trying to "manage" your weight? If so, how?
- What did you eat yesterday?
- How much do you think or worry about weight, shape, and food?

The answers can reveal a pattern of DE or a full ED. The inquiry indicates concern and awareness about this issue, making it easier for a woman to break through shame and denial and begin to discuss these problems.

We can no longer deny the weight women bear and the weight they find unbearable. Like the Dixie Chicks on their CD, *Dixie Chicks: Taking the Long Way* (New York, Columbia, 2006), *"I'm not ready to make nice."* I demand *"zero-tolerance"* of the health care system's ongoing dismissal of the impact of DE in adult women. It is time to start asking the questions.

BRIDGING THE GAP: FAMILY ISSUES

The Role of the Family Judith Brisman

Until recently, the family has unwittingly been shunted to the side when considering treatment and care of the ED patient. Every ED patient we see deserves a thoughtful and respectful consideration of the family's role with regard to prevention, treatment, and recovery. This perspective needs to accompany training in ED, at all levels of education, in an array of modalities.

Furthermore, research needs to explore how parents can effectively assist in the recovery process. How do varying clinical situations, differing developmental capacities, age, and prior clinical experience affect the way in which parents can best help their child? What family characteristics need be considered when recommending or discouraging parental intervention with food? Alongside research, clinical knowledge and experience may need to be more fully articulated, establishing a base of reasoning to guide family interventions.

Family support is needed throughout the treatment course. Parents need assistance dealing with their child, the treatment team, and third party payers. Insurance companies themselves need to be educated and supported, in considering the critical role of the family in treatment modalities, so that, in this culture of economic hardship, the financial support of the family is not aborted. Finally, more resources—supportive, clinical, and educational—are needed for parents and professionals alike so that the despair and fear, imbedded in work with ED, does not threaten to unhinge the embrace of care. The field must cultivate an atmosphere in which questioning and discovery override authoritative proof of cure. Only then will answers evolve in this field of so many unknowns.

Family Perspectives

Robbie Munn

The "chaos" of a prolonged and fluctuating recovery from an ED is still poorly understood by too many. This ignorance of what it takes to endure a journey with ED merely reflects the underestimation of its width and breadth. Individuals and families need not face these many challenges alone. There are excellent, resource-rich organizations available to help and support at any point, with the National Eating Disorders Association (NEDA, www.myneda.org; www.nationaleatingdisorders.org) and the Eating Disorders Coalition for Research, Policy, and Action (EDC, www.eatingdisorderscoalition.org) being the two best known. Clinicians need to encourage families to contact such groups. Throw them the lifelines already in place.

Doris and Tom Smeltzer

All degree programs in the therapeutic arts must provide students with the skills needed to detect and respond to their clients with ED. In my recent coursework in Counseling Psychology, we were provided with in-depth instructions on how to assess all first-time clients for depression, alcohol, and drug addictions, with extensive guidance in effective treatment modalities for the same. In this three-year master's program, ED were *never* mentioned. If, as clinicians, we are not assessing every client for DE behaviors, then our treatment protocols will always fall short of success.

Kitty Westin

There are countless ways to take action, and I learned early in my "career" as an advocate that one of the best ways is to have a "voice." I recognized the power of my voice when I joined the Eating Disorders Coalition for Research, Policy, and Action and began participating in Advocacy Training/Lobby Days in Washington. The EDC is a federal advocacy organization that trains grassroots activists to use their voices to educate Congress and policymakers about ED. I urge clinicians and researchers to add their voices to these efforts to recognize ED as a serious public health priority. We can change the world, one conversation at a time.

BRIDGING THE GAP: MIND, BODY, AND SPIRIT

Research and Training in Spirituality and ED Michael E. Berrett, Randy K. Hardman, and P. Scott Richards

Although the integration of spirituality into clinical treatment appears to be a growing trend in the ED treatment field, ED researchers have given this topic relatively little attention to date. The wide gulf that currently exists between practice and research regarding the subject of spirituality and ED is problematic. Given that approximately 95% of Americans profess a belief in God and 60% say that religion is very important to them, religion and/or spirituality are undoubtedly important to the majority of ED patients. Spirituality is only one component of rigorous clinical treatment, but it is vital, and its relationship with etiology, treatment, and recovery needs to be more fully understood.

There are many fascinating and important research questions that need investigation concerning the role of both Western (theistic), Eastern, and other forms of religion and spirituality in ED etiology, treatment, and recovery (Richards, Hardman & Berrett, 2007). We encourage ED researchers to give these topics more attention in the years ahead.

We also encourage ED professionals who have not yet received training in spiritual aspects of diversity to seek more education in this domain. Taking courses, reading books and articles, and watching videos about the world's religions, the psychology of religion, and spiritual approaches for psychotherapy, can be valuable. It is also worthwhile to consult with, and seek supervision from, professionals who have already developed expertise in spirituality and ED. Finally, we call upon ED organizations to more often sponsor continuing education workshops about religious and spiritual aspects of diversity and treatment.

BRIDGING THE GAP: FUTURE DIRECTIONS

A New Paradigm: Health as a Value Versus Size as a Goal Kathy Kater

Worry about weight affects almost everyone today, and at ever younger ages. If only it helped, it might be worth sacrificing a few to ED to keep most people healthier. But instead of improving choices, evidence now documents that *worry about weight* leads to poorer, or disordered, eating and fitness habits, diminished overall health, and *weight gain.* So it is that in the 50 years since weight was first framed as a "problem," with "weight control" cast as the solution, not only have ED emerged to threaten those who are vulnerable, but rates of obesity and associated health risks have risen exponentially. While other environmental changes have played a part, we cannot discount that the thinner we have tried to be, the fatter we have become. Clearly, *worry about weight* is not only ineffective, it is counterproductive.

Einstein noted, "*You cannot solve problems with the same thinking that created them.*" For those who view weight as a simple formula of calories in versus calories burned and believe that fatness causes diminished health, "obesity prevention" seems logical. But framing the goal as "eliminating fatness" is not only biologically naïve and incredibly prejudicial (given those whose lifestyles are optimal while their bodies remain fat), it fuels *worry about weight*, and is, therefore, iatrogenic.

Reframing how people think about and respond to weight is critical to preventing and treating both ED and health problems associated with added pounds. When all evidence is considered, it is clear that interventions should support a new paradigm that promotes *health as a value* versus *size as a goal*. This shift calls for a major change in campaigns and policies aimed at "obesity prevention" to promotion of wholesome eating and fitness for *everyone*, with acceptance of the diverse sizes and shapes that result.

Bridging the Research–Practice Gap Judith Banker and Kelly Klump

We must take swift, bold action to address the research–practice gap in the field of ED. This gap undermines the quality of our treatment, education, prevention, and research efforts as valuable research findings are not integrated into practice, and valuable practice experience and observations do not impact research directions. Although a divide between research and practice is commonplace in fields with an applied and a basic science component, it cannot be tolerated in the field of ED, where the mortality rate, the burden of illness, and the lack of funding for treatment and research worldwide critically impact the individuals and families who are suffering the effects of these devastating illnesses. Furthermore, serious holes exist in our knowledge base regarding effective treatment, education, and prevention that continue to expose these individuals and families to stigmatization and discrimination by the public and those responsible for health care policy and provision.

Given all that is at stake, we must immediately take steps to forge a partnership between researchers and practitioners. The Academy for Eating Disorders has embraced this mission, developing the AED Guidelines for Research–Practice Integration (www.aedweb.org) as a road map for research–practice integration that can be adapted and applied on both an organizational and individual level by researchers and clinicians alike.

We call upon practitioners, across disciplines, to play a pivotal role in strengthening the research–practice partnership by implementing measures to assess the effectiveness of their programs and clinical techniques and sharing their findings with the ED professional community. In doing so, practitioners can be instrumental in improving the quality of their treatment and in contributing vital clinical observations and hypotheses to influence the direction of future research and improve the lives of people with ED and their families.

A strong research–practice partnership will help produce unified public messages by fostering the mutual exchange of clinical/practice knowledge and research findings, thus integrating our perspectives, enhancing our knowledge base, and improving the quality of our work. Moreover, the strategies used to facilitate research–practice integration can also be applied to efforts to bridge the gap between the public's understanding of ED and the current knowledge in our field.

It seems self-evident that to enhance our knowledge base, it is essential to promote collaboration and interchange between ED researchers and practitioners. Forging a dialectic between research and practice will result in a synthesis of research and clinical evidence that could translate into significant breakthroughs for our patients and for our treatment, research, and prevention efforts. There are simple yet important steps individual practitioners and treatment facilities can take to integrate research and practice within their own settings.

References

Austin, S. B. (2001). Population-based prevention of eating disorders: An application of the Rose prevention model. *Preventive Medicine, 32,* 268—283.

Becker, A. E., Eddy, K. T., & Perloe, A. (2009). Clarifying criteria for cognitive signs and symptoms for eating disorders in DSM-V. *International Journal of Eating Disorders, 42*(7), 611—619.

Bell, L. (2003). What can we learn from consumer studies and qualitative research in the treatment of eating disorders? *Eating and Weight Disorders, 8,* 181—187.

Bulik, C. (2004). Genetic and biological risk factors. In J. K. Thompson (Ed.), *Handbook of eating disorders and obesity* (pp. 3—16). Hoboken, NJ: Wiley.

Bulik, C. (2009). *Eating disorders: The science you need to know.* Minneapolis, MN: Keynote presentation at National Eating Disorders Association.

Cachelin, F. M., & Striegel-Moore, R. H. (2006). Help seeking and barriers to treatment in a community sample of Mexican American and European American women with eating disorders. *International Journal of Eating Disorders, 39,* 154—161.

Haas, H. L., & Clopton, J. R. (2003). Comparing clinical and research treatments for eating disorders. *International Journal of Eating Disorders, 33,* 412—420.

Horn, S. D., & Gassaway, J. (2007). Practice-based evidence study design for comparative effectiveness research. *Medical Care, 45*(10 Suppl. 2), S50—257.

Irving, L. (1999). A bolder model of prevention: Science, practice, and activism. In N. Piran, M. P. Levine, & C. Steiner-Adair (Eds.), *Preventing eating disorders: A handbook of interventions and special challenges* (pp. 63—83). Philadelphia, PA: Brunner/Mazel.

Levine, M. P. (2009, April 18). *Are media an important medium for clinicians? Mass media, eating disorders, and the Bolder Model of treatment, prevention, and advocacy.* Presentation at the conference "Eating Disorders: State of the Art Treatment Symposium," sponsored by The Center for Eating Disorders at Sheppard Pratt, Towson, MD.

Levoy, G. (1997). *Callings: Finding and following an authentic life.* New York, NY: Harmony Books.

Lewis, D. M., & Cachelin, F. M. (2001). Body image, body dissatisfaction, and eating attitudes in midlife and elderly women. *Eating Disorders: The Journal of Treatment and Prevention, 9,* 29—39.

Lowe, M. R., Bunnell, D. W., Neeren, A. M., Chernyak, Y., & Greberman, L. (2010). Evaluating the real-world effectiveness of cognitive-behavior therapy efficacy research on eating disorders: A case study from a community-based clinical setting. *International Journal of Eating Disorders.* Advance online publication. doi: 10.1002/eat.20782

Miller, W., Wilbourne, P., & Hettema, J. (2003). What works? A summary of alcohol treatment outcome research. In R. Hester & W. Miller (Eds.), *Handbook of alcoholism treatment approaches: Effective alternatives* (3rd ed.). (pp. 13—63) Boston, MA: Allyn & Bacon.

Mussell, M. P., Crosby, R. D., Crow, S. J., Knopke, A. J., Peterson, C. B., Wonderlich, S. A., & Mitchell, J. E. (2000). Utilization of empirically supported psychotherapy treatments for individuals with eating disorders: A survey of psychologists. *International Journal of Eating Disorders, 27,* 230—237.

Richards, P. S., Hardman, R. K., & Berrett, M. E. (2007). *Spiritual approaches in the treatment of women with eating disorders.* Washington, DC: American Psychological Association.

Smolak, L., & Levine, M. P. (1994). Critical issues in the developmental psychopathology of eating disorders. In L. Alexander & B. Lumsden (Eds.), *Understanding eating disorders* (pp. 37—60). Washington, DC: Taylor & Francis.

Smolak, L., Levine, M. P., & Murnen, S. K. (2006, June). *The scientific status of sociocultural models for eating disorders: A close look at controversy, theory and data.* Workshop presented at the International Conference on Eating Disorders of the Academy for Eating Disorders, Barcelona, Spain.

Tobin, D. T., Banker, J. D., Weisberg, L., & Bowers, W. (2007). I know what you did last summer (and it was not CBT): A factor analytic model of international psychotherapeutic practice in the eating disorders. *International Journal of Eating Disorders, 40,* 754—757.

Index

Academy for Eating Disorders (AED)
  guiding principles for research–practice integration, 475–476, 489
  Research–Practice Committee, 461–462
  research–practice divide factors
    attitudinal factors, 463
    ineffective translation and transfer of knowledge, 464
    lack of strategies for research–practice partnership promotion, 464
    relational factors, 462–463
    systemic factors, 463–464
  research–practice integration strategies
    challenges, 472
    incorporating research methods and outcome measures in practice, 468–471
    networks, 465
    opportunities, 471–472
    research access, 465–467
    translating practice to research, 467–468
    translating research to practice, 467
ACC, see Anterior cingulate cortex
Acceptance and commitment therapy (ACT), see also Model for Healthy Body Image and Weight
  mindfulness, 391
  overview, 164–168
  processes
    classification, 167
    pain types, 169, 171
    willingness and acceptance versus experiential avoidance, 168
ACT, see Acceptance and commitment therapy
Activism, eating disorder professionals, 59–60
Adderal, abuse, 240
Adult eating disorders, see Midlife eating disorders; Pregnancy
AED, see Academy for Eating Disorders
Affect-regulation
  body in psychotherapy, 47
  chronic hypoactivation, 41
  left hemisphere function, 45–46
  neural development sequence, 39–40
  neural pathways, 39
  overview, 38–39
  right hemisphere and dysregulation, 43–44, 46–47
  stress and chronic hyperactivation, 40–41
  therapeutic implications, 44–47
Aggression, gender difference studies, 5
Alliance
  importance, 197–198
  therapist ingredients
    authenticity, 199–200
    empathy and trust, 200
    endurance and frustration tolerance, 200–201
    engagement, 199
    humbleness and transparency, 201
    non-possessive warmth, 198
    self-nurturing ability, 201–202
Amenorrhea, see Menstruation
Amygdala, gender differences, 6
AN, see Anorexia nervosa
Anorexia nervosa (AN)
  diagnostic characteristics, 111
  exercise in management, 430–432
  family perspective on treatment, 349–360
  holistic perspective, 408–409
  male, 303
  micronutrient supplementation, 114
  outpatient treatment following weight restoration
    body image, 188–189
    comorbidity management, 193–194
    exercise, 189
    family and friends in recovery, 186–188
    integrated treatment approaches, 191–193
    interpersonal and dynamic issues, 190–191
    team members
      medical providers, 184–185
      nutrition specialists, 185
      psychopharmacologists, 183–184
    transition from inpatient treatment, 182–183
  pharmacotherapy
    antidepressants, 112–113
    anxiolytics, 114
    atypical antipsychotics, 113–114
    combination therapy, 114, 144
    duration, 115
  psychiatric comorbidities, 112
  relapse, 181

Anterior cingulate cortex (ACC)
  affect-regulation, 40
  gender differences, 5
  size in anorexia nervosa, 38
Anticonvulsants, bulimia nervosa management, 117
Antidepressants
  anorexia nervosa management, 112–113
  bulimia nervosa management, 116–117
  eating disorder not otherwise specified management, 120
Anxiety disorders
  anorexia nervosa outpatient management, 193–194
  assessment in eating disorders, 80–81
Anxiolytics, anorexia nervosa management, 114
Apriprazole
  anorexia nervosa management, 114
  bulimia nervosa management, 118
Art therapy, eating disorders, 155–156
Attachment theory, eating disorders, 42–43
Attitudinal body image, *see* Body image

Baclofen, eating disorder not otherwise specified management, 121
BDD, *see* Body dysmorphic disorder
BDI, *see* Beck Depression Inventory
Beck Depression Inventory (BDI), 471
BED, *see* Binge eating disorder
BI, *see* Body image
Binge eating disorder (BED), *see also* Eating disorder not otherwise specified
  diet assessment, 75–76
  holistic perspective, 409–410
  variants, 119–120
BITE, *see* Bulimic Investigatory Test, Edinburgh
Blood pressure (BP), assessment, 99
BMD, *see* Bone mineral density
BN, *see* Bulimia nervosa
Body dysmorphic disorder (BDD), assessment in eating disorders, 81
Body image (BI)
  anorexia nervosa outpatients, 188–189
  attitudinal body image, 80
  eating disorder disturbances, 444–445
  feminist perspective, 10–11
  frequency of dissatisfaction in women, 285–286
  midlife women, 295
  perceptual body image, 80
  self-evaluation, 79–80
Body mass index, *see* Weight
Body talk
  body image and eating disorders, 444–445
  childhood story deconstructing, 451–453
  fat attack translation into metaphorical language, 450–451
  language of symbolism and metaphor, 447–450
  metaphor and healing, 445–447
  new lens for food and fat, 453–455
  overview, 443–444
  transformative power of metaphor, 455–456
Bolder Model, research–practice divide, 480–491
Bone mineral density (BMD), assessment, 105
Borderline personality disorder (BPD)
  clinical features, 218
  diagnosis
    clinical diagnosis, 221–222
    features, 217–218
    Gunderson criteria, 221–222
  eating disorder association
    risk factor in development, 222–223
    self-injury in eating disorders, 223
    symptom severity and outcomes, 223, 227
    treatment impact, 223
  epidemiology, 218–219
  family dysfunction, 220
  genetics, 219
  trauma in development, 219–220
  treatment
    eclectic treatments, 225–226
    quagmires, 226–227
    systematized treatments, 224–225
  triggers, 220
BP, *see* Blood pressure
BPD, *see* Borderline personality disorder
Bulimia nervosa (BN)
  diagnostic characteristics, 115
  exercise in management, 432
  family perspective on treatment, 360–364
  holistic perspective, 409
  micronutrient supplementation, 118
  pharmacotherapy
    anticonvulsants, 121
    antidepressants, 120
    combination therapy, 121
    duration, 119
  psychiatric comorbidities, 116
Bulimic Investigatory Test, Edinburgh (BITE), 469
Bupropion
  anorexia nervosa management, 112
  bulimia nervosa management, 116–117

Cardiac stress test, eating disorder findings, 101
CAT, *see* Cognitive analytic psychotherapy
CBT, *see* Cognitive behavioral therapy
Child abuse, assessment in eating disorders, 82

Cognitive analytic psychotherapy (CAT), anorexia nervosa outpatients, 191–193
Cognitive behavioral therapy (CBT)
　affect-regulation, 45–46
　bulimia nervosa management, 119, 143
　CBT-E model, 191
　post-traumatic stress disorder management in eating disorders, 254–256
　psychodynamic therapy integration, 153–154
　substance abuse management in eating disorders, 241
Continuing education, eating disorder professionals, 61–62
Corpus callosum, gender differences, 5
Couples therapy, substance abuse management in eating disorders, 242
Creatinine, eating disorder findings, 107

Dance/movement therapy (DMT)
　case examples, 415–417
　integration of concepts, 416–417
　kinesthetic awareness, 415
　kinesthetic empathy, 416
　rhythmic synchrony, 414–415
DBT, *see* Dialectical behavioral therapy
Denial, assessment in eating disorders, 81–82
Depression
　anorexia nervosa
　　comorbidity, 112
　　outpatient management, 194
　bulimia nervosa association, 116
　eating disorder not otherwise specified association, 120
Diagnostic Interview for Borderlines (DIB), 221
Dialectical behavioral therapy (DBT)
　mindfulness, 391
　post-traumatic stress disorder management in eating disorders, 256
　self-inflicted violence management in eating disorder, 275–276, 282
　substance abuse management in eating disorders, 243
DIB, *see* Diagnostic Interview for Borderlines
Diet
　assessment
　　diversity
　　　binge eating, 75–76
　　　grazing behavior, 76
　　quantity, 74–75
　　nutritional rehabilitation, *see* Nutritional rehabilitation
　protein deficiency, 97
　unbalanced nutrition effects
　　cognition, 128–130
　　emotion and behavior, 130

Dietary restraint, assessment, 74
Dietary restriction, assessment, 74
Diet pills, abuse, 239
Dissociation
　affect-regulation and chronic hypoactivation, 41
　eating disorder role, 41–42
　post-traumatic stress disorder mechanism, 41
Diuretics, abuse, 77–78, 238–239
DMT, *see* Dance/movement therapy
Dronabinol (THC), anorexia nervosa management, 114

EAT-26, *see* Eating Attitudes Test-26
Eating Attitudes Test-26 (EAT-26), 469
Eating disorder not otherwise specified (EDNOS)
　diagnostic characteristics, 119–120
　exercise in management, 432–433
　holistic perspective, 410
　male, 302–303
　mortality, 90
　pharmacotherapy
　　anticonvulsants, 121
　　antidepressants, 120
　　combination therapy, 121
　psychiatric comorbidities, 120
Eating Disorders Examination (EDE), 469
Eating Disorders Examination-Questionnaire (EDE-Q), 469
Eating Disorders Inventory-3 (EDI-3), 470
Eating Disorders Quality of Life Scale (EDQLS), 470
EBP, *see* Evidence-based practice
ECG, *see* Electrocardiography
Echocardiography, eating disorder findings, 101
EDE, *see* Eating Disorders Examination
EDE-Q, *see* Eating Disorders Examination-Questionnaire
EDI-3, *see* Eating Disorders Inventory-3
EDNOS, *see* Eating disorder not otherwise specified
EDQLS, *see* Eating Disorders Quality of Life Scale
Electrocardiography (ECG), eating disorder findings, 99–101
EMDR, *see* Eye movement desensitization and reprocessing
Emotion processing, deficits in eating disorders, 43
Enema, abuse, 77–78
Evidence-based practice (EBP), 459–460
Exercise
　anorexia nervosa outpatients, 189
　assessment of excessive exercise, 78–79, 426
　eating disorder management
　　anorexia nervosa, 430–432
　　bulimia nervosa, 432
　　critical themes, 433–434
　　eating disorder not otherwise specified, 432–433

Exercise (*Continued*)
    overview, 425–427
    quality versus quantity, 427–429
    mindful exercise
        cultivation, 434–436
        treatment team mindfulness, 436–437
    perspective in weight neutrality, 29
    prospects for study, 437–438
Eye movement desensitization and reprocessing (EMDR), post-traumatic stress disorder management in eating disorders, 26

Faith, *see* Spirituality
Family, *see also* Integrated relational/motivational approach
    anorexia nervosa recovery role, 186–188
    borderline personality disorder and dysfunction, 220
    functional assessment in eating disorders, 82
    history of study in eating disorders, 335–337
    intervention in eating disorder management, 157
    parental neglect, 337–338
    parenting challenges in adult eating disorders, 291–292
    perspectives on eating disorder treatment
        anorexia nervosa, 349–360
        bulimia nervosa, 360–364
    research–practice divide of role in eating disorders, 486–488
    systems theory, 337
Family therapy
    family-based treatment
        empirical support for eating disorders, 339
        factors affecting
            age and developmental capacity, 342–343
            parental involvement and interference, 343
            previous treatment, 343–344
        limitations of empirical research, 340–341
        parental intervention with food, 338–339
        prospects for study, 345–346
        spouses, parents, and siblings as patients, 344–345
        treatment direction assessment, 342
    post-traumatic stress disorder management in eating disorders, 256
    substance abuse management in eating disorders, 242
Fasting, assessment, 78
Fat, cultural meanings, 22–23
Food, *see* Diet

Gastroesophageal reflux disease (GERD), purging induction, 104
Gastrointestinal tract
    assessment
        purging complications, 104
        restriction, 103
    complications from eating disorders, 103
Gastroparesis, food restriction effects, 103
Gender competence, enhancement, 12–14
Gendering, biopsychosocial process, 6–7
GERD, *see* Gastroesophageal reflux disease
Gestalt therapy
    eating disorders, 156
    holistic medicine, 407–408

HAES, *see* Health at Every Size
Health at Every Size (HAES), model, 17–18, 25–26, 29–30
Heart rate (HR), assessment, 98–99
Hippocampus, gender differences, 5
History, *see* Medical history
Holistic medicine, *see also* Spirituality
    clinician resources, 407
    dance/movement therapy
        case examples, 415–417
        integration of concepts, 416–417
        kinesthetic awareness, 415
        kinesthetic empathy, 416
        rhythmic synchrony, 414–415
    eating disorder perspectives
        anorexia nervosa, 408–409
        binge eating disorder, 409–410
        bulimia nervosa, 409
        eating disorder not otherwise specified, 410
    Gestalt perspective, 407–408
    language
        action words, 419–420
        delivery rhythm and pacing, 417–418
        emotion, 418
        idiomatic phrases, 418
        trigger words, 420–421
        words, 418–419
    overview, 405–406
    prospects for study, 421–422
    shattered self in eating disorders, 406
    yoga, 410–413
HR, *see* Heart rate
Hypoglycemia, brain effects, 102–103
Hypothermia, assessment, 97–98

ICB, *see* Inappropriate compensatory behaviors
Identity, exploration in mindfulness, 397–398
Inappropriate compensatory behaviors (ICB), assessment, 76–79
Infection, eating disorder risks, 108
Insulin, abuse assessment, 79
Integrated relational/motivational approach
    prospects, 331–332

rationale, 324
relational-cultural theory
  eating disorders
    family factors, 323–324
    patient factors, 322–323
    research, 321–322
  stages of change theory
    eating disorders, 321
    motivational interviewing, 320–321
  theory, 319
  therapist
    stance, 329–331
    strategies and principles, 324–329
Interpersonal therapy (IPT)
  eating disorder management, 191
  post-traumatic stress disorder management in eating disorders, 256
  substance abuse management in eating disorders, 241–242
IPT, see Interpersonal therapy

Laboratory tests, eating disorder findings, 107
Language, see Body talk; Holistic medicine
Laxatives
  abuse, 74, 77–78, 238
  gastrointestinal effects, 104
Lithium, bulimia nervosa management, 117
Makeover mind, weight neutrality, 24–26
Male eating disorders
  classification and diagnosis, 302–303
  cultural and developmental factors, 304–305
  endocrine function, 304
  exercise, muscularity, and fitness, 304
  overview, 301
  prevalence, 302
  psychotherapy
    case example, 309
    countertransference, 312
    female therapists, 312–313
    male properties, 309–311
    male therapists, 313
    therapist guidelines, 311–312
  sexual orientation, 304
  socialization factors, 305–306
  treatment, 306–308

MAOIs, see Monoamine oxidase inhibitors
Mass media, see Media
Mathematical ability, gender difference studies, 5
MCM-III, see Millon Clinical Multiaxial Inventory-III
Media literacy
  assessment, 62
  office modeling, 62
  overview, 57–58
  promotion, 60
  therapy application, 61
Media
  advocacy by therapists, 63–64
  objectification theory, 8–9
  risk factor for negative body image and disordered eating, 55–59
Medical history
  checklist, 95–96
  differential diagnosis, 94–96
  gathering, 93–94
  weight history, 73
Medical illness, classification of eating disorders, 90–91
Memantine, eating disorder not otherwise specified management, 121
Menstruation, see also Puberty
  amenorrhea diagnostic utility, 73–74, 05
MET, see Motivational enhancement therapy
Metaphor, see Body talk
Methylphenidate
  abuse, 240
  bulimia nervosa management, 118
Midlife eating disorders
  classification, 287–288
  clinical issues, 293–295
  epidemiology, 285–287
  feminist framework, 294
  medical issues, 295–296
  obstacles to getting help, 296
  parenting challenges, 291–292
  transition risks, 288–289
Midlife eating disorders, research–practice divide, 486
Millon Clinical Multiaxial Inventory-III (MCM-III), 222
Mindfulness, see also Spirituality
  basis of psychotherapies, 390–391
  brain
    activation studies, 390
    circuitry, 391–392
  compassion mediation, 393
  countertransference, 400–402
  exercise
    cultivation, 434–436
    treatment team mindfulness, 436–437
  identity, perfectionism, and mindfulness, 397–398
  mindful eating and nutritional rehabilitation, 139–140, 396
  Model for Healthy Body Image and Weight, 173–175
  overview, 387–390
  prospects for study, 402

Mindfulness (*Continued*)
  radical acceptance, 398–399
  stress reduction, 391
  therapy initiation, 392–394
  transference, 399–400
  well-being skills, 394–395
Minimization, assessment in eating disorders, 81–82
Model for Healthy Body Image and Weight
  action alignment with core values, 176–179
  cognitive defusion versus fusion, 171–173
  mindfulness as intervening action, 173–175
  overview, 164
  processes
    classification, 167
    pain types, 169, 171
    willingness and acceptance versus experiential avoidance, 168
    self as context versus attachment to conceptualized self, 175–176
Monoamine oxidase inhibitors (MAOIs), bulimia nervosa management, 116
Mood disorders
  anorexia nervosa outpatient management, 193–194
  assessment in eating disorders, 81
Mood stabilizers, bulimia nervosa management, 117
Motivational enhancement therapy (MET), substance abuse management in eating disorders, 242
Munn, Genny, family perspective on anorexia nervosa treatment, 355–360, 487

Naltrexone, bulimia nervosa management, 118
NES, *see* Night eating syndrome
Neuropsychiatric assessment, eating disorder findings, 101–103
Night eating syndrome (NES), features, 120
Nutrition, *see* Diet
Nutritional rehabilitation
  duration, 131
  final phase, 139–140
  fullness management, 131–132
  interrupting symptom use
    duration of meal plans, 135
    eating plans, 132–134
    self-monitoring, 134–135
  middle phase
    flexibility, 136–137
    hunger and satiety, 136
    internal regulation, 138–139
    off-limits foods and challenging situations, 137
    shift from structure and monitoring, 137–138
  overview, 130–131
  research-clinical gap, 140

Obesity
  epidemiology, 19–20
  facts versus panic, 18–20
  weight neutral stance in therapy, 21–31
Objectification
  eating disorder professional opposition, 60
  feminist perspective, 10–11
  media, 8–9
  outcomes, 9–10
  self-objectification, 9
Obsessive-compulsive disorder (OCD)
  assessment in eating disorders, 80
  bulimia nervosa association, 116
OCD, *see* Obsessive-compulsive disorder
Olanzapine, anorexia nervosa management, 113–114
Ondansetron, bulimia nervosa management, 118
Oppression, internalization and eating disorders, 31
Orlistat, abuse, 239
Osteopenia, eating disorder findings, 105–106
Osteoporosis, eating disorder findings, 105–106

Pancreatitis, purging induction, 104
PDQ-4, *see* Personality Diagnostic Questionnaire-4
Perceptual body image, *see* Body image
Personality Diagnostic Questionnaire-4 (PDQ-4), borderline personality scale, 230–231
Physical activity, *see* Exercise
Physical examination
  blood pressure, 99
  heart rate, 98–99
  point of maximal impulse, 99
  skin, 104
  temperature, 97
  vital signs, 94, 96
  weight and body mass index, 96–97
PMI, *see* Point of maximal impulse
Point of maximal impulse (PMI), assessment, 99
Post-traumatic stress disorder (PTSD), *see also* Trauma
  affect-regulation, 41
  diagnosis
    criteria, 252
    women, 11
  dissociation mechanism, 41
  eating disorder association
    assessment, 253–254
    case studies, 259–266
    clinician recommendations, 258–259
    literature review, 253
    treatment, 254–257
  research–practice divide and eating disorders, 484–485
  self-inflicted violence role, 271
Pregnancy, eating disorders, 290–291

Psychiatric illness, classification of eating disorders, 90–91
Psychodynamic therapy, symptom management integration
　case studies, 144–145, 148–149, 151–154, 158
　countertransference monitoring, 151–158
　general considerations, 146
　overview, 143–144
　symptoms
　　discussing, 147–149
　　psychodynamic functions of eating disorder symptoms, 150
　　respecting, 146–147
　　transference relationship, 149, 151
　theory, 145–146
PTSD, *see* Post-traumatic stress disorder
Puberty
　age of onset trends for women, 4
　transition difficulty, 7–8
Purging, gastrointestinal complications, 104

Q-LES-Q, *see* Quality of Life Employment and Satisfaction Questionnaire
Quality of Life Employment and Satisfaction Questionnaire (Q-LES-Q), 470

Radical acceptance, mindfulness, 398–399
Recovery
　averting disaster, 202–203
　bridging experience and empiricism, 210–211
　client recipe, 202, 208
　outcome studies
　　coming out from the inside, 205
　　duration as defining factor in recovery, 204–205
　　insight and behavioral control in recovery, 204
　　recovery definition, 203–204
　qualitative research, 206–210
Refeeding syndrome (RFS), features and monitoring, 107–108
Relational-cultural theory, integrated relational/motivational approach
　eating disorders
　　family factors, 323–324
　　patient factors, 322–323
　research, 321–322
Religion, *see* Spirituality
Renal function, eating disorder findings, 106–107
Research–practice divide, *see also* Academy for Eating Disorders
　Bolder Model, 480–491
　diagnosis and treatment, 481–482
　extending frame of research and practice, 482–483
　family role in eating disorders, 486–488

　imagination to imaginaction, 483
　midlife eating disorders, 486
　overview, 479
　self-inflicted violence and eating disorders, 485
　spirituality and eating disorders, 488
　substance abuse and eating disorders, 484
　trauma and eating disorders, 484–485
　treatment implications from neurological perspective, 480
RFS, *see* Refeeding syndrome
Rosenberg's Self-Esteem Scale, 471

Selective serotonin reuptake inhibitors (SSRIs)
　anorexia nervosa management, 112, 115
　bulimia nervosa management, 116
　eating disorder not otherwise specified management, 120
　substance abuse management in eating disorders, 242–243
Self-Harm Inventory (SHI), 222, 231–232
Self-inflicted violence (SIV)
　definition, 269
　impact
　　patient and relationships, 273–274
　　treatment providers, 279–280
　recovery, 274–275
　research–practice divide and eating disorders, 485
　scope, 269–270
　treatment with comorbid eating disorder
　　dialectical behavioral therapy, 275–276, 282
　　milleu therapy, 278–279
　　unified approach, 277–278
　underpinnings
　　genetics, 271–272
　　psychosocial causes, 272
　　trauma, 271
Self Mutilators Anonymous (SMA), 274–275, 282
Sensory motor therapy, mindfulness, 390
Serotonin and norepinephrine reuptake inhibitors (SNRIs)
　anorexia nervosa management, 112
　bulimia nervosa management, 116
SHI, *see* Self-Harm Inventory
Sibutramine, eating disorder not otherwise specified management, 121
SIV, *see* Self-inflicted violence
Skin, eating disorder findings, 104
Sleep, quality in eating disorders, 29–30
SMAm, *see* Self Mutilators Anonymous
Smeltzer, Andrea, family perspective on bulimia nervosa treatment, 360–364, 487
SNRIs, *see* Serotonin and norepinephrine reuptake inhibitors

Spirituality, *see also* Holistic medicine; Mindfulness
   common spiritual issues in eating disorder patients, 370–371
   guidelines for treatment framework
      assessment of spirituality, 372
      implementation of interventions, 372–373
      spiritually safe therapeutic environment, 371
   health benefits, 367–368
   literature on eating disorder studies, 368
   pathways to eating disorder recovery
      gifts of love, 378–380
      language learning, 375–376
      listening to and following the heart, 373–374
      mindfulness, 376–377
      principled living, 377–378
      therapeutic mirror reflecting spiritual identity, 380–382
   research
      recommendations, 382
      research–practice divide in eating disorders, 488
   theory of benefits, 369–370
   training recommendations, 382–383
SSRIs, *see* Selective serotonin reuptake inhibitors
Stages of change theory
   eating disorders, 321
   motivational interviewing, 320–321
STAI, *see* State Trait Anxiety Inventory
State Trait Anxiety Inventory (STAI), 470
Stereotype management, skills, 31–32
Steroids, abuse of prescription drugs, 239–240
Storytelling, *see* Body talk
Stress, *see also* Post-traumatic stress disorder
   affect-regulation and chronic hyperactivation, 40–41
   gender difference studies of brain effects, 5
   mental health problem correlation in women, 3
   mindfulness-based stress reduction, 391
Substance abuse
   assessment in eating disorders, 81, 235–236
   bulimia nervosa association, 116
   comparison with eating disorders
      differences, 237–238
      similarities, 236–237
   diet pills, 239
   diuretics, 238–239
   laxatives, 238
   methylphenidate, 240
   orlistat, 239
   prevalence with eating disorders, 234–235
   research–practice divide and eating disorders, 484
   steroids, 239–240
   syrup of Ipecac, 239
   treatment in eating disorder
      challenges, 233–236

      cognitive behavioral therapy, 241
      couples therapy, 242
      dialectical behavioral therapy, 243
      empirically supported interventions, 240–241
      family therapy, 242
      integrated or sequential treatment, 243, 245
      interpersonal psychotherapy, 241–242
      motivational enhancement therapy, 242
      pharmacotherapy, 242–243
      placement indicators for levels of care, 243–244
      twelve-step programs, 243
Superior mesenteric artery syndrome, food restriction effects, 103
Sustainable eating, weight neutrality, 27–29
Symbolism, *see* Body talk
Syrup of Ipecac, abuse, 239

TCAs, *see* Tricyclic antidepressants
THC, *see* Dronabinol
Therapeutic relationship, *see* Alliance
Thin, cultural meanings, 22–23
Thyroid hormone, eating disorder findings, 105
Thyroid-stimulating hormone (TSH), eating disorder findings, 105
Topiramate
   bulimia nervosa management, 117
   eating disorder not otherwise specified management, 121
Trauma, *see also* Post-traumatic stress disorder; Self-inflicted violence
   assessment in eating disorders, 82
   borderline personality disorder development role, 219–220
   conceptualization, 252–253
   sharing with therapist, 251–252
Tricyclic antidepressants (TCAs)
   anorexia nervosa management, 112
   bulimia nervosa management, 116
   eating disorder not otherwise specified management, 120
TSH, *see* Thyroid-stimulating hormone

Valproic acid, bulimia nervosa management, 117
Verbal ability, gender difference studies, 5
Visuo-spatial ability, gender difference studies, 5
Vomiting, self-induced and assessment, 77

Weight neutrality
   cultural meanings of fat and thin, 22–23
   eating disorder professional promotion, 60
   evidence and beliefs, 20–21
   Health at Every Size model, 17–18, 25–26, 29–30
   intuitive eating, 27

makeover mind, 24—26
   self-care behaviors, 29—30
   stigmatization effects, 21—22
   sustainable eating, 27—29
Weight
   accurate measurement, 72—73
   body mass index determination, 97
   history from medical records, 73
   physiological markers, 73—74
   status assessment, 71—72

Wellbeing skills, mindfulness, 394—395
Westin, Anna Selina, family perspective on anorexia nervosa treatment, 349—355, 487—488

Yoga, eating disorder management, 410—413

Zonisamide, eating disorder not otherwise specified management, 121

FIGURE 9.1

FIGURE 9.2

FIGURE 25.1

FIGURE 25.2

CPSIA information can be obtained
at www.ICGtesting.com
Printed in the USA
LVHW060251120319
610283LV00001B/2/P